Lecture Notes in Computer Science 3524

Commenced Publication in 1973
Founding and Former Series Editors:
Gerhard Goos, Juris Hartmanis, and Jan van Leeuwen

Roman Barták   Michela Milano (Eds.)

Integration of AI and OR Techniques in Constraint Programming for Combinatorial Optimization Problems

Second International Conference, CPAIOR 2005
Prague, Czech Republic, May 30–June 1, 2005
Proceedings

 Springer

Volume Editors

Roman Barták
Charles University
Faculty of Mathematics and Physics
Malostranské nám. 2/25, 118 00 Prague 1, Czech Republic
E-mail: roman.bartak@mff.cuni.cz

Michela Milano
DEIS
University of Bologna
Viale Risorgimento 2, 40136 Bologna, Italy
E-mail: mmilano@deis.unibo.it

Library of Congress Control Number: 2005926642

CR Subject Classification (1998): G.1.6, G.1, G.2.1, F.2.2, I.2, J.1

ISSN	0302-9743
ISBN-10	3-540-26152-4 Springer Berlin Heidelberg New York
ISBN-13	978-3-540-26152-0 Springer Berlin Heidelberg New York

Springer is a part of Springer Science+Business Media

springeronline.com

© Springer-Verlag Berlin Heidelberg 2005
Printed in Germany

Typesetting: Camera-ready by author, data conversion by Scientific Publishing Services, Chennai, India
Printed on acid-free paper   SPIN: 11493853   06/3142   5 4 3 2 1 0

Preface

The Second International Conference on Integration of AI and OR Techniques in Constraint Programming for Combinatorial Optimization Problems (CPAIOR 2005) was held in Prague, Czech Republic, during May 31 - June 1, 2005.

The conference is intended primarily as a forum to focus on the integration and hybridization of the approaches of Constraint Programming (CP), Artificial Intelligence (AI), and Operations Research (OR) technologies for solving large scale and complex real life optimization problems. Therefore, CPAIOR is never far from industrial applications.

The interest of the research community in this conference is witnessed by the high number of submissions received this year, reaching almost 100 papers. From these submissions, we chose 26 to be published in full in the proceedings.

This volume includes summaries of the invited talks of CPAIOR: one from industry, one from the embedded system research community and one from the Operations Research community. The invited speakers were: Filippo Focacci from ILOG S.A. France, one of the leading companies in the field; Paul Pop, professor in the Embedded Systems Lab, in the Computer and Information Science Department, Linköping University; Paul Williams, full professor of Operations Research at the London School of Economics.

The day before CPAIOR, a Master Class has been organized by Gilles Pesant, where leading researchers gave introductory and overview talks in the area of Metaheuristics and Constraint Programming. The Master Class is intended for PhD students, researchers, and practitioners. We are very grateful to Gilles who brought this excellent program together.

For conference publicity we warmly thank Willem Jan van Hoeve and Petr Vilím who did a great job for the high number of submissions received. We are very grateful to Michel Rueher who took care of the non trivial task of finding funds for covering speakers expenses, proceedings, and student grants.

Many thanks to the program committee, who reviewed all the submissions in detail and discussed conflicting papers deeply. Due to the unexpected number of submissions, their load has been almost doubled and their effort has been paid with nothing more than a free dinner.

A special thank goes to Ondřej Čepek from Charles University and Milena Zeithamlová from Action M Agency who spent time in budgeting, planning, booking, and making it all work.

Finally, we would like to thank the sponsors who make it possible to organize this conference: the ARTIST, Network of Excellence sponsoring the talk by Paul Pop and making an interesting cross-fertilization possible; Carmen Systems, Sweden; CoLogNet, Network of Excellence; IISI (Intelligent Information Systems Institute, Cornell), USA; ILOG S.A., France; SICS, Sweden.

June 2005   Roman Barták and Michela Milano

Organizers

Charles University in Prague, Faculty of Mathematics and Physics
Action M Agency (local arrangements)

Executive Committee

Roman Barták, Charles University, Czech Republic (conference co-chair)
Michela Milano, Universitá di Bologna, Italy (conference co-chair)
Ondřej Čepek, Charles University, Czech Republic (organization chair)

Program Committee

Abderrahmane Aggoun, Cosytec, France
Philippe Baptiste, Ecole Polytechnique, France
Roman Barták, Charles University, Czech Republic (chair)
Chris Beck, University of Toronto, Canada
Mats Carlsson, SICS, Sweden
Ondřej Čepek, Charles University, Czech Republic
Hani El Sakkout, CISCO, UK
Bernard Gendron, CRT and Univ. of Montreal, Canada
Carmen Gervet, IC-Parc, UK
Carla Gomes, Cornell University, USA
John Hooker, Carnegie Mellon University, USA
Narendra Jussien, Ecole des Mines de Nantes, France
Stefan Karisch, Carmen Systems, Canada
Francois Laburthe, Bouygues, France
Andrea Lodi, Universitá di Bologna, Italy
Michela Milano, Universitá di Bologna, Italy (chair)
George Nemhauser, Univ. of Georgia Tech, USA
Gilles Pesant, CRT and Ecole Polytechnique de Montreal, Canada
Jean-Francois Puget, ILOG, France
Jean-Charles Régin, ILOG, France
Michel Rueher, Univ. of Nice-Sophia Antipolis, France
Meinolf Sellmann, Brown University, USA
Helmut Simonis, IC-Parc, UK
Sven Thiel, Max Planck Institute, German
Gilles Trombettoni, Univ. of Nice-Sophia Antipolis, France
Michael Trick, Carnegie Mellon University, USA
Pascal van Hentenryck, Brown University, USA
Mark Wallace, Monash University, Australia
Weixiong Zhang, Washington University, USA

Additional Referees

Carlos Ansotegui
Konstantin Artiouchine
Nicolas Beldiceanu
Hachemi Bennaceur
Thierry Benoist
Lucas Bordeaux
Ken Brown
Tom Carchrae
Alberto Caprara
Filipe Carvalho
David Daney
Pierre Deransart
Andrew Eremin
Xavier Gandibleux
Etienne Gaudin
Frédéric Goualard
Laurent Granvilliers
Jesper Hansen
Warwick Harvey

Vitaly Lagoon
Yahia Lebbah
Olivier Lhomme
Chu Min Li
Vassilis Liatsos
Tomas Liden
Ivana Ljubic
Ivan Luzzi
Roger Mailler
Michele Monaci
Bertrand Neveu
Stefano Novello
Ammar Oulamara
Nikos Papdakos
Thierry Petit
Ulrich Pferschy
Nikolai Pisaruk
Diego Fernandez Pons
Philippe Refalo

Guillaume Rochart
Andrea Roli
Benoit Rottembourg
Louis-Martin Rousseau
Jean-David Ruvini
Andrew Sadler
Paul Shaw
Barbara Smith
Stefano Smriglio
Peter Stuckey
Andrea Tramontani
Charlotte Truchet
Jean-Paul Watson
Quanshi Xia
Xiaolan Xie
Neil Yorke-Smith
Tallys Yunes
Alessandro Zanarini

Sponsors

ARTIST, Network of Excellence
Carmen Systems, Sweden
CoLogNet, Network of Excellence
IISI (Intelligent Information Systems Institute, Cornell), USA
ILOG S.A., France
SICS, Sweden

Table of Contents

Integration of Rules and Optimization in Plant PowerOps

Thomas Bousonville, Filippo Focacci, Claude Le Pape, Wim Nuijten,
Frederic Paulin, Jean-Francois Puget, Anna Robert, and Alireza Sadeghin

ILOG S.A, 9 rue de Verdun, 94253 Gentilly, France
{tbousonville, ffocacci, clepape, wnuijten, fpaulin, jfpuget,
anrobert, asadeghin}@ilog.fr

Abstract. Plant PowerOps (PPO) [9] is a new ILOG product, based
on business rules and optimization technology, dedicated to production
planning and detailed scheduling for manufacturing. This paper describes
how PPO integrates a rule based system with the optimization engines
and the graphical user interface. The integration proposed is motivated
by the need to allow business users to manage unexpected changes in
their environment. It provides a flexible interface for configuring, maintaining and tuning the system and for managing optimization scenarios.
The proposed approach is discussed via several use cases we encountered
in practice in supply chain management. Nevertheless, we believe that
most of the ideas described in this paper apply in almost any area of
optimization application.

1 Introduction

Most manufacturing companies are organized today around integrated programs
called Enterprise Resource Planning (ERP) systems. ERP systems provide the
information backbone needed to manage the day-to-day execution handling the
many transactions that document the activity of a company. Since the beginning of the new century, Advanced Planning and Scheduling (APS) systems have
been increasingly adopted to plan the production taking into account capacity
and material flow constraints in order to meet customer demand. APS systems
embed algorithms for planning and scheduling spanning from the application of
very simple priority rules to complex optimization algorithms depending on the
needs of each customer. Although rule-based scheduling and simulation-based
scheduling are still widely used, today the best APS systems offer scheduling algorithms based on Meta-heuristics, Constraint Programming and Mathematical
Programming.

The highly competitive marketplace on the one hand pushes to improve the
production efficiency; on the other hand it pushes to increase the flexibility
necessary to adapt to the continuous variations of customer demand. Today
manufacturing companies need to produce a higher variety of products and customized products. The increasing needs for flexibility are pushing today's APS

R. Barták and M. Milano (Eds.): CPAIOR 2005, LNCS 3524, pp. 1–15, 2005.

systems to their limits. Many companies are struggling with the limitation of the first generation of APS systems and are looking for new solutions.

A company that needs to implement advanced supply chain optimization tools has two possible choices: either it will implement an APS package or it will build it using optimization components and technology via often long and costly custom development. The drawback of buying an existing APS package is that it provides a generic optimization model which will not take into consideration all production constraints and policies characteristic of the company. Often the company is forced to fit into the predefined model. The bottom line can be a very high total cost of ownership combined with unhappy end users who, in some cases, replace the system with their previously developed Excel spreadsheet. The alternative of developing a custom solution is only viable for few companies (often with large OR departments). And even in this case the usability of custom development is not guaranteed. In both cases, changing the supply chain optimization system to follow the rapidly evolving business conditions is an issue.

The challenge for APS packages vendors is therefore to provide enough generality to avoid developing an optimization engine for each and every customer and to build a flexible and configurable enough system to meet the real needs of the customer. Ideally, such package should be configurable by its business users. These are the people that actually solve a business problem, such as producing the production plan for a plant. These business users usually do not possess the IT skills that are needed for adapting an APS package to the peculiarities of their plant.

There are many ways in which an IT package could be made more flexible. One could add a scripting language to it for instance. Unfortunately, even scripting languages are deemed to be too complex to be learned and used by business users. Another possibility could be to use tools that business users use, such as a spreadsheet. However, a spreadsheet interface is not powerful enough to express complex use cases such as business policies. A third approach to flexibility is emerging nowadays. It is called business rules. This approach let business users make statements about their business in a friendly way. These rules are then used to preprocess (or postprocess) the input (or output) of an optimization application. This use of rules is quite different than the so called rule inference systems that were used in expert systems in the 80's. Indeed, rules aren't used here to solve a problem. Rather, they are used to state what the problem is really about. This difference is at the root of the current success of business rules in the market place.

We are convinced that advances and the increased popularity of Business Rules create the opportunity to provide the flexibility the lack of which was limiting the applicability of APS systems. In this paper we describe how Plant PowerOps [9] takes advantage of this technology and we claim that the proposed rules interface can be generalized to the vast majority of optimization applications. Indeed, although this paper does not propose any advance in either rule based systems or operations research, it presents a new, extremely pragmatic, way of applying and integrating rule based systems to optimization models and algorithms.

The structure of the paper is as follows: in section 2 we present an overview of Plant PowerOps briefly describing the types of problems solved by PPO and its architecture. Section 3 is devoted to present different use cases where the integration of rules and optimization is demonstrated to be a powerful combination to overcome the limits of today's optimization software. Section 4 explains the reasons behind some of the design decisions taken in the development of the proposed integration, discusses open questions and future work. Section 5 presents some related approaches. Section 6 concludes the paper.

2   Plant PowerOps Overview

Plant PowerOps (PPO) is a new *Advanced Planning and Scheduling (APS)* system dedicated to production planning and detailed scheduling for manufacturing. It enables users to plan the production taking into account capacity and material flow constraints to meet customer demand.

Although Plant PowerOps provides production planning, lot sizing, and detailed scheduling engines, due to space limitations, we will concentrate the examples to the detailed scheduling features of PPO. The scheduling engine is used to schedule production activities (such as chemical reactions, mixing, forming, assembling or separating), and setup activities (such as cleaning or preparing) on different machines or production lines in order to efficiently produce quality finished products in a timely manner, while satisfying customer demand for the finished product.

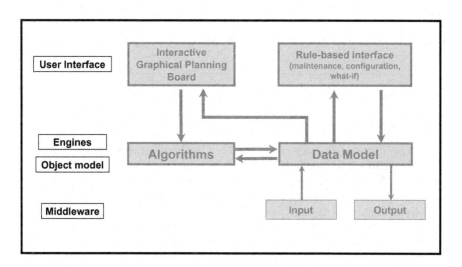

Fig. 1. Architecture

As shown in figure 1, Plant PowerOps provides

- a pre-defined data model to capture intricacies of the manufacturing operations. The pre-defined data model represents, for example, production recipes, production orders, calendars (e.g. breaks, shifts, productivity profiles), resources, customer demands.
- effective optimization models and algorithms based on Mathematical Programming, Constraint Programming and Local Search. These algorithms automatically generate feasible, cost-effective detailed schedules minimizing a combination of objective functions such as total tardiness, total earliness, total setup cost, makespan, processing cost, etc.
- a graphical planning board to visualize, analyze, manually adjust and update production schedules.
- a rule-based customization interface to configure and maintain over time the graphical user interface and the parameters of the model and of the algorithm. The rule-based interface provides also the ability to define optimization scenarios and to modify problem data and solutions.
- an integration framework to connect PPO to a database, to an existing ERP (Enterprise Resource Planning), or to an existing MES (Manufacturing Execution System).

2.1   Rule-Based Interface

The integration between the business rules and the optimization model and between the business rules and the graphical user interface is loose. Rules apply either in a pre-processing step (i.e. before the execution of the engine) or in a post-processing step (i.e. after the execution of the engine).

The rule interface of PPO is based on *ILOG JRules* [8] which parses and interprets *production rules* and executes them in forward-chaining using the Rete algorithm [5].

The typical syntax of the production rules interpreted by PPO is the following:

```
    when
        conditions
    then
        actions
```

Conditions (or left hand side of the rule) are methods returning booleans on objects of the PPO data model. For example, a condition looking for all activities which are due on Jan 1st is translated as *evaluate(theActivity.getDueDate() equals "Jan 1, 2005")*. *Actions* (or right hand side of the rule) can be (i) any method (returning void) typically modifying the state of the objects, (ii) insertion in the working memory, (iii) retraction from working memory. We use actions to produce side effects on the model or on the solution. On top of the rule language, JRules provides a syntax in natural language like flavor that is used in all the examples shown in this paper.

The rule engine and the optimization engine are fully independent and communicate only through modification of the model and of the solutions. The Plant PowerOps user selects the rules she/he wants to apply in a given scenario before or after the optimization per se. Moreover, the optimization model is exposed via a high level *closed* interface. The interface is *closed* in the sense that we do not provide direct access to decision variables so that only predefined constraints are possible.

The left hand side of a rule checks conditions on the model and its right hand side produces side effects on the model if the conditions are met. Rules apply either in a pre-processing step (i.e. before the execution of the engine) or in a post-processing step (i.e. after the execution of the engine). Typically, a pre-processing rule can be seen as a way to transform a model (coming from the legacy system) into a new model upon which optimization is performed. The optimization engine optimizes the transformed model and has no knowledge of the rules that applied to generate it. Post-processing rules check the state of the model and the solutions after optimization has occurred and possibly modify the solutions. The advantage of the loose integration proposed relies on its simplicity and modularity.

Note that the way rules are used in PPO is very different from the way rules are used in expert systems. In expert systems rules are used to solve the problem at hand. This usually requires complex sequences of rules firing, and maintenance was a real concern. In PPO the number of rules we expect to be active is very limited; the rules are often independent from each other and rarely chained together. This simplifies the maintenance issue while still allowing business users to understand and manage them.

3   Use Cases

3.1   The Chocolate Factory

In order to demonstrate the interest of integrating business rules and optimization algorithms we will describe some use cases that could be faced by supply chain managers and production planners of an imaginary chocolate factory.

This imaginary factory produces chocolate confections. Many production steps and machines are required to complete the manufacturing process. The manufacturing process is driven by customer demands and production orders, these processes being driven by recipes and the materials they produce. Costly setup times are required, and multiple process modes (e.g. an activity may be performed in alterative machines) are possible and may be associated with different process costs. Also, activities have precedence constraints.

The factory produces chocolate eggs, rabbits and squirrels. These can be made of either dark chocolate or milk chocolate. Each shape of product –egg (E), rabbit (R), or squirrel (S)– made of either dark (D) or milk (M) chocolate, can be filled with coconut cream (C), hazelnut cream (H), or filled with nothing (N). The possible combinations are identified using three-letter acronym such as *DCE* for dark coconut egg, *MHR* for milk hazelnut rabbit, *DNS* for dark no filling squirrel, etc.

The chocolate factory is composed of two production lines located in two different cities in Switzerland. Each production line has the following production equipment: a cocoa grinding machine, a chocolate mixing machine, a nut grinding machine, a cream mixing machine, and a molding machine.

Recipes for the 12 products with cream filling consist of 5 activities (*Cocoa grinding, Chocolate mixing, Nut grinding, Cream mixing* and*Molding*). Recipes for the 6 products with no cream filling consist of only 3 activities (*Cocoa grinding, Chocolate mixing, Molding*). For each recipe there are precedence constraints such as *Cocoa grinding must precede chocolate mixing.*

There are 5 customer demands:

- One for 3 batches of MNE due Jan. 2, 2005 at 6:00 am
- One for 2 batches of MHR due Jan. 3, 2005 at 6:00 am
- One for 3 batches of DCE due Jan. 4, 2005 at 6:00 am
- One for 4 batches of MHR due Jan. 5, 2005 at 6:00 am
- One for 7 batches of DNE due Jan. 6, 2005 at 6:00 am

These demands result in a total of 19 production orders, one for each batch.

In December, 2004, you are trying to solve a scheduling problem to commence on January 1, 2005, 06:00 am. This example also includes setup times and setup costs, and processing costs. For example, the chocolate mixer requires cleaning when changing from mixing milk chocolate to mixing dark chocolate. The objective is that manufacturing activities should finish as close as possible to their ideal due dates. There are costs associated with late completion. There are also costs associated with the choice of alternative resources that activities are performed in, and with setup times required by those activities. These should also be kept to a minimum.

Typically, the production planner runs Plant PowerOps once a day in order to meet customer expectations and keep internal costs to a minimum. In addition, the production planner runs Plant PowerOps whenever unexpected events (such as a machine breakdown) occur, in order to repair the schedule adapting it to the new situation of the factory. The supply chain manager uses Plant PowerOps to run several simulation scenarios in order to adapt the supply chain business policies to modifications of the market.

We first describe the activities of the supply chain manager; we will successively move to the description of the activities of the production planner.

3.2   What-If Analysis

One of the most important tasks of the supply chain manager is to design and control the production system. A way to achieve supply chain efficiency is to simulate and study the impact of external events and production policies. A very first step is to modify the production data. This can clearly be done by hand on a local copy of the legacy system. A much more effective way to perform massive and complex data modifications is to describe the modification using a rule-based language. The examples of this section demonstrate how business rules can be used to simulate events; in the examples of section 3.3 we demonstrate how business rules can be used to define policies to be applied upon these events. In particular, example 3 represents a business policy applicable to example 1 and example 4 represents a business policy applicable to example 2.

Example 1. Resource Shutdown. Although resource breakdowns are quite infrequent, they may have very important consequences to the efficiency of the production system. Also the supply chain manager may consider the possibility to close part of the factory on a specific day (e.g. Jan 1st 2005). The simulation of a resource shutdown is a necessary first step to try several action plans (business policies) that will be executed upon such an event (see example 3).

Declarations
  for *the resource*, instance of resource
    where the name of *the resource* is Cream mixer 1,
      or the name of *the resource* is Chocolate mixer 1,
  for *the bucket*, instance of bucket
    where *the bucket* is between Jan 1, 2005 6:00am and Jan 2, 2005 6:00am
Then
    the resource is unavailable in *the bucket*

Note that the left hand side of this rule is expressed by the Declarations section which designates the matching objects. Note also that in this first example the resource shutdown can easily be coded using a graphical user interface instead of the rules interface. Section 4.3 discusses the relation between GUI and rules interface.

This rule is automatically translated into a production rule that has a side effect on the model:

```
when {
    the_resource:IloMSResource((getName() equals "Cream mixer 1")
        or (getName() equals "Chocolate mixer 1"));
    the_bucket:IloMSBucket(isBetween("Jan 1, 2005 6:00am",
        "Jan 2, 2005 6:00am"));
} then
    modify the_resource.setCapacity(0,the_bucket);
```

Example 2. Important Sales Agreement with a Customer. The conditions of a big sales agreement are going to be negotiated with an important customer of the company, which could result in doubling the business made with this company. During the Sales and Operations Planning meeting, the sales representative asks the supply chain manager to study the impact on the production that would be caused by the deal (e.g. on production capacity).

Declarations
    for *the customer order*, instance of demand
      where *this demand* is a customer order
If
    the name of the customer for *the customer order* is "Hane"
Then
    set the quantity of material requested by *the customer order* to
      the quantity of material requested by *the customer order* $\times$ 2

3.3 Business Policies

As mentioned before, simulation in terms of massive and complex data modification is only the first step for an effective management of the production system. Once we are able to simulate events, we want to define those business policies that enable us to best deal with the events.

Example 3. Reduce Safety Stock During a Resource Shutdown. Safety stocks are necessary to face unexpected events and stochastic data. A resource breakdown is one such event and it justifies the usage of safety stock. Combining the optimization algorithm and the business rules capability of Plant PowerOps the supply chain manager is able to find the following business policy:

Declarations
    for *the down bucket*, instance of bucket,
    for *any resource*, instance of resource,
    for *any material*, instance of material,
    for *the impacted bucket*, instance of bucket
    where the start time of *this bucket* is greater
      than the start time of *the down bucket*
      and the start time of *this bucket* is less than
      the start time of *the down bucket* + 15
If
    the capacity of *any resource* in *the down bucket* is 0
Then
    set the safety stock of *any material* in *the down bucket* to 0
    and set the safety stock of *any material* in *the impacted bucket* to 0

This rule accounts for the fact that not only during the shutdown time, but also during a given time that follows the resource unavailability it is appropriate to use the safety stock in order to fulfill the demand. The rule is telling the optimizer to accept a lower stock by reducing the desired safety stock level to 0. Note that by using the rules in example 1 and 3 the supply chain manager is able to define an appropriate policy to apply in case of resource breakdowns or decided shutdown.

Example 4. Gold Customer Production Policy. In order to obtain a pivotal selling agreement (see example 2), the CEO of the company has promised to never deliver late large orders coming from the gold customer. The supply chain manager has implemented the following business policy into the system.

> **Declarations**
>   for *the demand* , instance of demand
>   where *the demand* is a customer order
> **If**
>   the category of the customer for *the demand* is gold
> **Then**
>   set the tardiness variable cost for the due date of *the demand* to "high"

3.4   Model Preprocessing

Adding an APS on top of an ERP means stepping from pure transactional data processing to the more complex optimization tasks. It often turns out that the existing data in the ERP data base is not sufficient (i) to express all constraints that hold for the production problem, (ii) to incorporate implicit preferences of the planner, (iii) to balance between conflicting objectives in the evaluation of a solution.

While adding appropriate fields for static data to the legacy system is not a big issue, there are numerous cases where this data has to be calculated dynamically (optimization weights, load dependent preferences, etc.). Maintenance of these data and procedures can become a nightmare.

Preprocessing rules help to express explicitly the necessary transformation logic and avoid out-of-date and inconsistent data by creating it dynamically. They also provide an easy way to build a set of preferences the user wants to apply in a given context.

Example 5. Products for the Same Customer Demand are to be Produced on the Same Production Line. For some products a high degree in regularity is important. Let's assume a nearly identical molding quality can only be guaranteed when the chocolate is processed on the same line. To dispatch this constraint we can use the following rule:

> **Declarations**
>   for *the demand*, instance of demand,
>   for *order A*, instance of production order
>   where *the demand* is satisfied by *order A*,
>   for *order B*, instance of production order
>   where *the demand* is satisfied by *order B*,
>   for *activity 1*, instance of the activities generated from *order A*,
>   for *activity 2*, instance of the activities generated from *order B*
> **If**
>   the name of *activity 1* contains Molding
>   and the name of *activity 2* contains Molding
> **Then**
>   insert in the working memory a new activity compatibility constraint
>   so that *activity 1* and *activity 2* are processed on the same line

Note that activity compatibility constraints are part of the object model of Plant PowerOps [9]. This constraint forces two given activities to be executed in resources belonging to the same production line.

3.5   Tune the Engine

An effective plan is always a trade-off between conflicting objectives. For example, in order to minimize the setup and production costs we should produce long campaigns of similar products. Such a production policy will probably lead to poor customer satisfaction because there is a continuous demand for a mix of different products. After having classified its possible customers in three categories (*normal, silver, gold*), the supply chain manager decided to adapt the objectives of the optimization to the configuration of the customer demands to be satisfied. In case of large amounts of demands from gold customers, customer satisfaction should be privileged. Otherwise production efficiency should be more important.

Example 6. Emphasize Customer Satisfaction.

> **If**
>   the percentage of gold customers is less than 20
> **Then**
>   set the total setup cost weight to "high"
>   and set the total tardiness weight to "low"
> **Else**
>   set the total setup cost weight to "low"
>   and set the total tardiness weight to "high"

The last three following scenarios concern the use of the business rules interface during the activity of the production planner. The production planner uses Plant PowerOps for generating the day to day schedule of the factory and is not allowed to change the business policies defined by the supply chain manager. He/she is nevertheless able to use the rule based interface to configure the system for his/her daily activities and to run validation tests.

3.6   Data Validation

Example 7. Minimal Order Quantity. For technical reasons (or by mistake) the sales department may enter into the system customer orders with low quantity of finished products. To prevent these orders from being considered in the planning, the following rule enables the production planner to make sure that only orders with more than two batch units are scheduled.

> **Declarations**
>   for *the demand*, instance of demand
>   where *the demand* is a customer order
> **If**
>   the quantity of material requested by it the demand is less than 3
> **Then**
>   display the name of *the demand*
>   and display "requests less than 3 product units"

3.7 Solution Checking

In addition to built-in solution checking, PPO allows defining factory specific checking rules applied to solutions. It does not matter if the solution has been generated by the optimizer or by hand. This technique is well suited to check soft constraints or desired properties that are not directly expressed in the constraint model.

Example 8. Temporal Dispersion of Related Activities. The following rule keeps the planner informed when two activities belonging to the same production order have been scheduled far away from each other.

Declarations
    for *order A*, instance of production order,
    for *activity 1*, instance of the activities generated from *order A*,
    for *activity 2*, instance of the activities generated from *order A*,
    for *the solution* is the best scheduling solution
If
    the start time of *activity 2* in *the solution* is greater than
        the end time of *activity 1* in *the solution* + 15
Then
    add in the checker of the solution a violation "Dispersed activities"

3.8 Graphical Rules

Graphical actions include coloring, filtering and selection. While most of them are predefined (filtering types of resources, color late activities), others have more complex parameters.

Example 9. Select Late Activities That Belong to a Gold Customer. As we have seen above, some orders may have a higher priority than others. Therefore we would like to refine the information presented by selecting only the late orders that are produced for a gold customer. Using the rule interface this can be expressed as follows:

Declarations
    for *the customer order*, instance of demand
        where *this demand* is a customer order,
    for *the order*, instance of production order
        where *the customer order* is satisfied by *the order*,
    for *activity 1*, instance of the activities generated from *the order*
If
    the category of the customer for *the customer order* is gold
    and the tardiness cost of *activity 1* in the solution is greater than 0
Then
    add *activity 1* to selection
Else
    remove *activity 1* from selection

For *coloring* different types of color schemas make sense: customer type, order value, material properties, etc. Using a rule based configuration interface the user can establish a series of commonly used coloring schemas without coding. The gain against a call for application extension can be measured in money, time and autonomy [1].

All the scenarios presented demonstrate the flexibility of a rule-based interface on top of optimization algorithms. Note that these scenarios could not have been done on the any of the most popular APS in the market without a major development effort. In fact either they do not provide any form of scripting language (e.g. Oracle/APS), or the scripting language does not allow modifications of the optimization model (e.g. the ABAP language of SAP/APO). In some cases it is possible to write special purpose optimization algorithms replacing the ones available in the APS (for example, this is true for both Oracle and SAP). However this implies a large project, including writing transformation from and to business model and a brand new optimizer. It would be overkill to achieve one of the scenarios described by such custom development.

4　Open Questions and Future Work

4.1　Loose Integration or Tight Integration

Although conceptually interesting, we are convinced that, in general, a tighter integration where the rules and the optimization engines directly communicate, would be much more complex without bringing a sensible added value. A tighter integration would end up being yet another high level optimization language (based on business rules). Such an imaginary *rule-based optimization language* would be far from being practical as supply chain optimization tool dedicated to people with little optimization experience. Moreover, the interaction of optimization and rules engines would generate difficult robustness issues. On the contrary, in the proposed approach, the robustness of embedded heuristics is enforced by the closed model. The end user may enrich the model using predefined constraints and influence the search procedure, but not interact with it. Therefore, once we are able to deal with infeasible input data, we do not have to deal with issues such as rules that could make the optimization problem impossible to handle as this translates into infeasible input data. Somewhere in between the loose integration proposed in this paper and a tight integration is applied in [4] for optimization systems used in the airline and railway industries and is described in section 5.

---

[1] Note that the manufacturing object model of Plant PowerOps does not provide the concept of a customer category (*normal, silver, gold*). Plant PowerOps enables users to dynamically attach properties to objects. These properties can be used in the left hand side (the *if* statement) of business rules thus providing a powerful mechanism to extend the object model and to write constraints (rules) based on these extensions as shown in the examples 4 and 9.

4.2   Use Rules to Guide the Search Heuristics

Although we are convinced that a loose integration of rules and optimization is better than a tight cooperative integration, nevertheless we are aware that business rules may play an important role in a more sophisticated method to guide the engine towards desired solutions. The design of methods to guide the search based on business rules is subject of future work. We believe that the following types of interactions could be highly interesting. Interaction of rules and optimization in constructive search methods; definition of local moves via business rules; use business rules to describe how to repair an infeasible schedule, and finally definition of soft constraints (preferences) via business rules. The challenge of the design of rule-based methods to guide the search will be to keep the clear separation between the rule based interface and the optimization engines.

4.3   Rules and GUI

Nowadays business rules systems such as *ILOG JRules* [8] provide a user friendly interface to write rules in natural language (see all provided examples) or technical language (see example 1). Rules can be stored in rule repositories and saved. Moreover, *parametric rules* or *rule templates* can be defined to enable users to generate new rules by modifying (specializing) a given rule template. Despite all that, writing a rule is always a complex task compared to a sequence of clicks in a graphical user interface. Consider example 1 of section 3.2 where a rule defines a machine breakdown or shutdown. This is a typical case where a small graphical item could provide the very same functionality with a much simpler user interaction. In our experience, it is not always easy to decide which functionality should be provided as GUI items and which should be provided via a rules interface. Our current approach is to provide a set of pre-defined rules first, which may become part of the graphical user interface later upon request.

5   Related Work

The integration of rule-based systems and optimization has been widely investigated in the literature. For example, one of the first constraint-based scheduling system, SOJA [10], used rules both to select the activities to schedule over the next day and to heuristically guide the constraint-based search. A more systematic approach proposing integration of constraints and rules can be found in the programming language LAURE [3] [1]. Caseau and Koppstein propose a multi-paradigm object-oriented language integrating rule-based and constraint-based technology. LAURE supports forward chaining production rules and backward chaining. In LAURE rule-based programming provides deductive capabilities that is merged with constraint satisfaction for improving the efficiency of constraint satisfaction. The integration of rules and optimization in LAURE is tight, and the rule technology is part of the optimization language used to solve the problems. The programming language LAURE evolved in a new programming language called CLAIRE [2] which packages the features proved useful in LAURE in a much simpler language. The backward chaining functionality of LAURE was removed, and the forward chaining functionality was basically used to build propagation algorithms.

A different integration of rule technology and problem solving can be found in the vast literature on *Constraint Handling Rules* (see e.g. [6]). Constraint Handling Rules is a high-level rule-based language for writing constraint solvers and reasoning systems. Again, the spirit of the integration of rules and optimization is very different from the loose integration proposed in this paper.

A rule-based front end to optimization is available in the crew pairing optimization system of *Carmen Systems* ([4], [7]) where the rule language *Rave* is used to define feasible pairings. In the airline and railway industries, legal pairings must satisfy a large number of governmental and collective agreements which vary from an airline to another. Such rules are not hardwired, but rather specified by the user using the specific rule language *Rave*. The interaction between the optimization engine and the rule engine is tighter than the one proposed in this paper as the rule engine is called to validate possible pairings during column generation. It is still a loose integration in the sense that the rule engine behaves as a black box for the optimization engine and provides simply a yes/no answer on the feasibility of possible pairings. The advantages of the integration of rules and optimization of *Carmen System* are that the rules can be easily changed and maintained by users and it is easy to perform what-it analysis.

6   Conclusions

We have proposed a new, pragmatic, approach for the integration of business rules and optimization engines. The proposed integration provides the flexibility, adaptability and extensibility that was missing in today's supply chain optimization systems. Besides the presentation of the integration framework, one goal was to present a categorization of pertinent rules for optimization applications. This classification was done based on the rule purpose in the application context: what-if analysis, business policies, model preprocessing, engine tuning, data validation, graphical actions and solution checking. Although the proposed approach is described on supply chain optimization, we believe it can be applied to most optimization applications. For example, similar investigation is conducted at ILOG in the area of transportation. We hope that the flexibility provided by the interaction of rules and optimization removes many obstacles in the adoption of Advanced Planning and Scheduling systems.

References

1. Y. Caseau and P. Koppstein. A Rule-based approach to a Time-Constrainted Traveling Salesman Problem. In *Proceedings of Symposium of Artificial Intelligence and Mathematics*, 1992.
2. Y. Caseau and F. Laburthe. CLAIRE: Combining objects and rules for problem solving. In T. Ida M.T. Chakravarty, Y. Guo, editor, *Proceedings of the JICSLP'96 workshop on multi-paradigm logic programming*, 1996.

3. Yves Caseau and Peter Koppstein. A cooperative-architecture expert system for solving large time/travel assignment problems. In *Database and Expert Systems Applications*, pages 197–202, 1992.
4. N. Kohl E. Andersson, E. Housos and D. Wedelin. *Crew pairing optimization*, pages 228–258. Kluwer Academic Publishers, 1990. G. Yu, editor.
5. C.L. Forgy. Rete: a fast algorithm for the many pattern/many object pattern match problem. *Artificial Intelligence*, pages 17–37, 1982.
6. Thom Frühwirth. Theory and practice of constraint handling rules. *Journal of Logic Programming, Special Issue on Constraint Logic Programming*, 37(1-3):95–138, October 1998.
7. Curt A. Hjorring and Jesper Hansen. Column generation with a rule modelling language for airline crew pairing. In *Proceedings of the 34th Annual Conference of the Operational Research Society of New Zealand*, 1999.
8. ILOG. *ILOG JRules 5.0 User's Manual and Reference Manual*.
9. ILOG. *ILOG Plant PowerOps 1.0 User's Manual and Reference Manual*.
10. C. Le Pape. Soja: A daily workshop scheduling system. soja's system and inference engine. In *Proceedings of the Fifth Technical Conference of the British Computer Society Specialist Group on Expert Systems, Warwick, United Kingdom*, 1985.

Embedded Systems Design: Optimization Challenges

Paul Pop*

Computer and Information Science Department,
Linköping University, Sweden
http://www.ida.liu.se/~paupo

Embedded systems are everywhere: from alarm clocks to PDAs, from mobile phones to cars, almost all the devices we use are controlled by embedded systems. Over 99% of the microprocessors produced today are used in embedded systems, and recently the number of embedded systems in use has become larger than the number of humans on the planet.

The complexity of embedded systems is growing at a very high pace and the constraints in terms of functionality, performance, low energy consumption, reliability, cost and time-to-market are getting tighter. Therefore, the task of designing such systems is becoming increasingly important and difficult at the same time.

New automated design optimization techniques are needed, which are able to: successfully manage the complexity of embedded systems, meet the constraints imposed by the application domain, shorten the time-to-market, and reduce development and manufacturing costs.

In this talk, the presenter will introduce several embedded systems design problems, and will show how they can be formulated as optimization problems. Solving such challenging design optimization problems are the key to the success of the embedded systems design.

---

* Member of ARTIST2 Network of Excellence.

R. Barták and M. Milano (Eds.): CPAIOR 2005, LNCS 3524, p. 16, 2005.
© Springer-Verlag Berlin Heidelberg 2005

Models for Solving the Travelling Salesman Problem

H.P. Williams

Department of Operational Research,
The London School of Economics,
Houghton Street, London WC2A 2AE, UK
http://personal.lse.ac.uk/williahp/

The Travelling Salesman Problem is a classic problem of Combinatorial Optimisation and involves routing around a number of cities in order to cover the minimum total distance. It is notoriously difficult to solve practical sized instances optimally. The classical Integer Programming formulation involves an exponential number of constraints.

Alternative formulations will be given which use less constraints (a polynomial number). These rely on (often ingenious) ways of introducing extra variables with a variety of real-life interpretations. The purpose of this talk will be to suggest the use of 'lateral' thinking in creating new formulations. These extra variables can be incorporated in extra 'logical constraints' which can help the solution process. The compactness of the formulations and the existence of extra variables which can be exploited in search strategies suggests they might be valuable if a Constraint Programming approach is adopted. This aspect is still to be investigated in detail.

Basically three distinct ideas are used in the different formulations.

Firstly sequence variables are introduced representing the sequence in which cities are visited. These can be used to prevent subtours by using $O(n^2)$ constraints (instead of the exponential number needed in the classical formulation). These extra variables also allow one to specify extra relations which help the solution process.

Secondly flow variables are introduced together with material balance constraints. These force the tour to be connected. $O(n^2)$ constraints are needed. If, however, the flows are split into distinct quantities, leading to a multicommodity flow formulation then $O(n^3)$ extra constraints are needed but the Linear Programming Relaxation of the model is of equal strength to that of the classical (exponential) formulation.

Thirdly staged variables are used with a third index representing the stage at which an arc is traversed. There are a number of ways of incorporating these variables into constraints to prevent subtours leading to models with $O(n)$ (remarkably) and $O(n^2)$ constraints. Again the existence of these variables allows one to specify extra conditions which could aid the solution process.

These formulations can be compared by projecting the polytopes of the Linear Programming relaxations into the same space. Remarkably the resultant polytopes are proper subsets of each other. The hierarchy of sizes of the poly-

R. Barták and M. Milano (Eds.): CPAIOR 2005, LNCS 3524, pp. 17–18, 2005.

topes is independent of problem instance allowing one to rank the quality of the formulations in terms of the strength of the Linear Programming relaxation. This does not, however, mean that the relative qualities of the formulations will be the same if other Search and bounding procedures are used other than Linear Progaramming.

Finally the possibility of arriving at better formulations syntactically (as opposed to semantically) will be discussed.

Set Variables and Local Search*

Magnus Ågren, Pierre Flener, and Justin Pearson

Department of Information Technology,
Uppsala University, Box 337,
SE – 751 05 Uppsala, Sweden
{agren, pierref, justin}@it.uu.se

Abstract. Many combinatorial (optimisation) problems have natural models based on, or including, set variables and set constraints. This was already known to the constraint programming community, and solvers based on constructive search for set variables have been around for a long time. In this paper, set variables and set constraints are put into a local-search framework, where concepts such as configurations, penalties, and neighbourhood functions are dealt with generically. This scheme is then used to define the penalty functions for five (global) set constraints, and to model and solve two well-known applications.

1 Introduction

Many combinatorial (optimisation) problems have natural models based on, or including, set variables and set constraints. Classical examples include set partitioning and set covering, and such problems also occur as sub-problems in many real-life applications, such as airline crew rostering, tournament scheduling, time-tabling, and nurse rostering. This was already known to the constraint programming community, and constructive search (complete) solvers for set variables have been around for a long time now (see for example [11, 15, 19, 2]).

Complementary to constructive search, local search [1] is another common technique for solving combinatorial (optimisation) problems. Although not complete, it usually scales very well to large problem instances and often compares well to, or outperforms, other techniques. Historically, the constraint programming community has been mostly focused on constructive search and has only recently started to apply its ideas to local search. This means that concepts such as high declarativeness, global constraints with underlying incremental algorithms, and high-level modelling languages for local search have been introduced there (see [12, 25, 22, 16, 10, 7, 13, 23, 14, 6] for instance).

In this paper, we introduce set variables and (global) set constraints to constraint-based local search. More specifically, our *contributions* are as follows:

- We put the local-search concepts of penalties, configurations, and neighbourhood functions into a *set-variable framework*. (Section 2)

---

* This paper significantly extends and revises Technical Report 2004-015 of the Department of Information Technology, Uppsala University, Sweden.

R. Barták and M. Milano (Eds.): CPAIOR 2005, LNCS 3524, pp. 19–33, 2005.

- In order to be able to use (global) set constraints generally in local search, we propose a *generic penalty scheme*. We use it to give the *penalty definitions of five (global) set constraints*. Other than their well-known *modelling* merits, we show that (global) set constraints provide opportunities for a hardwired global reasoning while *solving*, which would otherwise have to be hand-coded each time for lower-level encodings of set variables, such as integer variables for the characteristic functions of their set values. (Section 3)
- In order to obtain efficient solution algorithms, we propose methods for the *incremental penalty maintenance* of the (global) set constraints. (Section 4)
- The (global) set constraints are used to *model and solve two well-known problems*, with promising results that motivate further research. (Section 5)

After this, Section 6 discusses related and future work and concludes the paper.

2   Local Search on (Set) Constraint Satisfaction Problems

A *constraint satisfaction problem (CSP)* is a triple $\langle V, D, C \rangle$, where V is a finite set of variables, D is a finite set of finite domains, each $D_v \in D$ containing the set of possible values for the corresponding variable $v \in V$, and C is a finite set of constraints, each $c \in C$ being defined on a subset of the variables in V and specifying their valid combinations of values.

The definition above is very general and may be used with any choice of finite-domain variables. The variables in V may, for example, range over sets of integers (integer variables), strings, or, as in our case, sets of values of some type (set variables, defined formally below). Of course, a CSP may also contain variables with several kinds of domains. As an example, consider a CSP $\langle V, D, C \rangle$ in which some variables $\{i_1, \ldots, i_k\} \subset V$ are integer variables, and some other variables $\{s_1, \ldots, s_k\} \subset V$ are set variables. These could for instance be connected with constraints stating that the cardinality of each s_j must not exceed i_j.

In this paper, we assume that *all* the variables are set variables, and that all the constraints are stated on variables of this kind. This is of course a limitation, since many models contain both set variables and integer variables. However, mixing integer variables and set variables makes the constraints harder to define, and we consider this to be future work. Fortunately, interesting applications, such as the two in this paper, are already possible to model.

Definition 1 (Set Variable and its Universe). *Let* $P = \langle V, D, C \rangle$ *be a CSP. A variable* $s \in V$ *is a set variable if its corresponding domain* $D_s = 2^{U_s}$, *where* U_s *is a finite set of values of some type, called the* universe of s.

Note that this definition does not allow the indication of a non-empty set of required values in the universe of a set variable, hence this must be done here by an explicit constraint. This is left as future work, as not necessary for our present purpose.

Definition 2 (Configuration). *Let* $P = \langle V, D, C \rangle$ *be a CSP. A configuration for* P *is a total function* $k : V \to \bigcup_{s \in V} D_s$ *such that* $k(s) \in D_s$ *for all* $s \in V$.

Definition 3 (Delta of Configurations). *Let $P = \langle V, D, C \rangle$ be a CSP and let k and k' be two configurations for P. The* delta *of k and k', denoted delta(k, k'), is the set $\{(s, v, v') \mid s \in V$ & $v = k(s) - k'(s)$ & $v' = k'(s) - k(s)$ & $v \neq v'\}$, where $-$ stands for the set difference.*

Example 1. Consider a CSP $P = \langle \{s_1, s_2, s_3\}, \{D_{s_1}, D_{s_2}, D_{s_3}\}, C \rangle$ where $D_{s_1} = D_{s_2} = D_{s_3} = 2^{\{d_1, d_2, d_3\}}$ (hence $U_{s_1} = U_{s_2} = U_{s_3} = \{d_1, d_2, d_3\}$). One possible configuration for P is defined as $k(s_1) = \{d_3\}, k(s_2) = \{d_1, d_2\}, k(s_3) = \emptyset$, or equivalently as the set of mappings $\{s_1 \mapsto \{d_3\}, s_2 \mapsto \{d_1, d_2\}, s_3 \mapsto \emptyset\}$. Another configuration for P is defined as $k' = \{s_1 \mapsto \emptyset, s_2 \mapsto \{d_1, d_2, d_3\}, s_3 \mapsto \emptyset\}$. Now, the delta of k and k' is $delta(k, k') = \{(s_1, \{d_3\}, \emptyset), (s_2, \emptyset, \{d_3\})\}$.

Definition 4 (Neighbourhood Function). *Let K denote the set of all possible configurations for a CSP P and let $k \in K$. A* neighbourhood function *for P is a function $\mathcal{N} : K \rightarrow 2^K$. The* neighbourhood *of P with respect to k and $\mathcal{N}$ is the set of configurations $\mathcal{N}(k)$.*

Example 2. Consider P and k from Example 1. A possible neighbourhood of P with respect to k and some neighbourhood function $\mathcal{N}$ for P is the set $\mathcal{N}(k) = \{k_1 = \{s_1 \mapsto \emptyset, s_2 \mapsto \{d_1, d_2, d_3\}, s_3 \mapsto \emptyset\}, k_2 = \{s_1 \mapsto \emptyset, s_2 \mapsto \{d_1, d_2\}, s_3 \mapsto \{d_3\}\}\}$. This neighbourhood function moves the value d_3 in s_1 to variable s_2 or variable s_3, decreasing the cardinality of s_1 and increasing the one of s_2 or s_3.

We will use two *general neighbourhoods* in this paper, which are defined next. For both, let $s \in V$, $S \subseteq V - \{s\}$, and let $k \in K$ be a configuration for a CSP $P = \langle V, D, C \rangle$, where K is the set of all configurations for P. The first one, called *move*, is defined by the neighbourhood function with the same name:

$$move(s, S)(k) = \{k' \in K \mid \exists d \in k(s) : s' \in S \ \& \ d \in U_{s'} - k(s') \ \& \\ delta(k, k') = \{(s, \{d\}, \emptyset), (s', \emptyset, \{d\})\}\}$$

This neighbourhood, given k, is the set of all neighbourhoods k' that differ from k in the definition of two distinct set variables s and s', the difference being that there exists exactly one $d \in k(s)$ such that $d \in k(s) \Leftrightarrow d \notin k'(s)$ and $d \notin k(s') \Leftrightarrow d \in k'(s')$. Hence, d was moved from s to s'.

The second one, called *swap*, is defined by the neighbourhood function:

$$swap(s, S)(k) = \{k' \in K \mid \exists d \in k(s) : \exists d' \in U_s - k(s) : s' \in S \ \& \ d' \in k(s') \\ \& \ d \in U_{s'} - k(s') \ \& \\ delta(k, k') = \{(s, \{d\}, \{d'\}), (s', \{d'\}, \{d\})\}\}$$

This neighbourhood, given k, is the set of all neighbourhoods k' that differ from k in the definition of two distinct set variables s and s', the difference being that there exists exactly one pair $(d \in k(s), d' \in U_s - k(s))$ such that $d \in k(s) \Leftrightarrow d \notin k'(s)$ and $d \notin k(s') \Leftrightarrow d \in k'(s')$, and the opposite for d'. Hence, d and d' were swapped between s and s'.

We will now define the notion of penalty of a constraint, which, informally, is an estimate on how much a constraint is violated. Below is a general definition, followed by a generic scheme for balancing the penalties of different constraints, which is then specialised for each constraint in Section 3.

Definition 5 (Penalty). *Let $P = \langle V, D, C \rangle$ be a CSP and let K denote the set of all possible configurations for P. A penalty of a constraint $c \in C$ is a function $penalty(c) : K \rightarrow \mathbb{N}$. The penalty of P with respect to k is the sum $\sum_{c \in C} penalty(c)(k)$.*

Example 3. Consider once again P from Example 1 and let c_1 and c_2 be the constraints $s_1 \subseteq s_2$ and $d_3 \in s_3$ respectively. Let the penalty functions of c_1 and c_2 be defined as: $penalty(c_1)(k) = |k(s_1) - k(s_2)|$, and $penalty(c_2)(k) = 0$, if $d_3 \in k(s_3)$, or 1, *otherwise* . Now, the penalties of P with respect to the different configurations in the neighbourhood of Example 2 are $penalty(c_1)(k_1) + penalty(c_2)(k_1) = 1$, and $penalty(c_1)(k_2) + penalty(c_2)(k_2) = 0$ respectively.

In order for a constraint-based local-search approach to be effective, different constraints should have balanced penalty definitions [6]: i.e. for a set of constraints C, no $c \in C$ should be easier in general to satisfy compared to any other $c' \in C$. This may be application dependent, in which case weights could be added to tune the penalties, see [13] for example. For set constraints, we believe that one such penalty definition is to let (by extension of the integer-variable ideas in [10]) the penalty of a set constraint c be the length of the shortest sequence of atomic set operations (defined below) that must be performed on the variables in c under a configuration k in order to satisfy c.

Definition 6 (Atomic Set Operations). *Let $P = \langle V, D, C \rangle$ be a CSP, let k be a configuration for P, and let $s \in V$. An atomic set operation on $k(s)$ is one of the following changes to $k(s)$:*

1. *Add a value d to $k(s)$ from its complement $U_s - k(s)$, denoted $Add(k(s), d)$.*
2. *Remove a value d from $k(s)$, denoted $Remove(k(s), d)$.*

Note that no value-replacement operation is considered here; its inclusion would imply a reduction of some of the penalties in Section 3.

Example 4. Performing $\Delta = [Add(k(s), d), \ Remove(k(s), b), \ Add(k(s'), b)]$ on $k(s) = \{a, b, c\}$ and $k(s') = \emptyset$ will yield $\Delta(k(s)) = \{a, c, d\}$ and $\Delta(k(s')) = \{b\}$.

Definition 7 (Operation-Based Penalty for Set Constraints). *Let $P = \langle V, D, C \rangle$ be a CSP and let K be the set of all configurations for P. Let $c \in C$ be a constraint defined on a set of set variables $S \subseteq V$. The penalty of c, $penalty(c) : K \rightarrow \mathbb{N}$, is the length of the shortest sequence of atomic set operations that must be performed in order to satisfy c given a specific configuration k.*

From this definition it follows that $penalty(c)(k) = 0$ if and only if c is satisfied with respect to k. Also, as will be seen, to find a penalty that complies with this definition for a given set constraint is not always obvious.

3   (Global) Set Constraints and Their Penalties

We now present five (global) set constraints and define their penalties. Throughout this section, we assume that k is a configuration for a CSP $P = \langle V, D, C \rangle$, and that $c \in C$.

3.1   AllDisjoint

The global constraint $AllDisjoint(S)$, where $S = \{s_1, \ldots, s_n\}$ is a set of set variables, expresses that all distinct pairs in S are disjoint, i.e. that $\forall i < j \in 1 \ldots n : s_i \cap s_j = \emptyset$. The penalty of an $AllDisjoint(S)$ constraint under k is equal to the length of the shortest sequence Δ of atomic set operations of the form $Remove(k(s), d)$ that must be performed in order for $\forall i < j \in 1 \ldots n : \Delta(k(s_i)) \cap \Delta(k(s_j)) = \emptyset$ to hold. We define the penalty as:

$$penalty(AllDisjoint(S))(k) = \left(\sum_{s \in S} |k(s)| \right) - \left| \bigcup_{s \in S} k(s) \right| \tag{1}$$

Indeed, we need to remove all repeated occurrences of any value, and their number equals the difference between the sum of the set sizes and the size of their union. Hence the following proposition:

Proposition 1. *The penalty (1) is correct with respect to Definition 7.*

3.2   Cardinality

The constraint $Cardinality(s, m)$, where s is a set variable and m a natural-number constant, expresses that the cardinality of s is equal to m, i.e. that $|s| - m$. This constraint would of course be more powerful if we allowed m to be an integer variable. However, as was mentioned earlier, the penalty would be more complicated if we did this, and we see this as future work.

The penalty of a $Cardinality(s, m)$ constraint under k is equal to the length of the shortest sequence Δ of atomic set operations of the form $Add(k(s), d)$ or $Remove(k(s), d)$ that must be performed in order for $|\Delta(k(s))| = m$ to hold. The penalty below expresses this:

$$penalty(Cardinality(s, m))(k) = abs(|k(s)| - m) \tag{2}$$

where $abs(e)$ denotes the absolute value of the expression e. Indeed, we need to add (remove) exactly as many values to (from) $k(s)$ in order to increase (decrease) its cardinality to m. Hence the following proposition:

Proposition 2. *The penalty (2) is correct with respect to Definition 7.*

3.3   MaxIntersect

The global constraint $MaxIntersect(S, m)$, where $S = \{s_1, \ldots, s_n\}$ is a set of set variables and m a natural-number constant, expresses that the cardinality of the intersection between any distinct pair in S is at most m, i.e. that $\forall i < j \in 1 \ldots n : |s_i \cap s_j| \leq m$. This constraint expresses the same as an $AllDisjoint(S)$ constraint when $m = 0$. However, as will be seen, keeping the $AllDisjoint$ constraint is useful for this special case. Again, allowing m to be an integer variable would make the constraint more powerful and is future work.

The penalty of a $MaxIntersect(S, m)$ constraint under k is equal to the length of the shortest sequence Δ of atomic set operations of the form $Remove(k(s), d)$ that must be performed such that $\forall i < j \in 1 \dots n : |\Delta(k(s_i)) \cap \Delta(k(s_j))| \leq m$ holds. In fact, finding a closed form for the exact penalty of a $MaxIntersect$ constraint with respect to Definition 7 turns out not to be that easy. The following expression gives an upper bound on this penalty, namely the sum of the excesses of the intersection sizes:

$$penalty(MaxIntersect(S, m))(k) \leq \sum_{1 \leq i < j \leq n} max(|k(s_i) \cap k(s_j)| - m, 0) \quad (3)$$

Example 5. Assume that $k(s_1) = \{d_1, d_2, d_3\}$, $k(s_2) = \{d_2, d_3, d_4\}$, $k(s_3) = \{d_1, d_3, d_4\}$, and that $c = MaxIntersect(\{s_1, s_2, s_3\}, 1)$. The penalty of c according to (3) is $2 + 2 + 2 = 3$. Indeed, we may satisfy c by performing the sequence of 3 operations $[Remove(k(s_1), d_1), Remove(k(s_2), d_2), Remove(k(s_3), d_3)]$. However, this is not the shortest sequence that achieves this, since after performing $[Remove(k(s_1), d_3), Remove(k(s_2), d_3)]$, the constraint c is also satisfied.

Proposition 3. *The bound of (3) is an optimal upper bound w.r.t. Definition 7.*

Proposition 4. *The upper bound of (3) is zero iff $MaxIntersect(S, m)$ holds.*

However, the upper bound of (3) is not correct with respect to Definition 7 when $m = 0$. Consider $s_1 = \{d_1, d_2\}$, $s_2 = \{d_2, d_3\}$, and $s_3 = \{d_2, d_3\}$. The penalty under (3) of $MaxIntersect(\{s_1, s_2, s_3\}, 0)$ is $1 + 1 + 2 = 4$ whereas the one of $AllDisjoint(\{s_1, s_2, s_3\})$ correctly is $6 - 3 = 3$ under (1).

We may also obtain a lower bound, by using a lemma due to Corrádi [8].

Lemma 1 (Corrádi). *Let $s_1, \dots, s_n$ be r-element sets and U be their union. If $|s_i \cap s_j| \leq m$ for all $i \neq j$, then $|U| \geq \frac{r^2 \cdot n}{r + (n-1) \cdot m}$.*

This lemma can be applied for n ground sets that do not necessarily all have the *same* cardinality r, but rather with r being the *maximum* of their cardinalities, as is the case with $MaxIntersect(S, m)$ and $|S| = n$. It suffices to apply the corrective term $\delta = n \cdot r - \sum_{s \in S} |k(s)|$ when using the lower bound for a configuration k where $r = max_{s \in S} |k(s)|$. Note that δ is the amount of distinct new elements (from a sufficiently large fictitious universe disjoint from $\bigcup_{s \in S} k(s)$) that one must add to the sets in $\{k(s) \mid s \in S\}$ in order to make them all be of size r.

We now have the following lower bound on the penalty of a $MaxIntersect(S, m)$ constraint under a configuration k (where $|S| = n$ and $r = max_{s \in S} |k(s)|$):

$$penalty(MaxIntersect(S, m))(k) \geq \quad (4)$$
$$\left\lceil \frac{r^2 \cdot n}{r + (n-1) \cdot m} \right\rceil - \left(n \cdot r - \sum_{s \in S} |k(s)| \right) - \left| \bigcup_{s \in S} k(s) \right|$$

Example 6. Recall Example 5, where $m = 1$ and the $n = 3$ sets are of the same size $r = 3$, hence $\delta = 0$, and have a union of 4 elements. We get $penalty(c)(k) \geq \lceil \frac{27}{5} \rceil - 0 - 4 = 2$, which is correct with respect to Definition 7.

Now, the following proposition follows from Lemma 1:

Proposition 5. *The bound of (4) is an optimal lower bound w.r.t. Definition 7.*

The next proposition establishes what happens when $m = 0$, in which case *MaxIntersect*(S, m) is equivalent to *AllDisjoint*(S):

Proposition 6. *The lower bound of (4) is correct wrt Definition 7 when $m = 0$.*

Proof. When $m = 0$, then $\left\lceil \frac{r^2 \cdot n}{r + (n-1) \cdot m} \right\rceil = r \cdot n$ and the lower bound of (4) simplifies into the penalty expression (1). Hence it is correct, by Proposition 1.

Unfortunately, the lower bound is sometimes zero even though the constraint is violated. Consider $n = 10$ sets, all of size $r = 3$ (hence $\delta = 0$), that should have pairwise intersections of at most $m = 1$ element and that have a union of 8 elements. Then (4) gives 0 as lower bound on the penalty, but the constraint is violated as there are no such 10 sets, hence m would have to be at least 2.

However, we may still use (4) for the *MaxIntersect* constraint, but it would have to be in conjunction with (3), with the condition that if the lower bound of (4) is zero, then one uses the upper bound of (3) instead. In our experience, the lower bound of (4) is frequently correct. This also argues for keeping the explicit constraint *AllDisjoint*, since for that constraint (4) gives the correct penalty.

An often tighter upper bound than the one of (3) can be obtained by Algorithm 1. It obtains an estimate of the penalty by returning the length of a sequence of atomic set operations constructed in the following way: (i) Start with the empty sequence. (ii) Until the constraint is satisfied, add an atomic set operation removing a value that belongs to a set variable that takes part in the largest number of violating intersections. The algorithm uses the upper bound of (3) as the exit criterion, as it is zero only upon satisfaction of the constraint, by Proposition 4.

Algorithm 1 Calculating the penalty of a *MaxIntersect* constraint

---

function *penalty_max_intersect*$(S, m)(k)$
   $l \leftarrow 0$
   while *penalty*(*MaxIntersect*$(S, m))(k) > 0$ **do**   ▷ According to (3)
      choose $d \in \bigcup_{s \in S} k(s)$ s.t. $|\{(i, j) \mid i < j$ & $d \in k(s_i) \cap k(s_j)$ & $|k(s_i) \cap k(s_j)| > m\}|$ is maximised.
      choose $s_i \in S$ s.t. $|\{s_j \in S \mid i \neq j$ & $d \in k(s_i) \cap k(s_j)$ & $|k(s_i) \cap k(s_j)| > m\}|$ is maximised.
      $l \leftarrow l + 1$ ▷ i.e. an imaginary *Remove*$(k(s_i), d)$ operation was added
      Replace the binding for s_i in k by $s_i \mapsto k(s_i) - \{d\}$
   return l

In the current implementation of the *MaxIntersect* constraint, we use the upper bound given by (3). As we have seen, this is not always a good estimate on the penalty with respect to Definition 7. In the future, we plan to use (4) in conjunction with (3) or (an incremental variant of) Algorithm 1.

3.4   MaxWeightedSum

The constraint $MaxWeightedSum(s, w, m)$, where s is a set variable, $w : U_s \to \mathbb{N}$ is a weight function from the universe of s to the natural numbers, and m is a natural-number constant, expresses that $\sum_{d \in s} w(d) \leq m$. Note that we do not allow negative weights nor m to be an integer variable. Allowing these would need a redefinition of the penalty below.

The penalty of a $MaxWeightedSum(s, w, m)$ constraint under k is equal to the length of the shortest sequence Δ of operations of the form $Remove(k(s), d)$ that must be performed in order for $\sum_{d \in \Delta(k(s))} w(d) \leq m$ to hold. We define the following penalty:

$$penalty(MaxWeightedSum(s, w, m))(k) = \tag{5}$$
$$min\_card\left(\left\{s' \subseteq k(s) \mid \sum_{d' \in s'} w(d') \geq \left(\sum_{d \in k(s)} w(d)\right) - m\right\}\right)$$

where $min\_card(Q)$ denotes the cardinality of a set $q \in Q$ such that for all $q' \in Q$, $|q| \leq |q'|$, or 0 if $Q = \emptyset$. Indeed, we must remove at least the smallest set of values from $k(s)$ such that their weighted sum is at least the difference between the weighted sum of all values in $k(s)$ and m. Hence the following proposition:

Proposition 7. *The penalty (5) is correct with respect to Definition 7.*

3.5   Partition

The global constraint $Partition(S, q)$, where $S = \{s_1, \ldots, s_n\}$ is a set of set variables and q is a ground set of values, expresses that the variables in S are all disjoint, i.e. that $\forall i < j \in 1 \ldots n : s_i \cap s_j = \emptyset$, and that their union is equal to q, i.e. that $\bigcup_{s \in S} s = q$. Note that this definition of a partition allows one or more variables in S to be empty, which is useful in some applications, such as the progressive party problem below. The set q, called the *reference set*, could be generalised to be a set variable. The applications we currently look at do not expect this but this may change in the future. In that case, the penalty function below would have to be changed to take this into account.

The penalty of a $Partition(S, q)$ constraint under k is equal to the length of the shortest sequence Δ of atomic set operations that must be performed in order for $\forall i < j \in 1 \ldots n : \Delta(k(s_i)) \cap \Delta(k(s_j)) = \emptyset$ & $\bigcup_{s \in S} \Delta(k(s)) = q$ to hold. The following penalty expresses this:

$$penalty(Partition(S, q))(k) = \left(\sum_{s \in S} |k(s)|\right) - \left|\bigcup_{s \in S} k(s)\right| + \left|q - \bigcup_{s \in S} k(s)\right| \tag{6}$$

Indeed, the first two terms are those in (1) for *AllDisjoint* and the third term expresses that all unused elements of the reference set must be added to some set of the partition for the union to hold. Hence the following proposition:

Proposition 8. *The penalty (6) is correct with respect to Definition 7.*

Note that this penalty could be reduced by allowing replacement operations.

4   Incrementally Maintaining Penalties

This section presents how the penalties are maintained for two of the presented constraints, *AllDisjoint* and *MaxIntersect*. For the other three, *Partition* is similar to *AllDisjoint*, while *Cardinality* and *MaxWeightedSum* are rather straightforward to maintain. Since in local search one may need to perform many iterations, and since each iteration usually requires searching through a large neighbourhood, it is crucial that the penalty of a neighbouring configuration is computed efficiently. In order to do this, it is important to use *incremental algorithms* that, given a current configuration k, do not recompute from scratch the penalty of a neighbouring configuration k', but rather compute the penalty with respect to the penalty of k and the difference between k and k'.

This technique is used, for instance, in [12, 22] where invariants are used to get efficient incremental algorithms from high-level, declarative descriptions. In this paper, the incrementality is achieved explicitly for each constraint, and we consider it to be future work to implement this in a more general and elegant way. The aim of this paper is to explore the usefulness of the proposed framework and penalty definitions for set constraints.

4.1   Incrementally Maintaining *AllDisjoint*

Recall the penalty (1) for an *AllDisjoint* constraint in Section 3.1. In order to maintain this incrementally, we use a table *count* of integers, indexed by the values in $U = \bigcup_{s \in S} U_s$, such that $count[d]$ is equal to the number of variables that contain d. Now, the sum in (1) is equal to $\sum_{d \in U}(count[d] - 1)$ as it suffices to remove a value $d \in \bigcup_{s \in S} k(s)$ from all but one of the set variables in $\{s \in S \mid d \in k(s)\}$ in order to satisfy the constraint. This is easy to maintain incrementally given an atomic set operation.

4.2   Incrementally Maintaining *MaxIntersect*

Recall the penalty bound of (3) for a *MaxIntersect* constraint. In order to maintain this incrementally, we use the following two data structures: (i) A table *variables* indexed by the values in $U = \bigcup_{s \in S} U_s$, such that $variables[d]$ is the set of variables that d is a member of; (ii) for each variable s_i, a table $s_i.intersects$ indexed by the values in $\{i+1, \ldots, n\}$ such that $s_i.intersects[j] = |k(s_i) \cap k(s_j)|$.

The sum in (3) is then equal to $\sum_{1 \leq i < j \leq n} max(s_i.intersects[j] - m, 0)$ and all this may be maintained incrementally in the following way, given an atomic set operation o. If $o = Add(k(s_i), d)$ then (i) add s_i to $variables[d]$; (ii) for each variable s_j in $variables[d]$ such that $j > i$: if $s_i.intersects[j] \geq m$ then increase the sum in (3) by 1; and (iii) for each variable s_j in $variables[d]$ such that $j > i$: increase $s_i.intersects[j]$ by 1. If $o = Remove(k(s_i), d)$ then (i) remove s_i from $variables[d]$; (ii) for each variable s_j in $variables[d]$ such that $j > i$: if $s_i.intersects[j] > m$ then decrease the sum in (3) by 1; and (iii) for each variable s_j in $variables[d]$ such that $j > i$: decrease $s_i.intersects[j]$ by 1.

Implementing these ideas with respect to the lower bound of (4) and Algorithm 1 is future work.

5 Applications

This section presents two well-known applications for constraint programming: the *Progressive Party Problem* and the *Social Golfers Problem*. They both have natural models based on set variables. They have previously been solved both using constructive and local search. See, for instance, the references [21, 10, 25, 13, 6, 24] and [3, 20, 18, 9], respectively. The constraints in Section 3 as well as the search algorithms were implemented in OCaml and the experiments were run on an Intel 2.4 GHz Linux machine with 512 MB memory.

5.1   The Progressive Party Problem (PPP)

The problem is to timetable a party at a yacht club. Certain boats are designated as hosts, while the crews of the remaining boats are designated as guests. The crew of a host boat remains on board throughout the party to act as hosts, while the crew of a guest boat together visits host boats over a number of periods. The crew of a guest boat must party at some host boat each period (constraint c_1). The spare capacity of any host boat is not to be exceeded at any period by the sum of the crew sizes of all the guest boats that are scheduled to visit it then (constraint c_2). Any guest crew can visit any host boat in at most one period (constraint c_3). Any two distinct guest crews can visit the same host boat in at most one period (constraint c_4).

A Set-Based Model. Let H be the set of host boats and let G be the set of guest boats. Furthermore, let $capacity(h)$ and $size(g)$ denote the spare capacity of host boat h and the crew size of guest boat g, respectively. Let $periods$ be the number of periods we want to find a schedule for and let P be the set $\{1, \ldots, periods\}$. Now, let $s_{(h,p)}$, where $h \in H$ and $p \in P$, be a set variable containing the set of guest boats whose crews boat h hosts during period p. Then the following constraints model the problem:

$(c_1) : \forall p \in P : Partition(\{s_{(h,p)} \mid h \in H\}, G)$
$(c_2) : \forall h \in H : \forall p \in P : MaxWeightedSum(s_{(h,p)}, size, capacity(h))$
$(c_3) : \forall h \in H : AllDisjoint(\{s_{(h,p)} \mid p \in P\})$
$(c_4) : MaxIntersect(\{s_{(h,p)} \mid h \in H \text{ \& } p \in P\}, 1)$

Solving The PPP. If we are careful when defining an initial configuration and a neighbourhood for the PPP, we may be able to exclude some of its constraints. For instance, it is possible to give the variables $s_{(h,p)}$ an initial configuration and a neighbourhood that respect c_1. We can do this (i) by assigning random disjoint subsets of G to each $s_{(h,p)}$, where $h \in H$, for each period $p \in P$, making sure that each $g \in G$ is assigned to some $s_{(h,p)}$ and (ii) by using a neighbourhood specifying that guests from a host boat h are moved to another host boat h' in the same period, and nothing else.

Algorithm 2 is the solving algorithm we used for the PPP. It takes the constant sets P, G, H, and the functions $capacity$ and $size$ as defined above as parameters, specifying an instance of the PPP, and returns a configuration k for a CSP with respect to that instance. *MaxIter* and *MaxNonImproving* are additional arguments as described below. If $penalty(\langle V, D, C\rangle)(k) = 0$, then a solution was found within *MaxIter* iterations. The algorithm uses the notion of *conflict of a variable* (line 10), which, informally, is an estimate on how much a variable contributes to the total penalty of a set of constraints with respect to a configuration.

Algorithm 2 Solving the PPP

1: **procedure** $solve\_progressive\_party(P, G, H, capacity, size)$
2:      Initialise $\langle V, D, C\rangle$ w.r.t. P, G, H, $capacity$, and $size$ to be a CSP $\in$ PPP
3:      $iteration \leftarrow 0$, $non\_improving \leftarrow 0$, $best \leftarrow \infty$
4:      $k \leftarrow \emptyset$, $tabu \leftarrow \emptyset$, $history \leftarrow \emptyset$
5:      **for all** $p \in P$ **do**   ▷ Initialise s.t. c_1 is respected
6:          Add a random mapping $s_{(h,p)} \mapsto G'$, where $G' \subset G$, for each $h \in H$ to k
7:          s.t. $penalty(Partition(\{s_{(h,p)} \mid h \in H\}, G))(k) = 0$
8:      **while** $penalty(\langle V, D, C\rangle)(k) > 0$ & $iteration < MaxIter$ **do**
9:          $iteration \leftarrow iteration + 1$, $non\_improving \leftarrow non\_improving + 1$
10:          choose $s_{(h,p)} \in V$ s.t. $\forall s' \in V : conflict(s_{(h,p)}, C)(k) \geq conflict(s', C)(k)$
11:          $N \leftarrow move(s_{(h,p)}, \{s_{(h',p)} \mid h' \in H$ & $h' \neq h\})(k)$
12:          choose $k' \in N$ s.t. $\forall k'' \in N : penalty(\langle V, D, C\rangle)(k') \leq$
              $penalty(\langle V, D, C\rangle)(k'')$
13:          and $((s_{(h',p)}, d, iteration) \notin tabu$ or $penalty(\langle V, D, C\rangle)(k') < best)$,
14:          where $delta(k, k') = \{(s_{(h,p)}, \{d\}, \emptyset), (s_{(h',p)}, \emptyset, \{d\})\}$
15:          $k \leftarrow k'$, $tabu \leftarrow tabu \cup \{(s_{(h',p)}, d, iteration + rand\_int(5, 40))\}$
16:          **if** $penalty(\langle V, D, C\rangle)(k) < best$ **then**
17:              $best \leftarrow penalty(\langle V, D, C\rangle)(k)$, $non\_improving \leftarrow 0$,
18:              $history \leftarrow \{k\}$, $tabu \leftarrow \emptyset$
19:          **else if** $penalty(\langle V, D, C\rangle)(k) = best$ **then**
20:              $history \leftarrow history \cup \{k\}$
21:          **else if** $non\_improving = MaxNonImproving$ **then**
22:              $k \leftarrow$ a random element in $history$
23:              $non\_improving \leftarrow 0$, $history \leftarrow \{k\}$, $tabu \leftarrow \emptyset$
24:      **return** k

The algorithm starts by initialising a CSP for the PPP, necessary counters, bounds, and sets (lines $2 - 4$), as well as the variables of the problem (lines $5 - 7$). As long as the penalty is positive and a maximum number of iterations has not been reached, lines $8 - 23$ explore the neighbourhood of the problem in the following way. (i) Choose a variable $s_{(h,p)}$ with maximum conflict (line 10). (ii) Determine the neighbourhood of type *move* for $s_{(h,p)}$ with respect to the other variables in the same period (line 11). (iii) Move to a neighbour k' that minimises the penalty (lines $12 - 14$).

In order to escape local minima it also uses a tabu list and a restarting component. The tabu list *tabu* is initially empty. When a move from a configuration k to a configuration k' is performed, meaning that for two variables $s_{(h,p)}$ and $s_{(h',p)}$, a value d in $k(s_{(h,p)})$ is moved to $k(s_{(h',p)})$, the triple $(s_{(h'p)}, d, iteration + t)$ is added to *tabu*. This means that d cannot be moved to $s_{(h',p)}$ again for the next t iterations, where t is a random number between 5 and 40 (empirically chosen). However, if such a move would imply the lowest penalty so far, it is always accepted (lines $13 - 15$). By abuse of notation, we let $(s, d, t) \notin tabu$ be false iff $(s, d, t') \in tabu$ & $t \le t'$.

The restarting component (lines $16 - 23$) works in the following way. Each configuration k such that $penalty(\langle V, D, C \rangle)(k)$ is at most the current lowest penalty is stored in the set *history* (lines $16 - 20$). If a number *MaxNonImproving* of iterations passes without any improvement to the lowest overall penalty, then the search is restarted from a random element in *history* (lines $21 - 23$). A similar restarting component was used in [13, 24] (saving one best configuration) and [6] (saving a set of best configurations), both for integer-domain models of the PPP.

5.2   The Social Golfers Problem (SGP)

In a golf club, there is a set of golfers, each of whom play golf once a week (constraint c_1) and always in ng groups of size ns (constraint c_2). The objective is to determine whether there is a schedule of nw weeks of play for these golfers, such that there is at most one week where any two distinct players are scheduled to play in the same group (constraint c_3).

A Set-Based Model. Let G be the set of golfers and let $s_{(g,w)}$ be a set variable containing the players playing in group g in week w. Then the following constraints model the problem:

$(c_1) : \forall w \in 1 \dots nw : Partition(\{s_{(g,w)} \mid g \in 1 \dots ng\}, G)$
$(c_2) : \forall g \in 1 \dots ng : \forall w \in 1 \dots nw : Cardinality(s_{(g,w)}, ns)$
$(c_3) : MaxIntersect(\{s_{(g,w)} \mid i \in 1 \dots ng$ & $w \in 1 \dots nw\}, 1)$

Solving The SGP. Similar to the PPP, we need to define an initial configuration and a neighbourhood for the SGP. This, and a slightly changed tabu list, are the only changes in the algorithm compared to the one we used for the PPP, hence the algorithm for the SGP is not shown.

We choose an initial configuration k and a neighbourhood that respects the constraints c_1 and c_2, i.e. that each golfer plays every week and that each group is of size ns. We do this (i) by assigning random disjoint subsets of size ns of G to each $s_{(g,w)}$ where $g \in 1 \dots ng$ for each week $w \in 1 \dots nw$ and (ii) by choosing the neighbourhood called *swap*, specifying the swap of two distinct golfers between a given group g and another group g' in the same week. Given such a swap of golfers between two different groups $s_{(g,w)}$ and $s_{(g',w)}$, what is now inserted in the tabu list are both $(s_{(g,w)}, d, t)$ and $(s_{(g',w)}, d, t)$ with t being as for the PPP.

5.3   Results

Tables 1 and 2 show the experimental results for the PPP and SGP, respectively. For both, each entry in the table is the mean value of successful runs out of 100.

Table 1. Run times in seconds for the PPP. Mean run time of successful runs (out of 100) and number of unsuccessful runs (if any) in parentheses

H/periods (fails)	6	7	8	9	10
1-12,16			1.2	2.3	21.0
1-13			7.0	90.5	
1,3-13,19			7.2	128.4 (4)	
3-13,25,26			13.9	170.0 (17)	
1-11,19,21	10.3	83.0 (1)			
1-9,16-19	18.2	160.6 (22)			

Table 2. Run times in seconds for the SGP. Mean run time of successful runs (out of 100) and number of unsuccessful runs (if any) in parentheses

ng-ns-nw	time (fails)	ng-ns-nw	time (fails)	ng-ns-nw	time (fails)
6-3-7	0.4	6-3-8	215.0 (76)	7-3-9	138.0 (5)
8-3-10	14.4	9-3-11	3.5	10-3-13	325.0 (35)
6-4-5	0.3	6-4-6	237.0 (62)	7-4-7	333.0 (76)
8-4-7	0.9	8-4-8	290.0 (63)	9-4-8	1.7
10-4-9	2.5	6-5-5	101.0 (1)	7-5-5	1.3
8-5-6	8.6	9-5-6	0.9	10-5-7	1.7
6-6-3	0.2	7-6-4	1.2	8-6-5	18.6
9-6-5	1.0	10-6-6	3.7	7-7-3	0.3
8-7-4	4.9	9-7-4	0.8	10-7-5	3.4
8-8-3	0.5	9-8-3	0.6	10-8-4	1.4
9-9-3	0.7	10-9-3	0.8	10-10-3	1.1

The numbers in parentheses are the numbers of unsuccessful runs, if any, for that instance. We empirically chose $MaxIter = 500,000$ and $MaxNonImproving = 500$ for both applications. For the PPP, the instances are the same as in [25, 6, 24] and for the SGP, the instances are taken from [9]. For both applications, our results are comparable to, but not quite as fast as, the current best results ([6, 24] and [9] respectively) that we are aware of. We believe that they can be improved by using more sophisticated neighbourhoods and meta-heuristics, as well as by implementing the ideas in Section 3.3 for the $MaxIntersect$ constraint.

6   Conclusion

We have proposed to use set variables and set constraints in local search. In order to do this, we have introduced a generic penalty scheme for (global) set constraints and used it to give incrementally maintainable penalty definitions for five such constraints. These were then used to model and solve two well-known combinatorial problems.

This research is motivated by the fact that set variables may lead to more intuitive and simpler problem models, providing the user with a richer set of tools, as well as more preserved structure in underlying solving algorithms such as the incremental algorithms for maintaining penalties: (global) set constraints provide opportunities for hard-wired global reasoning that would otherwise have to be hand-coded each time for lower-level encodings of set variables.

In terms of related work, Localizer [12, 22], by Michel and Van Hentenryck, was the first modelling language to allow the definition of local search algorithms in a high-level, declarative way. It introduces invariants to obtain efficient incremental algorithms. It also stresses the need for globality by making explicit the invariants *distribute* and *dcount*.

In [10], Galinier and Hao use a similar scheme to ours for defining the penalty of a constraint in local search: they define as the penalty of a (global) constraint c the minimum number of variables in c that must change in order for it to be satisfied. Note, however, that this work is for integer variables only. Nareyek uses global constraints in [16] and argues that this is a good compromise between low-level CSP approaches, using only simple (e.g., binary) constraints, and problem-tailored local search approaches that are hard to reuse.

Comet [13], also by Van Hentenryck and Michel, is an object-oriented language tailored for the elegant modelling and solving of combinatorial problems. With Comet, the concept of differentiable object was introduced, which is an abstraction that reconciles incremental computation and global constraints. A differentiable object may for instance be queried to evaluate the effect of local moves. Comet also introduced abstractions for controlling search [23] and modelling using constraint-based combinators such as logical operators and reification [24]. Both Localizer and Comet support set invariants, but these are not used as variables directly in constraints.

Generic penalty definitions for constraints are useful also in the soft-constraints area. Petit *et al.* [17] use a similar penalty definition to the one of Galinier and Hao [10] as well as another definition where the primal graph of a constraint is used to determine its cost. This definition of cost is then refined by Petit and Beldiceanu in [5], where the cost is expressed in terms of graph properties [4]. Bohlin [6] also introduces a scheme built on the graph properties in [4] for defining penalties, which is used in his Composer library for local search. To our knowledge, none of these approaches considers set variables and set constraints.

Open issues exist as well. Other than fine-tuning the performance of our current prototype implementation, further (global) set constraints should be added. What impact will a change to the penalty of *MaxIntersect* with respect to Section 3.3 have? In what way should the penalties of the (global) set constraints in this paper be generalised to allow problems containing variables with several kinds of domains? For instance, it would be useful to be able to replace m with an integer variable in the *Cardinality*, *MaxIntersect*, and *MaxWeightedSum* constraints, to allow negative weights in the latter, and to have a variable reference set in the *Partition* constraint.

Overall, our results are already very promising and motivate such further research.

Acknowledgements. This research was partially funded by Project C/1.246/ HQ/JC/04 of EuroControl. We thank the referees for their useful comments.

References

1. E. Aarts and J. K. Lenstra, editors. *Local Search in Combinatorial Optimization.* John Wiley & Sons Ltd., 1997.
2. F. Azevedo and P. Barahona. Applications of an extended set constraint solver. In *Proc. of the ERCIM / CompulogNet Workshop on Constraints,* 2000.
3. N. Barnier and P. Brisset. Solving the Kirkman's schoolgirl problem in a few seconds. In *Proc. of CP'02,* volume 2470 of *LNCS,* pages 477–491. Springer, 2002.
4. N. Beldiceanu. Global constraints as graph properties on a structured network of elementary constraints of the same type. In *Proc. of CP'00,* volume 1894 of *LNCS,* pages 52–66. Springer-Verlag, 2000.
5. N. Beldiceanu and T. Petit. Cost evaluation of soft global constraints. In *Proc. of CPAIOR'04,* volume 3011 of *LNCS,* pages 80–95. Springer-Verlag, 2004.
6. M. Bohlin. Design and Implementation of a Graph-Based Constraint Model for Local Search. PhL thesis, Mälardalen University, Västerås, Sweden, April 2004.
7. P. Codognet and D. Diaz. Yet another local search method for constraint solving. In *Proc. of SAGA'01,* volume 2264 of *LNCS,* pages 73–90. Springer-Verlag, 2001.
8. K. Corrádi. Problem at Schweitzer competition. *Mat. Lapok,* 20:159–162, 1969.
9. I. Dotú and P. Van Hentenryck. Scheduling social golfers locally. In *Proc. of CPAIOR'05, LNCS,* Springer-Verlag, 2005.
10. P. Galinier and J.-K. Hao. A general approach for constraint solving by local search. In *Proc. of CP-AI-OR'00.*
11. C. Gervet. Interval propagation to reason about sets: Definition and implementation of a practical language. *Constraints,* 1(3):191–244, 1997.
12. L. Michel and P. Van Hentenryck. Localizer: A modeling language for local search. In *Proc. of CP'97,* volume 1330 of *LNCS.* Springer-Verlag, 1997.
13. L. Michel and P. Van Hentenryck. A constraint-based architecture for local search. *ACM SIGPLAN Notices,* 37(11):101–110, 2002. *Proc. of OOPSLA'02.*
14. L. Michel and P. Van Hentenryck. Maintaining longest paths incrementally. In *Proc. of CP'03,* volume 2833 of *LNCS,* pages 540–554. Springer-Verlag, 2003.
15. T. Müller and M. Müller. Finite set constraints in Oz. In *Proc. of 13th Workshop Logische Programmierung,* pages 104–115, Technische Universität München, 1997.
16. A. Nareyek. Using global constraints for local search. In *Constraint Programming and Large Scale Discrete Optimization,* volume 57 of *DIMACS: Series in Discrete Mathematics and Theoretical Computer Science,* pages 9–28. AMS, 2001.
17. T. Petit, J.-C. Régin, and C. Bessière. Specific filtering algorithms for over constrained problems. In *Proc. of CP'01,* volume 2293 of *LNCS,* Springer, 2001.
18. S. Prestwich. Supersymmetric modeling for local search. In *Proc. of 2nd International Workshop on Symmetry in Constraint Satisfaction Problems, at CP'02.*
19. J.-F. Puget. Finite set intervals. In *Proc. of CP'96 Workshop on Set Constraints.*
20. J.-F. Puget. Symmetry breaking revisited. In *Proc. of CP'02,* volume 2470 of *LNCS,* pages 446–461. Springer-Verlag, 2002.
21. B. M. Smith *et al.* The progressive party problem: Integer linear programming and constraint programming compared. *Constraints,* 1:119–138, 1996.
22. P. Van Hentenryck and L. Michel. Localizer. *Constraints,* 5(1–2):43–84, 2000.
23. P. Van Hentenryck and L. Michel. Control abstractions for local search. In *Proc. of CP'03,* volume 2833 of *LNCS,* pages 65–80. Springer-Verlag, 2003.
24. P. Van Hentenryck, L. Michel, and L. Liu. Constraint-based combinators for local search. In *Proc. of CP'04,* volume 3258 of *LNCS.* Springer-Verlag, 2004.
25. J. P. Walser. *Integer Optimization by Local Search: A Domain-Independent Approach,* volume 1637 of *LNCS.* Springer-Verlag, 1999.

The Temporal Knapsack Problem
and Its Solution

Mark Bartlett[1], Alan M. Frisch[1], Youssef Hamadi[2], Ian Miguel[3],
S. Armagan Tarim[4], and Chris Unsworth[5]

[1] Artificial Intelligence Group, Dept. of Computer Science,
Univ. of York, York, UK
[2] Microsoft Research Ltd., 7 J J Thomson Avenue,
Cambridge, UK
[3] School of Computer Science, University of St Andrews,
St Andrews, UK
[4] Cork Constraint Computation Centre, Univ. of Cork,
Cork, Ireland
[5] Department of Computing Science,
University of Glasgow, UK

Abstract. This paper introduces a problem called the temporal knapsack problem, presents several algorithms for solving it, and compares their performance. The temporal knapsack problem is a generalisation of the knapsack problem and specialisation of the multidimensional (or multiconstraint) knapsack problem. It arises naturally in applications such as allocating communication bandwidth or CPUs in a multiprocessor to bids for the resources. The algorithms considered use and combine techniques from constraint programming, artificial intelligence and operations research.

1   Introduction

This paper defines the temporal knapsack problem (TKP), presents some algorithms for solving it and compares the performance of the algorithms on some hard instances. TKP is a natural generalisation of the knapsack problem and a natural specialisation of the multi-dimensional knapsack problem. Nonetheless, it is—as far as we know—a new problem.

In the TKP a resource allocator is given bids for portions of a timeshared resource — such as CPU time or communication bandwidth — or a shared-space resource — such as computer memory, disk space, or equivalent rooms in a hotel that handles block-booking. Each bid specifies the amount of resource needed, the time interval throughout which it is needed, and a price offered for the resource. The resource allocator will, in general, have more demand than capacity, so it has the problem of selecting a subset of the bids that maximises the total price obtained.

We were initially drawn to formulating the TKP from our interest in applying combinatorial optimisation techniques in the context of grid computing.

R. Barták and M. Milano (Eds.): CPAIOR 2005, LNCS 3524, pp. 34–48, 2005.

Applications that use a grid simultaneously require different resources to perform large-scale computations. An advanced-reservation system will be used to guarantee a timed access to resources through some service level agreement [1].

The agreement is reached via negotiation, where end users present reservation bids to resource providers. Each bid specifies the resource category, start time, end time and required quality of service (e.g., bandwidth, number of nodes) [2]. If end users offer a price they are willing to pay for the resource, advanced reservation allocation in a grid infrastructure becomes equivalent to the TKP.

Our algorithms are designed to be used by a resource provider to select efficiently the right subset of customers with respect to resource requirements and the provider's utility. So far advanced reservation is not used in grid infrastructures, which still use specialised *fifo* scheduling policies inherited from high-performance computing. However, the convergence between web services and grid computing, combined with the arrival of the commercial grid, make the efficient use of valuable resources critical [3]. Our algorithms fit well into next-generation grids and represent the first attempts towards efficient grid resource schedulers.

2   The Temporal Knapsack Problem

A formal statement of the TKP is given in Figure 1. Here, and throughout, $bids(t)$ is $\{b \in bids | t \in duration(b)\}$. It is important to notice that TKP is *not* a scheduling problem.

Figure 2(a) illustrates an instance of TKP that has seven bids, $b_1, \ldots, b_7$, and 10 times, $t_1, \ldots, t_{10}$, which are displayed on the x-axis. The instance has a uniform capacity of 10, which is not shown. The optimal solution to this instance is to accept bids b_1, b_4, b_5, and b_6, yielding a total price of 22.

The traditional knapsack problem, as overviewed by Martello and Toth [4], is a special case of TKP in which there is only a single time. Since the knapsack problem, which is NP-hard, is a special case of TKP, TKP is also NP-hard.

Given: *times*, a finite, non-empty set totally ordered by $\leq$
            for each $t \in times$, $capacity(t)$, a positive integer
            bids, a finite set
            for each $b \in bids$,
                $price(b)$, a positive integer
                $demand(b)$, a positive integer
                $duration(b) = [start(b), end(b)]$, a non-empty interval of *times*
Find: a set $accept \subseteq bids$
Such that: $\forall t \in times, \sum_{b \in (accept \cap bids(t))} demand(b) \leq capacity(t)$
Maximising: $\sum_{b \in accept} price(b)$

Fig. 1. Definition of the temporal knapsack problem

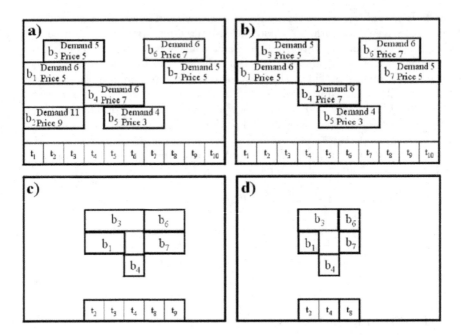

Fig. 2. An instance of the temporal knapsack problem to which the reduce operator is applied

The multidimensional knapsack problem (MKP, also known as the multi-constraint knapsack problem), as overviewed by Fréville [5], is a generalisation of TKP. Each time and bid in the TKP corresponds, respectively, to a dimension and an item in MKP. If $t_1, \ldots, t_n$ are the times in TKP, then every bid b corresponds to an item in MKP whose size is an n-dimensional vector of the form $\langle 0 \cdots 0 \; demand(b) \cdots demand(b) \; 0 \cdots 0 \rangle$ and the MKP capacity is the n-dimensional vector $\langle capacity(t_1) \cdots capacity(t_n) \rangle$. Since TKP polynomially-reduces to MKP, which is NP-easy, TKP is also NP-easy.

TKP readily reduces to integer linear programming. An instance of TKP generates the integer linear program that has a 0/1 variable X_b for each bid b:

Maximise: $\displaystyle\sum_{b \in bids} price(b) \cdot X_b$

Subject to: $\displaystyle\sum_{b \in bids(t)} demand(b) \cdot X_b \leq capacity(t)$, for each $t \in times$

$X_b \in \{0, 1\}$, for each $b \in bids$

Each solution to this integer linear program corresponds to an optimal solution to the TKP in which bid b is accepted if and only if X_b is assigned 1.

It shall often be convenient to switch between a TKP formulation, as given in Figure 1, and its linear programming formulation. To do this readily, it is important to notice that each time t produces one linear constraint; we shall refer to this as the constraint at t.

3   The Decomposition Algorithm

A decomposition algorithm for solving the TKP is presented here in three stages of increasing detail. Sec. 3.1 gives the algorithm as a non-deterministic procedure, which implicitly defines a search tree. Sec. 3.2 presents two methods for searching the tree for an optimal solution. Finally, Sec. 3.3 explains how linear programming is used to compute cuts, upper bounds, lower bounds and variable assignments, all of which are used to prune the search space.

3.1   A Non-deterministic Algorithm

Starting with all bids unlabelled, the decomposition algorithm records the solution by labelling each bid either "accept" or "reject." The algorithm employs three basic operations: reduce, which simplifies a problem instance; branch, which generates one branch in which an unlabelled bid is labelled "accept" and another branch in which that bid is labelled "reject"; and split, which decomposes a problem instance into smaller, independent problem instances. Let us begin by exemplifying the three operators.

The reduce operator performs two kinds of simplification. The first kind removes from the problem instance any bid b whose demand exceeds the capacity available at some time t in the duration of the bid. The operator labels all removed bids with "reject." This simplification corresponds to the constraint programming operation of achieving bound consistency on the constraint at t by removing the value 1 from the domain of X_b. For example, in the instance of Figure 2(a), bid b_2 has demand 11 at t_1, t_2 and t_3, yet the capacity at these times is only 10. So, the reduce operator removes b_2 from the instance and labels it "reject," which results in the instance shown in Figure 2(b).

The second kind of simplification removes unnecessary times from the instance. In the instance of Figure 2(b), at times t_1, t_5, t_6, t_7 and t_{10} the total demand does not exceed the capacity. Hence the constraints at these four times are satisfied in all assignments to the variables; in constraint programming terminology, they are entailed. Hence these times can be removed from the instance and from the durations of all bids. The result of this is that the duration of bid b_5 is now the empty set, meaning that b_5 participates in no constraints. Hence, b_5 is accepted in all optimal solutions. (In constraint programming terminology, any feasible solution that rejects b_5 is dominated by another feasible solution that is identical except that it accepts b_5.) Thus, the reduce operator removes b_5 from the instance and labels it "accept," resulting in the instance displayed in Figure 2(c).

There is a second method by which the reduce operator removes times from an instance. It is often the case that two adjacent times have the same bids. In such a case the time with the larger capacity imposes a weaker constraint and therefore can be removed; if the two times have the same capacity, then either time can be removed. In the instance of Fig. 2(c), t_2 and t_3 impose the same constraint as do t_8 and t_9. Thus, t_3 and t_9 can be removed from the instance, resulting in the instance of Figure 2(d).

The split operator decomposes a problem instance into subproblems that can be solved independently. A split can be performed between any two adjacent times, t and t', such that $bids(t) \cap bids(t') = \emptyset$. In the instance of Fig. 2(d), a split can be made in between t_4 and t_8 resulting in two subproblems: one comprising times t_2 and t_4 and bids b_1, b_3 and b_4; and a second comprising time t_8 and bids b_6 and b_7. Splitting is rarely used in constraint programming, though two recent exceptions are the work of Walsh [6] and of Marinescu and Dechter [7].

The branch operator is the familiar one from constraint programming, artificial intelligence and operations research. A bid b is selected, it is then removed from the problem and two branches are generated: one in which b is labelled "reject" and the other in which it is labelled "accept." On the accept branch, $demand(b)$ must be subtracted from the capacity at all times in $duration(b)$.

The decomposition algorithm, which applies the previous operations, is shown as Algorithm 1. This performs some initialisation and then calls the recursive procedure Solve.

Algorithm 1: Decomposition

Input: P, an instance of TKP;
Reduce(P);
Split(P) into set of problems S;
for $(s \in S)$ **do** Solve(s);

Procedure Solve(P); (where P is an instance of TKP)
if $bids = \emptyset$ **then** return;
else
| Select a bid b from $bids$;
| Non-deterministically do one of;
| (1) RejectBid(b);
| Reduce(P);
| Split(P) into set of problems S;
| **for** $s \in S$ **do** Solve(s)
| (2) AcceptBid(b);
| Reduce(P);
| Split(P) into set of problems S;
| **for** $s \in S$ **do** Solve(s)

Let us now turn our attention to the four procedures used in the decomposition algorithm: RejectBid, AcceptBid, Reduce and Split. The algorithms for these support procedures are given in Algorithm 2. In this discussion, and throughout, let $Demand(t)$ be the total demand at time t—that is

---

Algorithm 2: Support Procedures for the Decomposition Algorithm

---

Procedure RejectBid($b : bids$);
label b reject;
remove b from bids;

Procedure AcceptBid($b : bids$);
label b accept;
remove b from bids;
for $t \in duration(b)$ **do** subtract $demand(b)$ from $capacity(t)$;

Procedure Reduce(P);
for $b \in bids$ **do** TestForcedReject(b);
for $t \in times$ **do if** $Demand(t) \leq capacity(t)$ **then** RemoveTime(t);
if $|times| \geq 2$ **then**
$\quad$ set t_a to $min(times)$;
$\quad$ **while** $t_a \neq max(times)$ **do**
$\quad\quad$ set t_b to $next(t_a)$;
$\quad\quad$ **if** $bids(t_a) = bids(t_b)$ **then**
$\quad\quad\quad$ **if** $capacity(t_a) \geq capacity(t_b)$ **then** RemoveTime(t_a); set t_a to t_b
$\quad\quad\quad$ **else** RemoveTime(t_b)
$\quad\quad$ **else** set t_a to t_b

Procedure Split(P);
Let t_a and t_b be two times such that $next(t_a) = t_b$ and $bids(t_a) \cap bids(t_b) = \emptyset$;
if *no such times exist* **then** return(P);
else
$\quad$ Let P_1 be the TKP instance with times $\{t|t \leq t_a\}$ and bids $\{b|end(b) \leq t_a\}$;
$\quad$ Let P_2 be the TKP instance with times $\{t|t \geq t_b\}$ and bids $\{b|start(b) \geq t_b\}$;
$\quad$ return(Split(P_1) $\cup$ Split(P_2))

Procedure RemoveTime($t : times$);
remove t from times;
for $b \in bids(t)$ **do**
$\quad$ remove t from $duration(b)$;
$\quad$ **if** $duration(b) = \emptyset$ **then** AcceptBid(b)

Procedure TestForcedReject(b:bids);
if *for some* $t \in duration(b)$, $demand(b) > capacity(t)$ **then** RejectBid(b)

$\sum_{b \in bids(t)} demand(b)$. Also let $next(t)$ be the smallest time strictly greater than t; $next(t)$ is undefined if t is the largest time.

RejectBid(b) and AcceptBid(b) are simple; both label bid b appropriately and remove it from the problem instance. In addition, AcceptBid(b) must subtract the demand of the bid from the resource capacities available. Recall that reduce performs two kinds of simplification: (1) removing bids that must be rejected (forced rejects) and (2) removing unnecessary times, which may lead to removing bids that must be accepted (forced accepts). This is achieved by performing (1) to completion and then performing (2) to completion. Once this is done, there is no need to perform (1) again; doing so would not force any more rejects. In fact, it can be shown (though space precludes doing so here) that every time Solve is invoked it is given an instance such that

- $\forall t \in times \; \forall b \in bids(t) \; demand(b) \leq capacity(t)$,
- $\forall t \in times \; capacity(t) < Demand(t)$,
- $\forall b \; duration(b) \neq \emptyset$ and
- $\forall t, t' \in times \; t' = next(t) \implies bids(t) \neq bids(t')$.

An instance that has these properties is said to be *reduced* since performing the Reduce operator on the instance would have no effect. For each the two occurrences of Reduce in Solve we have implemented a significant simplification of the operator by considering the context in which it occurs.

As presented, the decomposition algorithm defines an AND/OR search tree.[1] Each node consists of a TKP instance. The root node is an AND node and comprises the initial problem instance. Every leaf node comprises a TKP instance with no bids. The children of an AND node are the (one or more) instances generated by applying the Split operator to the AND node. The set of feasible solutions to an AND node is the cross product of the feasible solutions of its children. Each child of an AND node is an OR node. Each OR node, other than the leaves, has two children generated by the branching in the algorithm—one child in which a selected bid is accepted and one in which that bid is rejected. The set of feasible solutions of an OR node is the union of the feasible solutions of its children.

The Decomposition Algorithm is correct in that the feasible solutions of the AND/OR search tree include all optimal solutions. The feasible solutions of the tree generally contain non-optimal solutions, which is obvious once one notices that the non-deterministic algorithm does not use the price of the bids. The next section considers how to explore the tree to find an optimal solution and how to use bounds on the objective function to prune the tree during the exploration.

3.2   The Search Strategy

This section explains how the standard branch-and-bound framework for OR trees can be adapted to handle AND/OR trees, such as those of the previous subsection. We refer this adapted framework as AOBB. The framework does not specify how the algorithm should choose the next node to be expanded. To do this, we currently employ two strategies: the AO* algorithm described by Nilsson [9] (which is itself based on an algorithm of Martelli and Montanari [10]), and a depth-first algorithm. The AO* search strategy is an extension of the A* algorithm to AND/OR search spaces, and retains two important properties of A*: the first feasible solution found is guaranteed to be an optimal one, and no

---

[1] The idea that a non-deterministic program implicitly defines an AND/OR tree was used in the very first paper published in the journal *Artificial Intelligence* [8].

algorithm that is guaranteed to find an optimal solution expands fewer nodes than AO* [11]. The drawback of AO* is that it requires a large amount of memory; the number of nodes in memory is $\Omega(2^{|bids|})$. In contrast, depth-first search stores only $\Omega(|bids|)$ nodes. However, in general, depth-first search explores more nodes than necessary to determine an optimal solution.

As with other branch-and-bound algorithms, AOBB stores at each search node an upper and lower bound on the objective function value for the TKP instance at that node. The AOBB algorithm repeatedly (1) selects an unexpanded OR node, (2) expands the OR node and then its children, and (3) propagates new bounds through the tree and uses these bounds to prune the tree. It performs this sequence of three stages until the tree contains no nodes to expand, at which point the result of the pruning is that the tree contains nothing but an optimal solution. Notice that whenever an OR node is expanded its children (which are AND nodes) are immediately expanded, producing OR nodes. Thus, the search tree's new leaves are always OR nodes. The only time an AND node is a leaf is at the start when the tree contains only the root node. Let us now consider the three major stages in more detail.

Stage 1: The node to process next is found. This is where AO* and depth-first differ. AO* selects a leaf node by descending the search tree, starting at the root and taking the child with the highest upper bound at an OR node. AO* allows any child to be taken at an AND node; our implementation takes the child with the largest spread between its upper and lower bounds. It is important to notice that successive descents can take different paths from a node since the node's bounds may change between the descents.

In contrast, when depth-first search expands a node its children are ordered from left to right and this ordering is fixed throughout the execution. Depth-first descends from the root node by always taking the left-most child that has unexpanded descendants. Depth-first search allows the children to be ordered in any manner. Our implementation orders the children of an OR node from left to right so that their upper bounds are non-decreasing and the children of an AND node so that the spread between their upper and lower bounds is non-decreasing.

Stage 2: The node is processed and expanded. The node found by the above stage will be either an OR node or the root node.

In the case of the root node, the resulting problem is reduced and split to form a set of child nodes, which are OR nodes, each containing independent subproblems. For each of these OR nodes, an upper and lower bound on its objective function value is obtained through solving the linear relaxation of the TKP at the node. As explained in the next subsection, at each OR node this linear program is also used to calculate three cuts (implied constraints in constraint programming terminology): a Gomory mixed-integer cut, a reduced costs constraint and a reversed-reduced-cost constraint. The Gomory cut is added to the linear programming form of the problem at the node and at all of its future descendents. This cut reduces the feasible region of the linear program without removing any integer solutions. Bounds consistency is enforced on all three cuts, which might determine the value of certain variables, i.e., whether a bid should be accepted or rejected. The RejectBid and AcceptBid operators are performed as appropriate, followed by the Reduce operator.

In the case of the node to expand being an OR node, an unlabelled bid is chosen to branch on, and two child AND nodes are created, one in which the bid is labelled "accept" and one in which it is labelled "reject." Both of these AND nodes are then processed and expanded in the same way as described for the root node. By expanding an OR node and both its children in a single stage in this way, the algorithm gains efficiency.

Stage 3: The new bound values are propagated and the tree is pruned. Starting at the OR nodes just created and working up the tree to the root, the value of the upper bound (ub) and the lower bound (lb) are updated for each node as follows.

$$ub(n) = \begin{cases} \max_{n' \in children(n)}(ub(n') + a(n, n')) & \text{if } n \text{ is an OR node} \\ \sum_{n' \in children(n)}(ub(n') + a(n, n')) & \text{if } n \text{ is an AND node} \end{cases}$$

$$lb(n) = \begin{cases} \max_{n' \in children(n)}(lb(n') + a(n, n')) & \text{if } n \text{ is an OR node} \\ \sum_{n' \in children(n)}(lb(n') + a(n, n')) & \text{if } n \text{ is an AND node} \end{cases}$$

where $a(n, n')$ is the sum of the prices of all bids accepted in moving from node n to node n', and $children(n)$ is the set of nodes that are children of n.

As this stage assigns and reassigns bounds, it checks to see if any OR node has one child whose upper bound does not exceed the lower bound of the other child. In such a case the best solution from the first child can be no better than that of the second child, so the first child and all its descendants are removed from the tree.

Having seen how the three stages operate, the last search issue that must be addressed is that of how bids are chosen for branching in Stage 2. We have tried two strategies for this. The *demand* strategy, chooses a bid with the highest demand. The intuition behind this is that labelling a bid with high demand is likely to lead to more propagation than one with low demand. The second strategy, called *force-split* is designed to yield nodes that can be split, preferably near the center. This strategy searches the middle half of the times for a pair of adjacent times, t and t', that minimises the cardinality of $S = bids(t) \cap bids(t')$; ties are broken in favour of the time nearest the center. The algorithm then branches on each bid in S in non-increasing order of their demands. This sequence of branches generates leaf nodes that can each be split between t and t'.

3.3   Generating Cuts and Bounds

This section explains how we use linear programming to generate cuts and bounds on the objective function value. The presentation assumes the reader is familiar with the basic theory underlying linear programming, such as that which is presented by Chvatal [12].

As shown in Section 2, a TKP instance can be represented as an integer linear program. This enables us to generate cuts that reduce the size of the feasible region for TKP without eliminating any potential integer solutions. To enhance the given search strategy, we employ the well-known Gomory mixed-integer cut (GMIC), which is considered one of the most important classes of cutting planes (see [13]).

Using the results from the linear relaxed TKP model ($0 \leq X_b \leq 1$, $\forall b \in bids$) and an objective function value z of a known feasible integer solution, a valid GMIC [14] for the TKP model can be written as

$$\lfloor z_U \rfloor - z \geq \sum_{\substack{f_i \leq f_0 \\ i \in N_1}} \lfloor -r_i \rfloor X_i + \sum_{\substack{f_j \leq f_0 \\ j \in N_2}} \lfloor r_j \rfloor (1 - X_j) + \sum_{\substack{f_i > f_0 \\ i \in N_1}} \left(\lfloor -r_i \rfloor + \frac{f_i - f_0}{1 - f_0} \right) X_i +$$

$$\sum_{\substack{f_j > f_0 \\ j \in N_2}} \left(\lfloor r_j \rfloor + \frac{f_j - f_0}{1 - f_0} \right) (1 - X_j) + \sum_{\substack{f_k \leq f_0 \\ k \in S}} \lfloor -r_k \rfloor s_k + \sum_{\substack{f_k > f_0 \\ k \in S}} \left(\lfloor -r_i \rfloor + \frac{f_k - f_0}{1 - f_0} \right) s_k$$

where, N_1 (N_2) is the set of indices for non-basic variables at their lower (upper) bounds; S, the set of slack variables s; r, the reduced costs; and z_U, the objective value of the linear relaxation model. It is clear that z_U provides an upper bound for the linear mixed integer TKP. In this notation, $f_0 = z_U - \lfloor z_U \rfloor$; $f_i = -r_i - \lfloor -r_i \rfloor$, $\forall i \in N_1$; and $f_j = r_j - \lfloor r_j \rfloor$, $\forall j \in N_2$.

The generated GMICs are added to the linear relaxed TKP instance and all its descendants in the search tree. Each added GMIC removes some of the non-integer solutions from the relaxed feasible region, but none that are integer. This also helps to improve the upper bound z_U.

We also employ the "reduced costs constraints" (RCC) and "reverse-reduced costs constraints" (R-RCC) discussed by Oliva et al. [15]. Following their work, the "pseudo-utility criterion" is used to obtain a reasonably good feasible solution. This criterion is computationally cheap, especially once the solution to the linear relaxation is known and the optimal values of the dual variables, λ, are determined. In this criterion all X_b are sorted in non-increasing order of $price(b)/(demand(b) \cdot \sum_{t \in duration(b)} \lambda_t)$ and the demands are satisfied in this order, as long as there is enough capacity. The resulting objective function value, denoted by z_L, yields a lower bound for the optimum z. From

$$z - \sum_{i \in N_1} r_i X_i + \sum_{j \in N_2} r_j (1 - X_j) - \sum_{k \in S} r_k s_k = z_U,$$

one can devise two useful constraints: RCC on the right, and R-RCC on the left.

$$z_U - z_{U+} \leq -\sum_{i \in N_1} r_i X_i + \sum_{j \in N_2} r_j (1 - X_j) - \sum_{k \in S} r_k s_k \leq z_U - z_L$$

where z_{U+} represents an upper bound which is stronger than the one provided by the linear relaxation (z_U). Such an upper bound can be obtained by using a "surrogate relaxation". In our case, this relaxation consists of adding together all of the constraints weighted with their associated dual values.

The aforementioned cuts and constraints are used for propagation purposes. By enforcing bounds consistency, the domains of decision variables including slacks are filtered and in certain cases are reduced to a singleton. Bounds consistency on a $RCC/R\text{-}RCC$ constraint of the form $a \leq \sum_i b_i x_i + \sum_j c_j (1 - x_j) \leq d$ gives the following additional bounds on the domains of the variables $x_p \in N_1 \cup S$ and $x_q \in N_2$:

$$\left\lceil \frac{a - \sum_i b_i - \sum_j c_j + b_p}{b_p} \right\rceil \leq x_p \leq \left\lfloor \frac{d}{b_p} \right\rfloor \qquad \text{and}$$

$$1 - \left\lfloor \frac{d}{c_q} \right\rfloor \leq x_q \leq 1 - \left\lceil \frac{a - \sum_i b_i - \sum_j c_j + c_q}{c_q} \right\rceil .$$

The GMIC works in a manner similar to RCC.

4   Performance Comparison

This section compares the effectiveness of three algorithms at solving a range of randomly-generated TKP instances. The algorithms considered are:

- the decomposition algorithm using AO* search with the forced-split variable-selection strategy;
- the decomposition algorithm using depth-first search with the forced-split variable-selection strategy; and
- the integer linear program solver provided by CPLEX version 8.1 with the default settings. The solver uses a branch-and-cut algorithm — branch-and-bound augmented by the use of cuts. After an initial "presolve" phase, which removes redundant constraints and attempts to tighten the bounds on the variables, the solver creates a tree, whose root contains the linear relaxation of the problem, and proceeds to expand nodes of this tree until an optimal integer solution has been found. At each node, the linear relaxation of the problem at that node is solved. If this leads to a solution in which some variables have fractional values, a selection of cuts are generated and added to the problem. The problem is then solved again, and if some variables are still non-integer, one is chosen to branch on, producing one child with the chosen variable set to 1 and another with it set to 0.

Preliminary experiments showed that the decomposition algorithm consistently performed better with forced-split selection than with demand selection. This is the case for both AO* search and depth-first search. Consequently, extensive experiments were not performed for the demand strategy.

Our method for randomly generating TKP instances is controlled by six parameters: *ntimes*, *max_length*, *max_demand*, *ucapacity*, *max_rate* and *nbids*. These are all integers, except *max_rate*, which is a floating point number. Given these parameters, an instance is generated that has *times* = $\{t_1, \ldots t_{ntimes}\}$, ordered in the obvious way, a uniform capacity of *ucapacity*, and *nbids* bids.

Each bid within an instance is generated by randomly choosing its start time, end time, demand and rate from a uniform distribution over the following ranges:

$$start(b) \in [1, ntimes],$$
$$end(b) \in [start(b), \max(ntimes, start(b) + max\_length - 1)],$$
$$demand(b) \in [1, max\_demand],$$
$$rate(b) \in [1, max\_rate].$$

All of these are integers except that $rate(b)$ is a floating point number. From these values, we set $price(b)$ to $round(rate(b) \cdot demand(b) \cdot (end(b) - start(b) + 1))$.

Performance was assessed on a set of randomly-generated instances in which most factors affecting complexity were kept static,

$$ntimes = 2880,$$
$$max\_length = 100,$$
$$max\_demand = 50,$$
$$ucapacity = 400,$$

while varying the values of two parameters: $nbids$ and max_rate. The value 2880 corresponds to the number of 15 minute slots in 30 days. By varying max_rate from 1.0 to 2.0 in increments of .2, and then further increasing its value to 4, 8 and 16, and by varying $nbids$ from 400 to 700 in increments of 50, we generated 63 problem suites, each containing 20 instances. By using these parameter values, we have focussed the majority of our experiments on instances generated with max_rate between 1 and 2, but also consider instances with larger values of max_rate in order to show the effect this parameter has over a greater range.

Figure 3 shows the mean solution time taken by each algorithm for all generated instances in which there are a given number of bids and $max\_rate \leq 2$, The graph reveals that for the problems with the lowest number of bids, both decomposition algorithms outperform the CPLEX solver. However, as the problem size increases, the performance of the decomposition algorithms deteriorates faster than that of CPLEX. The AO* algorithm is unable to solve some instances

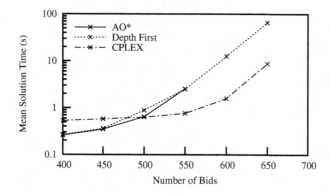

Fig. 3. Mean time taken to solve instances with a given number of bids

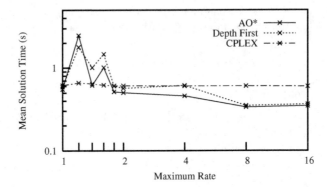

Fig. 4. Mean time taken to solve instances with a given range of rates

with 600 bids as it encounters memory problems and thrashes badly. The depth-first algorithm and CPLEX are both capable of solving all instances with up to and including 650 bids, but encounter problems on some instances with 700 bids; CPLEX through encountering memory-shortage problems and depth-first search through the exceptionally long time required to solve some problem instances.

Figure 4, shows the performance of the three algorithms on all instances in which max_rate has a given value and $nbids \leq 550$. The graph shows that for smaller values of max_rate, on average the CPLEX solver performs best, followed by the AO* algorithm, with the depth-first exploration proving worse. However for larger values of max_rate the AO* and depth-first algorithms clearly outperform the CPLEX solver. It is worth noting that the performance of CPLEX, unlike the other two programs, is barely affected by the value of max_rate.

While we have reported mean solution time throughout, it should be mentioned that these are influenced strongly by the extremely long run times required for a few of the instances. For most problem suites, the algorithms solve most instances in very short times; however for a few instances substantially longer is taken, resulting in these instances having a large influence on the mean. Despite this, we report mean times rather than median times (which would not be affected by these extreme values) as the frequency of the very hard instances increases with both increasing numbers of bids and decreasing max_rate, and their frequency is a significant component of the difficulty of a particular problem suite.

5   Conclusion and Future work

This paper has defined the temporal knapsack problem and identified it as a formalisation of some problems that naturally arise in making advanced reservations. The TKP specialises the multidimensional knapsack problem by imposing a temporal structure.

We have designed a special-purpose algorithm for solving the TKP. Its novel feature is that it exploits the temporal structure to decompose problem instances into subproblems that can be solved independently. It also uses a branching method that is designed to increase the frequency with which decompositions are made. The decomposition algorithm combines techniques from constraint programming (e.g., bound consistency, entailed constraints), artificial intelligence (e.g., AND/OR search spaces and the AO* and depth-first methods for searching them) and operations research (e.g., linear relaxations, cuts, branch and bound).

The TKP readily reduces to integer linear programming, which can be solved with an off-the-shelf system such as CPLEX.

Our experiments compared the time it takes to solve randomly-generated instances of TKP with three algorithms: CPLEX and the decomposition algorithm with both AO* search and depth-first search. CPLEX and decomposition with AO* search are effective on instances with approximately 650 and 550 bids, respectively, but encounter space problems on larger instances. Decomposition with depth-first search is effective on instances with approximately 650 bids but runs slowly on larger instances, though it does not encounter space problems.

In comparing these solution programs one must consider that the algorithms and implementation of CPLEX have been refined over decades, whereas those of our decomposition algorithm have been refined over months. With this in mind, we speculate that with further development—such as that outlined below—the decomposition algorithm could handle larger instances than CPLEX. We also have come to appreciate that beating CPLEX requires significant effort.

We see many ways in which believe that the decomposition algorithm and its implementation could be improved. The most important improvement would be to employ a search algorithm that takes the middle-ground between time-efficient, space-hungry AO* and time-hungry, space-efficient depth-first search. Such an algorithm could be developed by generalising one of the memory-bounded versions of A*, such as SMA* [16], to operate on AND/OR search trees. It also is likely that the decomposition algorithm would benefit from a better heuristic for choosing where to force splits. We conjecture that a better heuristic could be developed by carefully trading off the advantage of splitting into equal-sized subproblems and the advantage of minimising the amount of branching required to force a split. Finally, the algorithm would surely benefit from further development of its data structures. In particular, it should be possible to efficiently identify a greater number of redundant times.

Acknowledgements

We thank Michael Trick, Michel Vasquez and Lucas Bordeaux for helpful discussions. This project has been partly funded by Microsoft Research. Ian Miguel is supported by a UK Royal Academy of Engineering/EPSRC Post-doctoral Research Fellowship and S. Armagan Tarim by Science Foundation Ireland under Grant No. 03/CE3/I405 as part of the Centre for Telecommunications Value-Chain-Driven Research (CTVR) and Grant No. 00/PI.1/C075.

References

1. Foster, I., Kesselman, C., Tuecke, S.: The anatomy of the Grid: Enabling scalable virtual organization. The International Journal of High Performance Computing Applications **15** (2001) 200–222
2. Roy, A., Sander, V.: Advanced reservation API. GFD-E5, Scheduling Working Group, Global Grid Forum (GGF) (2003)
3. Foster, I., Kesselman, C., Nick, J., Tuecke, S.: The physiology of the grid: An open grid services architecture for distributed systems integration (2002)
4. Martello, S., Toth, P.: Knapsack Problems: Algorithms and Computer Implementations. Wiley, New York (1990)
5. Fréville, A.: The multidimensional 0-1 knapsack problem: An overview. European Journal of Operational Research **155** (2004) 1–21
6. Walsh, T.: Stochastic constraint programming. In: Proc. of the Fifteenth European Conf. on Artificial Intelligence, IOS Press (2002) 1–5
7. Marinescu, R., Dechter, R.: AND/OR tree search for constraint optimization. In: Proc. of the 6th International Workshop on Preferences and Soft Constraints. (2004)
8. Manna, Z.: The correctness of nondeterministic programs. Artificial Intelligence **1** (1970) 1–26
9. Nilsson, N.J.: Principles of Artificial Intelligence. Tioga (1980)
10. Martelli, A., Montanari, U.: Additive AND/OR graphs. In: Proc. of the Fourth Int. Joint Conf. on Artificial Intelligence. (1975) 345–350
11. Chang, C.L., Slagle, J.: An admissible and optimal algorithm for searching AND/OR graphs. Artificial Intelligence **2** (1971) 117–128
12. Chvatal, V.: Linear Programming. W.H.Freeman, New York (1983)
13. Bixby, R.E., Fenelon, M., Gu, Z., Rothberg, E., Wunderling, R.: MIP: Theory and practice — closing the gap. In Powell, M.J.D., Scholtes, S., eds.: System Modelling and Optimization: Methods, Theory, and Applications. Kluwer Academic Publishers (2000) 19–49
14. Wolsey, L.A.: Integer Programming. John Wiley and Sons, New York (1998)
15. Oliva, C., Michelon, P., Artigues, C.: Constraint and linear programming : Using reduced costs for solving the zero/one multiple knapsack problem. In: Proc. of the Workshop on Cooperative Solvers in Constraint Programming (CoSolv 01), Paphos, Cyprus. (2001) 87–98
16. Russell, S.J.: Efficient memory-bounded search methods. In: Proc. of the Tenth European Conf. on Artificial Intelligence, Vienna, Wiley (1992) 1–5

Simplifying Diagnosis Using LSAT: A Propositional Approach to Reasoning from First Principles

Andreas Bauer

Institut für Informatik,
Technische Universität München,
D-85748 Garching b. München, Germany
baueran@in.tum.de

Abstract. In face of the unwieldiness of non-monotonic logic engines, or Prolog/CLP meta interpreters as they are commonly used for model based reasoning and diagnosis, this paper proposes a simple, but effective improvement for performing the complex diagnostic task. The chosen approach is twofold: firstly, the problem of contradicting first order system descriptions with a set of observations is reduced to propositional logic using the notion of symptoms, and secondly, the determination of conflict sets and minimal diagnoses is mapped to a problem whose technical solution has experienced a sheer boost over the past years, namely k-satisfiability using state-of-the-art SAT-solvers. Since the involved problems are (mostly) $\mathcal{NP}$-complete, the ideas for additional improvements for a more diagnosis-specific SAT-solver are also sketched and their implementation by means of a non-destructive solver, LSAT, evaluated.

Keywords: model based reasoning, model based diagnosis, SAT-solving, system monitoring, formal specification.

1 Introduction

Ever since Reiter's seminal work on diagnosis from first principles [19], the automated reasoning and model based diagnosis communities have spawned a lot of work on the implementation and improvement of the proposed as well as on related ideas. Amongst these, probably the most influential ones have been Reiter's own *default logic* [19,4], the concept of *abduction* [20], or even McCarthy's notion of *circumscription* [16]. While reference implementations such as the *General Diagnostic Engine* [6] (GDE) realise some of these ideas there is — at least from a practical point of view — still a lot left to be desired in terms of time and space complexity of such implementations mainly due to the sheer complexity of the underlying decision problems.

Given a system S, the diagnostic task is to identify those parts or components $c_i \in S$ which are assumed to be faulty in order to explain an observed behaviour of S. If no such set of components can be isolated, then the system is assumed

R. Barták and M. Milano (Eds.): CPAIOR 2005, LNCS 3524, pp. 49–63, 2005.

to work according to its specification, i. e. is correct. According to the logic- and consistency-based approaches to diagnosis, this task is performed by detecting a contradicting behaviour of a system S when compared to its expected behaviour which is captured by a system description $SD \subset S$. Such a contradiction is then expressed in terms of a set of conflicts or diagnoses.

Moreover, in order to determine practically useful sets of conflicts and diagnoses that would allow hinting to specific, faulty parts of a system, all the proposed diagnosis methods involve a subsequent task minimising the solutions which, in itself, is a computationally complex undertaking. In Reiter's case, for instance, this task is not separable from the initial determination of conflicts. However, his method relies on the availability of a suitable first order theorem prover for finding at least a single conflict set to execute an algorithm for finding minimal *hitting sets* of conflicts that constitute the diagnoses.

The hitting set problem, on the other hand, also known as the transversal problem, is one of the key problems in the combinatorics of finite sets and the theory of diagnosis per se. It turns out to be a hard problem which also helps to explain the continuing hesitation of a broader industrial application of model based diagnosis techniques. Further, partly empirical, results from other authors regarding the wieldiness of implementations for non-monotonic reasoning (i. e. default logic, circumscription, etc.) second this conclusion (see §5, for further details on non-monotonic reasoning).

Contribution

This paper will show that the recent achievements in solving the k-satisfiability problem with heuristic search algorithms and pruning using state-of-the-art SAT-solvers can also be used for consistency based diagnosis and even for those cases where the task is not related to merely boolean circuits and the likes. More so, empirical results will show that, using a SAT based approach to diagnosis, one can handle several thousand variables (i. e. abstract system components) at ease and — when constrained to an appropriate n-fault assumption (see §4.1) — even tens and hundreds of thousands. Clearly, this is much more than non-monotonic reasoning engines can currently handle in reasonable time and space.

Therefore the contribution of this paper is to introduce an alternative methodology (based on simple propositional logic) for diagnosing technical systems, regardless as to whether these are software- or hardware-based, or both. Specifically, LSAT is presented which is a prototype SAT solver tailored to perform system's diagnosis.

Outline

After a brief overview over the theory of consistency- and logic-based diagnosis using system models in §2, the transformation of the (non-monotonic) diagnostic reasoning from first principles to propositional logic is then described in §3. LSAT, the main implementation vehicle for the concepts presented in this paper, is outlined in greater detail in §4, and an evaluation of the deployed algorithms

w. r. t. processing large combinatorical benchmarks is given in § 4.2. A section on related work (§ 5) describes other recent results in complexity measures regarding non-monotonic reasoning which seconds an important claim of this paper; that is, diagnosis should be tackled using modern SAT-solvers with specific, problem-oriented heuristics. Finally, § 6 presents some conclusions.

2   Consistency Based Diagnosis

In model and consistency based diagnosis[1], the system to be diagnosed, S, is determined by a tuple $(SD, COMP)$, where SD constitutes a finite set of first order sentences comprising a system description, and $COMP$ a finite set of components in S. The set of components can be of almost arbitrary granularity; depending on the properties of the system to be diagnosed, $COMP$ may refer to, say, Java threads, user session objects within a web application, or even physical entities such as sensors, actuators, or entire nodes of a computer network. The overall system behaviour is then defined in terms of the components' behaviours and their causal dependencies, represented as shared variables/predicates.

2.1   Definitions

In this section, let us recall some notions, notations and terminology used later in the paper.

Definition 1 (Observation). *An observation for a system $S = (SD, COMP)$ is a finite set of first order sentences each comprising a mapping of in- and outputs of S to actual/observed values: $input_i, output_i : c \in COMP \to Num$. The index i denotes the i-th input (resp. output) for component c, whereas Num represents the class of all numerical sorts. An observation OBS for S is denoted by $(SD, COMP, OBS)$.*

Example 1. The example system S depicted in Fig. 1 contains two multiplication components, M_1 and M_2, and two summation components, A_1, A_2. We use the more compact representation of an n-tuple $\langle i_1, \ldots, i_4, m_1, m_2, o_1, o_2 \rangle$ to capture the model's observed in- and output values in $\mathbb{N}$. Hence, $S = (SD, \{M_1, M_2, A_1, A_2\})$, with $OBS = \langle 2, 3, 4, 5, 6, 20, 26, 26 \rangle$. Without going much into further detail at this point, SD basically captures the behaviour and causality of each component.   ◊

Diagnosis can be understood as the process of finding and isolating differences between a system's model, i. e. the intended behaviour, and reality, i. e. observed behaviour. Typically, in order to reason about system models, at least one predicate needs to be introduced for the mere purpose of representing "normal" and

---

[1] From this point forward, the terms consistency-, logic-, and model-based are used synonymously due to the similarities of these approaches and their common problems w. r. t. complexity of their realisations.

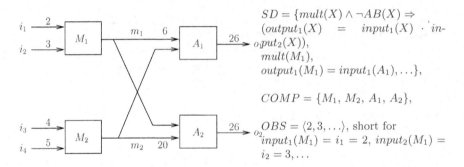

Fig. 1. A simple system description for a multiplier and adder example

"abnormal" parts of the system: $\neg AB(c)$ denotes a component which works according to its specification, while $AB(c)$ denotes an abnormal component. A diagnosis can now be defined w.r.t. to these predicates which help explain an observed behaviour.

Definition 2 (Diagnosis). *A diagnosis for a system $S = (SD, COMP)$ is a minimal set $\Delta \subseteq COMP$ such that*

$$SD \cup OBS \cup \{AB(c) \mid c \in \Delta\} \cup \{\neg AB(c) \mid c \in COMP \backslash \Delta\}$$

is consistent.

Proposition 1. *$\emptyset$ is a diagnosis (and the only diagnosis) for $(SD, COMP, OBS)$, iff*

$$SD \cup OBS \cup \{\neg AB(c) \mid c \in COMP\}$$

is consistent, i.e. iff the observation does not conflict with what the system should do if all its components were behaving correctly. (For a proof, see [19, § 3].)

Using this definition and continuing with what is presented in Ex. 1, it is self evident that substituting o_1 with anything but 26 will lead to the conclusion $\Delta = \{A_1\}$, i.e. $\neg AB(M_1)$, $\neg AB(M_2)$, $\neg AB(A_2)$, and $AB(A_1)$.

Definition 3 (Conflict Set). *A conflict set for $(SD, COMP, OBS)$ is a set $\{c_i, \ldots, c_j\} \subseteq COMP$ with $1 \leq i \leq j$ such that*

$$SD \cup OBS \cup \{\neg AB(c_i), \ldots, \neg AB(c_j)\}$$

is inconsistent.

Hence, $\{M_1, M_2, A_1, A_2\}$, would be a conflict set for our example, given $o_1 \neq 26$. Further, a conflict set for $(SD, COMP, OBS)$ is called *minimal*, iff no proper subset of it is a conflict set for $(SD, COMP, OBS)$ at the same time. That is, $\{A_1\}$ is a minimal conflict set. Diagnoses are then minimal conflict sets.

2.2   Reasoning with Incomplete Information

For those cases where obviously only a single component is at fault, finding minimal conflict sets seems a straightforward task under the gross assumption that OBS contains all relevant in- and outputs of the system, or its respective diagnosis model. However, in practice, one cannot always rely on the availability of complete information, but rather — and more realistically — on a *black* or *grey box* view yielding partial information.

Example 2. If the network depicted in Fig. 1 is modified according to Fig. 2, there is already a significantly larger amount of possible conflict sets — implicitly depending on the values of $\{m_1, m_2\} \not\subseteq OBS$:

$$\{M_1, M_2\}, \{M_1, A_1\}, \{M_1\}, \{M_2, A_1\}, \{M_2\}, \{A_1\}, \{M_1, M_2, A_1\}.$$

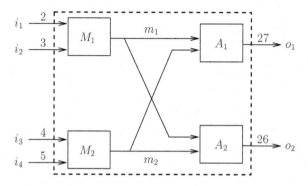

Fig. 2. Example of a black/grey box view where the values for m_1 and m_2 cannot be observed. According to qualitative measures, o_1 is faulty

Obviously, incomplete information creates a lot of diagnosis candidates, but as the above example demonstrates, these are entangled with assumptions regarding the missing elements of OBS. In other words, $\{M_2\}$ is a conflict set, iff the assignment for at least one unobservable connection contradicts the system specification according to the set SD, e. g. $m_2 \neq 20$.   ◇

3   A Propositional Solution

Of course, the traditional diagnosis approaches such as applications of default logic, or abduction, although specifically tailored for dealing with incomplete and inconsistent information, face increasing difficulties the more in- and outputs remain unobservable. Reducing this problem to propositional logic, however, suits this situation perfectly, given a number of prerequisites are fulfilled.

3.1   Combining Qualitative and Logical Measures

Let us assume we have two different models of the system under consideration, *a)* a qualitative model for reasoning about in- and output values of components, and *b)* an abstract model representing only causality; we can then use predicates that indicate whether an observation has been correct, or in error: $ok(m_1)$, for instance, would indicate that a result of M_1 is correct according to the qualitative assertion. Such "micro-evaluations" are realistic in many real-world scenarios, e.g. where sensor values are checked and compared, sometimes even multiple times to rule out tampered results due to jitter.

Qualitative assertions are typically made by dedicated *monitors* which continuously interpret a system component's in- and output values w.r.t. aberrations from the specification. However, monitors are not part of SD themselves, but rather constitute safety properties which ought to be fulfilled by single components $c_i \in COMP$, respectively.

In our case, these monitors are represented as predicates. That is, if the predicate holds, an observed value is assumed correct, otherwise it hints to existing aberrations. For example, the result of the boolean monitor $\beta(output_1(M_1)) = ok(output_1(M_1))$ would indicate conformance of the observed value $output_1(M_1)$ to its specification.

Definition 4 (Symptom). *Let* $S = (SD, COMP, OBS)$ *be a system under diagnosis. The ok-predicate is then defined over a subset of all in- and outputs, of a component* $c \in COMP$. *A negative evaluation of* $ok(i \in OBS)$ *is then called a symptom for an error in* S.

In other words, a negative result of a boolean monitor does not necessarily indicate that the monitored component is at fault. It merely hints to the fact that *some* component is faulty as captured in the following proposition:

$$\neg ok(i \in OBS) \Rightarrow \exists_{c \in COMP} \neg AB(c)^2.$$

Unlike the AB-predicate which is defined only over $c_i \in COMP$, ok is defined w.r.t. to observable system values. This notion of a symptom is then used to contradict the merely causal system specification and in order to distil a finite set of negative AB-predicates; that is, faulty components.

3.2   Reduction of SD

An alternative and, foremost, only causal first order system description for Ex. 1 and 2 could be expressed as follows:

$$SD = \{ok(i_1) \wedge ok(i_2) \wedge \neg AB(M_1) \Rightarrow ok(m_1),$$
$$ok(i_3) \wedge ok(i_4) \wedge \neg AB(M_2) \Rightarrow ok(m_2),$$

---

[2] Of course, c may be the monitored component, but this reasoning is part of the deductive diagnosis process.

$$ok(m_1) \wedge ok(m_2) \wedge \neg AB(A_1) \Rightarrow ok(o_1),$$
$$ok(m_1) \wedge ok(m_2) \wedge \neg AB(A_2) \Rightarrow ok(o_2)\}$$

Assuming we can evaluate and thus know when at least some observables w. r. t. ok hold, we can rewrite SD in terms of boolean variables and without any predicates:

$$SD' = \{ok\_i_1 \wedge ok\_i_2 \wedge \neg AB\_M_1 \Rightarrow ok\_m_1,$$
$$ok\_i_3 \wedge ok\_i_4 \wedge \neg AB\_M_2 \Rightarrow ok\_m_2, \ldots\}$$

Notice, both system descriptions SD and SD' are now comprising only causal and structural information, contrary to the literature of consistency based diagnosis, where SD always includes the behavioural part which we have "sourced out" in a separate, qualitative model which the monitors are based upon. Notice, monitor generation itself is an active field of research and not directly scope of this paper (see § 5).

Obviously, the mapping, $\Phi : pred \in (PL(\Sigma) \rightarrow \mathbb{B}) \rightarrow var \in \mathbb{B}$, where $PL(\Sigma)$ denotes a first order predicate logic formula defined over a signature Σ, is straightforward: each respective predicate, $pred$, is mapped to exactly one distinctive variable var_i of type $\mathbb{B}$.

3.3   Hitting Sets and Minimality

Finding conflict sets due to contradictions between expected and observed behaviour is crucial for failure diagnosis. However, in accordance to Definition 2, only the *minimal* diagnoses are of real, practical value. For this purpose, Reiter proposes a hitting set algorithm which constructs a so called *HS-tree* [19] that carries the minimal diagnoses, such that no diagnosis which is already included as a subset of a previously found diagnosis is chosen.

Formally, the problem addressed by Reiter's algorithm is as follows: a collection of non-empty sets $\mathcal{C} = \{C_1, \ldots, C_n\}$ of a set C, representing conflicts, is given. A hitting set (or transversal) of $\mathcal{C}$ is a subset $H \subseteq C$ that meets every set in the collection $\mathcal{C}$. We call a hitting set minimal, if no proper subset of H is a hitting set. More in depth information on the algorithm, specific optimisations, and analyses may also be found in § 5.

Despite well known improvements [8, 19, 14], the minimal hitting set problem remains generally $\mathcal{NP}$-complete, and it is practically undesirable to perform a thorough analysis when applied to diagnosis. Therefore, the chosen approach of this paper is to define a maximum "failure threshold" instead which is reflected in the diagnosis algorithm laid out in § 4.1. A diagnosis is then determined based upon *minimal cardinality* of occurring AB-predicates in the solution set, rather than upon the theory of set inclusion. Essentially, this allows us to improve on the determination of conflicts and still yields practically relevant diagnoses using the same algorithm; no subsequent procedure for minimalisation is required.

4   LSAT

In the previous section we have demonstrated how diagnosis related tasks such as fault isolation and conflict set minimalisation are basically reducible to propositional logic, under the premise that a qualitative model, which can be used to monitor symptoms, is available. Although the complexity of the involved decision problems has not shrunk to polynomial time, propositional (system) models exhibit the advantage of being manageable by using SAT-solvers.

The purpose of a SAT-solver is to accept a formula, P, in clause normal form (CNF), and to return a variable assignment $\{\alpha(p_1 \in P), \ldots, \alpha(p_n \in P)\}$, such that P evaluates to *true*. If no such assignment can be found, P is not satisfiable.

In recent years, the area of SAT-solving has advanced dramatically: CNF formulas of hundreds of thousands or even millions of literals can now be handled by state-of-the-art solvers, such as (z)Chaff [17], or SATO [24] to name just two of the most popular solvers. These programs more or less are based on the Davis-Putnam-Logemann-Loveland (DPLL) algorithm [5] which constructs semantic trees of CNF formulas. Normally, DPLL is a "destructive" algorithm in a sense that it recursively splits the semantic tree — based on some heuristics — and descends until a valid assignment has been found; most SAT-solvers indeed work this way.

In contrast, LSAT[3] is a SAT-solver which uses a non-destructive implementation of DPLL based on mutually linked lists of atoms and clauses (see Fig. 3) where variable assignments are not recursively pushed on the runtime stack, but are encoded in the global data structure itself. This way, LSAT can come up with more than one truth assignment, if applicable.

Having a single global data structure further improves on the space complexity of the algorithm: rather than keeping the entire semantic tree on the stack, only linear space for the main data structure is required.

The flag 'abnormal' denotes a component $c \in COMP$, and 'inact' contains a pointer to the variable which inactivated the current clause (NULL otherwise). Initially all clauses are active, i. e. no truth assignment has been made. 'pos-clauses' (resp. 'neg-clauses') is a list of pointers to clauses where the variable occurs with positive sign. 'pos-literals' (resp. 'neg-literals') is a list of pointers to variables which occur with positive sign in the clause. The rest is self explaining; further details can be found in the implementation.

4.1   Computing Single and Multiple Fault Diagnoses

Due to its non-destructive nature and the linear space complexity, LSAT is well suited for performing the diagnosis task based on propositional models. More so, if LSAT is able to conclude one diagnosis for a model, it is able to conclude

---

[3] LSAT has been released under the GPL open source license and is available in terms of C++ code from the author's home page at http://home.in.tum.de/~baueran/lsat/.

Variable

| index:int |
| value:int |
| abnormal:bool |
| pos-clauses:List<Clause> |
| neg-clauses:List<Clause> |

Clause

| pos-literals:List<Variable> |
| neg-literals:List<Variable> |
| inact:Variable |
| literals:int |

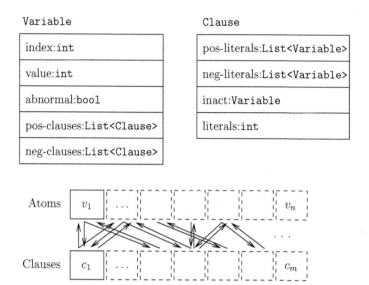

Fig. 3. LSAT's internal CNF representation visualised with v_i:Variable and c_i:Clause

all possible diagnoses, i. e. there is no need for a subsequent access to a theorem prover in order to determine conflict sets, or the likes.

In contrast, the diagnosis related literature frequently proposes the so called *single fault assumption* for two reasons: *a)* it is often realistic to assume merely single components at fault, rather than a total failure of a whole set of components, and *b)* most operations in non-monotonic reasoning approaches are sufficiently expensive such that the occurrence of a single fault is often used as (premature) exit condition. Naturally, the counterpart of the single fault assumption is the *multiple fault assumption* [6, 12].

Generally speaking, LSAT supports the n-fault assumption (where $n \geq 0$) and uses the 'abnormal' flag in the data type Variable in order to determine which diagnoses are useful, i. e. which are of a minimal cardinality w. r. t. the number of faulty components. That is, 'abnormal' determines the set of potential faults that are not symptoms (see Definition 4).

If defined, LSAT adheres to the n-fault assumption by keeping track of the positively assigned "AB-atoms/predicates" and by cutting off the semantic tree iff $AB(c_i)+\ldots+AB(c_j) > n$. Trivially, $n = 1$ selects the single fault assumption, $n > 1$ the multiple fault assumption.

Example 4. Given a system and observations $S = (SD, COMP, OBS)$, as it is depicted in Fig. 2, and an according propositional logic system description similar to Ex. 3, we may use the presented concepts so far to explain $o_1 \in OBS = 27$ as follows:

$$SD'' = \{\neg[1/ok\_i_1] \vee \neg[2/ok\_i_2] \vee [9/AB\_M_1] \vee [5/ok\_m_1],$$
$$\neg[3/ok\_i_3] \vee \neg[4/ok\_i_4] \vee [10/AB\_M_2] \vee [6/ok\_m_2],$$
$$\neg[5/ok\_m_1] \vee \neg[6/ok\_m_2] \vee [11/AB\_A_1] \vee [7/ok\_o_1],$$
$$\neg[5/ok\_m_1] \vee \neg[6/ok\_m_2] \vee [12/AB\_A_2] \vee [8/ok\_o_2], \ldots\}$$

SD'' is obtained by applying $\Phi(SD)$ and performing a subsequent CNF conversion which can be achieved in polynomial time [18]. Hence, SD'' represents the causal dependencies as well as possible states of our system using only natural numbers for each respective variable; 9, 10, 11, and 12 represent components in S. Although of no particular semantic value, the substitutions with natural numbers will be necessary for expressing SD'' in terms of an extended DIMACS format.

2-fault assumption: A snapshot of the semantic tree decision procedure for our example is depicted in Fig. 4. In each "step", the variables of $(SD, COMP, OBS)$ are assigned and the set of models expanded. Here, $\alpha(AB(11)) = 1 \equiv \alpha(11) = 1$ violates the 2-fault assumption since $\alpha(9) = 1$ and $\alpha(10) = 1$ on the same branch. The rounded arrow indicates one backtracking step in order to continue the algorithm using an alternative assignment, $\alpha(11) = -1$, i.e. $\neg AB(11)$. In other words, the n-fault assumption is the pruning criterion for the semantic tree procedure.

Using LSAT's extended DIMACS format, this example could be encoded and automatically solved as follows:

01	p cnf	12 18	Standard DIMACS header.
02	9	-1 -2 5	$\underline{SD}$: causal dependencies of S.
03	10	-3 -4 6	$(10 \vee -3 \vee -4 \vee 6)$
04	11	-5 -6 7	$\bigwedge$ $(11 \vee -5 \vee -6 \vee 7)$

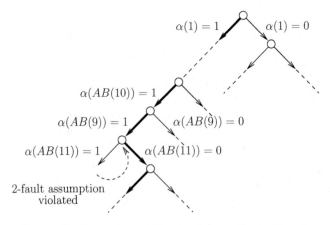

Fig. 4. Semantic tree for the system of Ex. 2: $\alpha(x)$ are the truth assignments; bold arrows indicate a valid assignment path, while the left-most path shows a violation of the 2-fault assumption

```
05      12  -6 -5 8      ∧ ...
06      -5  -9
07      -6 -10
08      -7 -11
09      -8 -12
10   a   9  10 11 12     COMP: the directive a defines the components in S.
11       1              OBS: ok(1)
12       2
13       3              ...
14       4
15      -7              Symptom: ¬ok(7). (Notice, {5, 6} ⊄ OBS.)
16       8
17      -9  9           Our hypotheses, i.e. all components may either be
18     -10 10           normal, or abnormal.
19     -11 11
20     -12 12
```

Similarly to the procedure shown in Fig. 4, LSAT is then able to determine all models for S with at most two faulty components; symptoms are underlined, "real" faults framed:

```
01   9  10  -11 -12 8 -7 -6 -5 4 3 2 1
02   9  -10 11  -12 8 -7  6 -5 4 3 2 1
03   9  -10 -11 -12 8 -7  6 -5 4 3 2 1
04  -9  10  11  -12 8 -7 -6  5 4 3 2 1
05  -9  10  -11 -12 8 -7 -6  5 4 3 2 1
06  -9 -10  11  -12 8 -7  6  5 4 3 2 1
```

Each of these six results encodes one valid truth assignment for S, such that the contradicting observation, i.e. $o_1 = 27$ can be explained under the assumption that no more than two components are responsible for the failure. Result #1, for instance, assumes $AB(9) \wedge AB(10) \wedge \neg AB(11) \wedge \neg AB(12)$.   ◊

4.2   Evaluation

Clearly, the emphasis of LSAT is on model based reasoning and diagnosis, rather than trying to outperform programs like SATO or (z)Chaff. However, LSAT does contain a couple of optimisations such as *unit propagation* and an implementation of the *purity rule* in order to deal with far bigger examples than shown in this paper so far. In order to elaborate on the feasibility of the chosen approach, a number of more or less standard benchmarks originating from the area of circuit design are shown in this section. This ISCAS set of benchmark circuits is widely used by the ECAD community for testing several digital design tools. The respective combinatorical tests range from several hundred to ca. 20,000 "components" and ca. 60,000 clauses.

Table 1 summarises the results of applying a selection of tests to LSAT on a Pentium 4 architecture with 512 MB of RAM using either the 5-fault assumption, or no restriction at all. If a test could not be finished within 60 seconds, it was considered to be a timeout.

Not surprisingly, the numbers substantiate the appropriateness of the presented concepts. Four tests could not be finished using the ∞-fault assumption, while LSAT had no problems solving these tasks when constrained to five faults. The variance between CPU time and the performed number of algorithmic steps can be explained by the heuristic approach and variantly efficient accessor functions in the LSAT tool.

Table 1. Modified ISCAS'89 benchmarks under the n-fault assumption

Name:	#$COMP$:	#Var.:	#Cl.:	∞-fault #Steps:	CPU:	5-fault #Steps:	CPU:
s208.1	66	122	389	84	0.17 sec	60	0.25 sec
s298	75	136	482	27	0.11 sec	58	0.32 sec
s444	119	205	714	20	0.18 sec	105	0.91 sec
s526n	140	218	833	−	timeout	295	0.23 sec
s820	256	312	1,335	−	timeout	562	0.59 sec
s1238	428	540	2,057	38	0.97	262	0.21 sec
s13207	2,573	8,651	27,067	−	timeout	17	0.57 sec
s15850	3,448	10,383	33,189	−	timeout	41	0.17 sec
s35932	12,204	17,828	60,399	2,339	11.16 sec	29	0.21 sec

5 Related Work

There are recent works which have used SAT-solvers in order to diagnose and debug the design of digital circuits, e.g. [1, 23]. However, research in this area treats the solvers foremost as mere tools, disregarding *a)* the cardinality of the models, and *b)* the adaption of the underlying algorithms and concepts. More so, in the "digital world", there is no need to reduce first order problems to propositional logic, since a circuit is already digestible by the solver as is.

Closer to the ideas presented in this paper is research undertaken by Baumgartner et al. such as [3], for instance. They describe the DRUM-2 system which is based on *hypertableaux* and can be used for model based reasoning. Their system is capable of finding minimal diagnoses based on "abnormal components", similar to LSAT. But the (system) descriptions used by Baumgartner remain first order, and results suggest that the hypertableaux do not seem to scale to the same extent as SAT-solving does, especially when considering the most recent developments in this area. LSAT, although still in a prototypic condition outperforms hypertaubleaux, even without fine tuning any heuristics which help finding "suitable" free variables in a given CNF formula.

Of course, first order systems are more expressive than LSAT models, but often less efficient for the reasoning algorithms. This is especially striking in non-monotonic logic: we call a logic monotonic if the truth of a proposition does not change when new information, i. e. axioms, are added to the knowledge base. In contrast, a logic is non-monotonic, if the truth of a proposition may change when new information is added to or old information is deleted from the base. Abduction, default logic, and the *closed world assumption* of Prolog, and many CLP systems are examples for applications of non-monotonic logic.

Recent complexity results for abductive reasoning and default logic as originally considered for diagnosis by McCarthy and Reiter (amongst others) indicate that only specific subsets and "sub-problems" can be dealt with in an efficient manner [7, 4]. Diagnosis based on these prominent concepts remains subject to restrictions which do not exist when using a reduced (but less expressive) propositional model of a diagnosable system along with suitable monitoring mechanisms.

Both diagnosability, i. e. strategic placement of sensors, and generation of monitors — often called observers — are active fields of research today; see, for example, [21], [11], [10], and [9]. Hence, the diagnosis approach shown in this paper should be considered an addition to these activities in order to complement the reasoning about system failure and corresponding causes for it.

6   Conclusions

This work has presented an efficient approach to model based diagnosis based on k-satisfiability and models of minimal cardinality. The beauty of the proposed solution lies in its simplicity, because it combines the advantages of state-of-the-art SAT-solving and deals with large designs by pruning the search space according to user defined criteria, i. e. n-fault assumption on AB-predicates.

The evaluation in § 4.2 hints to the scalability of this approach and its potential impact on system diagnosis and monitoring. More so, the algorithms used require at most polynomial space and can be examined and tested in detail using the freely available implementation of the presented LSAT program (see p. 56).

Unlike many other SAT-solvers, LSAT is able to come up with *all* models (under the n-fault assumption) yielding sensible conflict sets, hence diagnoses. Although this notion of minimality is not correlated to Reiter's original HS-tree [19], thus set theory, it provides for technically useful diagnoses.

Depending on the system under consideration, the generation of monitors, however, may vary dramatically. Possible realisations may be purely in software, e. g. monitoring threads and middleware, or mostly in hardware, e. g. intelligent sensors as increasingly used in the automotive domain, for instance. Although this paper has focussed mainly on aspects of diagnosis, the combination with monitoring techniques provides potential for a broader application of these techniques, especially in scenarios where quality and safety properties are increasingly important, e. g. embedded systems.

Acknowledgements

The author thanks his colleagues Gernot Stenz and Reinhold Letz for insightful discussions regarding SAT-solving and for their comments on the technicalities of the implementation. Martin Leucker and Martin Wildmoser both provided valuable feedback on an earlier version of this paper.

References

1. M. Ali, A. Veneris, S. Safarpour, M. Abadir, R. Drechsler, and A. Smith. Debugging Sequential Circuits Using Boolean Satisfiability. In *5th International Workshop on Microprocessor Test and Verification (MTV'04)*, Austin, 2004.
2. P. Baumgartner, P. Fröhlich, U. Furbach, and W. Nejdl. Semantically Guided Theorem Proving for Diagnosis Applications. pages 460–465.
3. P. Baumgartner, P. Fröhlich, U. Furbach, and W. Nejdl. Tableaux for Diagnosis Applications. Technical Report 23–96, Universität Koblenz-Landau, Institut für Informatik, Rheinau 1, D-56075 Koblenz, 1996.
4. R. Ben-Eliyahu-Zohary. Yet some more complexity results for default logic. *Artificial Intelligence*, 139(1):1–20, 2002.
5. M. Davis, G. Logemann, and D. Loveland. A machine program for theorem-proving. *Communications of the ACM*, 5(7):394–397, 1962.
6. J. de Kleer and B. C. Williams. Diagnosing multiple faults. *Artificial Intelligence*, 32(1):97–130, 1987.
7. T. Eiter and T. Lukasiewicz. Complexity results for explanations in the structural-model approach. *Artif. Intell.*, 154(1-2):145–198, 2004.
8. R. Greiner, B. A. Smith, and R. W. Wilkerson. A correction to the algorithm in Reiter's theory of diagnosis. *Artificial Intelligence*, 41:79–88, 1989.
9. J. Håkansson, B. Jonsson, and O. Lundqvist. Generating online test oracles from temporal logic specifications. *STTT*, 4(4):456–471, 2003.
10. K. Havelund and G. Rosu. Monitoring Java Programs with Java PathExplorer. *Electronic Notes Theoretical Computer Science*, 55(2), 2001.
11. K. Havelund and G. Rosu. Synthesizing Monitors for Safety Properties. In *Tools and Algorithms for Construction and Analysis of Systems*, pages 342–356, 2002.
12. Y. C. Kim, K. K. Saluja, and V. D. Agrawal. Multiple faults: Modeling, simulation and test. In *Proceedings of the 2002 conference on Asia South Pacific design automation/VLSI Design*, page 592. IEEE Computer Society, 2002.
13. C. M. Li and Anbulagan. Look-ahead versus look-back for satisfiability problems. In *CP*, pages 341–355, 1997.
14. L. Li and Y. F. Jiang. Computing minimal hitting sets with genetic algorithms. *Algorithmica*, 32(1):95–106, 2002.
15. I. Lynce and J. P. Marques-Silva. Efficient data structures for backtrack search SAT solvers. In *Proceedings of the 5th International Symposium on the Theory and Applications of Satisfiability Testing (SAT)*, May 2002.
16. J. McCarthy. Circumscription — a form of non-monotonic reasoning. *Artificial Intelligence*, 13:27–39, 1980.
17. M. W. Moskewicz, C. F. Madigan, Y. Zhao, L. Zhang, and S. Malik. Chaff: Engineering an Efficient SAT Solver. In *Proceedings of the 38th Design Automation Conference (DAC'01)*, 2001.

18. A. Nonnengart and C. Weidenbach. Computing small clause normal forms. In A. Robinson and A. Voronkov, editors, *Handbook of Automated Reasoning*, volume I, chapter 6, pages 335–367. Elsevier Science B.V., 2001.
19. R. Reiter. A theory of diagnosis from first principles. *Artificial Intelligence*, 32(1):57–95, 1987.
20. P. Torasso, L. Console, L. Portinale, and D. T. Dupré. On the role of abduction. *ACM Comput. Surv.*, 27(3):353–355, 1995.
21. Y. L. Traon, F. Ouabdesselam, C. Robach, and B. Baudry. From diagnosis to diagnosability: axiomatization, measurement and application. *J. Syst. Softw.*, 65(1):31–50, 2003.
22. F. Vatan. The complexity of the diagnosis problem. Technical Support Package (TSP) NPO-30315, NASA Jet Propulsion Laboratory, Apr. 2002.
23. A. Veneris. Fault Diagnosis and Logic Debugging Using Boolean Satisfiability. In *Fourth International Workshop on Microprocessor Test and Verification Common Challenges and Solutions*, Austin, Texas, May 2003.
24. H. Zhang. SATO: an efficient propositional prover. In *Proceedings of the International Conference on Automated Deduction (CADE'97)*, volume 1249 of *LNAI*, pages 272–275, 1997.

The *tree* Constraint

Nicolas Beldiceanu[1], Pierre Flener[2,3], and Xavier Lorca[1]

[1] École des Mines de Nantes, LINA FREE CNRS 2729,
FR-44307 Nantes Cedex 3, France
{Nicolas.Beldiceanu, Xavier.Lorca}@emn.fr
[2] Department of Information Technology, Uppsala University,
Box 337, SE-751 05 Uppsala, Sweden
Pierre.Flener@it.uu.se
[3] The Linnaeus Centre for Bioinformatics, Uppsala University,
Box 598, SE-751 24 Uppsala, Sweden

Abstract. This article presents an arc-consistency algorithm for the *tree* constraint, which enforces the partitioning of a digraph $\mathcal{G} = (\mathcal{V}, \mathcal{E})$ into a set of vertex-disjoint anti-arborescences. It provides a necessary and sufficient condition for checking the *tree* constraint in $\mathcal{O}(|\mathcal{V}| + |\mathcal{E}|)$ time, as well as a complete filtering algorithm taking $\mathcal{O}(|\mathcal{V}| \cdot |\mathcal{E}|)$ time.

1 Introduction

Graph partitioning constraints were already considered from an early stage of constraint programming research as natural shortcuts for expressing constraints on a graph. This was for instance the case of the *Hamiltonian circuit* and *spanning tree* constraints of ALICE [11]. Later on, this was also the case for the *cycle* [3] and *path* constraints [5, 14, 15], which were respectively introduced in some later version of CHIP [7] and Ilog Solver [12]. But curiously, despite its study within the Operations Research and algorithm design communities [6, 13], the problem of partitioning a digraph into a set of vertex-disjoint anti-arborescences[1] was so far ignored by the constraint programming community. This problem has a lot of practical applications, for instance in VLSI circuit design. The application that motivated us is the construction of a supertree from given trees with overlapping leaf sets, such that the ancestor relationships of the given trees are preserved. This is an important issue in phylogeny and has applications in molecular biology and linguistics [1], such as the construction of the Tree of Life [4]. See the description of future work in Section 4 for how this phylogenetic problem relates to the problem described here.

---

[1] A digraph $\mathcal{A}$ is an *anti-arborescence* with *anti-root* r iff there exists a path from all vertices of $\mathcal{A}$ to r and the undirected graph associated with the digraph $\mathcal{A}$ is a tree.

R. Barták and M. Milano (Eds.): CPAIOR 2005, LNCS 3524, pp. 64–78, 2005.

This paper addresses the mentioned digraph partitioning problem from a constraint programming perspective.[2] We should stress that, as usual within constraint programming, our goal is not partitioning a given digraph $\mathcal{G}$ into vertex-disjoint anti-arborescences, but rather first to find out whether this is possible at all or not, and second to detect those arcs of $\mathcal{G}$ that do not belong to any partitioning. Throughout this article, we use for simplicity the term *tree* rather than the term *anti-arborescence*.

The *tree* constraint has the form *tree*(NTREE, VERTICES), where NTREE is a integer variable[3] and VERTICES is a collection of n items, each item consisting of the following attributes:

- **index** is an integer between 1 and n.
- **father** is an integer variable whose domain is a subset of the values of the interval $[1, n]$.

The i-th item of the VERTICES collection is denoted VERTICES[i]. Furthermore, VERTICES[i].attr represents the value of attribute attr of VERTICES[i]. A collection of n items, each having p attributes $a_1, a_2, ..., a_p$ is denoted by:

$$\{(a_1 - v_{11}, ..., a_p - v_{1p}), (a_1 - v_{21}, ..., a_p - v_{2p}), ..., (a_1 - v_{n1}, ..., a_p - v_{np})\}$$

In order to define the *tree* constraint we first introduce the digraph associated with any instance of the *tree* constraint. We then define the meaning of the *tree* constraint as a graph property that must hold on the digraph associated with a ground[4] instance of the *tree* constraint.

Definition 1. *Digraph associated with a tree constraint*
To any tree(NTREE, VERTICES) *constraint we associate the digraph* $\mathcal{G} = (\mathcal{V}, \mathcal{E})$, *where:*

- *To each item* VERTICES[i], $(1 \leq i \leq n)$, *of the* VERTICES *collection corresponds a vertex of* $\mathcal{V}$ *denoted by* v_i. *Observe that* $|\mathcal{V}| = n$.
- *For every pair of items* (VERTICES[i], VERTICES[j]), *where i and j are not necessarily distinct, there is an arc from v_i to v_j in* $\mathcal{E}$ *(i.e.,* $(v_i, v_j) \in \mathcal{E})$ *if* $j \in$ dom(VERTICES[i].father). *Let:*

$$m = |\mathcal{E}| = \sum_{i=1}^{n} |dom(\text{VERTICES}[i].\text{father})|$$

Observe that each vertex of the digraph $\mathcal{G}$ associated with a ground instance of the *tree* constraint has exactly one successor.

---

[2] The term "tree constraint" exists in the constraint programming community, but the *tree processing* problem defined in [1] assembles a set of trees in one single tree according to some *dominance* constraints.

[3] A integer variable V is a variable that ranges over a finite set of integers denoted by $dom(V)$. $min(V)$ and $max(V)$ respectively denote the minimum and the maximum value of V.

[4] An instance such that all its integer variables are fixed.

Definition 2. *A ground instance of a* tree(NTREE, VERTICES) *constraint holds if* VERTICES[i].index = i , (1 ≤ i ≤ n), *and if its associated digraph* $\mathcal{G} = (\mathcal{V}, \mathcal{E})$ *verifies the two following conditions:*

- *$\mathcal{G}$ consists of* NTREE *connected components.*
- *Each connected component of $\mathcal{G}$ does not contain any circuit involving more than one vertex.*

The index and the father attributes of an item can be respectively interpreted as the unique identifier of that item and as the successor of that item in the partionning into trees.

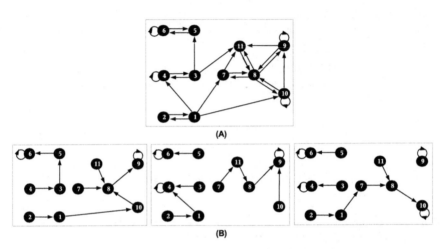

(A)

(B)

Fig. 1. (A) A digraph $\mathcal{G}$ and **(B)** three possible vertex-disjoint tree partitionings of $\mathcal{G}$

Example 1. For the digraph depicted by part (A) of Figure 1, a *tree* constraint is stated as *tree*(NTREE, VERTICES) where:

$$\begin{aligned} \text{VERTICES} = \{&(\text{index} - 1, \text{father} - F_1), (\text{index} - 2, \text{father} - F_2), \\ &(\text{index} - 3, \text{father} - F_3), (\text{index} - 4, \text{father} - F_4), \\ &(\text{index} - 5, \text{father} - F_5), (\text{index} - 6, \text{father} - F_6), \\ &(\text{index} - 7, \text{father} - F_7), (\text{index} - 8, \text{father} - F_8), \\ &(\text{index} - 9, \text{father} - F_9), (\text{index} - 10, \text{father} - F_{10}), \\ &(\text{index} - 11, \text{father} - F_{11})\}, \end{aligned}$$

$dom(\text{NTREE})$, $dom(F_1)$, $dom(F_2)$, $dom(F_3)$, $dom(F_4)$, $dom(F_5)$, $dom(F_6)$, $dom(F_7)$, $dom(F_8)$, $dom(F_9)$, $dom(F_{10})$, $dom(F_{11})$ respectively are $\{1, 2, 3, 4, 5\}$, $\{2, 4, 7, 10\}$, $\{1\}$, $\{4, 5, 11\}$, $\{3, 4\}$, $\{6\}$, $\{5, 6\}$, $\{8, 11\}$, $\{7, 9, 10, 11\}$, $\{8, 9, 11\}$, $\{8, 9, 10\}$, $\{8\}$.

Part (B) of Figure 1 shows three possible solutions of the vertex-disjoint partitioning of $\mathcal{G}$ with respectively 2, 3 and 4 trees. Observe that, as stated by the second condition of Defintion 2, there is no circuit involving more than one vertex. In order to achieve arc-consistency we have to prune NTREE as well as $F_1, F_2, \ldots, F_{11}$ in the following way:

- We want to find out that 1 and 5 are not feasible numbers of trees for partitioning $\mathcal{G}$, then $dom(\text{NTREE}) = \{2, 3, 4\}$.
- According to the previous restriction of $dom(\text{NTREE})$, we restrict the domains of F_1, F_6, F_8 respectively to $dom(F_1) = \{4, 7, 10\}$, $dom(F_6) = \{6\}$ and $dom(F_8) = \{9, 10\}$.

Example 1 will be used throughout this article in order to illustrate the different propositions.

The *tree* constraint was introduced within a catalogue of global constraints [2, page 74] but no filtering algorithm was known. The contribution of this article is an $\mathcal{O}(n \cdot m)$ arc-consistency filtering algorithm for the *tree* constraint.

The rest of the article is organised as follows: Section 2 provides a necessary and sufficient condition for partitioning the digraph $\mathcal{G}$ associated with a *tree* constraint according to a given set $dom(\text{NTREE})$ of potential numbers of trees. Section 3 shows how to exploit this necessary and sufficient condition in order to prune NTREE as well as the **father** variables. Finally, Section 4 concludes this article and outlines future work.

2   Checking Feasibility

This section first gives a necessary and sufficient condition for the *tree* constraint to hold. Second, it sketches an $\mathcal{O}(n + m)$ algorithm for evaluating that condition. Before presenting it, we introduce some terminology regarding the digraph $\mathcal{G} = (\mathcal{V}, \mathcal{E})$ associated with a *tree* constraint, as well as a lower and upper bound on the number of trees needed for partitioning $\mathcal{G}$:

- To each instance of a *tree*(NTREE, VERTICES) constraint we associate the *reduced digraph* $\mathcal{G}_r$ derived from $\mathcal{G}$ in the following way: to each strongly connected component of $\mathcal{G}$ we associate a vertex of $\mathcal{G}_r$; to each arc of $\mathcal{G}$ that connects different strongly connected components corresponds an arc in $\mathcal{G}_r$. A strongly connected component of $\mathcal{G}$ that corresponds to a sink of $\mathcal{G}_r$ is called a *sink component*.
- A vertex v of $\mathcal{G} = (\mathcal{V}, \mathcal{E})$ such that $(v, v) \in \mathcal{E}$ is called a *potential root*. The arc (v, v) is called a *loop*. A strongly connected component of $\mathcal{G}$ that contains at least one potential root is called a *rooted component*.
- A vertex u of $\mathcal{G} = (\mathcal{V}, \mathcal{E})$ is a *door* of the strongly connected component associated with u iff there exists $(u, v) \in \mathcal{E}$ such that u and v do not belong to the same strongly connected component of $\mathcal{G}$.
- A *connecting arc* (u, v) of $\mathcal{G} = (\mathcal{V}, \mathcal{E})$ is an arc of $\mathcal{E}$ such that u and v do not belong to the same strongly connected component. Similary, a *nonconnecting arc* (u, v) of $\mathcal{E}$ is an arc such that u and v belong to the same strongly connected component.
- A vertex v of $\mathcal{G} = (\mathcal{V}, \mathcal{E})$ is a *winner* if v is a *door* or if $(v, v) \in \mathcal{E}$, i.e., a potential root.
- *Enforcing an arc* (u, v) of $\mathcal{G}$ corresponds to removing from $\mathcal{G}$ all arcs (u, w) such that $w \neq v$.

Example 2. Figure 2 illustrates the previous terms according to the digraph introduced in part (A) of Figure 1. In part (A), the *winners* correspond to the *doors* and *potential roots*. The *connecting arcs* and the *loops* are depicted by a black line, while the other arcs are depicted by a dotted line. S_2, S_3, S_4 are *rooted components* while S_3, S_4 are *sink components*. Part (B) depicts the reduced digraph associated with $\mathcal{G}$. To each strongly connected component $\mathcal{S}_i$ of $\mathcal{G}$ corresponds a vertex $\mathcal{R}_i$ of $\mathcal{G}_r$. Observe that R_3 and R_4 represent *sink* vertices.

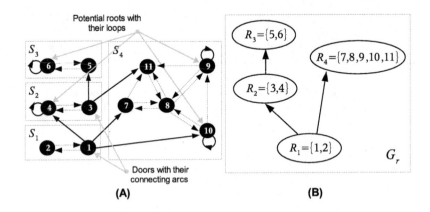

Fig. 2. (A) The digraph $\mathcal{G}$ and its strongly connected components S_1, S_2, S_3, S_4. (B) The reduced digraph $\mathcal{G}_r$ associated with $\mathcal{G}$

We now present a lower and upper bound on the number of trees that can possibly cover a given digraph $\mathcal{G}$ associated with a *tree* constraint. For this purpose, we name by MINTREE the number of sinks of $\mathcal{G}_r$ and by MAXTREE the number of potential roots of $\mathcal{G}$.

Proposition 1. *A lower bound on the minimum number of trees for partitioning the digraph $\mathcal{G}$ associated with a* tree *constraint is the number of sinks in $\mathcal{G}_r$ (i.e.,* MINTREE*).*

Proof. Proposition 1 stems from the fact that there is no path between two vertices that belong to two distinct sink components of $\mathcal{G}$.   □

Proposition 2. *An upper bound on the maximum number of trees for partitioning the digraph $\mathcal{G}$ associated with a* tree *constraint is the number of potential roots of $\mathcal{G}$ (i.e.,* MAXTREE*).*

Proof. Since each tree has a distinct root, we cannot have more trees than the number of potential roots.   □

We now state the necessary and sufficient condition to verify on a *tree* constraint in order to have at least one solution.

Proposition 3.   *Necessary and sufficient condition for a* tree *constraint*

A constraint tree(NTREE, VERTICES) *has at least one solution iff the following two conditions both hold:*

> *(1) All sink components of $\mathcal{G}$ are rooted components,*
> *(2) dom(NTREE) $\cap$ [MINTREE, MAXTREE] $\neq \emptyset$.*

Proof. We first prove that the conjunction of (1) and (2) is a necessary condition. If a sink component of $\mathcal{G}$ is not a rooted component, then there will be at least one circuit in $\mathcal{G}$ among a subset of vertices associated with this component, and the *tree* constraint cannot hold. Moreover, if *dom*(NTREE) $\cap$ [MINTREE, MAXTREE] $= \emptyset$ then *max*(NTREE) $<$ MINTREE or *min*(NTREE) $>$ MAXTREE. And we know, from Propositions 1 and 2, that the *tree* constraint then has no solutions. Secondly, we prove that the conjunction of (1) and (2) is sufficient. For this purpose, we show, in a two step construction, that for each value t in [MINTREE, MAXTREE], there exists at least one vertex-disjoint partitioning of $\mathcal{G}$ into t distincts trees. Step 1 selects t root vertices and chooses for each strongly connected component of $\mathcal{G}$ the vertex that will be the root of a tree or that will be attached to another component. Step 2 constructs for each strongly connected component a spanning forest.

STEP 1
- We choose one potential root r for each sink component of $\mathcal{G}$ and we enforce the loop (r, r) on r. Let $\mathcal{R}_1$ denote the set of thus selected roots.
- If $t >$ MINTREE then we choose a set $\mathcal{R}_2$ of $t -$ MINTREE potential roots in $\mathcal{G}$, distinct from $\mathcal{R}_1$, and we enforce a loop for each vertex of $\mathcal{R}_2$.
- For all strongly connected components for which we did not enforce a loop, we choose one vertex v that is a door and we enforce a connecting arc starting from v. Let $\mathcal{R}_3$ denote the set of thus selected doors.

STEP 2
For a given strongly connected component $\mathcal{S} = (\mathcal{V}_\mathcal{S}, \mathcal{E}_\mathcal{S})$ of $\mathcal{G}$:
- Let $\mathcal{H}_\mathcal{S} = \mathcal{V}_\mathcal{S} \cap (\mathcal{R}_1 \cup \mathcal{R}_2 \cup \mathcal{R}_3)$,
- Let $\mathcal{L}_\mathcal{S} = \mathcal{V}_\mathcal{S} - \mathcal{H}_\mathcal{S}$.

For each strongly connected component $\mathcal{S}$ of $\mathcal{G}$, we call the function introduced in Lemma 1 (see the Appendix), TreeCovering($\mathcal{S}, \mathcal{H}_\mathcal{S}, \mathcal{L}_\mathcal{S}, \emptyset$), in order to build a vertex-disjoint partitioning of $\mathcal{S}$ with $|\mathcal{H}_\mathcal{S}|$ trees and having their roots in $\mathcal{H}_\mathcal{S}$.

Thus, we have shown how to build a vertex-disjoint partitioning of $\mathcal{G}$ with t trees, for all $t \in$ [MINTREE, MAXTREE]. And, since *dom*(NTREE)$\cap$[MINTREE, MAXTREE] $\subseteq$ [MINTREE, MAXTREE], we know that there exists at least one solution for the *tree* constraint. $\qquad\square$

Proposition 4. *The worst-case complexity for checking the necessary and sufficient condition for a* tree *constraint (i.e., Proposition 3) is $\mathcal{O}(n + m)$ time.*

Proof. Evaluating the worst-case complexity for implementing Proposition 3 is done by analysing the following items:

1- Computing the strongly connected components of $\mathcal{G}$ takes $\mathcal{O}(n + m)$ time with Tarjan's algorithm [16].
2- Checking that each sink component of $\mathcal{G}$ contains at least one potential root takes $\mathcal{O}(n)$ time.
3- Checking that $dom(\text{NTREE}) \cap [\text{MINTREE}, \text{MAXTREE}] \neq \emptyset$ takes $\mathcal{O}(1)$ time.

Observe that the worst-case complexity makes the following hypotheses on the complexity of the primitives that access the domains of the variables:

- In item 3, we assume that we can get the minimum and maximum values of a integer variable in $\mathcal{O}(1)$ time.
- Since item 1 uses depth-first search, we need to iterate over the successors of a vertex of $\mathcal{G}$. This is done by iterating through the potential values of a father variable. In order to achieve $\mathcal{O}(n+m)$ time, getting the next successor (i.e., the next value of a father variable) needs to be done in $\mathcal{O}(1)$ time. Therefore we assume that a domain is represented by a list of intervals.   □

3   Domain Filtering

This section first shows how to prune the domains of the father variables $F_1, F_2, \ldots, F_n$ and of the variable NTREE from the digraph $\mathcal{G}$ associated to a *tree* constraint. All the pruning rules are derived from the necessary and sufficient condition given by Proposition 3. Then it proves the completeness of the previous pruning rules and finally sketches an $\mathcal{O}(n \cdot m)$ arc-consistency filtering algorithm.

The pruning rules remove some arcs of the digraph $\mathcal{G}$ associated with a *tree* constraint. Observe that since there is a one to one correpondence between the arcs of $\mathcal{G}$ and the father variables and their respective domains, removing an arc (u, v) from $\mathcal{G}$ is equivalent to removing value v from the domain of variable F_u.

3.1   Filtering for a *tree* Constraint

We first present a proposition that restricts the domain of NTREE according to condition (2) of Proposition 3.

Proposition 5. *The domain of* NTREE *is restricted by the two following rules:*

- *If* $max(\text{NTREE}) > \text{MAXTREE}$ *then* $max(\text{NTREE}) = \text{MAXTREE}$ (3)
- *If* $min(\text{NTREE}) < \text{MINTREE}$ *then* $min(\text{NTREE}) = \text{MINTREE}$ (4)

Proof. Conditions 3 and 4 are respectively derived from Propositions 2 and 1.

□

Example 3. We illustrate how to prune the domain of NTREE according to the digraph $\mathcal{G}$ depicted by part (A) of Figure 1. As $\mathcal{G}$ contains 2 sink components and 4 potential roots, MINTREE and MAXTREE are respectively equal to 2 and 4. Therefore, assuming that $dom(\text{NTREE}) = \{1, 2, 3, 4, 5\}$, Proposition 5 removes the values 1 and 5 from $dom(\text{NTREE})$.

Before presenting the next proposition, we need to introduce the notion of *strong articulation point* given by Gondran and Minoux in [9, page 175].

Definition 3. *A strong articulation point of a strongly connected digraph $\mathcal{G}$ is a vertex such that if we remove it, $\mathcal{G}$ is broken into at least two strongly connected components.*

The withdrawal of a strong articulation point p, in a strongly connected component $\mathcal{S}$ of the digraph $\mathcal{G}$ associated with a *tree* constraint, creates two types of strongly connected components:

- let $\Delta^p_{out} = \{\mathcal{S}^1_{out}, \ldots, \mathcal{S}^l_{out}\}$ be the possibly empty set of new strongly connected components from which we can reach, by a path that does not contain p, a *winner* of $\mathcal{S}$.
- let $\Delta^p_{in} = \{\mathcal{S}^1_{in}, \ldots, \mathcal{S}^q_{in}\}$ be the possibly empty set of new strongly connected components from which we cannot reach, by a path that does not contain p, a *winner* of $\mathcal{S}$.

Property 1. Let p be a strong articulation point of a strongly connected component of $\mathcal{G}$. Then p belongs to all paths from any vertex of Δ^p_{in} to any vertex of Δ^p_{out}.

Proposition 6. *An outgoing arc (p, v) of a strong articulation point p that reaches a vertex v of a strongly connected component of Δ^p_{in} never belongs to any solution of a* tree *constraint.*

Proof. For any outgoing arc (p, v) of a strong articulation point p, if v belongs to a strongly connected component of Δ^p_{in}, then, by Property 1, every path from v to any vertex of a strongly connected component of Δ^p_{out} contains p. Thus, enforcing (p, v) creates a circuit with some vertices of Δ^p_{in} and p. Therefore, (p, v) never belongs to any solution of a *tree* constraint. □

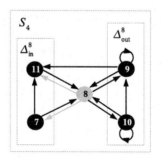

Fig. 3. Pruning according to a strong articulation point

Example 4. Figure 3 illustrates Proposition 6 on the strongly connected component S_4 of the digraph $\mathcal{G}$ depicted by part (A) of Figure 1. Vertex 8 is a strong articulation point since its removal breaks S_4 into four strongly connected components. 9 and 10 are potential root vertices, and since 7 and 11 have neither loops nor connecting arcs, we have $\Delta_{out}^8 = \{\{9\}, \{10\}\}$ and $\Delta_{in}^8 = \{\{7\}, \{11\}\}$. Then, from Proposition 6, the arcs $(8, 7)$ and $(8, 11)$ (drawn in gray in Figure 3) are infeasible.

We now introduce a final proposition that allows us to prune according to the fact that we have to build a vertex-disjoint partitioning compatible with $dom(\texttt{NTREE})$.

Proposition 7. *Let* $C = dom(\texttt{NTREE}) \cap [\texttt{MINTREE}, \texttt{MAXTREE}]$. *For each strongly connected component* $\mathcal{S}$ *of* $\mathcal{G}$:

1. *If* $\mathcal{S}$ *is a sink component of* $\mathcal{G}$ *that contains one single potential root* r, *then all the outgoing arcs of* r, *except the loop* (r, r), *are infeasible.*
2. *Otherwise:*
   2.1. *If* $C = \{\texttt{MAXTREE}\}$ *then, for each potential root* r *of* $\mathcal{S}$, *all the non-loop arcs* (r, v) $(v \neq r)$ *are infeasible.*
   2.2. *If* $C = \{\texttt{MINTREE}\}$ *and* $\mathcal{S}$ *is a non-sink component then all the loops of* $\mathcal{S}$ *are infeasible.*
   2.3. *If there exists a single winner* w *in* $\mathcal{S}$, *which is a door, then all the non-connecting arcs* (w, v) *are infeasible.*

Proof. For item 1, let r be the potential root of $\mathcal{S}$ and assume that we enforce an outgoing arc (r, v) $(v \neq r)$. Then, as $\mathcal{S}$ does not contain any doors, we cannot leave $\mathcal{S}$ and thus create a circuit involving at least two vertices. Item 2.1 (respectively 2.2) is a direct consequence of Proposition 2 (respectively Proposition 1). For item 2.3, assume that we have a single door w in $\mathcal{S}$ and consider that no potential root belongs to $\mathcal{S}$. If we do not take a connecting arc of w, then we can never leave $\mathcal{S}$ and therefore we create a circuit in $\mathcal{S}$ involving at least two vertices. Thus, the *tree* constraint cannot hold. □

Example 5. In order to illustrate Proposition 7 on the digraph depicted by part (A) of Figure 1, we consider the following three cases:

- First, we do not assume any restriction on $dom(\texttt{NTREE})$. Then, in this context, item 1 removes the arc $(6, 5)$, while item 2.3 removes $(1, 2)$.
- If $\texttt{NTREE}$ is equal to $\texttt{MAXTREE}$ (i.e., 4) then, in addition to the arcs removed by items 1 and 2.3, item 2.1 removes the arcs $(4, 3)$, $(9, 8)$, $(9, 11)$, $(10, 8)$ and $(10, 9)$ since a loop is enforced for each of the vertices 4, 9 and 10.
- If $\texttt{NTREE}$ is equal to $\texttt{MINTREE}$ (i.e., 2) then, in addition to the arcs removed by items 1 and 2.3, item 2.2 removes the arc $(4, 4)$.

3.2 Arc-Consistency

Now, we prove that Propositions 5, 6 and 7 characterise all the arcs that do not belong to any solution of a *tree* constraint. For this purpose, we assume that the necessary and sufficient condition of Proposition 3 holds.

Proposition 8. *If Proposition 3 holds then the two following equivalences lead to the completeness of the pruning rules:*

- *Let $t \in dom(\text{NTREE})$, then t is incompatible with a* tree *constraint if and only if t is pruned by Proposition 5.*
- *Let $(u, v) \in \mathcal{E}$, then (u, v) is incompatible with a* tree *constraint if and only if (u, v) is removed by at least one proposition among Propositions 6 and 7.*

Proof. We first prove that Proposition 5 removes all infeasible values in the domain of NTREE. Indeed, we have completeness since Proposition 3 enforces for each $t \in [\text{MINTREE}, \text{MAXTREE}]$ the existence of a vertex-disjoint partitioning of $\mathcal{G}$ with t trees.

Second, we prove that Propositions 6 and 7 remove all infeasible values for the father variables. Now, for each arc (u, v) that was not pruned by Propositions 6 or 7, we show how to build a vertex-disjoint partitioning of $\mathcal{G}$ with t trees, where $t \in dom(\text{NTREE}) \cap [\text{MINTREE}, \text{MAXTREE}]$.

STEP 1
Let $dom(\text{NTREE}) \cap [\text{MINTREE}, \text{MAXTREE}] = [\text{mintree}, \text{maxtree}]$,
A1 **Selecting a root in each sink component of $\mathcal{G}$:**
For each sink component $\mathcal{S}$ of $\mathcal{G}$, if u is a potential root of $\mathcal{S}$ and $u = v$ then (u, v) has to be enforced. Otherwise, we select a potential root r of $\mathcal{S}$ different from u and we enforce the loop (r, r). Observe that item 1 of Proposition 7 garanties us to find a potential root different from u. Let $\mathcal{R}_1$ denote the set of thus selected roots in the different sink components of $\mathcal{G}$.
A2 **Completing the set of roots in order to get mintree or mintree+1 trees:**
    CASE 1: mintree > MINTREE
    Since we have to build at least mintree trees, we choose to build exactly mintree trees. Therefore, we enforce a set $\mathcal{R}_2$ of mintree − MINTREE potential roots distinct from $\mathcal{R}_1$. Observe that if $u = v$ and u does not belong to any sink component then u must belong to $\mathcal{R}_2$.
    CASE 2: mintree = MINTREE
    • If $u = v$ and u does not belong to any sink component then MINTREE < MAXTREE and we have to enforce the loop (u, u), and $\mathcal{R}_2 = \{u\}$. Therefore, we choose to build mintree + 1 trees.
    • Otherwise, we build mintree trees and $\mathcal{R}_2 = \emptyset$.
A3 **Selecting a *door* in the strongly connected components that do not contain a vertex of $\mathcal{R}_1 \cup \mathcal{R}_2$.**
For all the strongly connected components $\mathcal{S} = (\mathcal{V}_{\mathcal{S}}, \mathcal{E}_{\mathcal{S}})$ for which no loops are enforced (i.e., $\mathcal{V}_{\mathcal{S}} \cap (\mathcal{R}_1 \cup \mathcal{R}_2) = \emptyset$):
    • If $u \in \mathcal{V}_{\mathcal{S}}$ and (u, v) is a connecting arc, then (u, v) is enforced. Observe that if u is the only door of $\mathcal{S}$ then $v \notin \mathcal{V}_{\mathcal{S}}$ by item 2.3 of Proposition 7.
    • Otherwise, if $u, v \in \mathcal{V}_{\mathcal{S}}$ or $u \notin \mathcal{V}_{\mathcal{S}}$ then a door w, different from u, is chosen and we enforce one of its connecting arcs.

Let $\mathcal{R}_3$ denote the set of thus selected doors.

STEP 2

For a given strongly connected component $\mathcal{S} = (\mathcal{V}_\mathcal{S}, \mathcal{E}_\mathcal{S})$:

- Let $\mathcal{H}_\mathcal{S} = \mathcal{V}_\mathcal{S} \cap (\mathcal{R}_1 \cup \mathcal{R}_2 \cup \mathcal{R}_3)$.
- Let $\mathcal{L}_\mathcal{S} = \mathcal{V}_\mathcal{S} - \mathcal{H}_\mathcal{S}$.
- Let $\mathcal{A}_\mathcal{S} = \{(u,v)\}$ if $u, v \in \mathcal{V}_\mathcal{S}$, otherwise $\mathcal{A}_\mathcal{S} = \emptyset$.

For each strongly connected component $\mathcal{S}$ of $\mathcal{G}$, we call the function introduced in Lemma 1 (see the Appendix), `TreeCovering`$(\mathcal{S}, \mathcal{H}_\mathcal{S}, \mathcal{L}_\mathcal{S}, \mathcal{A}_\mathcal{S})$, in order to build a vertex-disjoint partitioning of $\mathcal{S}$ with $|\mathcal{H}_\mathcal{S}|$ trees that includes (u,v) if $u, v \in \mathcal{V}_\mathcal{S}$.

Thus, for each arc $(u,v) \in \mathcal{E}$ that is not pruned by Propositions 6 or 7, we have shown how to build a vertex-disjoint partitioning of $\mathcal{G}$ with t trees, where $t \in dom(\texttt{NTREE}) \cap [\texttt{MINTREE}, \texttt{MAXTREE}]$. □

3.3   Polynomial Arc-Consistency Algorithm

We show how to process all the pruning rules in $\mathcal{O}(n \cdot m)$ time. Two parts are distinguished, the first one only considers the pruning according to the strong articulation points (Proposition 6), the second one considers the pruning of `NTREE` (Proposition 5) and the pruning related to Proposition 7.

Then, in the first part, we are interested in Proposition 6 and we propose the `TreeFiltering` algorithm below. For this purpose we have to detect all the strong articulation points of a strongly connected component of $\mathcal{G}$:

- Finding the strong articulation points takes at the maximum $\mathcal{O}(n \cdot m)$ time because we have not found a more efficient algorithm than withdrawing a vertex and testing if the remaining subgraph is strongly connected or not.
- Detecting the arcs to be pruned takes $\mathcal{O}(n \cdot m)$ time because for each of the strongly connected components $\mathcal{S}$ we have to withdraw each strong articulation point p detected:
  - we have to search Δ_{in}^p, thanks to a depth-first search beginning from the winners of $\mathcal{S}$.
  - we mark the vertices reachable from at least one winner as vertices of Δ_{out}^p.
  - we remove the arcs, that reach Δ_{in}^p vertices from p, according to Proposition 6.

Now, in the second part, it is straightforward that the pruning related to Proposition 5 is carried out in constant time. Moreover, the pruning related to the general Proposition 7 consists of four steps:

- Items 1 and 2.3 of Proposition 7: the time complexity of these steps lies in the construction of the reduced digraph, which takes $\mathcal{O}(m + n)$ time.
- Item 2.1 of Proposition 7: all the potential roots have to be detected, that takes $\mathcal{O}(n)$ time.

- Item 2.2 of Proposition 7: we have to detect all the non-sink components of $\mathcal{G}$; then the time complexity lies in the construction of the depth-first search in $\mathcal{O}(n + m)$ time.

TreeFiltering$(\mathcal{G}) : \mathcal{R}$.
Input : the digraph $\mathcal{G}$.
Output : $\mathcal{R}$ the set of prunable arcs of $\mathcal{G}$.
$\mathcal{R} \leftarrow \emptyset$; \\ *the set of arcs pruned.*
$W \leftarrow \{u \mid u \text{ is a winner}\}$;
For each vertex scc **of** $\mathcal{G}$ **do**
$\quad\Psi_{scc} \leftarrow \emptyset$; \\ *the set of the strong articulation points of scc.*
$\quad$\\ *we detect the strong articulation points (s.a.p.) of scc.*
$\quad$**For each vertex** u **of** scc **do**
$\quad\quad$**if** scc **without** u **is not strongly connected then** $\Psi_{scc} \leftarrow \Psi_{scc} \cup \{u\}$;
$\quad$\\ *we search infeasible arcs.*
$\quad$**For each vertex** u **of** Ψ_{scc} **do** PruneArcs$(u, scc, W, \mathcal{R})$;

PruneArcs$(u, scc, W, \mathcal{R}) : \mathcal{R}$
Input : a strongly connected component scc, a strong articulation point u of scc, the set W of winners and the set $\mathcal{R}$.
Output : $\mathcal{R}$ the set of prunable arcs, increased by those discovered in scc.
For each vertex v **of** scc **do** $reach[v] \leftarrow false$;
\\ *detecting the blocks obtained by the withdrawal of each s.a.p. of scc.*
$\Sigma^u_{scc} \leftarrow \{C^u_1, ..., C^u_m\}$;
\\ *we search in each block the infeasible arcs.*
For each $C^u_i \in \Sigma^u_{scc}$ **do**
$\quad search[C^u_i] \leftarrow false$;
$\quad$**For each** $w \in W$ **such that** $(w \in C^u_i \wedge \neg reach[w]) \vee (w = u \wedge \neg search[C^u_i])$ **do**
$\quad\quad$Visit$(w, u, reach[\ ])$;
$\quad search[C^u_i] \leftarrow true$;
$\quad$\\ *withdrawal of infeasible outgoing arcs of u.*
$\quad$**If** $\exists (u, v) \in \mathcal{E}$ **such that** $v \in C^u_i \wedge \neg reach[v]$ **then**
$\quad\quad W \leftarrow W \cup \{u\}$;
$\quad\quad \mathcal{R} \leftarrow \mathcal{R} \cup \{(u, v)\}$;

Visit$(v, u, reach[\ ]) : reach[\ ]$
Input : a *winner* v, a strong articulation point u and a boolean table $reach[\ ]$.
Output : the table $reach[\ ]$.
$reach[v] \leftarrow true$;
For each $v \neq u$ **and** $w \in pred[v]$ **such that** $\neg reach[w]$ **do** Visit$(w, u, reach[\ ])$;

TreeFiltering algorithm

Example 6. We present a trace of TreeFiltering according to the strongly connected component depicted by Figure 3:

- In TreeFiltering: $\mathcal{R} \leftarrow \emptyset$; $W \leftarrow \{9, 10\}$; $\Psi_{\mathcal{S}_4} \leftarrow \{8\}$; PruneArcs$(8, \mathcal{S}_4, W, \mathcal{R})$.
- In PruneArcs: for each $u \in \{7, 8, 9, 10, 11\}$ do $reach[u] \leftarrow false$; $\Sigma^8_{\mathcal{S}_4} \leftarrow \{\{7\}, \{9\}, \{10\}, \{11\}\}$. We process a depth-first search with the recursive function Visit(). Finally, vertices 7 and 11 are not reachable from 9 or 10, then $\Delta^8_{in} = \{\{7\}, \{11\}\}$ and $\Delta^8_{out} = \{\{9\}, \{10\}\}$, thus $\mathcal{R} \leftarrow \{(8, 7), (8, 11)\}$ according to Proposition 6.

4   Conclusion and Perspectives

This article provides an arc-consistency algorithm description and a necessary and sufficient condition for the *tree* constraint.

On the one hand, the necessary and sufficient condition for the *tree* constraint consists in two conditions checked in $\mathcal{O}(n + m)$ time (Proposition 3). On the other hand, the key point of the arc-consistency algorithm is the detection and processing (Proposition 6) of the strong articulation points. Unfortunately, to our knowledge, the existence of an $\mathcal{O}(m)$ algorithm is an open problem, thus the current complexity is $\mathcal{O}(n \cdot m)$ time. Furthermore, note that it would be possible to get a relaxed $\mathcal{O}(n + m)$ time algorithm using a subset of the strong articulation points. A natural choice for such a subset is the set of articulation points[5] provided without the orientation of the arcs of the digraph associated with the *tree* constraint. An implementation of this relaxed algorithm was carried out in *Choco* [10] with the version 1.0 available to http://choco.sf.net.

Future work will address the phylogenetic problem of constructing a supertree from given trees with overlapping leaf sets, such that the ancestor relationships of the given trees are preserved. This problem can be modelled in terms of a generalisation of the *tree* constraint, where the VERTICES collection has been augmented with an optional attribute giving the direct ancestors of the considered vertex in the given trees, but with NTREE = 1. Notice that this takes a number of integer variables that is *linear* in the number of vertices, and hence linear in the number of species (the leaves of the given trees), rather than a number quadratic in the number of species as in a previous constraint-programming approach to supertree construction [8]. The advantages of deploying constraint programming on this phylogenetic problem, as opposed to the purely algorithmic approaches advocated so far (see [4] for instance), are that any combination of biological side constraints (on branch lengths, speciation dates, nested species, etc) to the otherwise purely combinatorial problem can be incorporated without having to design a new algorithm each time, that *all* the supertrees can be enumerated without having to generalise an otherwise deterministic algorithm, and that an explanation of why the given trees are incompatible can be provided when no supertree is found. In the latter case, the supertree problem can also be re-cast as an optimisation problem, and constraint programming will facilitate experiments with emerging cost functions.

---

[5] An *articulation point* [9, page 16] of an undirected graph $\mathcal{G}$ is a vertex v of $\mathcal{G}$ such that if we remove it, the number of connected components of $\mathcal{G}$ deprived of v increases.

Acknowledgements. This project was partially funded by grant HPRI-CT-2001-00153 within the *Human Research Potential and Socio-Economic Knowledge Base: Access to Research Infrastructures* (ARI) programme of the European Commission, when the first author visited the second author at The Linnaeus Centre for Bioinformatics. Finally, the implementation of the constraint would not have been possible without the relevant advice of Hadrien Cambazard and Guillaume Rochart regarding *JChoco*.

References

1. E. Althaus, D. Duchier, A. Koller, K. Mehlhorn, J. Niehren, and S. Thiel. An efficient graph algorithm for dominance constraints. *Journal of Algorithms*, 48(1):194–219, May 2003. Special Issue of SODA 2001.
2. N. Beldiceanu. Global constraints as graph properties on structured network of elementary constraints of the same type. Technical report, SICS T2000:01, Sweden, January 2000.
3. N. Beldiceanu and E. Contejean. Introducing global constraint in CHIP. *Mathl. Comput. Modelling*, 20(12):97–123, 1994.
4. O. R. Bininda-Emonds, editor. *Phylogenetic Supertrees: Combining Information to Reveal the Tree of Life*. Kluwer, 2004.
5. H. Cambazard and E. Bourreau. Conception d'une contrainte globale de chemin. *JNPC*, pages 107–120, 2004. In French.
6. Z.-Z. Chen. Efficient algorithms for acyclic colorings of graphs. *Theoretical Computer Science*, 230(1-2):75–95, 2000.
7. M. Dincbas, P. V. Hentenryck, H. Simonis, T. G. A. Aggoun, and F. Berthier. The Constraint Logic Programming Language CHIP. In *Int. Conf. on Fifth Generation Computer Systems (FGCS'88)*, pages 693–702, Tokyo, Japan, 1988.
8. I. P. Gent, P. Prosser, B. M. Smith, and W. Wei. Supertree construction using constraint programming. In F. Rossi, editor, *Proceedings of CP'03*, volume 2833 of *LNCS*, pages 837–841. Springer-Verlag, 2003.
9. M. Gondran and M. Minoux. *Graphes et algorithmes*. Eyrolles, Paris, 2nd edition, 1985. In French.
10. F. Laburthe and the OCRE group. CHOCO: implementing a CP kernel. In *CP00 Post Conference Workshop on Techniques for Implementing Constraint programming Systems (TRICS)*, Singapore, Sept. 2000.
11. J.-L. Laurière. A language and a program for stating and solving combinatorial problems. *Artificial Intelligence*, 10:29–127, 1978.
12. J.-F. Puget. A C++ Implementation of CLP. In *Second Singapore International Conference on Intelligent Systems (SPICIS)*, pages 256–261, Singapore, November 1994.
13. A. Roychoudhury and S. Sur-Kolay. Efficient algorithms for vertex arboricity of planar graphs. In *Proceedings of the 15th Conference on Foundations of Software Technology and Theoretical Computer Science*, volume 1026 of *LNCS*, pages 37–51. Springer-Verlag, 1995.
14. M. Sellmann. *Reduction techniques in Constraint Programming and Combinatorial Optimization*. PhD thesis, University of Paderborn, 2002.

15. M. Sellmann. Cost-based filtering for shortest path constraints. In *9th international Conference on the Principles and Practice of Constraint Programming (CP)*, volume 2833 of *LNCS*, pages 694–708. Springer-Verlag, 2003.

16. R. Tarjan. Depth-first search and linear graph algorithms. In *SIAM J. Comput.*, volume 1, pages 146–160, 1972.

Appendix

Lemma 1 is used in the proofs of Propositions 3 and 8. It presents a constructive method for building a vertex-disjoint partitioning into anti-arborescences of a particular digraph depicted by the assumptions 1, 2 and 3.

Lemma 1. *Let* $\mathcal{G} = (\mathcal{V}, \mathcal{E})$ *be a digraph such that:*

(1) Let $\mathcal{H}, \mathcal{L} \subseteq \mathcal{V}$ *such that* $\mathcal{H} \cup \mathcal{L} = \mathcal{V}$ *and* $\mathcal{H} \cap \mathcal{L} = \emptyset$.
(2) For each $v \in \mathcal{L}$ *there exists a path from* v *to at least one vertex of* $\mathcal{H}$.
(3) Let $\mathcal{A} \subset \mathcal{E}$ *such that* $|\mathcal{A}| \leq 1$ *and if* $|\mathcal{A}| = 1$ *then* $\mathcal{A} = \{(u, v)\}$.

The following algorithm computes $|\mathcal{H}|$ *vertex-disjoint anti-arborescences and having their roots in* $\mathcal{H}$:

> TreeCovering$(\mathcal{G}, \mathcal{H}, \mathcal{L}, \mathcal{A}) : \mathcal{F}$
> **Input** $: \mathcal{G}, \mathcal{H}, \mathcal{L}, \mathcal{A}$.
> **Output** $: \mathcal{F}$, *the set of vertex-disjoint anti-arborescences.*
>
> $\mathcal{F} \leftarrow \emptyset$;
> **While** $\mathcal{L} \neq \emptyset$ **do**
> | **If** $\mathcal{A} \neq \emptyset$ **and** $\exists h \in \mathcal{H}$ **such that** $(v, h) \in \mathcal{E}$ **then**
> | | $\mathcal{F} \leftarrow \mathcal{F} \cup \{(v, h), (u, v)\}$;
> | | $\mathcal{H} \leftarrow \mathcal{H} \cup \{u, v\}$;
> | | $\mathcal{L} \leftarrow \mathcal{L} - \{u, v\}$;
> | **Else**
> | | **Let** $w \in \mathcal{L}$ **and** $h \in \mathcal{H}$ **such that** $(w, h) \in \mathcal{E}$;
> | | $\mathcal{F} \leftarrow \mathcal{F} \cup \{(w, h)\}$;
> | | $\mathcal{H} \leftarrow \mathcal{H} \cup \{w\}$;
> | | $\mathcal{L} \leftarrow \mathcal{L} - \{w\}$;

Proof. By assumption (2), we know that from every vertex $w \in \mathcal{L}$ there exists a path to at least one vertex of $\mathcal{H}$. Thus we make sure that $\mathcal{L}$ will become an empty set, i.e., that all vertices of $\mathcal{V}$ are covered. □

Filtering Algorithms for the NVALUE Constraint

Christian Bessiere[1], Emmanuel Hebrard[2], Brahim Hnich[3], Zeynep Kiziltan[4], and Toby Walsh[2]

[1] LIRMM, CNRS/University of Montpellier, France
[2] NICTA and UNSW, Sydney, Australia
[3] 4C and UCC, Cork, Ireland
[4] University of Bologna, Italy
bessiere@lirmm.fr, {ehebrard, tw}@cse.unsw.edu.au,
brahim@4c.ucc.ie, zkiziltan@deis.unibo.it

Abstract. The constraint NVALUE counts the number of different values assigned to a vector of variables. Propagating generalized arc consistency on this constraint is NP-hard. We show that computing even the lower bound on the number of values is NP-hard. We therefore study different approximation heuristics for this problem. We introduce three new methods for computing a lower bound on the number of values. The first two are based on the *maximum independent set problem* and are incomparable to a previous approach based on intervals. The last method is a linear relaxation of the problem. This gives a tighter lower bound than all other methods, but at a greater asymptotic cost.

1 Introduction

The NVALUE constraint counts the number of distinct values used by a vector of variables. It is a generalization of the widely used ALLDIFFERENT constraint [12]. It was introduced in [4] to model a musical play-list configuration problem so that play-lists were either homogeneous (used few values) or diverse (used many). There are many other situations where the number of values (e.g., resources) used at the same time are limited. In such cases, a NVALUE constraint can aid both modelling and solving.

Enforcing generalized arc consistency (GAC) on the NVALUE constraint is NP-hard [3]. One way to deal with this intractability is to identify a tractable decomposition or approximation method. The NVALUE constraint can be decomposed into two other global constraints: the ATMOSTNVALUE and the ATLEAST-NVALUE constraints. Unfortunately, while enforcing GAC on the ATLEAST-NVALUE constraint is polynomial, we show that enforcing GAC on the AT-MOSTNVALUE constraint is also NP-hard. We will therefore focus on various approximation methods for propagating the ATMOSTNVALUE constraint.

We introduce three new approximations. Two are based on graph theory while the third exploit a linear relaxation encoding. We compare the level of filtering achieved with a previous approximation method due to Beldiceanu based on intervals that runs in $O(n \log(n))$ [1] for finding a lower bound on N, and linear

R. Barták and M. Milano (Eds.): CPAIOR 2005, LNCS 3524, pp. 79–93, 2005.

for pruning values. We show that the two new algorithms based on graph theory are incomparable with Beldiceanu's, though one is strictly tighter than the other. Both algorithms, however, have a $O(n^2)$ time complexity. We also show that the linear relaxation method dominates all other approaches in terms of the filtering, but with a higher computational cost. Finally, we demonstrate how all of these methods can be used in a filtering algorithm for the NVALUE constraint.

2   Formal Background

2.1   Constraint Satisfaction Problems

A constraint satisfaction problem (CSP) consists of a set of variables, each with a finite domain of values, and a set of constraints that specify allowed combinations of values for subsets of variables. We use upper case for variables, X_i, or vectors of variables, $\bar{X}$, and lower case for values, v, or assignments, $\bar{v}$. The domain of a variable X_i, $D(X_i)$ is a set of values. A full or partial assignment $\bar{v} = \langle v_1, \ldots, v_m \rangle$ of $\bar{X} = \langle X_1, \ldots, X_m \rangle$ is a vector of values such that $v_i \in D(X_i)$. A solution to a CSP is a full assignment of values to the variables satisfying the constraints. The minimum (resp. maximum) value in the domain of a variable X_i is $min(X_i)$ (resp. $max(X_i)$). The cardinality of an assignment $\bar{v}$ is $card(\bar{v})$, the number of distinct values used. For instance if $\bar{v} = \langle a, b, a, b, c \rangle$, $card(\bar{v}) = 3$. The maximum (resp. minimum) cardinality of a vector of variables $\bar{X}$, $card\uparrow(\bar{X})$ (resp. $card\downarrow(\bar{X})$) is the largest (resp. smallest) cardinality among all possible assignments.

Constraint solvers typically explore partial assignments enforcing a local consistency property using either specialized or general purpose propagation algorithms. Given a constraint C on the variables $\bar{X}$, a *support* for $X_i = v_j$ on C is a partial assignment $\bar{v}$ of $\bar{X}$ containing $X_i = v_j$ that satisfies C. A value $v_j \in D(X_i)$ without support on a constraint is *arc inconsistent*. A variable X_i is *generalized arc consistent (GAC)* on C iff every value in $D(X_i)$ has support on C. A constraint C is GAC iff each constrained variable is GAC on C. A *bound support* on C is a support where the interval $[min(X_i), max(X_i)]$ is substituted for the domain of each constrained variable X_i. A variable X_i is *bound consistent (BC)* on C if $min(X_i)$ and $max(X_i)$ have bound support on C. A constraint is BC iff all constrained variables are BC on C.

In line with [11], we say that a local consistency property Φ on C is as strong as Ψ (written $\Phi \succeq \Psi$) iff, given any domains, if Φ holds then Ψ holds; Φ is stronger than Ψ (written $\Phi \succ \Psi$) iff $\Phi \succeq \Psi$ but not $\Psi \succeq \Phi$; Φ is equivalent to Ψ (written $\Phi \equiv \Psi$) iff $\Phi \succeq \Psi$ and $\Psi \succeq \Phi$; and that they are incomparable otherwise (written $\Phi \bowtie \Psi$).

2.2   Graph Theoretic Concepts

Given a family of sets $\mathcal{F} = \{S_1, \ldots, S_n\}$ and a graph $G = (V, E)$ with the set of vertices $V = \{v_1, \ldots, v_n\}$ and set of edges E, G is the *intersection graph of* $\mathcal{F}$ iff

$$\forall i, j \ \langle v_i, v_j \rangle \in E \leftrightarrow S_i \cap S_j \neq \emptyset$$

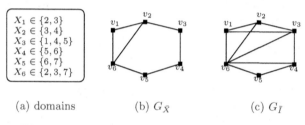

(a) domains   (b) $G_{\bar{X}}$ (c) $G_{\bar{I}}$

Fig. 1. Domains, intersection graph and interval graph

For any graph G, there exists a family of sets $\mathcal{F}$ such that the intersection graph of $\mathcal{F}$ is G. Thus, the class of intersection graphs is simply the class of all undirected graphs [8]. The class of graphs obtained by the intersection of *intervals*, instead of sets, is known as *interval graphs*.

Given a vector of variables $\bar{X} = \langle X_1, \ldots, X_m \rangle$, we use $G_{\bar{X}} = (V, E)$ for the induced intersection graph, i.e the graph where $V = \{v_1, \ldots, v_m\}$ and $\forall i, j \cdot \langle v_i, v_j \rangle \in E \leftrightarrow D(X_i) \cap D(X_j) \neq \emptyset$. Similarly, we use $\bar{I}$ for the same vector of variables, where all domains are seen as intervals instead, i.e., for each i, $D(X_i) = [min(X_i), max(X_i)]$. $G_{\bar{I}}$ is the induced interval graph, defined like $G_{\bar{X}}$, but on the intervals instead. For instance, the domains in Figure (1,a) induce the intersection graph in (1,b) and the interval graph in (1,c).

Finally, we recall that an *independent set* is a set of vertices with no edge in common. The *independence number* $\alpha(G)$ of a graph G, is the number of vertices in an independent set of maximum cardinality. A *clique* is the dual concept: a set of vertices such that any pair has an edge between. A *clique cover* of G is a partition of the vertices into cliques. The cardinality of the *minimum clique cover* is $\theta(G)$. For instance, the interval graph of Figure (1,c) has $\{\{v_1, v_2, v_3\}\{v_4, v_5, v_6\}\}$ as a minimal clique cover, hence $\theta(G_{\bar{I}}) = 2$. Similarly, the intersection graph of Figure (1,b) has $\{v_1, v_3, v_5\}$ as a maximal independent set, hence $\alpha(G_{\bar{X}}) = 3$.

3   The NVALUE Constraint

In this section we define the NVALUE constraint and we show that it can be decomposed into two simpler constraints. Whereas one of these constraints is polynomial to propagate using a maximum matching algorithm, the second is NP-hard so we look at approximate methods.

Definition 1. NVALUE$(N, \langle X_1, \ldots, X_m \rangle)$ *holds iff* $N = |\{X_i |\ 1 \leq i \leq m\}|$

Enforcing GAC on the NVALUE constraint is NP-hard in general [3]. We can, however, decompose it into two simpler constraints: the ATLEASTNVALUE and the ATMOSTNVALUE constraints.

Definition 2. ATLEASTNVALUE$(N, \langle X_1, \ldots, X_m \rangle)$ *holds iff* $N \leq |\{X_i |\ 1 \leq i \leq m\}|$. ATMOSTNVALUE$(N, \langle X_1, \ldots, X_m \rangle)$ *holds iff* $N \geq |\{X_i |\ 1 \leq i \leq m\}|$.

We can identify precisely when the decomposition of a NVALUE constraint does not hinder propagation.

Theorem 1. *If* ATLEASTNVALUE *and* ATMOSTNVALUE *are GAC and* $|D(N)| \neq 2$ *or* $min(N) + 1 = max(N)$, *then* NVALUE *is GAC.*

Proof. Suppose that the decomposition is GAC. Then we have $card\!\downarrow(\bar{X}) \leq min(N)$ and $card\!\uparrow(\bar{X}) \geq max(N)$. Thus, by Lemma 1 (see Appendix), N is GAC for NVALUE. Furthermore, we know that if $|D(N)| > 2$, or $D(N)$ contains a value v such that $card\!\downarrow(\bar{X}) < v < card\!\uparrow(\bar{X})$ then all variables in $\bar{X}$ are GAC (see Lemma 2 in Appendix). Therefore we only need to cover three cases:

- $D(N) = \{card\!\uparrow(\bar{X})\}$. Let v be an arc inconsistent value in $\bar{X}$. There is no assignment whose cardinality is greater than $card\!\uparrow(\bar{X})$, therefore v is arc inconsistent because it participates only in assignments of cardinality below N. Hence v is arc inconsistent for ATLEASTNVALUE, which contradicts the hypothesis.
- $D(N) = \{card\!\downarrow(\bar{X})\}$. Analogous to the last case.
- $D(N) = \{card\!\downarrow(\bar{X}), card\!\uparrow(\bar{X})\}$: If $card\!\downarrow(\bar{X}) + 1 = card\!\uparrow(\bar{X})$ then NVALUE is GAC (see Lemma 2 in Appendix). Otherwise, there is a gap between the bounds. This is the only case where the decomposition is GAC but NVALUE may not be. For instance, consider the domains: $X_1 \in \{1, 2, 3\}, X_2 \in \{1, 2\}, X_3 \in \{1\}, N \in \{1, 3\}$. Whilst enforcing GAC on NVALUE($N, \langle X_1, X_2, X_3 \rangle$) will prune $X_1 = 2$, these domains are GAC for the decomposition. □

If the domain of N contains only $card\!\downarrow(\bar{X})$ and $card\!\uparrow(\bar{X})$, and these two values are not consecutive, then NVALUE may not be GAC even though AT-MOSTNVALUE and ATLEASTNVALUE are GAC. However, as we show in section 7, we can make GAC on the decomposition equivalent, by performing an extra pruning in this situation.

3.1   The ATLEASTNVALUE Constraint

We first have a brief look at the ATLEASTNVALUE constraint. It is known [1] that $card\!\uparrow(\bar{X})$ is the cardinality of the maximal matching of the bipartite graph with a class of vertices representing the variables, another the values, and where an edge links two vertices if and only if it corresponds to a valid assignment. Indeed, this is the basic idea behind Régin's algorithm for enforcing GAC on the ALLDIFFERENT constraint [12]. We can easily derive a propagation procedure for ATLEASTNVALUE using the polynomial algorithm for the SOFTALLDIFF constraint [10] that counts the number of variables that need to be reassigned to satisfy the constraint. We can use this algorithm to compute $card\!\uparrow(\bar{X})$. Moreover, we can use this same algorithm to prune the values in $\bar{X}$ that do not belong to a maximal matching. This nearly provides us with an algorithm for enforcing GAC on ATLEASTNVALUE. One difference is that we do not always want to prune the values that do not participate in a maximal matching. We shall see how this algorithm can be used when pruning the variables in $\bar{X}$ in section 7.

We refer the reader to [10] for more details about this algorithm, and we focus on the constraint ATMOSTNVALUE for the rest of the paper.

3.2   The ATMOSTNVALUE Constraint

We adapt the proof of NP-hardness for NVALUE [3] to also show that enforcing GAC on an ATMOSTNVALUE constraint alone is intractable.

Theorem 2. *Enforcing GAC on a* ATMOSTNVALUE$(N, \langle X_1, \ldots, X_m \rangle)$ *constraint is NP-hard, and remains so even if N is ground.*

Proof. We use a reduction from 3SAT. Given a formula in k variables and n clauses, we construct the ATMOSTNVALUE$(X_1, \ldots, X_{k+n}, N)$ constraint in which $D(X_i) = \{i, \neg i\}$ for all $i \in [1, k]$, and each X_i for $i > k$ represents one of the n clauses. If the jth clause is $x \vee \neg y \vee z$ then $D(X_{k+j}) = \{x, \neg y, z\}$. By construction, the variables will consume k distinct values, hence if $N = k$, the constructed ATMOSTNVALUE constraint has a solution iff the original 3SAT problem has a satisfying assignment. The completeness is easy to see as the support is a polynomial witness. Hence testing a value for support is NP-complete, and enforcing GAC is NP-hard.   □

Note that this proof is a reduction of 3SAT into the problem of propagating GAC on $\bar{X}$ when N is ground. This means that pruning $\bar{X}$ alone is NP-hard. Indeed, even computing just the lower bound on N, given $\bar{X}$ is no easier.

Theorem 3. *Computing the value of $card\!\downarrow(\bar{X})$ is NP-hard.*

Proof. Computing $card\!\downarrow(\bar{X})$ is equivalent to finding the cardinality of a *minimum hitting set* of $\bar{X}$ seen as a family of sets. A *hitting set* of a family of sets $\mathcal{F}$, is a set that intersects each member of $\mathcal{F}$. Computing the cardinality of the smallest possible hitting set is NP-hard [9]. If we have one variable X_i in $\bar{X}$ for each set $S_i \in \mathcal{F}$, and $D(X_i) = S_i$, then $card\!\downarrow(\bar{X})$ is equal to the cardinality of a minimum hitting set of $\mathcal{F}$.   □

4   Existing Algorithm for the ATMOSTNVALUE Constraint

We first recall Beldiceanu's algorithm, then we introduce a graph theoretic view of his method. We shall refer to Beldiceanu's algorithm as OI, for *ordered intervals*. The first step is to order the domains by increasing lower bound. Then the following procedure (algorithm 1) can be applied, the value returned ($N_{distinct}$) is a lower bound on $card\!\downarrow(\bar{X})$.

The intervals are explored one at a time, and a new group, i.e. a clique of the interval graph, is completed when an interval is found that does not overlap with all previous ones in the group. The time complexity is $O(nlog(n))$ for sorting, and then the algorithm itself is linear, the loop visits each domain at most twice (when this domain is distinct from the previous). Hence, the worst case time complexity is dominated by $O(nlog(n))$. This algorithm is proved correct, that

---

Algorithm 1: OI: The interval-based algorithm introduced in [1]

> **Data** : $\bar{X} = [X_1, \ldots X_m]$
> **Result** : $N_{distinct}$
> $N_{distinct} \leftarrow 1$; $reinit \leftarrow 1$; $i \leftarrow 1$; $low \leftarrow -\infty$; $up \leftarrow \infty$;
> **while** $i < m$ **do**
> > $i \leftarrow i + 1 - reinit$;
> > **if** $reinit$ or $(low < min(X_i))$ **then** $low \leftarrow min(X_i)$;
> > **if** $reinit$ or $(up > max(X_i))$ **then** $up \leftarrow max(X_i)$;
> > $reinit \leftarrow (low > up)$;
> > $N_{distinct} \leftarrow N_{distinct} + reinit$;
>
> return $N_{distinct}$;

---

is, it returns a valid lower bound, by noticing that the intervals with smallest maximum value for each group are pairwise disjoint. Consequently, at least as many values as groups, that is, $N_{distinct}$, have to be used. As there was no proof given in [1], we present one here:

Proposition 1 (given in [1] without proof). *Let $\{C_1, \ldots, C_k\}$ be a partition of the intervals, output of OI. If $\bar{I} = \langle I_1, \ldots, I_k \rangle$ is the vector of intervals where I_i is the element of C_i with least maximum value, then all elements of $\bar{I}$ have empty pairwise intersections.*

Proof. OI scans all intervals by increasing lower bound, partitioning into groups on the way. When the algorithm ends, we have k groups $C_1, \ldots, C_k$. For any group C_i, consider the interval I_1 with least upper bound. This interval does not intersect any interval in any group C_j such that $j > i$. Suppose it was the case, i.e, there exists $I_2 \in C_j$ which intersects with I_1, since the intervals are ordered by increasing lower bound, I_2 cannot be completely *below* any interval in C_i. It must then be either completely *above* or overlapping. However, since I_1 has the least upper bound and intersects I_2, all intervals in C_i must also intersect I_2. It follows that I_2 should belong to C_i hence the contradiction. The set containing the interval with least upper bound of every group is then pairwise disjoint, and is of cardinality k.   □

Moreover, it is easy to see that, when the domains are indeed intervals, this bound can be achieved. If, for each group, we assign all the variables of this group to one of the common values, then we obtain an assignment of cardinality $N_{distinct}$. This argument is used in [1] to show that OI achieves BC on N.

Now, recall that $G_{\bar{X}}$ is the intersection graph of the variables in $\bar{X}$, whereas $G_{\bar{I}}$ is the interval graph of the same variables. It is easy to see that OI computes at once a clique cover and an independent set of $G_{\bar{I}}$. Moreover, since for any graph $\alpha(G) \leq \theta(G)$, if a graph G contains an independent set *and* a clique cover of cardinality n, we must conclude that $n = \alpha(G) = \theta(G)$. Indeed, interval graphs belong to the class of perfect graphs, for which, by definition, the independence number is equal to the size of the minimum clique cover. Therefore, we know that the output of OI, i.e., $N_{distinct}$ is equal to $\alpha(G_{\bar{I}})$ and also to $\theta(G_{\bar{I}})$. It can be shown that, in this case, the cardinality of the minimum clique cover on the interval graph is equal to the cardinality of the minimum hitting set on $\bar{I}$ itself. This is due to the fact that a set of intervals that pairwise intersect always share a common interval, any element of this interval hitting all of them. To summarize, in the special case where the domains of all variables in $\bar{X}$ are intervals (denoted $\bar{I}$), the following equality holds: $\alpha(G_{\bar{I}}) = \theta(G_{\bar{I}}) = card\!\downarrow\!(\bar{I})$

As a consequence, the value $N_{distinct}$ is exact, hence OI achieves bound consistency on N for the constraint ATMOSTNVALUE. However, considering domains as intervals may be in some case a very crude approximation. If we consider the intersection graph $G_{\bar{X}}$ instead of the interval graph $G_{\bar{I}}$, the relation becomes: $\alpha(G_{\bar{X}}) \leq \theta(G_{\bar{X}}) \leq card\!\downarrow\!(\bar{X})$

Any of those three quantities is a valid lower bound, though they are NP-hard to compute. They are, on the other hand, tighter approximations that do not consider domains as intervals. Indeed, since $G_{\bar{X}}$ has more edges than $G_{\bar{I}}$, it follows immediately that: $\alpha(G_{\bar{X}}) \geq \alpha(G_{\bar{I}})\ (= \theta(G_{\bar{I}}) = card\!\downarrow\!(\bar{I}))$

5   Three New Approaches

We present two algorithms approximating $\alpha(G_{\bar{X}})$ and a linear relaxation approximating directly the minimum hitting set problem, and hence $card\!\downarrow\!(\bar{X})$.

5.1   A Greedy Approach

We have seen that OI approximates the lower bound on N by computing the exact value of $\alpha(G_{\bar{I}})$, the independence number of the interval graph induced by $\bar{X}$. Here the idea is to compute the independence number of $G_{\bar{X}}$, $\alpha(G_{\bar{X}})$.

Whilst computing the exact value of $\alpha(G_{\bar{X}})$ is intractable for unrestricted graphs, some efficient approximation schemes exist for that problem. We use here a very simple heuristic algorithm for computing the independence number of a graph, referred to as "the natural greedy algorithm". We denote it MD, for *minimum degree* from now on. It consists in removing the vertices of minimum degree as well as their neighborhood in turn. The number of iterations i is such that $i \leq \alpha(G)$. This algorithm is studied in detail in [6]. If we suppose that the intersection graph is constructed once and maintained during search, then a careful implementation can run in $O(n+m)$ where n is the number of vertices and m is the number of edges (linear in the size of the graph). However, computing the intersection graph requires $n(n + 1)/2$ tests of intersection. Each of those may require at most d equality checks, where d is the size of the domains in $\bar{X}$. Notice that efficient data structures, such as bit vectors, are often used to represent domains and thus allow intersection checks in almost constant time in practice. This suggests an implementation where the graph is never actually computed, but an intersection check is done each time we need to know if an edge

---

Algorithm 2: MD: A greedy algorithm approximating the maximum independent set of a graph

Data : $G = (V, E)$	
Result : $N_{distinct}$	
if $G = \emptyset$ **then** return 0;	
choose $v \in V$ such that $d(v)$ is minimum;	
return 1+MD($G(V \setminus (\Gamma(v) \cup \{v\}))$);	

links two nodes. The worst case time complexity is then $O(dn^2)$ if intersection is linear in the size of the sets or $O(n^2)$ if it is constant. We denote $\Gamma(v)$ the neighborhood of v, $\Gamma(v) = \{w | vw \in E\}$.

5.2   Turán's Approximation

Alternatively, we can use an even simpler approximation. Turán proposed a lower bound of $\frac{n^2}{2m+n}$ for $\alpha(G)$ in [13], where n is the number of vertices and m the number of edges. Therefore assuming that m is computed once, and revised whenever a domain changes or whenever the constraint is called again, this formula gives a lower bound in constant time. The worst case time complexity is the same as MD's (because of the initialization). However, this heuristic can be much more efficient in practice. We refer to this method as Turan.

5.3   A Linear Relaxation Approach

We have shown the following inequalities: $\alpha(G_{\bar{I}}) \leq \alpha(G_{\bar{X}}) \leq card{\downarrow}(\bar{X})$.

We have seen that the cardinality of the minimum hitting set problem where the family of sets is formed by the domains of the variables in $\bar{X}$ is equal to the lower bound on N, that is, $card{\downarrow}(\bar{X})$. One difficulty is that approximation algorithms proposed in the literature for minimum hitting set return a set which may be too large, and so do not provide a valid lower bound. However, we consider here a linear relaxation that can be solved in polynomial time that gives a lower bound on the minimum hitting set cardinality, and thus, of N. We refer to this method as LP.

Given a vector of variables $\bar{X} = \langle X_1, \ldots X_m \rangle$, let $V = \bigcup_{v \in \bar{X}} D(x)$ be the total set of values. Then let $\{y_v | \ v \in V\}$ be a set of linear variables, and LP is as follows:

$$min \sum_{v \in V} y_v \quad subject \ to \quad \sum_{v \in D(X_i)} y_v \geq 1 \ \ \forall X_i \in \bar{X}$$

where $y_v \geq 0$ forall $v \in V$.

The best polynomial linear program solvers based on the interior point methods run in $O(v^3 L)$ where v is the number of variables and L is the number of bits in the input. The number of variables in our linear program is nd ($d = |D(X_i)|$) and we have $n = |\bar{X}|$ inequalities of size d. Therefore, the worst case time complexity is $O(n^4 d^4)$. In practice, the simplex method may behave better even though it has an exponential worst case time complexity.

6   Theoretical Analysis

We will compare local consistency properties applied to the ATMOSTNVALUE constraint. In our case, the ATMOSTNVALUE constraint holds iff the lower bound returned by the propagation algorithm (consistency) does not exceed $max(N)$. Thus, given consistency properties Φ and Ψ, $\Phi \succeq \Psi$ means that the lower bound on N returned by the algorithm enforcing Φ is greater than or equal to the lower bound returned by the algorithm enforcing Ψ.[5] We consider comparing the level of consistency achieved by the following algorithms: OI, MD, Turan, and LP.

Note that, since only the lower bound on N is considered in this comparison, OI is then equivalent to BC. We do not compare with generalized arc consistency either, as this is NP-hard to enforce and all our algorithms are polynomial and strictly weaker.

Theorem 4. MD $\succ$ Turan

Proof. For a proof that MD is as strong as Turan see [6]. Moreover, it is easy to find an example showing that MD is strictly stronger. For instance consider the following domains: $X_1 \in \{1,2,3,4,5,6,7,8\}$, $X_2 \in \{1,2\}$, $X_3 \in \{3,4\}$, $X_4 \in \{5,6\}$, *and* $X_5 \in \{7,8\}$.

The induced intersection graph is as follows:

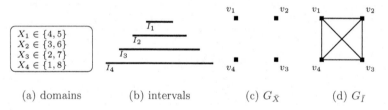

When applying MD, we obtain an independent set of size 4. However, Turan returns: $\lceil \frac{n^2}{2m+n} \rceil = \lceil \frac{25}{13} \rceil = 2$. and we deduce a lower bound of 2 for N.   □

Theorem 5. Turan $\bowtie$ OI

Proof. To see that Turan is not as strong as OI, consider the example used in the proof of Theorem 4. The domains being intervals, we know that OI computes the exact lower bound, 4. However the Turán heuristics gives us 2.

To see that OI is not as strong as Turán, consider the domains in Figure 2. The induced intersection graph $G_{\bar{X}}$ has 4 vertices ($n = 4$) and no edges ($m = 0$), thus Turan returns 4. However, the interval graph $G_{\bar{I}}$ induced by the same domains is a clique and then OI returns 1.   □

| $X_1 \in \{4,5\}$ |
| $X_2 \in \{3,6\}$ |
| $X_3 \in \{2,7\}$ |
| $X_4 \in \{1,8\}$ |

(a) domains   (b) intervals   (c) $G_{\bar{X}}$ (d) $G_{\bar{I}}$

Fig. 2. Example for OI $\not\succeq$ MD($G_{\bar{X}}$) and for OI $\not\succeq$ Turan

Theorem 6. MD $\bowtie$ OI

Proof. To see that MD $\not\succeq$ OI, consider the interval graph in Figure (3,a) induced by the intervals of Figure (3,b). The exact independence number is 4 (for instance $\{v_2, v_3, v_8, v_9\}$ is an independent set of cardinality 4), and thus OI returns 4.

---

[5] We refer to the level of local consistency achieved by an algorithm A as A as well.

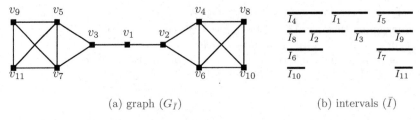

(a) graph $(G_{\bar{I}})$ (b) intervals $(\bar{I})$

Fig. 3. Example for MD $\not\succeq$ OI

However, the vertex with minimal degree is v_1, and no independent set of cardinality 4 involves v_1, therefore MD is not as strong as OI.

To see that OI $\not\succeq$ MD$(G_{\bar{X}})$, see Figure 2. It is easy to construct domains where the interval graph can have arbitrarily more edges than the intersection graph. For instance the domains in Figure 2,a induce a complete interval graph, or an unconnected intersection graph. Therefore OI is not as strong as MD. $\quad\square$

Theorem 7. LP $\succ$ MD, LP $\succ$ OI *and* LP $\succ$ Turan.

Proof. We first show that the value returned by LP is greater or equal to $\alpha(G_{\bar{X}})$. Consider a maximum independent set A of the intersection graph. We know that any two variables in A have no value in common. However for each variable $X_i \in A$ we have: $\sum_{v \in D(X_i)} y_v \geq 1$. Since the domains of those variables are disjoint, we have:

$$\sum_{v \in \bigcup_{X_i \in A} D(X_i)} y_v \geq |A| = \alpha(G_{\bar{X}})$$

And thus the total sum to minimize is greater than or equal to $\alpha(G_{\bar{X}})$. However, recall that OI, MD and Turan all approximate $\alpha(G_{\bar{X}})$ by giving a lower bound. Therefore LP is as strong as OI, MD and Turan. Moreover, the variables $X_1 \in \{1,2\}, X_2 \in \{2,3\}, X_3 \in \{1,3\}$ constitute an example showing that LP is strictly stronger, as the optimal sum for LP is 1.5, whilst $\alpha(G_{\bar{X}}) = 1$. $\quad\square$

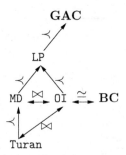

Fig. 4. Relations between algorithms and consistencies on the ATMOSTNVALUE constraint

7   A Propagation Algorithm for the NVALUE Constraint

A template for an approximate propagation algorithm for NVALUE is given in Algorithm 3. In this template, one may use any of the methods described in the previous sections. The pruning on N is straightforward (Line 1 and 2). When we have $max(N) < min(N)$ (Line 3), there is clearly an inconsistency, and the algorithm fails. In the following subsections, we consider the cases of Lines 4, 5 and 6, where some filtering may be achieved. All other cases (Line 7) satisfy the preconditions of Lemma 2 (see Appendix) and the constraint is GAC. Therefore, either the constraint is GAC, or we are unable to deduce any inconsistency because the lower bound lb for $card\downarrow(\bar{X})$ is not tight enough.

Algorithm 3: Algorithm for propagating the NVALUE constraint

    Data   : $\bar{X}$, N
    Result   :
    $ub \leftarrow card\uparrow(\bar{X}))$;
    $lb \leftarrow approx(card\downarrow(\bar{X})))$;
1 $max(N) \leftarrow min(max(N), ub)$;
2 $min(N) \leftarrow max(min(N), lb)$;
3 **if** $(max(N) < min(N))$ **then** fail;
4 **case** *(ub = min(N) = max(N) $\neq$ lb)* : pruning from below;
5 **case** *(lb = min(N) = max(N) $\neq$ ub)* : pruning from above;
6 **case** *(|D(N)| = 2 and min(N) + 1 < max(N))* : pruning from within;
7 **otherwise** return;

7.1   Pruning from Below

This pruning is triggered when $card\uparrow(\bar{X}) = min(N)$ and $card\downarrow(\bar{X}) < min(N)$. In this situation, we know that some assignments may have too small cardinality, and therefore some values may not participate in assignments of cardinality $card\uparrow(\bar{X}) = min(N)$, which is the only cardinality satisfying the constraint. Making ATLEASTNVALUE GAC is then sufficient to make the whole constraint GAC as this corresponds to the first of the three possible cases discussed in the proof of Theorem 1. In this situation, we can use a polynomial procedure for enforcing GAC on the SOFTALLDIFF constraint which counts the number of variables that have to be reassigned in order to be all different [10].

7.2   Pruning from Above

This is the dual case, we know that some assignments may have too large cardinality, and therefore some values may participate only in assignments of cardinality above $max(N)$ (and we assume $max(N) = card\downarrow(\bar{X})$). This corresponds to the second case of the proof of Theorem 1. Making ATMOSTNVALUE GAC is then sufficient to make the whole constraint GAC. Note that here we are not sure to achieve GAC.

In [1] (p. 6), two observations are made in order to prune $\bar{X}$ which are relevant here when using MD to compute $min(N)$. We reformulate these observations consistently to the graph notations we used. First, let A be a set of variables that form an independent set of the intersection graph, and let $X_i \in (\bar{X} \setminus A)$ be assigned to a value v which does not belong to any domain in A. It follows that the minimum number of values required will be at least $\alpha(G_{\bar{X}}) + 1$. Hence we can prune the value v from the domain of X_i when N is equal to $\alpha(G_{\bar{X}})$. This way of pruning the variables can be used with MD as well as with OI. There are no further difficulties when going from interval graphs to intersection graphs. Consequently, given an independent set A, we can propagate the following constraint: $\forall X_i \in \bar{X}, \exists X_j \in A \ s.t. \ X_i = X_j$. Second, suppose that A' is another distinct independent set. Thus, we have: $\forall X_i \in \bar{X}, \exists X_{j_1} \in A, \exists X_{j_2} \in A' \ s.t. \ (X_i = X_{j_1} \wedge X_i = X_{j_2})$. Therefore, one can prune values in $\bar{X}$ by finding a set of independent sets $\mathcal{A} = \{A_1, \ldots A_k\}$. The set of consistent value $\mathcal{V}$ is defined as follows: $\forall A \in \mathcal{A}, U_A = \bigcup_{X_i \in A} D(X_i), \quad \mathcal{V} = \bigcap_{A \in \mathcal{A}} U_A$. It may be difficult to compute *all* independent sets of cardinality equal to N. One must therefore find a set which is as large as possible. In [1] from the first one found with OI, each independent set that differs by only one vertex is deduced. This can be computed in linear time, without increasing the algorithm's complexity. As a result this way $\bar{X}$ is pruned, the algorithm described in [1] does not enforce BC on ATMOSTNVALUE. The following domains are a counter example: $X_1 \in \{1,2\}$, $X_2 \in \{2,3\}$, $X_3 \in \{3,4\}$, $X_4 \in \{4,5\}$, *and $N \in \{2\}$.* Only the values 2 for X_1, X_2 and 4 for X_3, X_4 are bound consistent. However, the independent sets considered will be $\{X_1, X_3\}$ and $\{X_1, X_4\}$. Therefore, the values that are consistent are $\{1, 2, 4\}$. This way of pruning can make holes in domains. Therefore the level of consistency achieved on ATMOSTNVALUE is incomparable with bound consistency. Although they are not equivalent, one can easily derive a procedure to enforce BC from OI. To check the (say lower) bound of a variable X_i, we assign this bound to X_i and compute N again. If $card\!\downarrow(\bar{X})$ after this assignment is greater than N, this bound is not BC.

With algorithms that do not compute independent sets in order to get a lower bound on N, like the linear relaxation method or the Turán heuristic, we are in a different situation. However, we can simply wait until $min(N) > max(N)$ and fail in this case, without pruning any variable in $\bar{X}$. Alternatively we could compute a new lower bound for each value v of each X_i, that is, $O(nd)$ times. We set $y_v = 1.$ and if the objective function fails to be lower than or equal to N, then v is arc inconsistent. Since the pruning on $\bar{X}$ happens in a limited number of situations, it may be cost effective to use this complete method.

7.3 Pruning from Within

This pruning is triggered when $card\!\downarrow(\bar{X}) = min(N)$, $card\!\uparrow(\bar{X}) = max(N)$ and $card\!\downarrow(\bar{X}) + 1 < card\!\uparrow(\bar{X})$. This is the last of the three cases in the proof of Theorem 1. In this case ATMOSTNVALUE and ATLEASTNVALUE can be GAC whilst NVALUE is not. However, we can use a conditional constraint to do some pruning in this particular case. The idea is, when these conditions are met, to trigger the following constraint to perform this extra filtering:

$$Min = card{\downarrow}(\bar{X}) \wedge Max = card{\uparrow}(\bar{X})\wedge$$
$$(\text{ATMOSTNVALUE}(Min, \bar{X})\vee \text{ATLEASTNVALUE}(Max, \bar{X}))$$

Min and Max are two extra variables. We have the following theorem:

Theorem 8. *If $D(N) = \{card{\downarrow}(\bar{X}), card{\uparrow}(\bar{X})\}$ and $card{\downarrow}(\bar{X}) + 1 < card{\uparrow}(\bar{X})$ then $\text{NVALUE}(N, \bar{X})$ is GAC iff the decomposition and (when the conditions are met) the conditional constraint are GAC.*

Proof. ($\Rightarrow$) The case where $D(N) = \{card{\downarrow}(\bar{X}), card{\uparrow}(\bar{X})\}$ or $card{\downarrow}(\bar{X}) + 1 < card{\uparrow}(\bar{X})$ does not hold is covered by Theorem 1. Now suppose this condition holds, and there is a value $v_i \in D(X_i)$ which is not GAC for NVALUE. By definition, this implies that any assignment such that the ith element is v_i has a cardinality different from $card{\downarrow}(\bar{X})$ and from $card{\uparrow}(\bar{X})$, since these values are in $D(N)$. Moreover, there is no assignment with cardinality above $card{\uparrow}(\bar{X})$ or below $card{\downarrow}(\bar{X})$. Therefore we deduce that any assignment $\bar{v}$ involving v_i is such that $card{\downarrow}(\bar{X}) < card(\bar{v}) < card{\uparrow}(\bar{X})$. Hence, if $Min = card{\downarrow}(\bar{X}) \wedge Max = card{\uparrow}(\bar{X})$ holds, then v_i would be inconsistent for both $\text{ATMOSTNVALUE}(Min, \bar{X})$ and $\text{ATLEASTNVALUE}(Max, \bar{X})$).

($\Leftarrow$) If a value v_i belongs to a support, i.e., an assignment whose cardinality is either $card{\downarrow}(\bar{X})$ or $card{\uparrow}(\bar{X})$, then either $\text{ATMOSTNVALUE}(Min, \bar{X})$ or $\text{ATLEASTNVALUE}(Max, \bar{X})$) or both are GAC.   □

Hence, we simply assign $card{\downarrow}(\bar{X})$ to N, then we compute B_1, the set of values inconsistent for ATMOSTNVALUE. Similarly, we assign $card{\uparrow}(\bar{X})$ to N and compute B_2, the set of values inconsistent for ATLEASTNVALUE, In both cases, we use the methods described in section 7.1 and 7.2. Once this is done, we restore the domain of N, and prune all values in $B_1 \cap B_2$. Notice that B_1 may be underestimated, hence we do not achieve GAC.

8   Related Work

Two algorithms, on the same line as OI, yet achieving BC, have been introduced in [2]. In this technical report, the authors also extend the constraint to deal with weights on values. Observe that filtering on the weighted version of the constraint can easily be done with the linear relaxation method. Indeed, the weights on values can be represented as coefficients in the linear equations.

The maximum independent set is a well known problem in graph theory and a number of approximation algorithms have been proposed. We used two simple and intuitive algorithms for the sake of simplicity and because MD is successful in practice. However, algorithms with better approximation ratio exist, for instance see [7]. Any such algorithm may replace MD into the propagation algorithm.

We have seen that the linear programming approach is always stronger, even than a complete method for finding a maximum independent set. It is difficult to identify where the linear relaxation for the minimum hitting was first introduced, as it is such a simple model. It is certainly given in [5]. One weakness of the linear programming approach is that it is difficult to deduce which values to prune when $min(N) = max(N)$.

9   Conclusion

Propagating generalized arc consistency on the NVALUE constraint is NP-hard. In order to filter inconsistent values, one has to obtain tight bounds on the number of distinct values used in assignments. Whilst the upper bound can be obtained in polynomial time with a maximal matching procedure, the lower bound alone is NP-hard to compute. Therefore, our focus is on methods which achieve lesser levels of consistency. A procedure proposed by Beldiceanu considers domains as intervals, which allows the independence number of the induced interval graph to be computed in polynomial time. The independence number of this graph is a valid lower bound on the number of distinct values. We introduce three new methods for approximating this lower bound. The first two approximate the independence number of the intersection graph. However, these algorithms have a quadratic worst case time complexity, and do not guarantee a tighter lower bound. The last and most promising approach is to use a linear relaxation of the minimum hitting set problem. The cardinality of the minimum hitting set is a tight lower bound on the number of distinct values. This always finds a tighter lower bound than the approaches based on the maximum independent set problem. In our future work, we will compare these methods experimentally.

Acknowledgements

Brahim Hnich is currently supported by Science Foundation Ireland under Grant No. 00/PI.1/C075. Toby Walsh and Emmanuel Hebrard are supported by National ICT Australia.

References

1. N. Beldiceanu. Pruning for the *minimum* constraint family and for the *Number of Distinct Values* constraint family. In *Proceedings CP-01*, 2001.
2. N. Beldiceanu, M. Carlsson, and S. Thiel. Cost-Filtering Algorithms for the two sides of the *Sum of Weights of Distinct Values* Constraint. SICS technical report, 2002.
3. C. Bessiere, E. Hebrard, B. Hnich, and T. Walsh. The complexity of global constraints. In *Proceedings AAAI-04*, 2004.
4. P. Roy F. Pachet. Automatic generation of music programs. In *Proceedings CP-99*, 1999.
5. S. Shahar G. Even, D. Rawitz. Hitting sets when the vc-dimension is small, (submitted to a journal publication) 2004.
6. M. Halldórsson and J. Radhakrishnan. Greed is good: Approximating independent sets in sparse and bounded-degree graphs. In *Proceedings STOC-94*, pages 439–448, 1994.
7. V. Th. Paschos M. Demange. Improved approximations for maximum independent set via approximation chains. *Appl. Math. Lett.*, 10:105–110, 1997.
8. E. Marzewski. Sur deux propriétés des classes s'ensembles. Fund. Math., 33:303–307, 1945.

9. D.S. Johnson M.R. Garey. *Computers and Intractability: A Guide to the Theory of NP-completeness.* W.H. Freeman and Company, 1979.

10. T. Petit, J.C. Regin, and C. Bessiere. Specific filtering algorithms for over-constrained problems. In *Proceedings CP-01*, 451-463.

11. R. Debruyne C. Bessiere. Some practicable filtering techniques for the constraint satisfaction problem. In *Proceedings IJCAI-97*, 1997.

12. J.C. Régin. A filtering algorithm for constraints of difference in CSPs. In *Proceedings AAAI-94*, pages 362–367, 1994.

13. P. Turán. On an extremal problem in graph theory. In *(in Hungarian)*, Mat. Fiz. Lapok, pages 48:436–452, 1941.

Appendix: Conditions When NVALUE Is GAC

In order to show when enforcing GAC on the decomposition of the NVALUE constraint enforces GAC on NVALUE, we used two lemmas. These identify when the variables in the NVALUE constraint are GAC. First, N is GAC iff its bounds are between $card{\downarrow}(\bar{X})$ and $card{\uparrow}(\bar{X})$.

Lemma 1. *Any value in $D(N)$ is GAC for NVALUE as long as it is lower than or equal to $card{\uparrow}(\bar{X})$ and greater than or equal to $card{\downarrow}(\bar{X})$.*

Proof. Let S be any assignment of $\bar{X}$. Consider assigning $\bar{X}$ as in S, one variable at a time. Let $\bar{X}_k$ be $\bar{X}$ at step k, that is, with k ground variables. Hence, since $\bar{X}$ involves m values, $\bar{X}_m$ corresponds to *Sol* At a step k, the value of $card{\downarrow}(\bar{X}_k)$ (resp. $card{\uparrow}(\bar{X}_k)$) increases (resp. decreases) by at most one with respect to step $k-1$. Moreover, when every variable is assigned, $card{\downarrow}(\bar{X}_m) = card{\uparrow}(\bar{X}_m) = card(S)$. Therefore, for any value p between $card{\downarrow}(\bar{X}_0)$ and $card{\uparrow}(\bar{X}_0)$, there exists k such that either $card{\downarrow}(\bar{X}_k) = p$ or $card{\uparrow}(\bar{X}_k) = p$. Consequently p has a support for a sub-domain $\bar{X}_k$ and is thus GAC.   □

Second, the variables in $\bar{X}$ are GAC if either $D(N) = [card{\downarrow}(\bar{X}), card{\uparrow}(\bar{X})]$ or there exists at least one value lower than $card{\uparrow}(\bar{X})$ and greater than $card{\uparrow}(\bar{X})$.

Lemma 2. *If either $D(N) = [card{\downarrow}(\bar{X}), card{\uparrow}(\bar{X})]$ or $card{\downarrow}(\bar{X})+1 < card{\uparrow}(\bar{X})$ and $[card{\downarrow}(\bar{X}) + 1, card{\uparrow}(\bar{X}) - 1] \cap D(N) \neq \emptyset$ then $\bar{X}$ is GAC.*

Proof. We first show the first part of the disjunction. Recall that $card{\downarrow}(\bar{X})$ (resp. $card{\uparrow}(\bar{X})$) is the cardinality of the smallest (resp. largest) possible assignment. Therefore, if the domain of N is equal to the interval $[card{\downarrow}(\bar{X}), card{\uparrow}(\bar{X})]$ it means that all assignments of $\bar{X}$ have a cardinality in $D(N)$.

For the second part, we use again the argument that assigning a single variable can affect the bounds by at most one. In other words, for all $X_i \in \bar{X}$, a value $v \in D(X_i)$ (without loss of generality) belongs to an assignment of cardinality either $card{\downarrow}(\bar{X})$, $card{\downarrow}(\bar{X}) + 1$, $card{\uparrow}(\bar{X})$ or $card{\uparrow}(\bar{X}) - 1$. Moreover, let $\bar{X}_{X_i=v}$ be $\bar{X}$ where the domain of X_i is reduced to $\{v\}$. We have $card{\downarrow}(\bar{X}_{X_i=v}) \leq card{\downarrow}(\bar{X}) + 1$ and $card{\uparrow}(\bar{X}_{X_i=v}) \geq card{\uparrow}(\bar{X}) - 1$. Hence, by assumption $D(N) \cap [card{\downarrow}(\bar{X}_{X_i=v}), card{\uparrow}(\bar{X}_{X_i=v})] \neq \emptyset$, and by applying Lemma 1, we know that there exists a tuple satisfying NVALUE with $X_i = v$.   □

Identifying and Exploiting Problem Structures Using Explanation-Based Constraint Programming

Hadrien Cambazard and Narendra Jussien

École des Mines de Nantes – LINA CNRS FRE 2729,
4 rue Alfred Kastler – BP 20722 – F-44307 Nantes Cedex 3, France
{hcambaza, jussien}@emn.fr

Abstract. Recent work have exhibited specific structure among combinatorial problem instances that could be used to speed up search or to help users understand the dynamic and static intimate structure of the problem being solved. Several Operations Research approaches apply decomposition or relaxation strategies upon such a structure identified within a given problem. The next step is to design algorithms that adaptatively integrate that kind of information during search. We claim in this paper, inspired by previous work on impact-based search strategies for constraint programming, that using an explanation-based constraint solver may lead to collect invaluable information on the intimate dynamic and static structure of a problem instance. We define several impact graphs to be used to design generic search guiding techniques and to identify hidden structures of instances. Finally, we discuss how dedicated OR solving strategies (such as Benders decomposition) could be adapted to constraint programming when specific relationships between variables are exhibited.

1 Introduction

Generic search techniques for solving combinatorial problems seems like the Holy Grail for both OR and CP communities. Several tracks are now explored: dynamically analyzing and adapting the way the solver actually solves a combinatorial problem, identifying specific structures in a given instance in order to speed up search, etc. The key point is to be able to identify, understand and use the intimate structure of a given combinatorial problem instance [7, 17, 18].

Refalo [16] recently defined impact-based solving strategies for constraint programming that dynamically use the structure of a solved problem. In this paper, we attempt to investigate the relationships between the variables of the problem. We intend to identify, differentiate and use both dynamic (created by the search algorithm) and static (relative to the instance) structures of the problem being solved. We focus on structures intended as subsets of variables that play a specific role within the problem. We define to this end several fine grained impact measures and induced impact graphs between variables to:

R. Barták and M. Milano (Eds.): CPAIOR 2005, LNCS 3524, pp. 94–109, 2005.

- identify hidden structures in problems;
- design generic search guiding techniques;
- pave the way of possible use of the impact analysis into decomposition based methods such as, for example, Benders decomposition.

Our new impact measures are made possible by the use of an explanation-based constraint solver that provides inside information about the solver embedded knowledge gathered from the problem.

The paper is organized as follows: Section 2 introduces the basis and motivations of our work. Several impact measures and associated graphs are presented in Section 3 distinguishing their respective ability to reflect dynamic and static structures on a concrete example. Finally, as we believe that the detection of hidden structures can be explicitly used into CP, we start to show the interest of those such structures as a guide for searching as well as the design of a dedicated resolution strategy inspired from a logic based decomposition.

2   The Idea of Exploiting Problem Structures Within Search Strategies

Efficient constraint programming search strategies exploit specific aspects or characteristics of a given (instance of a) problem. In Operation Research, relaxation or decomposition strategies exploit the fact that part of the problem can be treated as a *classical* problem (such as compatible or optimal flow problems, shortest path problems, knapsack problems, etc.). This is often called *structure* in the constraint programming community.

A problem is more generally said to be *structured* if its components (variables[1] and/or constraints) do not all play the same role, or do not have the same importance within the problem. In such a problem, the origin of the complexity relies on the different behavior (or impact) for specific components of the problem. One of the main difficulty in identifying structure in problems is that this structure is not always statically (at the instance level, before solving) present. The interplay between a given instance and the search algorithm itself may define or help to exhibit a hidden *structure* within the problem. We call it a *dynamic structure*. It is related to bad initial choices as well as new relationships due to the addition of constraints during the search. This does not make things easy when willing understand the complexity of a problem and use this information to speed up search techniques.

Backdoors, recently introduced in [18], are an interesting concept to characterize hidden structure in problem instances. They are informally defined as subsets of variables that encapsulate the whole combinatorics of a given instance of a problem: once this core part completely instantiated, the remaining subproblem can be solved very efficiently. Numerous search strategies are based, knowingly

---

[1] In the following, we will focus our study only on variables as components inducing a structure.

or not, upon this principle. The following two were the most influential for our work:

- Branching heuristics in CP attempt to early guide the search towards the backdoors variables as they try to perform choices that simplify the whole problem as much as possible. They are based on a simple idea: select a variable that lead the possibly smallest search space and that raises contradictions early as possible. This principle (often referred to the *first fail* principle [8]) is often implemented by taking the current domain and degree of constraindness (see [3] for variants) of the variables into account. More recently, [16] proposed to characterize the impact of a choice and a variable by looking at the search space reduction caused by this choice in average (another way of identifying a *backdoor*) and used this information as a guiding strategy.

- Benders decomposition [2] falls exactly within the range of backdoors techniques. It is a solving strategy based on a partition of the problem among its variables into two sets x, y. A master problem provides an assignment x^*, and a sub-problem tries to complete this assignment over the y variables. If this proves impossible, it produces a cut^2 (a constraint) added to the master problem in order to prune this part of the search space on the x side. The interesting cuts are those who are able to prune not only the current x^* solution from the search space (this is mandatory) but also the largest possible class of assignments that share common characteristics with x^* which make them suboptimal or inconsistent for the same reason. This technique is intended for problems with special structure. The master problem is based on a relevant subset of variables that generally verifies the two following assumptions:

  1. the resulting subproblem is easy. In practice, several small independent subproblems are used, making it easy to perform the required exhaustive search in order to produce the Benders cut.
  2. the Benders cut is accurate enough to ensure a quick convergence of the overall technique.

  In such a decomposition, the master problem can be considered as a *backdoors* because, thanks to condition 1, once completely instantiated the remaining problem can be solved efficiently. Moreover, if the remaining subproblem can be actually solved polynomially (this is referred as *strong* backdoors), a powerful cut based on the minimal conflict can often be computed.

For the latter technique, structure needs to be identified before search starts. Classical structure identification is made through an analysis of the constraint network. For example, it is common for solving graph coloring problems to look for maximal cliques in order to compute bounds or to add *all-different* constraint to tighten propagation on the problem. But, such an analysis only provides information on visible static structures. Nevertheless, hidden structure and dynamic

---

² This cut is often referred as the Benders cut.

one seems to be of very high interest for a lot of search strategies. Of course, their identification is at least as costly as solving the original problem. We believe that the propagation performed by the solver during search provides information that should lead to identify those hidden structures. One way of exploiting that information is to use explanations.

3 Identifying Problem Structure Using Explanations

Refalo [16] introduced an impact measure with the aim of detecting choices with the strongest search space reduction. He proposes to characterize the impact of a decision by computing the Cartesian product of the domains (an evaluation of the size of the search space) before and after the considered decision. We claim here that we can go a step further by analyzing where this propagation occurs and how past choices are involved. We extend those measures into an impact graph of variables, taking into account both the effects of old decisions and their effective involvement in each inference made during resolution.

Our objective being to identify variables that maximally constrain the problem as well as subsets of variables that have strong relationships and strong impact upon the whole problem (namely a *backdoors*). We have focused our study on the following points:

- the impact or influence of a variable on the direct search space reduction;
- the impact of a variable inside a chain of deductions made by the solver even a long time after the variable has been instantiated;
- the region of the problem under the influence of a variable and the precise links between variables.

Such information relies on the concept of explanation for CP [11].

3.1 Explanations for Constraint Programming

Explanations have been initially introduced to improve backtracking-based algorithms but have been recently used for many purposes including dynamic constraint satisfaction problems and user interaction.

Definition 1. *An explanation records some sufficient information to justify an inference made by the solver (domain reduction, contradiction, etc.). It is made of a set of constraints C' (a subset of the original constraints of the problem) and a set of decisions $dc_1, ..., dc_n$ taken during search. An explanation of the removal of value a from variable v will be written: $C' \land dc_1 \land dc_2 \land \cdots \land dc_n \Rightarrow v \neq a$.*

As the constraint solver always know (although may be not explicitly) *why* it removes a value from the domain of a variable, explanations can be computed within the solver[3] [12]. Thus, explanations computed by the solver account for

---

[3] Notice that when a domain is emptied (*i.e.* a contradiction is identified), an explanation for that situation is computed by uniting each explanation of each removal of value of the variable concerned.

the underlying logical chain of consecutive inferences made by the solver during propagation. In a way, explanations provide an accurate trace of the behavior of the solver as all operations are explained. In the following, E_i^{val} will denote the set of all explanations computed from the start of the search for all different removals of the value val of the variable i that have occurred throughout search.

3.2   Characterizing Impact

The impact of a decision $x_i = a$ can be expressed, according to the first fail principle, through the reduction of the search space implied in average by this decision. Nevertheless, this reduction does not only occur when the decision is posted to the problem but also when other (future) deductions that are partially based on the hypothesis $x_i = a$ are made.

The use of explanations can provide more information on the real involvement of the decision in the reduction. A past decision $x_i = a$ has an effective impact (in the solver's point of view) over a value val of variable x_j if it appears in the explanation justifying its removal.

We introduce now our new measures whose aim is to characterize the impact of a decision not only based on the immediate search space reduction. We denote $I_\alpha(x_i = a, x_j, val)$ the impact of taking decision $x_i = a$ on the value val of a variable x_j. α is an index used to distinguish our different measures.

Our first measure is expressed as the number of times a decision occurs in a removal explanation for value val from variable x_j. The size of the explanation is also taken into account as it reflects directly the number of hypothesis required to deduce the removal. Therefore, small explanations reveal strong relationships.

$$I_0(x_i = a, x_j, val) = \sum_{\{e \in E_j^{val}, x_i = a \in e\}} 1/|e|$$

From this basic measure, we introduced different impact measures based on the solver activity and the computation of explanations (measures I_1 and I_2) in order to exhibit dynamic structures. We also designed a search space reduction based measure (I_3) in order to capture static structures and to help guiding search. As search obviously direct propagation (and vice-versa), it seamed quite natural to normalize this basic measure according to search.

- The impact is here normalized according to the number of times a decision $x_i = a$ is taken during search: $|x_i = a|$. We simply intend here to distinguish frequent decisions (*i.e.* most likely recent ones) and hardly reconsidered ones (*i.e.* most likely quite old ones):

$$I_1(x_i = a, x_j, val) = \frac{\sum_{\{e \in E_j^{val}, x_i = a \in e\}} \frac{1}{|e|}}{|x_i = a|}$$

- Another way to normalize is to consider the age^4 a_e^d of a decision d when computing an explanation e with the aim of decreasing the impact of old decisions. This leads to:

$$I_2(x_i = a, x_j, val) = \sum_{\{e \in E_j^{val}, x_i = a \in e\}} \frac{1}{|e| \times a_e^{x_i = a}}$$

- The computation of impacts is spread within the whole resolution process as it is done during explanation computations. It is a quite different approach to [16] which analyses each decision separately to get its instantaneous impact. I_3 tries to identify recurrent search space reduction associated to a decision:

$$I_3(x_i = a, x_j, val) = \frac{\sum_{e \in E_j^{val}, x_i = a \in e} \frac{1}{|e|}}{|\{x_i = a \text{ active } \wedge val \in Dom(x_j)\}|}$$

$I_3(x_i = a, x_j, val)$ can be considered as the probability that the value val of x_j will be pruned if the decision $x_i = a$ is taken. It therefore considers the number of time a removal could have been done and the number of time it has been effectively done. This measure is therefore updated each time a new removal occurs and as long as $x_i = a$ is active. It takes into account the frequency as well as the proportion of the involvement of a decision within explanations of removals.

3.3   From Fine-Grained Impact Measures to Relations Between Variables

From the previous definitions, we can introduce different directed weighted graphs of impacts $GI(V, E)$ with weight $I(x, y)$ for any couple $(x, y) \in E = V \times V$. In order to define those weights, fine-grained impact measure introduced above are aggregated in the following way:

$$I(x_i = a, x_j) = \sum_{val \in D(x_j)} I(x_i = a, x_j, val)$$

where $I(x_i = a, x_j, val)$ can be replaced by any of the 4 measures introduced before $(\{I_\alpha \mid \alpha \in [0, 1, 2, 3]\})$.

We have a special case for I_3, as it intends to relate the impact to the domain reduction generated by a variable over another. Relating a variable and a decision is therefore normalized considering the domain initial size of the variable:

$$I(x_i = a, x_j) = (|D(x_j)| - \sum_{val \in D(x_j)} (1 - I_3(x_i = a, x_j, val))) / |D(x_j)|$$

In this context, $1 - I_3(x_i = a, x_j, val)$ corresponds to the probability of presence of the value val of the variable x_j after taking $x_i = a$.

---

[4] The distance in the search tree from the decision to the resulting removal.

The weight of an edge can now be computed in the following way :

$$I(x_i, x_j) = \sum_{v \in D(x_i)} I(x_i = v, x_j)$$

3.4   Overall Impact of a Given Decision

For measures I_1 and I_2, the overall impact of a decision is computed by accumulating impacts over variables of the problem:

$$I(x_i = a) = \sum_{x_j \in V} I(x_i = a, x_j)$$

As for measure I_3, focus is made on the average search space reduction. The current size P of the search space is the Cartesian product of the current domains of variables. Therefore, the overall impact of a decision for the whole problem is expressed through its effective search space reduction by considering the probable remaining space after the decision $((P_{before} - P_{after})/P_{before})$:

$$I(x_i = a) = (P - \prod_{x_j \in V} \sum_{val \in D(x_j)} (1 - I_3(x_i = a, x_j, val)))/P$$

3.5   An Illustrative Case

We take here a particular instance[5] of the benchmark problems introduced later in Section 4 in order to illustrate what kind of structures are isolated by our impact measures. Moreover, we will describe how the retrieved information may be used at the user level to investigate problems and instances.

We therefore consider a random binary problem in which a structure is inserted by increasing the tightness of some constraints in order to design several subsets of variables with strong relationships. Random instances are characterized by the tuple $< N, D, p1, p2 >$ (we use the classical B model [1]) where N is the number of variables, D the unique domain size, p_1 the density of the constraint network and p_2 the tightness of the constraints. Here we consider $N = 30$, $D = 10$, $p1 = 50\%$. We design three subsets of 10 variables whose tightness is $p_2 = 53\%$ while it is set to 3% in the remainder of the network.

The specific instance we chose here to illustrate our different measures is interesting because it seems harder to solve than expected for the mindom [8] classical variable selection heuristic. Using the different measures of impact introduced above, we would like to illustrate how several questions may be addressed when facing a problem:

- is it possible without any network analysis to identify the structure embedded within the instance ?
- why mindom is not performing as expected on this instance ? Is this due to the instance or to the heuristic itself ?

---

[5] One of the random generated problem with a given set of parameters.

Visualizing the Impact Graph. Figures 1 to 4 show the impact graph GI of the 30 variables involved in our instance. We use here a matrix-based representation [6]: variables are represented both on the rows and columns of the matrix. The cell at the intersection of row i and column j corresponds to the impact of the variable v_j on the variable v_i. The stronger the impact, the heavier the edge, the darker the cell. The matrix is ordered according to the order of the hidden kernel of variables[6].

Notice that we start search by applying a kind of singleton consistency propagation (every value of every variable is propagated [15]) to ensure that the impacts of variables are homogenously initialized. Although the graph is almost entirely connected, the matrix-based visualization depicted in Figure 1 makes it possible to see very clearly the structure of the problem, *i.e.* the three sets of variables having strong internal links, right after this first propagation step (we use here the generic impact measure I_0).

Fig. 1. The representation of the impact graph of variables at the end of the initialization phase using I_0 as measure of impact

Figure 2 depicts the impact graph after two minutes of search using `mindom` as variable selection heuristic (using both I_0 and I_3). One can notice how I_0 highly concentrates on dynamic structure (initial clusters are no longer visible compared to Figure 1) whereas I_3 is focused on the original static structure and interestingly forgets the weak links. The darker area for I_0 at the bottom left corner shows that the variables in the first two sets have an apparently strong influence on the variables belonging to the third set. This can be accounted for by the fact that bad decisions taken early on the variables of the first sets lead the solver into numerous try-and-fail steps on the variables of the third set.

---

[6] We are currently working on clustering algorithms [5] to discover this particular ordering from the impact graph alone.

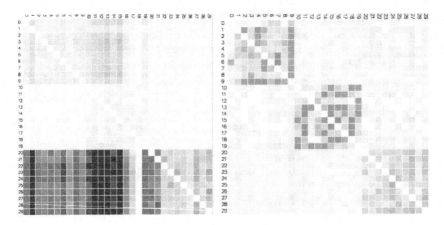

Fig. 2. The impact graph using I_0 (on the left) and I_3 (on the right) after two minutes of computation using the `mindom` heuristic

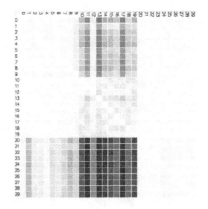

Fig. 3. A representation of the impact graph normalized according to the number of times a decision is taken (I_1)

Figure 3 represents a normalized representation of the same graph where the influence of a decision taken by the solver is divided by the number of times this decision occurred during the resolution (measure I_1). By doing so, we aim at refining the previous analysis by distinguishing two types of decisions: those having a great influence because they are repeated frequently, and those having a great influence because they guide the solver in some inconsistent branch of the search tree and appear in all inconsistency explanations. We can thus isolate early bad decisions that seem to involve the second set of variables.

Finally, Figure 4 represents the activity within the impact graph where the effect of old decisions is gradually discarded. As expected, it appears that the

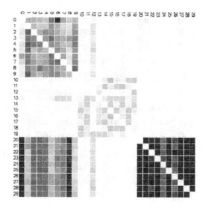

Fig. 4. A representation of the impact graph normalized according to the age of decisions (I_2)

solver keeps going back and forth between the first and third set of variables, with very negligible involvement of the second set. This must be related to poor decisions taken on the variables of the second set.

In order to further confirm this interpretation, we adapted our search heuristic so that it takes into account the impact of variables during the resolution and undoes immediately decisions whose influence increase outstandingly (because they appear in many explanations but do not provide any valuable pruning). The problem was then solved almost instantaneously.

4   Using Impacts to Improve Search

In this section, we illustrate how the impact measure introduced above can be used in order to improve search techniques. We use the impact measure in the branching heuristics within a tree search: Upon branching, first, we choose the variable x that maximizes $\sum_{a \in D(x)} I(x = a)$ and second, for that variable, we choose the value v that minimizes $I(x = a)$ in order to allow a maximum possible future assignments ($D(x)$ is here the current domain of x). Ties are randomly broken. As said earlier, impacts are initialized through the use of a singleton consistency-like propagation.

Our experiments were conducted on a Pentium 4, 3 GigaHz, running Windows XP. Our constraint solver is the most recent Java version of choco (choco. sf.net). Notice that as we are using explanations, our tree search is not limited to standard backtracking but we actually use the mac-cbj algorithm (getting higher in the search tree if it is possible upon encountering a contradiction). In practice, the behavior is very close to mac and behaves as it was only merely maintaining explanations. We considered three sets of benchmark problems:

1. The first set comes from experiments in [16]: a set of multiknapsack problems modelled with binary variables. For this set a time limit of 1500s is considered. We focused here on the number of developed nodes[7] as it is directly related the relevance of the measure. Moreover, as randomness is introduced in the problem solving when breaking ties, we report an average over 10 executions.
2. Our second set consists on random binary problems generated following the classical B model (see Section 3.5) with the following parameters: $< 50, 10, 30, p_2 >$. We considered here a time limit of 120s. We focus on the number of unsolved instances within the time limit for each value of p_2.
3. Our final set is made of random structured instances made as described in Section 3.5. A problem $< 45, 10, 35, p_2 >$ is structured with three kernels of 15 variables linked with an intra-kernel tightness p_2 and an inter-kernel tightness of 3%.

As for the impact measure, we compared three measures ($\{I_\alpha \mid \alpha \in [1, 2, 3]\}$) and our implementation of the measure introduced in [16] (denoted I_{ref}). As the measure completely specifies the search used, we will refer in the following to the I_α and I_{ref} strategies in the following.

4.1   First Benchmark: Multiknapsack Problems

On this first benchmark (whose results are reported on Table 1), I_{ref} appears to be the best search strategy. The use of explanations seems to provide good information but it is a long term learning (it requires a restart policy) and is much more costly (in time) so that it cannot solve the instance mknap1-6. I_3 is obviously too costly on this problem where near one million nodes need to be explored. The number of nodes of mindom is given here as a reference.

Table 1. Impacts on multiknapsack problems

	mindom	I_{ref}		I_3		I_3+restart	
	Nodes	Time	Nodes	Time	Nodes	Times	Nodes
mknap1-2	38	0	25.9	0	23.1	0	23.1
mknap1-3	385	0.1	188.7	0.3	354.1	0.3	255.2
mknap1-4	16947	0.7	982.7	4.2	2754	3.2	979.5
mknap1-5	99003	11.2	21439.6	229.1	110666.4	112.8	20237.4
mknap1-6	21532776	425.7	612068	> 1500		> 1500	

I_1 and I_2 are not accurate on these instances and maybe need a fine restart policy as they attempt to detect irrelevant first choices. As mentioned by Refalo, the use of restart only increases the overall computation time for I_{ref} but seems to be important for I_3. I_3 is indeed a fine-grained measure that maybe need more time to become accurate for the search.

---

[7] In the presented results, when a restart technique is used, only the number of nodes of the last execution are reported whereas the overall time is indicated.

Table 2. The number of unsolved instances (left) for each impact strategy on $< 50, 10, 30, p_2 >$ and the percentage of succesfully solved instances (right)

strategy	% success
I_{ref}	36 %
I_1	53.7 %
mindom	79 %
mindom + I_2	86 %
mindom + I_{ref}	92.9 %

4.2   Second Benchmark: Random Binary Problems

On this unstructured benchmark, the size of the domains (integer variables instead of binary ones) gives to mindom better chances to make good choices. the results (depicted in Figure 2) are not in favor of impact measures alone. However, their combination with mindom as a way of breaking equalities is much more powerful and allow to solve around 93 % of instances over the whole phase transition against 79 % for mindom alone. This combination avoids bad choices for I_{ref} which becomes the best technique whereas it was the worst one alone (I_1 could solve 17 % more instances than I_{ref}). The use of restart generally increases the overall computation time.

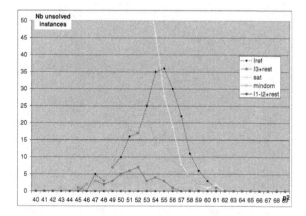

Fig. 5. Number of unsolved instances for different impact measures on random structured binary CSP only with the number of feasible instances (sat)

4.3   Third Benchmark: Structured Random Binary Problems

On this set of problems, the I_{ref} strategy seems to experience difficulties even compared to the classical `mindom` heuristic. Figure 5 reports the number of instances that were not successfully solved within the time limit of 120s. As restart does not help the I_{ref} strategy again on this problem but is effective for I_3, only the best results (*i.e.* with or without restart) of each technique are indicated. The more impressive results here are obtained again by focusing on the dynamic component of the inherent structure of the instances (*i.e.* using strategies I_1 and I_2). That is the only way that all the instances could be successfully solved. Notice that I_3 gives better results than I_{ref} despite its high cost.

The success of I_1 and I_2 may be due to the fact that the complexity of this benchmark does not reside purely in the instances but is more due to the level of the interaction with the search algorithm. The presence of such artificial structures favors from our point of view a kind of heavy tailed behavior and makes initial choices more critical. It can indeed be noticed on Figure 6 that I_1 is sometimes subject to *bad behavior* which does not only appear at the transition phase. The same phenomenon (on a larger scale) may be the cause of the poor performance of I_{ref}.

4.4   Impact-Based Heuristics: First Insights

I_1 and I_2 are strongly based on the solver activity during search (thus focusing on the dynamic component of the instance structure). It generally pays off using them on problems because (and that may be explains the relatively poor performance of I_{ref} in our benchmarks) they are able to detect past bad choices (those whose influence increases outstandingly throughout search without leading to solutions) do be undone.

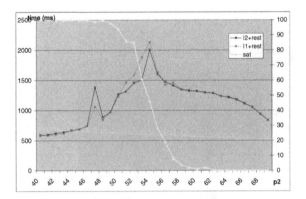

Fig. 6. Resolution time for I_1 and I_2 on random structured binary CSP with the number of feasible instances (sat)

I_3 is too costly (regarding time consumption) at the present time to be used as a default heuristic but some interesting compromise between I_{ref} and I_3 may be designed: taking advantage of the general robustness of I_{ref} while at the same time avoiding heavy-tailed behavior due to bad initial choices.

5   Perspective: Automated Logic Based Benders Decomposition in Constraint Programming

We are interested in Benders decomposition as it is intended for problems with a specific structure and specially, a master-slave relationship between variables. For us, the master set of variables could be restricted to a subset of variables exhibiting a strong overall impact over the whole problem.

Usually, classical Benders cuts are limited to linear programming and are obtained by solving the dual of the subproblem[8] and therefore requires that dual variables or multipliers to be defined to apply the decomposition. However, [9] proposes to overcome this limit and to enlarge the classical notion of *dual* by introducing an *inference dual* available for all kinds of subproblems. He refers to a more general scheme and suggests a different way of considering duality: a Benders decomposition based on *logic*.

However this inference dual must be implemented for each class of problems to derive accurate Benders cuts [10, 4]. One way of thinking the dual is to consider it as a certificate of optimality or an *explanation* (as introduced in Section 3.1) of inconsistency in our case. Our explanation-based constraint programming framework therefore provides in a sense an implementation of the logic based Benders decomposition in case of satisfaction problems [4]. One can notice here as the computation of explanations is *lazy*[9], the first explanation is taken whereas several explanations exist. One cannot look for the minimal explanation for evident scalability reasons. Therefore, such an inference dual provides an arbitrary[10] dual solution but not necessarily the optimal one. Obviously, the success of such an approach depends on the degree to which accurate explanations can be computed for the constraints of the subproblem.

Explanation-based constraint programming as used in algorithms like `mac-dbt` [13] or in `decision-repair` [14] kind of automatically focus on the master problem of such a decomposition but may be trapped by bad decisions and revert to a more conventional behaviour. The next step would be here to use the structure exhibited from the impact graphs presented above in order to apply a Benders decomposition scheme in a second phase of resolution. The identification of substructures once the master instantiated could guide the generation of cuts for the master to gather as much information as possible where lies the real combinatorics of the problem.

---

[8] Referring to linear programming duality.

[9] Not all possible explanations are computed when removing a value. Only the one corresponding to the solver actual reasoning is kept.

[10] This can also be accounted for linear duality where any dual solution is a bound for the primal problem.

6 Conclusion

In this paper, we introduced several indicators useful for both identification and use while searching of key structures at the heart of combinatorial problems. We focused our study on the relationship between variables and gave new perspectives on the design of generic search heuristics for constraint programming as well as search algorithms. We believe that the presence of *backdoors* or subset of variables exhibiting a strong impact over the whole problem could be explicitly used by *ad hoc* decomposition or relaxation strategies inspired from Operation Research. A concrete example is the Benders decomposition and its generic extension based on logic. It is indeed exactly a *backdoors* technique and could be applied in Constraint Programming as a nogood learning strategy.

References

1. D. Achlioptas, L. Kirousis, E. Kranakis, D. Krizanc, M. Molloy, and Y. Stamatiou. Random constraint satisfaction: a more accurate picture. In *Proceedings CP 1997*, pages 121–135, Linz, Austria, 1997.
2. J. F. Benders. Partitionning procedures for solving mixed-variables programming problems. *Numerische Mathematik*, 4:238–252, 1962.
3. F. Boussemart, F. Hemery, C. Lecoutre, and L. Sais. Boosting systematic search by weighting constraints. In *Proceedings ECAI'04*, pages 482–486, 2004.
4. Hadrien Cambazard, Pierre-Emmanuel Hladik, Anne-Marie Déplanche, Narendra Jussien, and Yvon Trinquet. Decomposition and learning for a real time task allocation problem. In *Proceedings CP 2004*, pages 153–167, 2004.
5. G. Cleuziou, L. Martin, and C. Vrain. Disjunctive learning with a soft-clustering method. In *ILP'03:13th International Conference on Inductive Logic Programming*, pages 75–92. LNCS, September 2003.
6. Mohammad Ghoniem, Narendra Jussien, and Jean-Daniel Fekete. VISEXP: visualizing constraint solver dynamics using explanations. In *Proceedings FLAIRS'04*, Miami, Florida, USA, May 2004.
7. Carla P. Gomes, Bart Selman, and Nuno Crato. Heavy-tailed distributions in combinatorial search. In *Proceeding CP 1997*, pages 121–135, Linz, Austria, 1997.
8. R. Haralick and G. Elliot. Increasing tree search efficiency for constraint satisfaction problems. *Artificial intelligence*, 14(9):263–313, 1980.
9. J.N. Hooker and G. Ottosson. Logic-based benders decomposition. *Mathematical Programming*, 96:33–60, 2003.
10. Vipul Jain and I. E. Grossmann. Algorithms for hybrid milp/cp models for a class of optimization problems. *INFORMS Journal on Computing*, 13:258–276, 2001.
11. Narendra Jussien. *The versatility of using explanations within constraint programming*. Habilitation thesis, Université de Nantes, France, 2003. also available as RR-03-04 research report at École des Mines de Nantes.
12. Narendra Jussien and Vincent Barichard. The PaLM system: explanation-based constraint programming. In *Proceedings of TRICS: Techniques foR Implementing Constraint programming Systems, a post-conference workshop of CP 2000*, pages 118–133, Singapore, September 2000.
13. Narendra Jussien, Romuald Debruyne, and Patrice Boizumault. Maintaining arc-consistency within dynamic backtracking. In *Proceedings CP 2000*, pages 249–261, Singapore, 2000. Springer-Verlag.

14. Narendra Jussien and Olivier Lhomme. Local search with constraint propagation and conflict-based heuristics. *Artificial Intelligence*, 139(1):21–45, July 2002.
15. P. Prosser, K. Stergiou, and T. Walsh. Singleton consistencies. In R. Dechter, editor, *Proceedings CP 2000*, pages 353–368, Singapore, 2000.
16. Philippe Refalo. Impact-based search strategies for constraint programming. In *Proceedings CP 2004*, pages 556–571, Toronto, Canada, 2004.
17. Ryan Williams, Carla Gomes, and Bart Selman. On the connections between backdoors and heavy-tails on combinatorial search. In *In the International Conference on Theory and Applications of Satisfiability Testing (SAT)*, 2003.
18. Ryan Williams, Carla P. Gomes, and Bart Selman. Backdoors to typical case complexity. In *Proceedings IJCAI 2003*, 2003.

A Hybrid Algorithm for a Class of Resource Constrained Scheduling Problems

Yingyi Chu and Quanshi Xia

IC-Parc, Imperial College London,
London SW7 2AZ, UK
{yyc, qx1}@imperial.ac.uk

Abstract. This paper presents a hybrid algorithm for a class of resource constrained scheduling problems based on decomposition. The general minimum completion time problem is considered, which has not been solved in a decomposed way by existing methods. The problem is first decomposed into an assignment master problem and a number of scheduling subproblems. The subproblem is formulated as both a constraint programming model and an integer programming model. The hybrid algorithm then combines constraint programming, integer programming and linear programming solvers in its three steps: the master problem solving, the subproblems solving and the cut generation. In particular, the cut generation method is based on the integer programming model, and in practice it is done by solving a linear program. Computational experiments have been carried out for the considered minimum completion time problems. The results show that the proposed algorithm could substantially reduce the solving time, compared with directly solving by mixed integer solvers.

1 Introduction

This paper studies an important class of resource constrained scheduling problems, where a set of jobs are assigned to and processed by a set of facilities, subject to resource constraints and release/due date constraints. Depending on the objective function, there are different versions of the problem, e.g. minimum cost problem, minimum completion time problem, etc.

These problems have attracted substantial research interests due to its importance in many application domains. Solution methods based on Benders decomposition have been proposed recently [12, 8, 13, 9] for some of the problems. A decomposition method partitions the problem into an assignment *master* problem and a number of independent scheduling *subproblems*, each for one facility. The master problem and the subproblems are solved iteratively and in particular the scheduling subproblems are often solved by constraint solvers as strong reasoning techniques are available. The key step is to generate valid cuts from the subproblems, guiding the search of the assignment solution in the master problem.

R. Barták and M. Milano (Eds.): CPAIOR 2005, LNCS 3524, pp. 110–124, 2005.
© Springer-Verlag Berlin Heidelberg 2005

For minimum cost problems, Jain and Grossmann (2001), Harjunkoski and Grossmann (2002) and Hooker (2004) employ the no good cut that excludes sets of incompatible jobs assigned to a facility. For minimum makespan problems, a valid cut is proposed by Hooker (2004), but under the assumption of same release dates for all jobs (and same due dates in the computational study). As is pointed out in [9, 10], all these cuts are based on the *logical explanation* of the individual solution processes for the specific problems, instead of the dual information from the linear integer formulation of the subproblems.

This paper presents a general approach for tackling the resource constrained scheduling problems. A hybrid method is proposed where the necessary cuts are generated based on the *dual information* from the subproblem's formulation. The subproblem is formulated as equivalent integer programming (IP) and constraint programming (CP) models. The IP formulation is used to generate the integer Benders cuts by exploiting the dual information, while the CP formulation is used to efficiently solve the scheduling subproblems using strong constraint propagations for cumulative scheduling. To be concrete, this hybrid approach is instantiated to a solution algorithm for the *general minimum completion time problem*, which is not solvable by previous decomposition methods, as different release/due dates are allowed.

The paper is organized as follows. Section 2 introduces the considered problems. Section 3 presents the hybrid method. Section 4 details the key step in the proposed method, i.e., the cut generation. Section 5 presents the computational experiments and results. Section 6 concludes the paper.

2   The Resource Constrained Scheduling Problems

This section introduces the considered scheduling problems. A formulation of the general minimum completion time problem is given, which is used for the subsequent algorithm development and the computational experiments.

Consider a set of jobs, denoted by $\mathcal{J}$, and a set of facilities (machines), denoted by $\mathcal{M}$. Each job $j \in \mathcal{J}$ has a release date r_j and a due date d_j. Each facility $m \in \mathcal{M}$ provides a fixed amount of resources specified by C_m. The processing time of job j on facility m is given by p_{jm}, and the job j consumes the amount C_{jm} of resources during its processing time on m.

Following [9], we employ a discrete time formulation (where times are discretized to integers), instead of the continuous time model used in [12, 8, 15], because the discrete time formulation is often easier to solve, especially when the cumulative constraint (instead of the simpler disjunctive constraint) is considered. Let $\mathcal{T}$ denote the whole set of discrete time points in the considered problem, $\{\min_j\{r_j\}, \cdots, \max_j\{d_j\}\}$.

Define binary variables x_{jmt} for any job j, any facility m and any time point t in $\mathcal{T}_{jm}$, where $\mathcal{T}_{jm} \equiv \{r_j, \cdots, d_j - p_{jm}\}$ represents the possible starting times of job j on facility m. The variables are used to indicate when and on which facility a job starts, i.e. $x_{jmt} = 1$ if and only if job j starts from the discrete time point t at facility m. Variable H denotes the overall completion time of all jobs. Using these variables, the following constraint states that each job must be processed by exactly one facility:

$$\forall j \in \mathcal{J}: \quad \sum_{m \in \mathcal{M}} \sum_{t \in \mathcal{T}_{jm}} x_{jmt} = 1 \tag{1}$$

and the resource constraints have to be observed on every facility at any time:

$$\forall t \in \mathcal{T}, \forall m \in \mathcal{M}: \quad \sum_{j \in \mathcal{J}} \sum_{t'=t-p_{jm}+1}^{t} C_{jm} x_{jmt'} \leq C_m \tag{2}$$

By definition, the completion time variable H satisfies:

$$\forall j \in \mathcal{J}, \forall m \in \mathcal{M}: \quad \sum_{t \in \mathcal{T}_{jm}} (t + p_{jm}) x_{jmt} \leq H \tag{3}$$

The minimum completion time problem is formulated as:

$$\mathbf{P}: \min_{x_{jmt}, H} H$$

$$s.t. \begin{cases} \sum_{m \in \mathcal{M}} \sum_{t \in \mathcal{T}_{jm}} x_{jmt} = 1 & \forall j \in \mathcal{J} \\ \sum_{j \in \mathcal{J}} \sum_{t'=t-p_{jm}+1}^{t} C_{jm} x_{jmt'} \leq C_m & \forall t \in \mathcal{T}, \forall m \in \mathcal{M} \\ \sum_{t \in \mathcal{T}_{jm}} (t + p_{jm}) x_{jmt} \leq H & \forall j \in \mathcal{J}, \forall m \in \mathcal{M} \\ x_{jmt} \in \{0, 1\} & \forall j \in \mathcal{J}, \forall m \in \mathcal{M}, \forall t \in \mathcal{T}_{jm} \\ H \in \mathbf{Z}[\max_j \{r_j + \min_m \{p_{jm}\}\}, \max_j \{d_j\}] \end{cases}$$

3   A Hybrid Algorithm Framework

3.1   Decomposition

To decompose the problem, we first reformulate the problem $\mathbf{P}$ by disaggregating the variables. Introducing the assignment variables y_{jm}, to indicate whether job j is assigned to facility m or not, we can rewrite the problem as:

$$\mathbf{P}': \min_{y_{jm}, x_{jmt}, H} H$$

$$s.t. \begin{cases} \sum_{m \in \mathcal{M}} y_{jm} = 1 & \forall j \in \mathcal{J} \\ \sum_{t \in \mathcal{T}_{jm}} x_{jmt} = y_{jm} & \forall j \in \mathcal{J}, \forall m \in \mathcal{M} \\ \sum_{j \in \mathcal{J}} \sum_{t'=t-p_{jm}+1}^{t} C_{jm} x_{jmt'} \leq C_m & \forall t \in \mathcal{T}, \forall m \in \mathcal{M} \\ \sum_{t \in \mathcal{T}_{jm}} (t + p_{jm}) x_{jmt} \leq H & \forall j \in \mathcal{J}, \forall m \in \mathcal{M} \\ y_{jm} \in \{0, 1\} & \forall j \in \mathcal{J}, \forall m \in \mathcal{M} \\ x_{jmt} \in \{0, 1\} & \forall j \in \mathcal{J}, \forall m \in \mathcal{M}, \forall t \in \mathcal{T}_{jm} \\ H \in \mathbf{Z}[\max_j \{r_j + \min_m \{p_{jm}\}\}, \max_j \{d_j\}] \end{cases}$$

A decomposition is based on a partition of variables. For the problem $\mathbf{P}'$, the variables y_{jm} and H are solved in an assignment master problem. If these variables are tentatively fixed, the rest of the problem, pertaining the variables x_{jmt}, further decomposes into $|\mathcal{M}|$ smaller subproblems, one for each facility m.

The master problem is written as:

$$\textbf{MP} : \min_{y_{jm}, H} H$$

$$s.t. \begin{cases} \sum_{m \in \mathcal{M}} y_{jm} = 1 & \forall j \in \mathcal{J} \\ \sum_{j \in \mathcal{J}} C_{jm} p_{jm} y_{jm} \leq C_m(H - \min_j\{r_j\}) & \forall m \in \mathcal{M} \\ \text{cuts generated from subproblems} \\ y_{jm} \in \{0,1\} \ \forall j \in \mathcal{J} \\ H \in \mathbb{Z}[\max_j\{r_j + \min_m\{p_{jm}\}\}, \max_j\{d_j\}] \end{cases}$$

The second constraint is a strengthening valid constraint, asserting that, for each facility, the total 'volume' of resources consumed by the assigned jobs cannot exceed the available 'volume' of resources up to the completion time H.

Based on the idea of Benders decomposition, the master problem is solved to obtain a tentative assignment $\bar{y}_{jm}$, and a tentative completion time $\bar{H}$. Fixing the tentative assignment, the subproblem is obtained, and it is immediately decomposed according to the facilities. The subproblems try to schedule the tentatively assigned jobs to minimize the completion time, subject to resource constraint and release/due date constraint. If at all facilities the jobs are indeed finished within $\bar{H}$, then the optimal solution is found. Otherwise, a cut is generated from each subproblem where a feasible schedule within $\bar{H}$ is impossible, and the master problem is resolved with the new cuts added. The algorithm iterates until the optimal solution is attained.

Given the tentative solution, the subproblem for each facility is a cumulative scheduling problem for the assigned jobs. In our method, the subproblems have to be formulated in two different ways, serving different functionalities in the hybrid scheme.

3.2   Subproblem Formulations

The most parsimonious formulation of the subproblems is the constraint programming formulation, where variables t_j are defined to denote the starting time of the job j and $\mathcal{J}_m$ denotes the set of assigned jobs $\{j | j \in \mathcal{J}, \bar{y}_{jm} = 1\}$.

$$\forall m : \textbf{SP}_{\textbf{CP}}^m(\bar{y}_{jm}, \bar{H}) : \bar{H} \geq \min_{t_j : j \in \mathcal{J}_m} \max_{j \in \mathcal{J}_m} \{t_j + p_{jm}\}$$

$$s.t. \begin{cases} \text{cumulative}([t_j : j \in \mathcal{J}_m], [p_{jm} : j \in \mathcal{J}_m], [C_{jm} : j \in \mathcal{J}_m], C_m) \\ t_j \in \mathbb{Z}[r_j, d_j - p_{jm}] \ \forall j \in \mathcal{J}_m \end{cases}$$

The subproblem minimizes the completion time on facility m subject to the cumulative constraint on the tentatively assigned jobs, and then the optimal value of it is compared with the tentative value $\bar{H}$.

In order to generate cuts based on Benders decomposition, the subproblems are also formulated as an integer programming model. In problem $\textbf{P}'$, by fixing the master problem variables, we obtain the following subproblems:

$$\forall m : \mathbf{SP}_{\mathbf{IP}}^m(\bar{y}_{jm}, \bar{H}) : \min_{x_{jmt}} 0$$

$$s.t. \begin{cases} \sum_{t \in \mathcal{T}_{jm}} x_{jmt} = \bar{y}_{jm} & \forall j \in \mathcal{J} \\ \sum_{j \in \mathcal{J}} \sum_{t'=t-p_{jm}+1}^{t} C_{jm} x_{jmt'} \leq C_m & \forall t \in \mathcal{T} \\ \sum_{t \in \mathcal{T}_{jm}} (t + p_{jm}) x_{jmt} \leq \bar{H} & \forall j \in \mathcal{J} \\ x_{jmt} \in \{0, 1\} & \forall j \in \mathcal{J}, \forall t \in \mathcal{T}_{jm} \end{cases}$$

In this case the subproblems are feasibility problems as the variables x_{jmt} do not contribute to the objective function in $\mathbf{P}'$.

Note that the subproblems are obtained based on formulation $\mathbf{P}'$, following the classic Benders decomposition procedure (ref. [2, 7]). The tentative master problem solution values only appear in the right hand sides. Yet the only difficulty is that the subproblems are now integer programs.

3.3   Hybrid Algorithm

The hybrid scheme is partitioned into three functional modules: the solution of the master problem, the solution of the subproblems, and the generation of cuts. This partition of functionality makes the *hybridization of solvers and formulations* possible. Different models and solvers are used for different modules: the master problem, given as an *IP model*, is solved by an *integer programming solver*; the subproblem is solved by a *constraint solver* using its *CP formulation*; the cut generation is based on the *IP formulation* of the subproblem and it is done by *solving a linear program*. The cut generation, which is the key step, will be detailed in Section 4, while this section presents the hybrid framework, and discusses the solution method of the master problem and the subproblems.

Algorithm 1 Hybrid Algorithm for Problem $\mathbf{P}'$

---

1. INITIALIZATION. Setup the initial master problem $\mathbf{MP}^{(0)}$ with no cut; set $k = 0$.

2. ITERATION.

(1) Master Problem Phase. Solve the integer linear program $\mathbf{MP}^{(k)}$ to obtain the tentative solution $\bar{y}_{jm}^{(k)}$ and $\bar{H}^{(k)}$; if $\mathbf{MP}^{(k)}$ is infeasible, then exit with the original problem infeasible.

(2) Subproblems Phase. For each facility m, solve the corresponding subproblem using the CP formulation $\mathbf{SP}_{\mathbf{CP}}^m(\bar{y}_{jm}^{(k)}, \bar{H}^{(k)})$; if all subproblems are feasible, then exit with the optimal solution found; otherwise continue to phase (3).

(3) Cut Generation Phase. For each subproblem that is infeasible, generate a Benders cut based on the IP formulation $\mathbf{SP}_{\mathbf{IP}}^m(\bar{y}_{jm}^{(k)}, \bar{H}^{(k)})$; add the new cuts to the master problem to construct $\mathbf{MP}^{(k+1)}$; set $k = k + 1$ and go back to phase (1).

The hybrid algorithm for the problem $\mathbf{P}'$ is summarized in the Algorithm 1. In step 2.(1), the master problem is solved by a standard mixed integer programming (MIP) solver, using the formulation $\mathbf{MP}$. In step 2.(2), the subproblem is solved by constraint programming. In particular, the edge-finding constraint propagation algorithm is applied to the cumulative constraint in $\mathbf{SP}_{CP}^{m}(\bar{y}_{jm}, \bar{H})$. The edge-finding algorithm reduces the domain of the t_j variables by finding the jobs that have to precede or succeed a set of other jobs, based on the volume of resources they consume (ref. [1]). Furthermore, the branching search is enhanced by a probe backtracking technique, which uses a forward probing method (in addition to the conventional forward local consistency checking) to prune and guide the search (ref. [6]). The constraint solving algorithms used here are provided as libraries of the ECL^{i}PS^e [11] platform. The step 2.(3) is unspecified in the Algorithm 1, but it will be completed at the end of Section 4.

4   Cut Generation

4.1   Benders Cuts from Integer Subproblems

The Benders cut generation from integer subproblems is developed in a general setting, and it is then applied to the considered scheduling problem. This approach is based on the earlier idea reported in [4], but here a more general method is developed and a more efficient way of cut generation is presented.

Consider the following generic program $\mathbf{P}_g$ in the general decomposed form.

$$\mathbf{P}_g : \min_{y, x_m} c^T y + d_1^T x_1 + d_2^T x_2 + \cdots + d_M^T x_M$$

$$s.t. \begin{cases} A_0 y & \geq b_0 \\ A_1 y + B_1 x_1 & \geq b_1 \\ A_2 y + \quad B_2 x_2 & \geq b_2 \\ \vdots \quad \vdots \qquad \ddots & \vdots \quad \vdots \\ A_M y + \qquad\qquad B_M x_M \geq b_M \\ y \in \mathcal{D}_y, \quad x_m \in \{0,1\}^{n_m} \end{cases}$$

where the vector y represents the master problem variables, and the subproblem variables are divided to the vectors x_m ($m = 1, \cdots, M$). In Benders decomposition, when y is fixed, the rest of the problem is decomposed into M subproblems. The master problem variables belong to a *finite* domain $\mathcal{D}_y$, while the subproblem variables are considered as binary.

The formulation $\mathbf{P}'$ for the scheduling problem fits into the above generic model. However we develop the cut generation method in a general setting using the problem $\mathbf{P}_g$.

According to Benders decomposition, the master problem and the subproblems are written as:

$$\mathbf{MP}_g : \min_{y, z_m} c^T y + z_1 + \cdots + z_M$$

$$s.t. \begin{cases} A_0 y \geq b_0 \\ \text{Benders cuts} \end{cases}$$

$$\forall m : \mathbf{SP}_g^m(\bar{\boldsymbol{y}}) : \min_{\boldsymbol{x}_m} \boldsymbol{d}_m^T \boldsymbol{x}_m$$

$$s.t. \begin{cases} \boldsymbol{B}_m \boldsymbol{x}_m \geq \boldsymbol{b}_m - \boldsymbol{A}_m \bar{\boldsymbol{y}} \\ \boldsymbol{x}_m \in \{0,1\}^{n_m} \end{cases}$$

where variables z_m represent the objective value of the subproblems. The master problem is solved to give a tentative solution $(\bar{\boldsymbol{y}}, \bar{z}_1, \cdots, \bar{z}_M)$. For each m, if the resulting $\mathbf{SP}_g^m$ is infeasible or the value $\bar{z}_m$ cannot be reached by $\mathbf{SP}_g^m$, then a Benders cut (over the master problem variables $\boldsymbol{y}$ and z_m) is generated. While there is a standard way of generating Benders cut when the subproblem is continuous, difficulties arise when the subproblem is an integer program. Next we focus on the Benders cut generation from the mth subproblem $\mathbf{SP}_g^m$.

It is essential that the Benders cut to be generated is *valid*, which means that it does not cut off any *feasible combination* of the values of $\boldsymbol{y}$ and z_m (i.e. the value of z_m can be reached by the subproblem parameterized by the value of $\boldsymbol{y}$). A valid Benders cut is often derived using the dual information from the subproblem. In order to extract dual information from the integer subproblem $\mathbf{SP}_g^m$, we define the fixed subproblems by fixing the integer variables $\boldsymbol{x}_m$ to a given value $\tilde{\boldsymbol{x}}_m$:

$$\forall \tilde{\boldsymbol{x}}_m \in \{0,1\}^{n_m} : \mathbf{SP}_g^m {}_{\mathbf{F}}(\bar{\boldsymbol{y}}, \tilde{\boldsymbol{x}}_m) : \min_{\boldsymbol{x}_m} \boldsymbol{d}_m^T \boldsymbol{x}_m$$

$$s.t. \begin{cases} \boldsymbol{B}_m \boldsymbol{x}_m \geq \boldsymbol{b}_m - \boldsymbol{A}_m \bar{\boldsymbol{y}} \\ \boldsymbol{x}_m = \tilde{\boldsymbol{x}}_m \\ \boldsymbol{x}_m \geq 0 \end{cases}$$

For each subproblem there are totally 2^{n_m} number of fixed subproblems. As $\boldsymbol{x}_m$ is fixed to an integer value, the integrality constraint is dropped. The fixed subproblems are dualized to $\mathbf{DSP}_g^m {}_{\mathbf{F}}(\bar{\boldsymbol{y}}, \tilde{\boldsymbol{x}}_m)$ in order to elicit dual values.

$$\forall \tilde{\boldsymbol{x}}_m \in \{0,1\}^{n_m} : \mathbf{DSP}_g^m {}_{\mathbf{F}}(\bar{\boldsymbol{y}}, \tilde{\boldsymbol{x}}_m) : \max_{\boldsymbol{u}, \boldsymbol{v}} (\boldsymbol{b}_m - \boldsymbol{A}_m \bar{\boldsymbol{y}})^T \boldsymbol{u} + \tilde{\boldsymbol{x}}_m^T \boldsymbol{v}$$

$$s.t. \begin{cases} \boldsymbol{B}_m^T \boldsymbol{u} + \boldsymbol{v} \leq \boldsymbol{d}_m \\ \boldsymbol{u} \geq 0, \quad \boldsymbol{v} : \text{free} \end{cases}$$

If $\mathbf{SP}_g^m {}_{\mathbf{F}}(\bar{\boldsymbol{y}}, \tilde{\boldsymbol{x}}_m)$ is infeasible (and thus $\mathbf{DSP}_g^m {}_{\mathbf{F}}(\bar{\boldsymbol{y}}, \tilde{\boldsymbol{x}}_m)$ is unbounded), then we use the homogeneous dual $\mathbf{HDSP}_g^m {}_{\mathbf{F}}(\bar{\boldsymbol{y}}, \tilde{\boldsymbol{x}}_m)$.

$$\forall \tilde{\boldsymbol{x}}_m \in \{0,1\}^{n_m} : \mathbf{HDSP}_g^m {}_{\mathbf{F}}(\bar{\boldsymbol{y}}, \tilde{\boldsymbol{x}}_m) : \max_{\boldsymbol{u}, \boldsymbol{v}} (\boldsymbol{b}_m - \boldsymbol{A}_m \bar{\boldsymbol{y}})^T \boldsymbol{u} + \tilde{\boldsymbol{x}}_m^T \boldsymbol{v}$$

$$s.t. \begin{cases} \boldsymbol{B}_m^T \boldsymbol{u} + \boldsymbol{v} \leq 0 \\ 0 \leq \boldsymbol{u} \leq 1, \quad -1 \leq \boldsymbol{v} \leq 1 \end{cases}$$

In the programs, $\boldsymbol{u}, \boldsymbol{v}$ are dual variables. As $\mathbf{SP}_g^m {}_{\mathbf{F}}(\bar{\boldsymbol{y}}, \tilde{\boldsymbol{x}}_m)$ is a *linear program*, strong duality property holds.

Much dual information can be extracted from the above dual programs. From an arbitrary feasible solution of any $\mathbf{DSP}_g^m {}_{\mathbf{F}}(\bar{\boldsymbol{y}}, \tilde{\boldsymbol{x}}_m)$, we can derive an *optimality inequality* over the master problem variables:

$$(\boldsymbol{b}_m - \boldsymbol{A}_m \boldsymbol{y})^T \tilde{\boldsymbol{u}} + \tilde{\boldsymbol{x}}_m^T \tilde{\boldsymbol{v}} \leq z_m \tag{4}$$

From an arbitrary feasible solution of any $\mathbf{HDSP}_g^m\mathbf{F}(\bar{y}, \tilde{x}_m)$, we can derive a *feasibility inequality* over the master problem variables:

$$(b_m - A_m y)^T \tilde{u} + \tilde{x}_m^T \tilde{v} \leq 0 \tag{5}$$

However, not all these inequalities are valid. The following lemmas identify the valid ones among them[1].

Lemma 1. *An optimality inequality (4) is valid if the following sign condition is satisfied:*

$$\begin{cases} \tilde{v}_i \leq 0 & \text{if } (\tilde{x}_m)_i = 1 \\ \tilde{v}_i \geq 0 & \text{if } (\tilde{x}_m)_i = 0 \end{cases} \quad \forall i = 1, \cdots, n_m \tag{6}$$

Lemma 2. *A feasibility inequality (5) is valid if the sign condition (6) is satisfied.*

Using the above condition, one can find valid Benders cuts from the large family of inequalities specified by (4) and (5).

While any valid cut can be added to the master problem, it is desirable to find one that is as tight as possible with respect to the tentative solution $(\bar{y}, \bar{z}_m)$. Formally, a *tightest* valid optimality cut with respect to $(\bar{y}, \bar{z}_m)$ is defined as a valid cut

$$(b_m - A_m y)^T \tilde{u}^* + \tilde{x}_m^{*T} \tilde{v}^* \leq z_m$$

such that

$$(b_m - A_m \bar{y})^T \tilde{u}^* + \tilde{x}_m^{*T} \tilde{v}^* = \max_{\tilde{x}_m, \tilde{u}, \tilde{v}} \{(b_m - A_m \bar{y})^T \tilde{u} + \tilde{x}_m^T \tilde{v} : \text{ s.t.}(6)\}$$

The *tightest* valid feasibility cut is defined similarly. In other words, it is a matter of choice of $\tilde{x}_m$ and $(\tilde{u}, \tilde{v})$, to maximize the left hand side value of the cut (with y instantiated to $\bar{y}$), giving a tightest cut with respect to $(\bar{y}, \bar{z}_m)$.

4.2   Cut Generation Programs

To elicit a valid optimality or feasibility cut, one needs to find out an assignment $\tilde{x}_m$ and a dual feasible value $(\tilde{u}, \tilde{v})$ such that the sign condition is satisfied. The sign condition (6) can be expressed as the following constraints:

$$\begin{cases} (\tilde{x}_m)_i \tilde{v}_i \leq 0 \\ (1 - \tilde{x}_m)_i \tilde{v}_i \geq 0 \end{cases} \quad \forall i \in 1, \cdots, n_m \tag{7}$$

To find a tightest cut, one could maximize the left hand side value with y instantiated to $\bar{y}$:

$$\max_{\tilde{x}_m, \tilde{u}, \tilde{v}} (b_m - A_m \bar{y})^T \tilde{u} + \tilde{x}_m^T \tilde{v} \tag{8}$$

---

[1] The proofs of all lemmas are given in the appendix.

Therefore, a tightest valid optimality cut can be generated using the dual constraint from $\mathbf{DSP}_g^m{}_\mathbf{F}(\bar{\boldsymbol{y}}, \tilde{\boldsymbol{x}}_m)$, the sign condition constraints (7) and the objective function (8):

$$\mathbf{CGP}_g^m(\bar{\boldsymbol{y}}) : \max_{\tilde{\boldsymbol{x}}_m, \tilde{\boldsymbol{u}}, \tilde{\boldsymbol{v}}} (\boldsymbol{b}_m - \boldsymbol{A}_m\bar{\boldsymbol{y}})^T \tilde{\boldsymbol{u}} + \tilde{\boldsymbol{x}}_m^T \tilde{\boldsymbol{v}}$$

$$s.t. \begin{cases} (\tilde{\boldsymbol{x}}_m)_i \tilde{\boldsymbol{v}}_i \leq 0 & \forall i \in 1, \cdots, n_m \\ (1 - \tilde{\boldsymbol{x}}_m)_i \tilde{\boldsymbol{v}}_i \geq 0 \; \forall i \in 1, \cdots, n_m \\ \boldsymbol{B}_m^T \tilde{\boldsymbol{u}} + \tilde{\boldsymbol{v}} \leq \boldsymbol{d}_m \\ \tilde{\boldsymbol{u}} \geq 0, \tilde{\boldsymbol{v}} : \text{free} \\ \tilde{\boldsymbol{x}}_m \in \{0, 1\}^{n_m} \end{cases}$$

A tightest valid feasibility cut can be generated using the dual constraint from $\mathbf{DSP}_g^m{}_\mathbf{F}(\bar{\boldsymbol{y}}, \tilde{\boldsymbol{x}}_m)$, the sign condition constraints (7) and the objective function (8):

$$\mathbf{HCGP}_g^m(\bar{\boldsymbol{y}}) : \max_{\tilde{\boldsymbol{x}}_m, \tilde{\boldsymbol{u}}, \tilde{\boldsymbol{v}}} (\boldsymbol{b}_m - \boldsymbol{A}_m\bar{\boldsymbol{y}})^T \tilde{\boldsymbol{u}} + \tilde{\boldsymbol{x}}_m^T \tilde{\boldsymbol{v}}$$

$$s.t. \begin{cases} (\tilde{\boldsymbol{x}}_m)_i \tilde{\boldsymbol{v}}_i \leq 0 & \forall i \in 1, \cdots, n_m \\ (1 - \tilde{\boldsymbol{x}}_m)_i \tilde{\boldsymbol{v}}_i \geq 0 & \forall i \in 1, \cdots, n_m \\ \boldsymbol{B}_m^T \tilde{\boldsymbol{u}} + \tilde{\boldsymbol{v}} \leq \boldsymbol{0} \\ \boldsymbol{0} \leq \tilde{\boldsymbol{u}} \leq \boldsymbol{1}, -\boldsymbol{1} \leq \tilde{\boldsymbol{v}} \leq \boldsymbol{1} \\ \tilde{\boldsymbol{x}}_m \in \{0, 1\}^{n_m} \end{cases}$$

However, the above nonlinear mixed integer programs can be simplified to linear cut generation programs. Define $\tilde{\boldsymbol{v}}_i^+ \equiv (1 - \tilde{\boldsymbol{x}}_m)_i \tilde{\boldsymbol{v}}_i$ and $\tilde{\boldsymbol{v}}_i^- \equiv (\tilde{\boldsymbol{x}}_m)_i \tilde{\boldsymbol{v}}_i$. Obviously, $\tilde{\boldsymbol{v}}_i = \tilde{\boldsymbol{v}}_i^+ + \tilde{\boldsymbol{v}}_i^-$. By this way the variables $\tilde{\boldsymbol{x}}_m$ and $\tilde{\boldsymbol{v}}$ can be eliminated. A tightest valid optimality cut is generated by the following cut generation program:

$$\mathbf{CGP}'^m_g(\bar{\boldsymbol{y}}) : \max_{\tilde{\boldsymbol{u}}, \tilde{\boldsymbol{v}}^+, \tilde{\boldsymbol{v}}^-} (\boldsymbol{b}_m - \boldsymbol{A}_m\bar{\boldsymbol{y}})^T \tilde{\boldsymbol{u}} + \boldsymbol{1}^T \tilde{\boldsymbol{v}}^-$$

$$s.t. \begin{cases} \boldsymbol{B}_m^T \tilde{\boldsymbol{u}} + \tilde{\boldsymbol{v}}^+ + \tilde{\boldsymbol{v}}^- \leq \boldsymbol{d}_m \\ \tilde{\boldsymbol{v}}^+ \geq \boldsymbol{0}, \tilde{\boldsymbol{v}}^- \leq \boldsymbol{0} \\ \tilde{\boldsymbol{u}} \geq \boldsymbol{0} \end{cases}$$

A tightest valid feasibility cut is generated by the following cut generation program:

$$\mathbf{HCGP}'^m_g(\bar{\boldsymbol{y}}) : \max_{\tilde{\boldsymbol{u}}, \tilde{\boldsymbol{v}}^+, \tilde{\boldsymbol{v}}^-} (\boldsymbol{b}_m - \boldsymbol{A}_m\bar{\boldsymbol{y}})^T \tilde{\boldsymbol{u}} + \boldsymbol{1}^T \tilde{\boldsymbol{v}}^-$$

$$s.t. \begin{cases} \boldsymbol{B}_m^T \tilde{\boldsymbol{u}} + \tilde{\boldsymbol{v}}^+ + \tilde{\boldsymbol{v}}^- \leq \boldsymbol{0} \\ \boldsymbol{0} \leq \tilde{\boldsymbol{v}}^+ \leq \boldsymbol{1}, -\boldsymbol{1} \leq \tilde{\boldsymbol{v}}^- \leq \boldsymbol{0} \\ \boldsymbol{0} \leq \tilde{\boldsymbol{u}} \leq \boldsymbol{1} \end{cases}$$

It is worth noticing that, to generate cuts, it is only necessary to construct the above cut generation programs, but not the fixed subproblems or their duals.

Although the generated Benders cut is valid, it may not be tight enough to cut off the current tentative master problem solution. This causes a problem for an iterative Benders algorithm such as the Algorithm 1. If, in some iteration, the tentative master problem solution is not cut off by any generated cut, then the subsequent iterations will stuck at the same tentative solution. Note that there is no such problem if a branch-and-cut based Benders algorithm, as is suggested in [14, 3], is used, where the master problem is only solved once by a branching procedure and the Benders cuts are accumulated to guide the search.

However, the problem for iterative algorithms can be remedied by excluding the tentative solution from subsequent master problems with a no-good cut. As long as the master problem variables have a finite domain $\mathcal{D}_y$, one can formulate a no-good cut that only excludes $(\bar{\boldsymbol{y}}, \bar{z}_m)$. For example, consider $\mathcal{D}_y = \{0, 1\}^{n_y}$. When the subproblem $\mathbf{SP}^m_g\mathbf{F}(\bar{\boldsymbol{y}})$ is infeasible, the following no-good cut excludes only $\bar{\boldsymbol{y}}$:

$$\sum_{i=1}^{n_y} \bar{\boldsymbol{y}}_i(1 - \boldsymbol{y}_i) + \sum_{i=1}^{n_y}(1 - \bar{\boldsymbol{y}}_i)\boldsymbol{y}_i \geq 1 \tag{9}$$

When the subproblem $\mathbf{SP}^m_g\mathbf{F}(\bar{\boldsymbol{y}})$ is feasible but cannot reach the tentative $\bar{z}_m$, the following no-good cut excludes $(\bar{\boldsymbol{y}}, \bar{z}_m)$:

$$z_m \geq \phi_{\mathbf{SP}} - (\phi_{\mathbf{SP}} - z^L_m)[\sum_{i=1}^{n_y} \bar{\boldsymbol{y}}_i(1 - \boldsymbol{y}_i) + \sum_{i=1}^{n_y}(1 - \bar{\boldsymbol{y}}_i)\boldsymbol{y}_i] \tag{10}$$

where $z^L_m \equiv \sum_{(\boldsymbol{d}_m)_i < 0}(\boldsymbol{d}_m)_i$ is a lower bound of the variable z_m in $\mathbf{MP}$, and $\phi_{\mathbf{SP}}$ is the subproblem's objective value in the current iteration.

4.3   Generating Cuts from $\mathbf{SP}^m_{\mathbf{IP}}(\bar{y}_{jm}, \bar{H})$

Applying the general method to problem $\mathbf{P}'$, we are now able to generate Benders cuts based on the subproblem formulations $\mathbf{SP}^m_{\mathbf{IP}}(\bar{y}_{jm}, \bar{H})$. In this case the subproblems $\mathbf{SP}^m_{\mathbf{IP}}(\bar{y}_{jm}, \bar{H})$ are feasibility problems, and therefore only feasibility cuts will be generated based on the homogeneous duals. Let $\mathcal{M}'$ denote the set of facilities where a cut needs to be generated. The corresponding cut generation programs can be formulated as is following. Note that in practice the IP formulations of the subproblems never need to be explicitly setup or solved.

$$\forall m \in \mathcal{M}' : \mathbf{HCGP}'^m_{\mathbf{IP}}(\bar{y}_{jm}, \bar{H}) :$$

$$\max_{\tilde{u}^A_j, \tilde{u}^B_t, \tilde{u}^C_j, \tilde{v}^+_{jt}, \tilde{v}^-_{jt}} \sum_{j \in \mathcal{J}} \bar{y}_{jm}\tilde{u}^A_j + \sum_{t \in \mathcal{T}} C_m\tilde{u}^B_t + \sum_{j \in \mathcal{J}} \bar{H}\tilde{u}^C_j + \sum_{j \in \mathcal{J}}\sum_{t \in \mathcal{T}_{jm}} \tilde{v}^-_{jt}$$

$$s.t. \begin{cases} \tilde{u}^A_j + \sum_{t'=t}^{t+p_{jm}-1} C_{jm}\tilde{u}^B_{t'} + (t + p_{jm})\tilde{u}^C_j + \tilde{v}^+_{jt} + \tilde{v}^-_{jt} \leq 0 & \forall j \in \mathcal{J}, \forall t \in \mathcal{T}_{jm} \\ -1 \leq \tilde{u}^A_j \leq 1, \ -1 \leq \tilde{u}^B_t \leq 0, \ -1 \leq \tilde{u}^C_j \leq 0 \\ 0 \leq \tilde{v}^+_{jt} \leq 1, \ -1 \leq \tilde{v}^-_{jt} \leq 0 \end{cases}$$

The subscript $_{\mathbf{IP}}$ is used to emphasize that it is derived based on the IP formulation of the subproblem. Variables $\tilde{u}^A_j$, $\tilde{u}^B_t$, $\tilde{u}^C_j$ represent the dual variables associated with the first, second and third constraint of $\mathbf{SP}^m_{\mathbf{IP}}(\bar{y}_{jm}, \bar{H})$ respectively.

Using the optimal values solved from the $\mathbf{HCGP}'^m_{\mathbf{IP}}(\bar{y}_{jm}, \bar{H})$, a Benders cut over the master problem variables can be generated:

$$\forall m \in \mathcal{M}' : \sum_{j \in \mathcal{J}} y_{jm}\tilde{u}^{A*}_j + \sum_{t \in \mathcal{T}} C_m\tilde{u}^{B*}_t + \sum_{j \in \mathcal{J}} H\tilde{u}^{C*}_j + \sum_{j \in \mathcal{J}}\sum_{t \in \mathcal{T}_{jm}} \tilde{v}^{-*}_{jt} \leq 0 \tag{11}$$

The above cut is valid according to Section 4.1 and Section 4.2, and it is eligible to be added to the master problem **MP**.

In case the cut (11) is not tight enough, a no-good cut needs to be formulated if the Algorithm 1 is used. If, on the mth facility, the tentatively assigned jobs cannot be scheduled in the facility at all (i.e. the optimization problem in $\mathbf{SP_{CP}^m}$ is already infeasible), then the no-good cut is given as:

$$\sum_{j \in \mathcal{J}_m} \bar{y}_{jm}(1 - y_{jm}) \geq 1 \tag{12}$$

If the assigned jobs can be scheduled but not within the tentative completion time $\bar{H}$, then the no-good cut is given as:

$$H \geq H_{\mathbf{SP}}^* - H_{\mathbf{SP}}^* \sum_{j \in \mathcal{J}_m} \bar{y}_{jm}(1 - y_{jm}) \tag{13}$$

where $H_{\mathbf{SP}}^*$ is the minimum completion time can be attained by the optimization problem in $\mathbf{SP_{CP}^m}$.

The cut generation phase (step 2.(3)) in the Algorithm 1 can now be specified as the Procedure 2, and the finite convergence of the algorithm is guaranteed.

Procedure 2 Cut Generation Phase for Problem **P'**

2.(3) Cut Generation Phase.
For each m belonging to $\mathcal{M}'$:
(a) construct the cut generation program $\mathbf{HCGP'}_{\mathrm{IP}}^m(\bar{y}_{jm}^{(k)}, \bar{H}^{(k)})$.
(b) generate a valid Benders cut (11) using the solution.
(c) if (11) does not cut off $(\bar{y}_{jm}^{(k)}, \bar{H}^{(k)})$, then generate a no-good cut (12) or (13).
Add the generated cuts to the master problem to construct $\mathbf{MP}^{(\mathbf{k+1})}$; set $k = k + 1$ and go back to phase (1).

Lemma 3. *The Algorithm 1, with its Cut Generation Phase instantiated by the Procedure 2, converges to the optimal solution of* **P'** *in a finite number of iterations.*

5   Computational Experiments

The proposed algorithm is implemented to solve the general minimum completion time problem. Problem instances are randomly generated by a similar way as in [9], but different release and due dates are allowed. The size of a problem instance is specified by the number of facilities M and the number of jobs J. For each problem size configuration, 10 problem instances are randomly generated. The capacity C_m of each facility is set to 10. The consumption of resources C_{jm} is drawn from a uniform distribution on $[1, 10]$. The processing time p_{jm} for each job j on a certain facility m is drawn from a uniform distribution on $[m, 20m]$, rounded to the nearest integer. Thus the average processing speeds of the facilities are different. As the average of $20m$ over all facilities is $10(M + 1)$, the total processing time for the jobs is roughly proportional to $10J(M + 1)/M$ per facility (ref. [9]). The release date r_j for each job is drawn from a uniform distribution on $[1, 10]$, rounded to the nearest integer. The due date d_j is calculated by r_j plus a time window w_j, which is calculated at one third of the value $10J(M + 1)/M$.

The algorithm is implemented in the ECL^{i}PS^e 5.8 [11] platform. The external solver used for solving the master problems and the cut generation programs is the XPRESS-MP 14.27 [5]. The cumulative scheduling subproblems are solved by constraint programming using the `ic_probe_for_scheduling` and `ic_edge_finding` libraries in the ECL^{i}PS^e. For comparison purpose, the test problems are also solved directly by the same external MIP solver. The formulation used for directly solving is $\mathbf{P}$, instead of $\mathbf{P'}$. To investigate the benefits of the Benders cuts of the form (11), we also implemented the hybrid algorithm with only the no-good cuts being generated. To show the effects of the CP component of the hybrid scheme, we implemented the decomposition algorithm with the subproblems solved by the MIP solver using the IP formulation. For all the algorithms, we set the timeout to 1800 seconds.

The computational results are summarized in Table 1. All numbers except for those in columns 'M', 'J' and '#TO' are average values. The first two columns record the problem size. The 'Optimal' shows the average optimal objective values. The computational results of the proposed algorithm are summarized under the heading 'Hybrid'. The solving times and numbers of iterations of the hybrid decomposition algorithm are shown in 'CPU' and '#Iter' respectively. In column 'CGT%', we give the percentage of solving time spent in the cut generation step. The results for the algorithm with no-good cuts only are recorded in 'Hybrid (NGC)'. The solving times and numbers of iterations are shown. The results for the non-hybrid algorithm without using CP for the subproblems (but still using the proposed Benders cuts) are shown in 'Non-hybrid'. As this variation of the algorithm dose not solve all instances within the time limit, we show the number of timeout cases (out of 10) in the column '#TO'. The solving times are given in 'CPU'. For comparison, the performance of directly MIP solving is summarized under 'Direct MIP'. Again as the MIP solver does not solve every problem instance within 1800 seconds, we show in column '#TO' the number of instances (out of 10) for which the MIP solving times out, and 'CPU' gives the solving time. Note that the values with a plus sign are computed using only the instances for which the corresponding solver does not time out. All other values are the average of 10 instances. The unit of times in the table is second.

The results show that the solving times of all methods increase as the problem scales, reflecting the growing complexity of the problem. Using the proposed hybrid method, all the tested problem instances are solved to optimality within the time limit, while directly MIP solving fails finding the optimality for some instances, and there are more timeout cases as the problem size increases. For smaller problems where both algorithms can prove optimality, the decomposition algorithm also spends much less solving time than the MIP solver. The results suggest that the proposed algorithm could be very useful in solving the minimum completion time problems efficiently. For other objectives, similar performance might be expected, but it is subject to further empirical study.

Table 1. Computational Results

M	J	Optimal	Hybrid			Hybrid (NGC)		Non-hybrid		Direct MIP	
			CPU	#Iter	CGT%	CPU	#Iter	#TO	CPU	#TO	CPU
2	8	31.2	0.87	10.7	13.3%	1.02	15.0	0	4.62	0	9.93
2	10	35.8	4.45	15.4	10.7%	6.03	25.3	0	53.06	1	213.29^+
2	12	38.9	131.11	48.0	6.3%	159.43	61.2	4	472.02^+	5	600.67^+
3	10	32.9	3.89	28.8	14.9%	4.03	33.6	0	19.24	0	115.25
3	12	36.2	19.89	53.0	11.5%	24.85	63.7	0	145.48	4	96.51^+
3	14	39.5	129.66	85.0	8.2%	153.59	102.7	3	448.52^+	6	735.87^+
4	12	31.6	4.28	19.6	12.6%	7.66	32.6	0	51.14	5	109.46^+
4	14	34.6	47.26	54.1	7.8%	135.43	83.3	2	427.69^+	8	132.25^+
4	16	34.0	148.97	78.8	7.4%	396.48	105.8	4	209.34^+	7	159.26^+

Next, the effects of incorporating the proposed Benders cuts are studied. Firstly, the results show that the overheads of cut generation account for a small portion of the total solving time, and the percentage decreases as the problem scales. Secondly, compared with the alternative algorithm with no-good cuts only, the algorithm that uses the proposed Benders cuts experiences less iterations. This difference becomes substantial in some larger problems. In terms of solving time, the algorithm also consistently outperforms the one with no-good cuts, in spite of the overheads incurred by solving the cut generation programs. The comparison indicates that the Benders cuts of the form (11) are indeed useful in improving the performance, yet without incurring too much overheads.

Finally, to show the effects of the CP component in the hybrid scheme, we compare the hybrid algorithm with an algorithm that uses MIP to solve both the master problem and the subproblems. The results show that the non-hybrid algorithm is substantially slower than the hybrid one, although its performance is still much better than that of directly MIP solving. Without CP, the algorithm even fails finding the optimal solution (within 1800 seconds) for a few instances. This difference could be attributed to the fact that the scheduling subproblems are often hard to solve, and that the employed CP methods, which are specially designed for single machine scheduling problems, are much more efficient than the MIP solver. The results show that the incorporation of CP solution methods indeed plays an important role in the proposed algorithm, in reducing the solving times for the considered class of problems.

6   Conclusions

This paper presents a hybrid method for the resource constrained scheduling problems. Different models and solvers are used in the three components of the hybrid scheme. In particular, the cut generation uses the dual information based on the integer programming model under a Benders decomposition framework. The approach has been instantiated to an algorithm for the minimum completion time problem. Computational results have shown that the proposed algorithm achieves substantial reduction of solving times, especially for larger problem instances.

References

1. P. Baptiste, C. Le Pape, and W. Nuijten. *Constraint-based Scheduling: Applying Constraint Programming to Scheduling Problems.* Kluwer, 2001.
2. J.F. Benders. Partitioning procedures for solving mixed-variables programming problems. *Numerische Mathematik*, 4:238–252, 1962.
3. A. Bockmayr and N. Pisaruk. Detecting infeasibility and generating cuts for mip using cp. In *Proc.of CPAIOR 03, Montreal, Canada*, 2003.
4. Y. Chu and Q. Xia. Generating benders cuts for a general class of integer programming problems. In M. Rueher J.-C. Regin, editor, *Lecture Notes in Computer Science 3011 – CPAIOR 04*, pages 127–141. Springer-Verlag, 2004.
5. Dash Inc. *Dash XPRESS-MP 14.27 User's Manual*, 2003.
6. H. El Sakkout and M. Wallace. Probe backtrack search for minimal perturbation in dynamic scheduling. *Constraints*, 5(4):359–388, 2000.
7. A.M. Geoffrion. Generalised benders decomposition. *Journal of Optimization Theory and Application*, 10:237–260, 1972.
8. Iiro Harjunkoski and I.E. Grossmann. Decomposition techniques for multistage scheduling problems using mixed-integer and constraint programming methods. *Comp. and Chem. Engineering*, 26:1533–1552, 2002.
9. J.N. Hooker. A hybrid method for planning and scheduling. In M. Wallace, editor, *Lecture Notes in Computer Science 3258 – Principles and Practice of Constraint Programming CP 04*, pages 305–316. Springer-Verlag, 2004.
10. J.N. Hooker and G. Ottosson. Logic-based benders decomposition. *Mathematical Programming*, 96:33–60, 2003.
11. Imperial College London. *ECLiPSe 5.8 User's Manual*, 2004.
12. V. Jain and I.E. Grossmann. Algorithms for hybrid milp/cp models for a class of optimisation problems. *INFORMS Journal on Computing*, 13:258–276, 2001.
13. R. Sadykov. A hybrid branch-and-cut algorithm for the one-machine scheduling problem. In M. Rueher J.-C. Regin, editor, *Lecture Notes in Computer Science 3011 – CPAIOR 04*, pages 409–415. Springer-Verlag, 2004.
14. E.S. Thorsteinsson. Branch-and-check: A hybrid framework integrating mixed integer programming and constraint logic programming. In T. Walsh, editor, *Lecture Notes in Computer Science 2239 – Principles and Practice of Constraint Programming CP 01*, pages 16–30. Springer-Verlag, 2001.
15. M. Turkay and I.E. Grossmann. Logic-based minlp algorithms for the optimal synthesis of process networks. *Comp. and Chem. Engineering*, 20:959–978, 1996.

Appendix: Proofs of Lemmas

Lemma 1:

Proof. To prove the validity, let $(\hat{\boldsymbol{y}}, \hat{z}_1, \cdots, \hat{z}_M)$ be any solution of $\mathbf{MP}_g$ such that the value of $\hat{z}_m$ can be reached by the subproblem parameterized by $\hat{\boldsymbol{y}}$ for each m. We prove that this solution is not cut off by the optimality inequality (4) Consider the mth subproblem. We have:

$$\hat{z}_m \geq \phi(\mathbf{SP}_g^m(\hat{\boldsymbol{y}})) = \min_{\boldsymbol{x}_m}\{\boldsymbol{d}_m^T\boldsymbol{x}_m : \boldsymbol{B}_m\boldsymbol{x}_m \geq \boldsymbol{b}_m - \boldsymbol{A}_m\hat{\boldsymbol{y}}, \boldsymbol{x}_m \in \{0,1\}^{n_m}\} \quad (14)$$

where $\phi(\cdot)$ is the value function of a program. Then there must exist a value $\hat{x}_m$ such that $\hat{z}_m \geq \phi(\mathbf{SP}_g^m{}_\mathbf{F}(\hat{y}, \hat{x}_m)) = \phi(\mathbf{DSP}_g^m{}_\mathbf{F}(\hat{y}, \hat{x}_m))$, i.e., $\hat{z}_m \geq (b_m - A_m\hat{y})^T\hat{u} + \hat{x}_m^T\hat{v}$, where $(\hat{u}, \hat{v})$ is an *optimal* solution of $\mathbf{DSP}_g^m{}_\mathbf{F}(\hat{y}, \hat{x}_m)$.

Note that the feasible region of $\mathbf{DSP}_g^m{}_\mathbf{F}(y, x_m)$ is independent of the value of y and x_m. Therefore, the dual value $(\tilde{u}, \tilde{v})$ (used by the optimality inequality (4)), which is feasible for $\mathbf{DSP}_g^m{}_\mathbf{F}(\bar{y}, \tilde{x}_m)$, is also a *feasible* solution of $\mathbf{DSP}_g^m{}_\mathbf{F}(\hat{y}, \hat{x}_m)$. This implies that

$$(b_m - A_m\hat{y})^T\tilde{u} + \hat{x}_m^T\tilde{v} \leq (b_m - A_m\hat{y})^T\hat{u} + \hat{x}_m^T\hat{v} \leq \hat{z}_m$$

Due to the assumption (6), we have $\tilde{x}_m^T\tilde{v} \leq \hat{x}_m^T\tilde{v}$ no matter which binary values the variables $\hat{x}_m$ take. Thus,

$$(b_m - A_m\hat{y})^T\tilde{u} + \tilde{x}_m^T\tilde{v} \leq (b_m - A_m\hat{y})^T\tilde{u} + \hat{x}_m^T\tilde{v} \leq \hat{z}_m$$

i.e. the optimality inequality (4) is satisfied by $(\hat{y}, \hat{z}_m)$.   □

Lemma 2:

Proof. Similar to the proof of Lemma 1, let $(\hat{y}, \hat{z}_1, \cdots, \hat{z}_M)$ be any feasible solution of $\mathbf{MP}_g$. We prove that it is not cut off by the feasibility inequality (5). Consider the mth subproblem. Since it is feasible, there must exist a value $\hat{x}_m$ such that $\mathbf{SP}_g^m{}_\mathbf{F}(\hat{y}, \hat{x}_m)$ is feasible, and therefore the corresponding homogeneous dual $\mathbf{HDSP}_g^m{}_\mathbf{F}(\hat{y}, \hat{x}_m)$ has a non-positive optimal value, i.e., $(b_m - A_m\hat{y})^T\hat{u} + \hat{x}_m^T\hat{v} \leq 0$, where $(\hat{u}, \hat{v})$ is an *optimal* solution of $\mathbf{HDSP}_g^m{}_\mathbf{F}(\hat{y}, \hat{x}_m)$. Then applying the same reasoning as in the proof of Lemma 1, the conclusion follows.   □

Lemma 3:

Proof. First we show that the algorithm terminates in a finite number of iterations. Due to the cut generation procedure, in each iteration except for the last, the current tentative solution from the master problem is cut off, and the value of $(\bar{y}_{jm}, \bar{H})$ is different in different iterations. Since the master problem variables have a finite domain, the algorithm has to terminate in finite iterations.

Next we show that the algorithm returns the optimal solution of the original program $\mathbf{P}'$. Note that the master problem $\mathbf{MP}^{(k)}$ is always a relaxation of $\mathbf{P}'$ as all the cuts added are valid. If the Algorithm 1 terminates in step 2.(1), then the original problem has to be infeasible as well. If it terminates in step 2.(2) of some iteration k, then the tentative assignment in this iteration renders feasible subproblems on every machine, and the value of $\bar{H}^{(k)}$ can be attained by the subproblems. As $\mathbf{MP}^{(k)}$ is a relaxation of $\mathbf{P}'$, $\bar{H}^{(k)}$ is always a lower bound of the optimal solution of $\mathbf{P}'$. Thus, $\bar{H}^{(k)}$ is a minimum completion time that can be achieved. The current assignment is the optimal solution for variables y_{jm}. The optimal starting times of the assigned jobs can be obtained from the subproblems.   □

On the Minimal Steiner Tree Subproblem and Its Application in Branch-and-Price

Wilhelm Cronholm[1], Farid Ajili[1], and Sofia Panagiotidi[2]

[1] IC-Parc, Imperial College London, London SW7 2AZ, UK
{w.cronholm, f.ajili}@imperial.ac.uk
[2] London e-Science Centre, Imperial College London, London, SW7 2AZ, UK
sp1003@imperial.ac.uk

Abstract. The minimal Steiner tree problem is a classical NP-complete problem that has several applications in the communication and transportation sectors. It has recently emerged as a subproblem in decomposition techniques such as column generation and Lagrangian schemes. This has set new computational challenges to the state of the art solving approaches. Our goal is to improve on existing branch-and-cut algorithms so that our approach successfully serves as a fast subproblem solver in a decomposition context. Compared with existing literature, our technical contributions include 1) a new preflow-push cutting strategy, revisiting a little known graph algorithm, that halves the runtime of the separation step, and 2) a branching scheme that fairly balances the search tree and speeds up the search. An evaluation in a multicast design application shows that the algorithm enhances a column generation hybrid. Moreover, our approach offers a significant speedup factor on a publicly available set of challenging Steiner tree benchmarks.

Keywords: networks, preflow-push algorithms, branch-and-cut, Steiner trees.

1 Introduction

The minimal Steiner Tree Problem (STP) consists of determining the least-cost tree that connects a given set of nodes in a graph. This NP-complete problem was originally formulated by Hakimi [10]. Since then, it has received considerable attention in the literature as it has a wide range of applications, e.g. network optimisation, distribution systems and VLSI layout [10, 18]. It is beyond the scope of this paper to cover all the STP literature. Surveys can be found in e.g. [10, 18]. Although useful in its own right, several network design models embed the STP as a substructure. It then emerges as a *subproblem* in decomposition algorithms such as column generation or Lagrangian relaxation.

If the aim is to compute fairly good solutions to the *stand-alone* Steiner tree problem within a reasonable time, then it has been recommended to use a variety of greedy "tree heuristics" [10, 18]. However, if the STP appears as a subproblem in a branch-and-price setting [6] or Lagrangian context [5, 19],

R. Barták and M. Milano (Eds.): CPAIOR 2005, LNCS 3524, pp. 125–139, 2005.
© Springer-Verlag Berlin Heidelberg 2005

existing algorithms require exact STP solvers that deliver an optimal tree. In branch-and-price for instance, finding the most profitable tree not only offers the best pricing strategy, but also enables exploiting the lower bound and achieving early termination and cost-based filtering at *any* column generation iteration. This is because the column generation bound plus the subproblem reduced cost is a valid lower bound even at *non-optimal* iterations.

Several authors have proposed approaches for solving the minimal STP, including Lagrangian relaxation [10] and branch-and-cut [3, 12, 15]. Previous empirical evaluation has suggested that the fastest approach is the branch-and-cut solver of Koch & Martin [12]. The authors showed that their solver is superior to other approaches, as it succeeded in solving almost all existing benchmarks for the first time. Several decomposition methods to solve the multicast network design have embedded [12] as a subproblem solver, e.g. [6] deploys column generation and [5, 19] exploit a special Lagrangian scheme.

A necessary ingredient for a decomposition method to be sufficiently competitive is an efficient subproblem solver. However, we have noticed that the hybrid branch-and-price algorithm of [6] spends nearly two thirds of the runtime in solving the Steiner tree subproblems in the pricing phase. Our performance analysis has demonstrated that the separation (i.e. cutting) phase of the branch-and-cut solver [12] is the largest computational bottleneck since the algorithm spends at least half of its runtime in identifying the inequalities whose addition cuts off the relaxed solution.

Another drawback of existing literature is that most exact STP algorithms, [3, 12, 15] included, use traditional Integer Programming (IP) branching on single arc variables. This is not well-suited, as it potentially leads to an unbalanced (binary) search tree: the weak branch forbidding an edge often barely changes the relaxation.

This paper proposes an innovative and non-trivial variation of branch-and-cut STP algorithms that successfully serves as a subproblem solver in decomposition methods. The solver also substantially improves the best known runtimes of a publicly available set of benchmarks as a stand-alone Steiner tree solver.

The contents of the paper are as follows. Section 2 outlines related work and our contributions. Section 3 overviews our branch-and-cut solver, in particular an enhanced branching scheme and a new cost-based filtering rule that exploits reduced costs. Section 4 discusses some separation issues and proposes a new cutting strategy. Section 5 presents a computational discussion of our approach under two use cases 1) run it as a stand-alone STP solver, 2) invoke it as a branch-and-price subproblem. Section 6 concludes the paper.

2   Related Work and Overview of Contributions

Exact algorithms for the minimal Steiner tree problem include Lagrangian relaxation and branch-and-cut, e.g. [10]. Because [12] is considered the most efficient algorithm for solving STP to optimality, multicast network design approaches [5, 6, 19] exploiting decomposition methods have all used it. In [6] it is reported that the invocations of [12] are by far the largest bottleneck on some challenging multicast design datasets. We also note that in the results of [19] around 45 minutes are spent at the root node of the search tree, even for quite small networks and reasonable number of multicast groups. This is likely spent in solving STP subproblems.

Most Steiner tree solvers rely heavily on reduction techniques [10, 12, 18] to reduce graph size (i.e. edge set and vertex set). These act as a preprocessing step that includes (resp. excludes) edges/nodes that do (resp. not) belong to at least one minimal Steiner tree.

The preprocessing, which often requires careful implementation, exploits special configurations in the graph. Since those tests often change the set of edges and/or nodes, they are difficult to accommodate in a decomposition-based search as they conflict with the much needed incrementality of the subproblem solver: we may have to re-setup a subproblem model/solver as soon as the graph changes. Indeed, some changes, such as node contraction, could result in deleting and adding variables and make several constraints invalid. In the decomposition methods [5, 6, 19], applying "non-logical" reduction tests every time the subproblem is solved can introduce a substantial overhead as the constraint part of the subproblem is kept unchanged throughout the search: only the cost vector changes.

Painful reduction techniques are not included in our algorithm. Because we are losing their strength, it is crucial to somehow compensate for this. It is done through 1) designing a new separation phase by carefully merging an algorithm proposed in [9] to compute a generalised form of a cut in a graph, 2) introducing a harmless cost-based pruning rule that reasons about optimality to exclude nodes from the solution. In fact, our focus on speeding up the separation phase began when our experimentation clearly revealed that it is very costly.

Despite the efficiency of [9] for computing the minimum unrestricted cut [14], this algorithm is not widely known in the literature. Koch & Martin [12] briefly mention that [9] is modified and used at the separation phase of their STP solver. However, they gave no details of how the algorithm is to be used within their solver and failed to provide a computational evaluation whether such an approach is beneficial. Moreover, recent publications [18, 19] do not seem to be aware of the idea of using [9].

From this regard, our contributions are two-fold. We demonstrate that using [9] as a black-box, or through a trivial adaptation, does not work and identify a straightforward counter-example. Our technical contribution is to detail a careful adaptation of [9]. In fact, the resulting preflow-push cutting strategy could be used in other areas such as survivable networks [11] and the travelling salesman methods, e.g. [16].

Strengthening the cutting strategy of the branch-and-cut was useful in enhancing the performance of a a branch-and-price method even with "easy" STP instances. Our evaluation in the stand-alone setting showed that the speedup factor is much clearer as soon as the instances get harder.

Yet another serious drawback of existing branch-and-cut approaches lies in the branching itself. Our runs showed that branching on a binary variable in the integer programming fashion tends to create search branches that cause unbalanced changes on the linear relaxation. To overcome this, we suggest a new branching scheme that leads to branches potentially having the same likelihood of containing the best solution.

3   Branch-and-Cut Approach

In this section we present our improved branch-and-cut approach. It builds on top of previous literature such as Koch & Martin [12]. We first introduce some notation and definitions.

Let $G = (V, A)$ be a directed graph, where V is its node set and A is its set of arcs. Let $n = |V|$ and $m = |A|$. An arc in A from node i to node j, is denoted (i, j). If H is a subset of nodes, then we denote by $\bar{H} = V \setminus H$ its set complement in V. A cut $(H, \bar{H})$ is a non-trivial partition of V. An $S - t$ cut is a cut $(H, \bar{H})$ such that $S \subseteq H$ and $t \in \bar{H}$. If $S = \{s\}$ then we use the shorter notation $s - t$ cut. Usually H is called the *source side* and $\bar{H}$ the *sink side* of the cut. The set of arcs in the cut is denoted by $\delta(H, \bar{H})$, so $\delta(H, \bar{H}) = \{(i, j) | i \in H, j \in \bar{H}\}$. If w is a set of weights attached to the arcs A we denote by $w(H, \bar{H})$ the weight of the cut, i.e. $w(H, \bar{H}) = \sum_{a \in \delta(H, \bar{H})} w_a$. The minimum $s - t$ cut problem is that of finding the minimum weight $s - t$ cut.

The minimal STP consists of determining the least-cost tree in a graph, that connects a given set of nodes. We use a similar *directed* version instead since it is known to give a tighter bound [4]. The directed version is named the *Steiner arborescence* problem. The transformation from the undirected Steiner tree problem to the directed problem is simple: replace, in the obvious way, every edge by two directed arcs with the same cost and select one terminal as the root. It is well known that this does not alter the solution set. For convenience we shall refer to an arborescence with the more general term tree throughout this document.

The Steiner tree problem can be formulated as follows: given a directed graph $G = (V, A)$, non-negative weights $w : A \to \mathbb{R}^+$, a non-empty subset $T \subseteq V$ of terminals, and a root $r \notin T$, find a directed subset of arcs R such that there is a path from r to every terminal and $\sum_{a \in R} c_a$ is minimised. A node not in $T \cup \{r\}$ is called a *nonterminal*.

We introduce decision variables: let x_a be a binary variable that is 1 if and only if arc a is used in the Steiner tree.

The solver uses a standard branch-and-cut approach outlined in Figure 1. The inner loop is a separation procedure which finds violated inequalities or proves that none exits. The algorithm calls the reduction technique described in Section 3.3. The LP is a continuous relaxation of the cut formulation of a tree. It is initialised with the flow inequalities described in [12]. The search separates Steiner cuts and adds them to the relaxation. A *Steiner cut* $(H, \bar{H})$ is a cut that separates the root r from *at least* one terminal: $r \in H$ and there exists a $t \in T$ such that $t \in \bar{H}$. To each Steiner cut $(H, \bar{H})$ in G is associated the following inequality:

Branch-and-Cut:
apply reduction techniques
initialise search
repeat
  select a leaf from the tree and consider the associated LP
  repeat
    solve the LP relaxation
    separate violated inequalities and add them to the LP
  until there are no violated inequalities
  branch if necessary otherwise remove the leaf from the tree
until there are no leaves in the branch-and-bound tree

Fig. 1. A general branch-and-cut algorithm

$$\sum_{a \in \delta(H, \bar{H})} x_a \geq 1 \quad . \tag{1}$$

The Steiner cut inequalities (1) ensure reachability from the root to every terminal. At each iteration of the separation algorithm described in Section 3.1, the LP-relaxation suggests a value $\bar{x}_{ij}$ for every variable x_{ij}. These values are then used to identify violated inequalities to be added to the LP to cut off the current solution. If no violated inequality exists the search branches as in Section 3.2.

3.1   Separation of Violated Inequalities

The separation algorithm is the most critical step in branch-and-cut. It finds, if any, relaxed inequalities (1) violated by the suggested solution $\bar{x}$ of the LP relaxation. The inequalities (1) can be separated as follows: for each $t \in T$ find the minimum cost $r - t$ cut in G using $\bar{x}$ as weights. This is commonly done with any max-flow/min-cut algorithm, see e.g. [1]. If, for some t, the cost w_{r-t}^* of the minimum $r - t$ cut in G is below 1 then the associated inequality is violated. If w_{r-t}^* exceeds 1 for each terminal t, then there is no violated inequalities and the search branches. Inequalities are added only if they are violated by a small tolerance. In fact, we use the "creep-flow" strategy [12] that favours least-cost $r - t$ cuts having minimum number of arcs. This makes the separation harder, but in practice is compensated for by the strength of the inequalities.

Note that all Steiner cuts are found in the input graph G. That is because the graph is not affected by the branching decisions of Section 3.2. Therefore the cuts generated are valid throughout the branch-and-cut search.

Section 4 describes a fast separation algorithm by adapting [9]. It considerably enhances the performance of the branch-and-cut search.

3.2   Branching Strategy

Branching is an essential part of the exact algorithm and has to be carefully designed to reach optimal performance. There are several requirements on the branching strategy. In particular it has to:

- substantially affect the relaxation and the LP cost
- yield a balanced search tree, with equal likelihood of the best solution being in the different branches.

In [12] IP branching is used; it branches on a single arc variable. Disallowing an arc is unlikely to change the problem significantly since most arcs will not be in the optimal tree anyway. On the other hand, including an arc has a larger impact on the LP since it severely restricts the number of feasible (shortest) trees. This results in an unbalanced search tree in practice.

To overcome this difficulty we apply a new hierarchical branching strategy. It commits decisions to several variables by first focusing on node membership. This is motivated by the fact that the difficulty in the STP is to decide which nodes are to appear in the minimal tree. Vertex oriented branching is also mentioned in [10] and used in [18]. However, it has not been used in the context of branch-and-cut.

For each node v, its "likelihood" h_v of being a member of the minimal Steiner tree is estimated by $h_v = \sum_{(j,v) \in A} \bar{x}_{(j,v)}$. In a primal solution, h_v is either 0 or 1. The nonterminal having the maximal integrality violation of h_v is selected, i.e. the node v that minimises $|0.5 - h_v|$ over the nonterminals for which $1 > h_v > 0$. The branching first includes v and excludes v on backtrack. It is accomplished by posting the following inequality with $b = 1$ on the forward branch and $b = 0$ on backtrack:

$$\sum_{(j,v) \in A} x_{jv} = b \quad . \tag{2}$$

The exclusion effectively disallows all arcs having one end in v as a result of flow inequalities [12]. Note that the branching decision is easily incorporated in the LP since each branching decision is the same as adding a cut. Clearly both branches cut off the relaxed solution $\bar{x}$ provided that there exists a nonterminal v such that h_v is fractional.

Whenever the node branching does not apply, $\bar{x}$ is often primal feasible. However, in rare cases it might not be. An example is when there are several optimal Steiner trees spanning the same nodes but crossing different arc sets. In such a case, we resort to IP branching for completeness by branching on the arc variable x_a that minimises $|0.5 - \bar{x}_a|$ over the set of fractional $\bar{x}_a$. The search sets $x_a = 1$ and $x_a = 0$ on backtrack.

3.3   A New Cost-Based Reduction Technique

Preprocessing that alters the graph is difficult to implement when the STP is a subproblem. However, there are still harmless reductions that can be exploited in a decomposition context. Here, the focus is on a reduction technique that only removes nodes/arcs that can not participate in *any* optimal Steiner tree. This type of reduction can be easily encoded by setting some arc variables to zero in the LP.

We now introduce a new reduction rule that exploits reduced costs extracted from solving the linear programming relaxation. We observed that applying standard reduced cost fixing is weak as the cut formulation of a tree is known to be highly degenerate. This same observation is supported by the results of [12].

Our technique is stronger than the standard reduced cost fixing. Its design has been inspired by our previous work in strengthening optimality reasoning in network routing [7]. We exploit the additivity of the reduced costs and the tree structure to infer an estimation of the cost incurred on the relaxation lower bound LB by including a node into the Steiner tree.

The incumbent cost of the branch-and-cut is denoted by UB. Let r_a denote the reduced cost of variable x_a; further let $r_a^+ = \max\{0, r_a\}$. Consider a nonterminal node v that the relaxed solution suggests not to be in the optimal tree, i.e. $h_v = 0$. If v was to be included in the minimal Steiner tree (i.e. the value of h_v switches from 0 to 1), then some path p from the root r to v must exist in any feasible tree. All arcs a in p such that $\bar{x}_a = 0$ have to be included. The cost of including a is r_a (resp. 0) if $\bar{x}_a = 0$ (resp. $\bar{x}_a > 0$). Thus, the incurred cost of ensuring x_a is 1 is r_a^+. The incurred cost of including the path p is the sum of r_a^+ for each a in p. A valid under-estimate of that is the length, say $\psi_v(r)$, of the shortest path from r to v in the graph where the weight of an arc a is r_a^+. This is captured by the following reduction test:

$$\text{if} \ \ LB + \psi_v(r) \geq UB \ \ \text{then} \ \sum_{(j,v)\in A} x_{jv} = 0 \tag{3}$$

This rule can be used every time the relaxation is re-optimised. The effect of (3) can be enhanced by using a CP-relaxation [5]. This would enable the inference of even more fixings. We have used the same constraint programming store as have been used in [5]. It makes some significant inference at the root, but less fixings afterwards. The reduction rule (3) has not been extensively tested and was therefore switched off in our experimentation.

4   A More Efficient Separation Algorithm

This section describes a new separation algorithm for the branch-and-cut Steiner tree solver. Normally, as described in Section 3.1, the separation is done by solving $|T|$ min-cut problems *independently*. Here we adapt ideas presented by Hao & Orlin [9] to be able to solve the separation problem much more efficiently in one go. The speed up is gained by re-using information from previous min-cut computations.

The rest of this section first briefly presents the Hao & Orlin algorithm (HO) in Section 4.1 and then explains in Section 4.2 the adaptation needed to find Steiner cuts.

4.1   Overview of the Hao and Orlin Algorithm

This section briefly describes HO [9]. It is assumed the reader is familiar with the preflow-push algorithm, first presented by Goldberg & Tarjan [8] to some extent. For general graph topics we refer the reader to [1]. Most of the materials in this section are from [9].

HO is a modified version of the algorithm by Goldberg & Tarjan [8]. It finds the minimum cut in a directed graph where only a set S of sources are specified to be in the source side. We denote this by a $S - *$ cut.

The unrestricted min-cut problem is to find the minimum weight cut in the graph without restrictions on any nodes. It has many applications in network reliability and is useful as a separation algorithm for the travelling salesman problem. This can easily be solved by $2(n-1)$ $s-t$ cut computations. However, for a directed graph, the unrestricted min-cut problem can be solved by two invocations of HO, the second where all arcs are reoriented. The total runtime is comparable to the time of one $s-t$ computation.

Recall that the preflow-push algorithm works with a *preflow*. A preflow is a flow except that the entering flow of a node v can exceed the leaving flow v. The *excess* is the difference between the last two. The preflow-push algorithm pushes flow from *active* nodes, nodes with positive excess, along estimated shortest paths in the *residual graph*, the "remaining flow graph". The shortest path estimation is done by *distance labels* of the nodes. The distance label of a node is a lower bound on the length of a path from the node to the sink in the residual graph. An overview of the preflow-push algorithm can be found on the right-hand side of Figure 2. During a min-cut computation of the preflow-push algorithm it is first assumed that all nodes, apart from the source node, are on the sink side W of the cut, but as the computation progresses nodes are transferred to the source side D. Specifically they are moved to D when there is no longer a path from the node to the sink in the residual graph. The algorithm maintains the invariant that there is no arc of the residual graph directed from any node in D to any node in W.

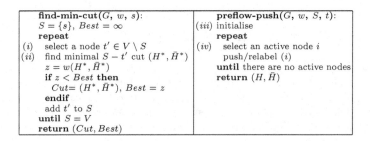

Fig. 2. Outline of the Hao and Orlin algorithm for computation of the minimum $S - *$ cut

HO exploits the preflow-push algorithm to find the least-cost $S - *$ cut. An overview can be found on the left-hand side of Figure 2. The preflow-push algorithm is invoked at (*ii*). The faster runtime of HO stems from the reuse of information from min-cut computations: the ending state of the last preflow-push computation is reused (at (*iii*)). Also, information from the min-cut computation is used to force an ordering in the selection of sinks at (*i*). Specifically, the distance labels are re-used, as well as information about nodes that have been transferred to the source side. These nodes are divided into "dormant" layers according to the time they where moved to the source side. These nodes will be moved back, "awakened", in reverse order at a later stage so all nodes will be selected at (i) during the execution. The new sink is selected as the node in the source side of the last computation with the smallest distance label. If there is no awake node then a layer of dormant nodes are woken up and the selection procedure is repeated. The algorithm terminates when all nodes are in S.

We now present the problems with adapting HO to the context of finding violated inequalities of form (1). In the light of that, we derive a sound adaptation.

4.2   Adapting HO for Computing Steiner Cuts

First, HO cannot be used as it is, since the cut found is not guaranteed to be a Steiner cut. Indeed, it is likely that the minimum cut will have weight 0 as is the case when the sink side consists of a set of nonterminals that are not suggested should be used. Clearly, this is not a Steiner cut.

A trivial modification is to only consider cuts that are found whenever the sink is a terminal. However, this does not guarantee finding a violated cut if there is one [17]. The reason is that the algorithm does not necessarily find the minimal $s - t$ cut when t is the sink. Instead, properties of the algorithm certify that the minimal $s - t$ cut has been found in previous iterations, if the current cut is not the one sought after. However, when considering Steiner cuts these properties do not hold since there is a difference between terminals and nonterminals. The example below illustrates this. In fact, it is not enough to consider *all* Steiner cuts generated during the execution of the algorithm since the optimal cut can be missed by moving a nonterminal to S too early.

Example 1. Consider Figure 3. Node s is the root, t is a terminal and v is a nonterminal. During the execution of HO, v is selected as the first sink. The minimal $r - v$ cut is found, which is $(\{s, t\}, \{v\})$ with a weight of 3. However, it is not a valid Steiner cut since there is no terminal on the sink side. Next, v is moved to the set of sources S and t is selected as the next sink. In this iteration the cut $(\{s, v\}, \{t\})$ is returned with a weight of 13. The cut is a valid Steiner cut since t is on the sink side. It is also the minimal Steiner cut found by the algorithm. However, as can easily be seen in Figure 3, the optimal Steiner cut is $(\{s\}, \{t, v\})$ with a weight of 4. This cut was missed since v was moved into S too early and the cut found when v was sink was not a Steiner cut. ∎

It follows from Example 1 that in order for the adaptation to work we need to make sure that every cut found is a Steiner cut. This is ensured by only selecting terminals as sinks at step (i). Then every minimum cut found is a valid Steiner cut. However, there is a problem with this approach. The problem relates to the distance labels. Recall that preflow-push algorithms use distance labels of the nodes to approximate the minimum number of arcs to reach the sink. When a terminal is selected as sink it might not have the minimum distance label, so other awake nodes may have smaller distance labels. This is a problem for two reasons.

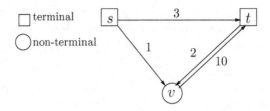

Fig. 3. Counter-example showing how the Hao and Orlin algorithm could miss the optimal Steiner cut. Arcs are labelled with costs

First, it interferes with the widely used and efficient gap-relabelling heuristic [2], which is included in the relabel procedure in [9]. The heuristic moves nodes that are known to be disconnected from the sink into the dormant set by detecting "gaps" in the distance labels. This is much quicker than the naive solution which only moves nodes to the dormant set as the distance label increases above a limit (usually n). However if nodes are allowed to have a smaller distance label than the sink, a gap does not guarantee that nodes are disconnected.

Second, the number of nodes that are not in the source set will be larger than the difference between n and the distance label of the current sink, if it is kept unchanged. This means that the aforementioned limit is hard to establish. The result is a bad performance since efficient data structures as well as the runtime complexity rely on a tight bound on the limit, see [17].

The interference with the gap-relabelling is easy to handle by taking the distance label of the current sink into account. However, the second problem still remains. We overcome both difficulties by setting the distance label of the new sink to 0 and resetting all the awake nodes' distances by the exact distance, in terms of number of arcs, to the new sink in the residual graph. This is called global relabelling and is an important feature in competitive preflow-push algorithms [2].

To speed up the computation even more, we always stop and select a new sink whenever we detect that the minimum cut for the current sink will not be the minimum one, for instance when we have already found a smaller cut. Also, we store all violated sub-optimal cuts found throughout the iterations, since they may benefit the LP.

To summarise: our change to HO is that we select the terminal in the awake set with the minimum distance label. We then reset the distance labels for that node to 0 and recalculate the exact distances for all other awake nodes. We transfer nodes that do not have a residual path to the new sink into the dormant set. If necessary we wake up layers from the dormant set when selecting a new sink until there is a terminal in the awake set, even if there are non-terminal nodes in the awake set.

It is easy to see that the algorithm is still correct with this modification. The dormant nodes do not play a role in finding the current minimum cut so changing the distance labels of the awake nodes does not alter the correctness. Also, whenever nodes are woken up from the dormant set, they are given a new distance label. We note that we might have to wake up several layers before a terminal is found.

5   Experimental Results

This section presents a computational discussion of our approach under two use cases 1) in a stand-alone context and 2) invoked in a branch-and-price solver. The aim is to assess the contribution of the new branching strategy and the improved separation. The techniques have been implemented in the Constraint Logic Programming language ECLiPSe.

We evaluate four parameter settings: n-ff, n-ho, a-ff and a-ho, where the first letter of the parameter setting describes the branching strategy: IP branching (a) or our branching (n). The second part describes the separation strategy: $|T|$ max-flow/min-cut computations (ff) or our strategy (ho).

5.1   Evaluation in Stand-Alone Context

Results for 42 instances from SteinLib [13] are reported. SteinLib is a set of publicly available benchmarks for the Steiner tree problem. The results can be seen in Table 1. SP-t designates the total separation time in seconds, LP-t is the total LP time in seconds, #N presents the size of the search tree and the upper bound UB is labelled with a star if the instance was proved optimal. First, we compare the contribution of the new separation strategy. Table 1 reveal that our cutting strategy is clearly superior. Indeed, the results show that the ho-strategy always enables significantly faster optimality proofs. Compared to a-ff, the improved algorithm a-ho finds a better primal solution on 2 instances, proves optimality on 2 instances when a-ff fails to do so. It also proves optimality on 1 instance when a-ff fails to find the optimal solution at all. In fact, the n-ho strategy manages to significantly improve the best known runtime on one of the most challenging instances (bipe2u) in SteinLib. The latter instance was only recently solved to optimality by [18]. It is not surprising that the ho strategy requires more separation iterations since it only generates the most violated Steiner cuts. The ff strategy generates almost one $r-t$ cut for each $t \in T$ at every iteration. However, the majority of the latter cuts are useless because for those examples where both cut strategies proved optimality ho required considerably less cuts in total. This supports the view that the most violated cut is the most beneficial.

The ho strategy more than halves the runtime in some cases. The faster separation enables ho to prove optimality quicker or explore a larger part of the search tree. This is especially apparent for the instances with a large number of terminals, suggesting that the gain from the ho strategy is clearer as the number of terminals increases.

We now turn our attention to the contribution of the branching. The new branching enables n-ho to prove optimality of 3 instances where a-ho fails to do so.

Table 1. Results for some SteinLib instances

Test	a-ff UB	a-ff Time	a-ff SP-t	a-ff LP-t	a-ff #cuts	a-ff #iter	a-ff #N	a-ho UB	a-ho Time	a-ho SP-t	a-ho LP-t	a-ho #cuts	a-ho #iter	a-ho #N	n-ff UB	n-ff Time	n-ff SP-t	n-ff LP-t	n-ff #cuts	n-ff #iter	n-ff #N	n-ho UB	n-ho Time	n-ho SP-t	n-ho LP-t	n-ho #cuts	n-ho #iter	n-ho #N
bipe2p	5766	5011	4916	22	137	6	2	5661	4923	2768	391	637	1250	885	5851	5008	4768	21	131	6	2	5653	4940	1706	647	561	744	460
bipe2u	55	5011	4796	23	142	6	2	54	4954	1258	3538	698	398	19	55	5009	4382	19	96	5	2	54*	2571	937	786	445	326	85
cc3-4p	2338	4978	1272	2061	8501	4877	3023	2338	4933	572	2372	7094	6700	3634	2338	4961	1537	2778	11251	3497	1600	2338	4915	515	3572	8110	5409	2165
cc3-4u	23*	1937	1551	1090	4864	818	57	23*	1038	179	765	3388	1251	109	23*	1255	1282	578	4295	722	43	23*	632	137	460	3105	987	21
cc3-5p	3676	4980	3400	1527	3242	471	6	3669	4939	766	3250	3435	1056	84	3681	4980	3551	1370	3481	488	5	3670	4912	831	3883	3946	1154	1
cc3-5u	36	4980	2944	1982	3225	431	1	36	4884	1151	3601	4938	1461	1	36	4980	2979	1954	3285	446	1	36	4865	1154	3586	4988	1487	29
cc6-3p	21608	5005	4510	73	599	11	1	21134	4885	4281	522	428	85	1	21608	5003	4467	171	599	11	1	21134	4863	3486	524	436	87	1
cc6-3u	206	5004	4791	140	749	9	1	209	4905	3486	1362	325	82	1	206	5002	4467	146	749	9	1	209	4897	511	1342	337	84	1
hc6p	4013	4981	4637	134	38390	8009	5126	4011	4758	3483	413	54478	55432	23647	4003*	3992	3732	146	42917	4918	1919	4003*	672	65	511	9901	7385	1767
hc6u	39	4983	4660	200	37539	4526	1496	39*	1511	1155	172	17573	15015	3143	39*	1749	1660	61	14936	1733	309	39*	283	34	221	3618	2762	305
hc7p	7944	4989	4448	297	11131	641	334	7938	4808	2364	983	10691	11228	6629	7905	4977	4516	313	11975	702	224	7905	4761	3234	1016	20941	13653	3109
hc7u	77	4984	4325	437	10101	318	22	77	4766	1937	2224	13158	6649	509	77	4982	4540	308	8724	346	9	77	4784	2084	2360	8385	496	496
160-041	1494*	50	41	3	41	8	1	1494*	16	6	4	24	11	1	1494*	50	41	3	41	8	1	1494*	15	6	4	24	11	1
160-042	1486*	34	27	4	17	4	1	1486*	9	2	49	10	11	1	1486*	34	27	6	17	2	1	1486*	9	0	47	10	4	1
160-043	1549*	901	833	43	248	59	13	1549*	109	49	38	164	83	13	1549*	1162	1056	72	287	67	17	1549*	119	48	45	150	82	11
160-044	1478*	10	8	2	6	2	1	1478*	7	0	22	2	2	1	1478*	10	16	18	6	2	1	1478*	7	0	21	2	2	1
160-045	1554*	481	446	18	186	43	9	1554*	85	47	5	138	68	7	1554*	443	405	18	186	44	11	1554*	84	45	21	132	68	9
160-111	2869*	18	16	1	66	7	1	2869*	56	40	11	31	11	1	2869*	18	16	8	66	7	1	2869*	7	5	31	31	11	1
160-112	2924*	182	166	10	295	41	11	2866*	28	24	3	97	29	1	2866*	153	143	8	128	13	3	2924*	28	23	10	168	57	3
160-113	2866*	49	46	2	128	13	1	2866*	28	24	3	97	29	1	2866*	48	45	8	128	13	1	2866*	28	23	9	97	29	1
160-114	2989*	177	167	4	280	27	1	2989*	69	55	11	191	60	1	2989*	175	165	12	280	27	1	2989*	67	54	10	191	60	1
160-115	2937*	1241	970	214	1304	183	59	2937*	638	319	242	1081	354	55	2937*	579	418	130	673	96	33	2937*	360	163	164	634	206	37
160-141	2549*	307	287	13	110	11	3	2549*	63	36	19	76	28	3	2549*	302	282	12	110	11	5	2549*	63	36	19	76	28	3
160-142	2562*	1272	1228	31	274	28	3	2562*	178	110	50	237	66	3	2562*	1113	1070	25	252	28	5	2562*	175	109	49	237	66	3
160-143	2557*	194	180	7	76	8	1	2557*	32	17	43	43	12	1	2557*	190	177	76	209	8	1	2557*	31	16	8	43	12	1
160-144	2617	5000	4660	282	737	72	5	2607*	1457	569	707	943	261	15	2607*	4155	3742	384	649	66	7	2607*	1523	561	834	924	260	15
160-145	2578*	746	746	27	209	20	7	2578*	114	69	35	141	41	21	2578*	768	732	26	209	20	21	2578*	113	68	34	141	41	13
160-211	5583*	4520	3780	609	2990	182	29	5583*	1877	1106	671	2259	534	15	5583*	3357	2763	506	2476	137	49	5583*	1369	810	513	1817	407	51
160-212	5643	4993	3724	998	3049	298	95	5643	4882	2206	2230	4046	1054	133	5643	4991	3522	1373	3350	225	13	5643	3516	1394	1993	3056	683	13
160-213	5647*	2730	2341	345	2217	139	23	5647*	1210	733	426	1433	355	25	5647*	2281	1984	272	1724	110	47	5647*	1471	767	657	1593	399	29
160-214	5720	4994	3831	961	3530	258	64	5720*	3505	1747	1528	3514	867	75	5720	4991	3627	1177	3173	213	58	5720*	4211	1717	2237	3905	924	73
160-215	5518*	1357	1273	69	1189	60	3	5518*	578	446	114	909	211	3	5518*	1278	1217	55	1144	58	3	5518*	551	433	100	904	210	3
320-011	2053*	242	225	11	174	29	3	2053*	77	52	17	146	44	3	2053*	249	233	11	181	30	3	2053*	77	52	16	145	4	3
320-012	1997*	14	10	1	21	5	1	1997*	6	2	2	7	4	1	1997*	14	10	9	21	5	1	1997*	6	2	18	4	4	1
320-013	2072*	292	277	10	161	24	3	2072*	116	89	20	192	64	3	2072*	286	272	9	161	24	1	2072*	114	88	48	192	64	7
320-014	2061*	1092	1021	53	446	72	7	2061*	324	228	78	475	145	5	2061*	525	468	47	320	53	5	2061*	192	131	68	279	99	7
320-015	2059*	754	692	49	355	59	7	2059*	188	121	121	263	91	5	2059*	517	478	54	272	45	5	2059*	192	108	68	247	86	7
320-041	1707*	923	812	54	76	13	1	1707*	175	57	57	47	18	7	1707*	914	804	54	76	4	7	1707*	169	55	56	47	18	1
320-042	1682*	242	168	21	19	4	1	1682*	90	12	12	11	5	1	1682*	239	166	21	19	4	1	1682*	85	11	55	11	5	1
320-043	1723	5040	4868	126	224	62	13	1723*	579	302	160	223	101	7	1726	5039	4703	114	203	38	5	1723*	630	282	222	220	94	7
320-044	1681*	314	238	23	26	5	1	1681*	107	20	20	15	28	1	1681*	310	235	22	26	5	1	1681*	101	19	28	15	9	1
320-045	1686*	96	14	30	6	2	1	1686*	87	30	30	1	30	1	1686*	95	14	30	6	2	1	1686*	82	1	30	1	2	1

The node branching is much faster on some instances and has roughly the same runtime on other ones. On the larger instances, except one, n-ho often finds a significantly better solution. This is a clear indication of the benefits of node branching. Moreover, the results show that the search tree is smaller for n-ho, compared to a-ho. This may be explained by the fact that the new branching evenly strengthens LB on both branches which result in stronger pruning of the search. Also, n-ho requires fewer cuts to prove a test optimal. This may be explained by the fact that when an arc is included it makes several previously computed cuts useless.

5.2   Branch-and-Price Evaluation

There are several difficulties that appear during the integration of the Steiner tree algorithm in a branch-and-price framework. We briefly mention some of them.

The Steiner cuts that are generated are valid not only within the branch-and-cut tree, but also throughout the branch-and-price tree itself. It is important that the implementation exploits that by maintaining a "cut pool" containing all previously generated Steiner cuts. These can be re-used at separation.

It is important to exclude trees that correspond to columns that already exist in the branch-and-price master. It is easily done with the solver described in this paper since we can add a linear inequality to the STP for each existing, and therefore non-improving, tree.

We have included the branch-and-cut solver into the branch-and-price approach of [6]. The solver was modified to return the first beneficial Steiner tree to enable fast progress of the column generation. Extra effort in optimising the subproblem is not necessarily beneficial for reaching global convergence fast. Two sets of instances occurring in a multicast network design application were considered. The first one, comprising 28 instances, considers less than 5 multicast commodities and the second one, with 12 instances, has around 19 commodities.

Curiously, there is little difference between the two branching strategies. Only in one case did n enable the search to find a slightly better solution. The explanation for this is probably that the emerged Steiner subproblems needed no branching: the root LP was integral. This is consistent with what appears to be the case in [19]. It is worth mentioning that commercial Internet topologies, like the ones we considered, are typically sparse: the number of edges is below $4|V|$. This explains why the branching is not beneficial for Internet-like topologies.

Table 2. Aggregated results for branch-and-price method [6] with instances

$|K|$	UB ratio			Time ratio			SP-t ratio			LP-t ratio			#cuts ratio			#bp-nodes		
	avg	min	max	avg	min	max	avg	min	max	avg	min	max	avg	min	max	avg	min	max
18–20	1.00	0.97	1.00	1.00	1.00	1.00	0.88	0.59	1.27	0.94	0.79	1.18	0.59	0.42	0.84	0.90	0.75	1.13
2–5	0.99	0.80	1.00	1.09	0.99	1.45	0.88	0.56	1.51	0.93	0.77	1.12	0.59	0.30	1.00	0.89	0.40	1.10

We now compare two variants n-ff and n-ho as subproblem solvers. Figure 2 presents an aggregated form of the results. The values are the ratio between n-ho and **n-ff**. For example, the average incumbent **n-ho** solution was 1% better than the one found by **n-ff** for small the instances. The column #bp-nodes describes the size of the branch-and-price tree.

Although it is marginally slower on average, the **ho** cutting strategy enables the branch-and-price algorithm to find slightly better solutions on average. The results also show that the separation is quicker and the number of cuts is drastically reduced. Again, this suggests that the most violated cuts are the important ones.

Interestingly, the **ho** cutting strategy enables the search to use fewer number of nodes. This may be because faster subproblem solving enables the encountering of primal solutions earlier and thus makes bound pruning possible.

6   Conclusion

Motivated by a multicast network design application, our work focuses on tailoring an existing Steiner tree approach so that it successfully serves as a fast subproblem solver in a branch-and-price framework. We adapt a little known preflow-push approach and demonstrate how to turn it into an effective separation algorithm. The search is also enhanced with a vertex oriented branching rule. We show that they both improve the performance in the decomposition context. We expect that our techniques will be more beneficial if the Steiner subproblems are *hard* to solve, unlike the reported multicast instances.

Even though our work originated in a decomposition framework, its benefits also unfold when solving stand-alone Steiner tree problems. In particular it succeeds in significantly improving the best known runtime for a test that was recently solved for the first time by [18].

There are many important applications which admit a natural formulation as a collection of cut-based covering inequalities similar to the Steiner tree problem. These include survivable networks e.g. [11]. Such applications often consist of finding several minimum $s - t$ cuts. It may be possible to adapt and exploit our cutting strategy in order to efficiently compute violated cuts.

References

1. R.K. Ahuja, T.L. Magnanti, and J.B. Orlin. *Network flows: theory, algorithms and applications*. Prentice-Hall, New Jersey, 1993.
2. B. V. Cherkassky and A. V. Goldberg. On implementing the push-relabel method for the maximum flow problem. *Algorithmica*, 19:390–410, 1997.
3. S. Chopra, E.R. Gorres, and M. R. Rao. Solving the Steiner tree problem on a graph using branch and cut. *Journal on Computing*, 4(3):320–335, 1992.
4. S. Chopra and M.R. Rao. The Steiner tree problem I: Formulations, compositions and extension of facets. *Mathematical Programming*, 64:209–229, 1994.
5. W. Cronholm and F. Ajili. Strong cost-based filtering for Lagrange decomposition applied to network design. In M. Wallace, editor, *CP'04*, number 3258 in LNCS, pages 726–730, Toronto, Canada, September 2004.

6. W. Cronholm and F. Ajili. Hybrid branch-and-price for multicast network design. In *Proceedings of the 2nd International Network Optimization Conference*, March 2005. To appear.

7. W. Cronholm, W. Ouaja, and F. Ajili. Strengthening optimality reasoning for a network routing application. In *Proceedings of the Fourth Workshop on Cooperative Solvers in Constraint Programming, COSOLV'04*, Toronto, Canada, September 2004.

8. A.V. Goldberg and R.E. Tarjan. A new approach to the maximum-flow problem. *Journal of the Association for Computing Machinery*, 35(4):921–940, October 1988.

9. J. Hao and J. B. Orlin. A faster algorithm for finding the minimum cut in a directed graph. *Journal of Algorithms*, 17:424–446, 1994.

10. F. K. Hwang, D. S. Richards, and P. Winter. *The Steiner Tree Problem*, volume 53 of *Annals of Discrete Mathematics*. Elsevier Science Publishers B. V., 1992.

11. H. Kerivin and A.R. Mahjoub. Separation of partition inequalities for the (1, 2)-survivable network design problem. *Operations Research Letters*, 30(4):265–268, August 2002.

12. T. Koch and A. Martin. Solving Steiner tree problems in graphs to optimality. *Networks*, 32(3):207–232, 1998.

13. T. Koch, A. Martin, and S. Voß. SteinLib: An updated library on steiner tree problems in graphs. Technical Report ZIB-Report 00-37, Konrad-Zuse-Zentrum für Informationstechnik Berlin, Takustr. 7, Berlin, 2000.

14. M.S. Levine. Experimental study of minimum cut algorithms. Master's thesis, Massachusetts Institute of Technology, 1995.

15. A. Lucena and J. E. Beasley. A branch and cut algorithm for the Steiner problem in graphs. *Networks, an International Journal*, 31(1):39–59, 1998.

16. M. Padberg and G. Rinaldi. An efficient algorithm for the minimum capacity cut problem. *Mathematical Programming*, 47:19–36, 1990.

17. S. Panagiotidi. Efficient implementation of a preflow push solver and its application to network design. Master's thesis, Imperial College London, September 2004.

18. T. Polzin. *Algorithms for the Steiner Problem in Networks*. PhD thesis, Faculty of Science and Technology, University of Saarlandes, May 2003.

19. M. Prytz and A. Forsgren. Dimensioning multicast-enabled communications networks. *Networks*, 39(4):216–231, 2002.

Constraint Programming Based Column Generation for Employee Timetabling

Sophie Demassey, Gilles Pesant, and Louis-Martin Rousseau

Centre for Research on Transportation (CRT), Université de Montréal,
C.P. 6128, succ. Centre-ville, Montreal, H3C 3J7, Canada

Abstract. The Employee Timetabling Problem (ETP) is a general class of problems widely encountered in service organizations (such as call centers for instance). Given a set of activities, a set of demand curves (specifying the demand in terms of employees for each activity for each time period) the problem consists of constructing a set of work shifts such that each activity is at all time covered by a sufficient number of employees. Work shifts can cover many activities and must meet work regulations such as breaks, meals and maximum working time constraints. Furthermore, it is often desired to optimize a global objective function such as minimizing labor costs or maximizing a quality of service measure. This paper presents variants of this problem which are modeled with the Dantzig formulation. This approach consists of first generating all feasible work shifts and then selecting the optimal set. We propose to address the shift generation problem with constraint satisfaction techniques based on expressive and efficient global constraints such as gcc and regular. The selection problem, which is modeled with an integer linear program, is solved by a standard MIP solver for smaller instances and addressed by column generation for larger ones. Since a column generation procedure needs to generate only shifts of negative reduced cost, the optimization constraint cost-regular is introduced and described. Preliminary experimental results are given on a typical ETP.

1 Introduction

Employee Timetabling Problems (ETP) form a wide class of optimization problems encountered in several industries and organizations. Generally, an ETP is the problem of designing valid employee schedules over a given time horizon that cover given workforce requirements. The timetabling attempts to optimize performance criteria such as the overall cost or the quality of service.

In a context where there are numerous possible activities, such as in call centers, a schedule refers to a sequence of activities performed during fixed time periods satisfying a given set of rules and regulations (e.g. a break of 15 minutes is necessary between two different work activities). Various constraints of that kind arise in real world ETP and their number and complexity quickly make these problems NP-hard. Furthermore, it is sometimes necessary to take into

R. Barták and M. Milano (Eds.): CPAIOR 2005, LNCS 3524, pp. 140–154, 2005.

account the individual preferences and skills of each employee, which significantly increases the complexity of the problem.

Because it is a broad modeling methodology, constraint programming (CP) is well suited to generate the valid individual schedules. Most of the constraints in this problem are defined in terms of allowed patterns of activities and limits on the amount of work that is scheduled. These constraints can be efficiently tackled in a constraint satisfaction model using global constraints such as gcc [8] and regular [7], together able to restrict the number of occurrences and to enforce patterns of values in a sequence of variables.

This paper describes a general algorithm based on such a CP framework for solving several variants of ETP. The basic ETP can be represented as a set covering problem, where each column is a valid schedule. Our solution method is based on an integer linear formulation of the whole optimization problem, where variables represent the different permitted schedules (which are pre-computed in CP). When the number of valid schedules becomes too large to be generated and stored, the linear relaxation of the integer program is solved by column generation. In this case, only a subset of the schedules (with negative reduced costs) are generated by CP and added iteratively to the master linear program. In order to generate only negative reduced cost schedules, the regular constraint is replaced by its extension cost-regular within the CP model.

The interest of this approach is its ability to handle variations of the problem without major modifications to the algorithm itself. Indeed, comparing with pure linear programming approaches that are generally developed for ETP, CP offers a more straightforward way to model complex constraints. Furthermore, the decomposition of the problem makes the processing of these hard constraints independent from the global optimization process. This hybrid constraint-linear programming approach also differs from pure constraint programming approaches by taking more efficiently into consideration the optimization criterion.

CP-based column generation approaches have been proposed for several problems more or less related to ETP. The general framework was first introduced in [5]. It has since been applied to airline crew scheduling [3, 12], vehicle routing [10], and cutting-stock problems [4]. The subproblems solved by CP have taken the form of constrained shortest path problems and constrained knapsack problems. In our case we use cost-regular, which bears some similarity to a constrained shortest path, as we shall see in Section 4. Recently other hybrid CP-LP algorithms have been advantageously applied to solve ETP. In particular, Benoist et al. [1] presented a CP-based Benders decomposition for solving the timetabling problem encountered in a large call center. They use CP including the flow global constraint to handle the underlying flow structure of the problem.

The paper is organized as follows. The next section gives definitions and notation for employee timetabling problems. A typical set of regulation constraints and the associated CP model is presented in Section 3. Section 4 presents the optimization constraint cost-regular, an extension of the regular global constraint. Section 5 describes the linear formulations for three different ETP as well as the column generation process for each of these formulations. Section 6 contains preliminary computational results. The final section presents our conclusion and perspectives.

2   Problem Statement

Employee Timetabling Problems come in many forms. The ones adressed here are essentially divided into two major classes: the anonymous and the personalized scheduling. In the anonymous version we are only interested in building a set of valid work schedules — employees are considered interchangeable. This problem is in fact equivalent to the shift scheduling problem. This paper mostly address the anonymous case, but gives some insight on how to deal with individual schedules.

The definitions and notation given in this section apply to ETP where the *planning horizon* (ex. one day) can be partitioned as a sequence of T consecutive time intervals called *periods* or *shifts*.

Activities. Different types of activities must be fulfilled by the employees on the planning horizon. Let W denote the set of these work activities. An estimation of the *workforce requirement* of each activity is given on the whole planning duration. For each activity a and each period t, r_{at} specifies the number of workers required to achieve activity a during period t. Eventually, c_{at} will denote the cost of assigning one employee to activity a at period t.

Besides work activities, we distinguish three other activities : break (p), rest (o) and lunch (l). These activities are not subjected to costs and workforce but they may be involved in specific constraints. Let $A = W \cup \{o, b, l\}$ denote the set of generalized activities.

Valid Schedules. A *schedule* is an assignment $s : [1..T] \longrightarrow A$ where $s(t)$ stands for the activity to perform at period t. Alternatively, schedule s can be expressed by a binary matrix $B^s = (b^s_{at})_{a \in A, t \in [1..T]}$ where $b^s_{at} = 1$ if $s(t) = a$ and $b^s_{at} = 0$ otherwise. $\mathcal{S}$ denotes the set of *valid schedules* that are individual schedules satisfying all regulation constraints.

Cost and Satisfaction. Usual objectives in ETP are the minimization of the overall cost or the maximization of employee satisfaction. These criteria can be formulated by considering the cost c^s for the company of allocating schedule s to an employee. Such a cost can also represent the degree of dissatisfaction for an employee being assigned to schedule s. The objective is then to minimize the sum of the costs of the schedules assigned to each employee. In this paper, c^s is computed as the sum of the costs of performing activity $s(t)$ at period t:
$$c^s = \sum_{t=1}^{T} c_{s(t)t}.$$

Regulation Constraints. Constraints describing permitted schedules come from legislation and contractual agreements. Three kinds of constraints are usually encountered:

- Some activities are not allowed to be performed at some times.
- Restrictions are imposed on the number of periods or on the number of consecutive periods an employee is assigned to a particular activity or to a group of activities during his schedule (e.g. at most 8 hours of work a day, at least one 15 minute break)
- Some given patterns specify the allowed sequences of activities for the workers (e.g. imposing a break before performing a new activity).

Overcoverage and Undercoverage. In some ETP formulations, the cost of the staff timetabling may include penalties due to overcoverage or undercoverage. For each activity a and period t, let $\hat{c}_{at}$ and $\check{c}_{at}$ be the additional cost when the timetabling covers the workforce demand r_{at} with, respectively, one more employee and one less employee.

Employees. When taking into consideration individual preferences and skills, the definition of a valid schedule becomes different for each employee (or team). Additional constraints restrict the assignment of a given employee to activities following his qualification or to periods following his availabilities.

For each employee $e \in \mathcal{E}$, let $\mathcal{S}_e$ be the set of valid schedules for employee e. The cost of a schedule s may then also differs between employees (for whom schedule s is permitted): let c^{es} be the cost of schedule $s \in \mathcal{S}_e$ when assigned to employee e.

3   Constraint Programming for Shift Scheduling

The subproblem (SP) of computing valid schedules can easily be modeled as a Constraint Satisfaction Problem with T decision variables $s_1, s_2, \ldots, s_T$ and finite discrete domains $D_1, D_2, \ldots, D_T$ all equal to A. There is an obvious one-to-one correspondence between complete instantiations of these variables and schedules $s : [1..T] \longrightarrow A$ by setting $s(t) = s_t$ for all periods t (i.e. $s_t = a$ means that activity a is performed at period t). The set $\mathcal{S}$ of valid schedules corresponds then to the set of all the solutions of this CSP including the regulation constraints.

The high expressiveness and modeling flexibility of constraint programming allows to formulate a wide variety of complex regulation constraints in terms of variables $s_1, \ldots, s_T$ (eventually with the help of additional variables). In particular, a number of global constraints well suited to model such constraints have been introduced in the constraint programming literature. Some of these global constraints are quickly described in Section 3.1. We give in Section 3.2 the example of a set of typical regulation constraints encountered for example in a large store where workers may be assigned to any sales activities.

3.1   Global Constraints for Shift Scheduling

A global constraint in constraint programming is both a way of modelling a specific substructure (common preferably to many decision problems) and a filtering algorithm dedicated to this substructure. In the context of ETP some global constraints of the literature are of great interest. These constraints mainly relate to the allowed values taken by a sequence of decision variables $X = (x_1, \ldots, x_n)$ (in domains $D_1 \times \cdots \times D_n$) together:

Global Cardinality Constraint [8]. This constraint does not consider the ordering of the variables but restricts the number of times each value is distributed on a set of variables. Formally,

$$\texttt{gcc}(< y_1, \ldots, y_m >, < v_1, \ldots, v_m >, X)$$

constrains variable y_j to be equal to the number of appearances of value v_j in the set of variables X. For the ETP, the $\texttt{gcc}$ constraint is helpful to give lower and upper bounds on the total amount of work performed over a day.

Stretch Constraint [6]. A $\texttt{stretch}$ refers to a subsequence $(x_i, x_{i+1}, \ldots, x_j)$ of variables all assigned to a same value v and that is maximal for this property in terms of inclusion (i.e. $x_{i-1} \neq v$ and $x_{j+1} \neq v$).

$$\texttt{stretch}(X, < v_1, \ldots, v_m >, < l_1^{\min}, \ldots, l_m^{\min} >, < l_1^{\max}, \ldots, l_m^{\max} >)$$

restricts the length of any stretch in X with value v_j to be at least equal to value $l_j^{\min}$ and at most to value $l_j^{\max}$.

For example in ETP, the $\texttt{stretch}$ constraint is helpful to indicate that activities must be assigned to a certain number of consecutive periods. Note that $\texttt{stretch}$ is more general and can also be applied to cyclic schedules.

Global Sequencing Constraint [9]. This constraint lies somewhere between the two preceding constraints since it looks like the $\texttt{stretch}$ constraint with the difference that values do not have to appear consecutively. It can be understood as a set of global cardinality constraints defined on every subsequence of X of a given length. It may occur in the ETP if restrictions on the amount of work performed are also given over a shorter duration, say every three hours.

Regular Constraint [6]. This constraint is able to express complex patterns in a sequence. Formally, given a deterministic finite automaton Π describing a regular language, constraint

$$\texttt{regular}(X, \Pi)$$

restricts the sequence of values taken by the variables of X to belong to the regular language associated to Π. For the ETP, it is useful to enforce sequencing rules for the activities.

3.2   Example of Regulation Constraints

We based our first experimentations on a mostly generic set of regulation constraints. These constraints as well as their formulation in a CSP using global constraints are presented below.

The problem consists of generating valid schedules for employees in a shop with different sales activities $a \in W$. The planning horizon (one day) is decomposed in periods of 15 minutes ($T = 96$). A valid schedule is an assignment $s : [1..T] \longrightarrow A$, where A includes also activities p (break), o (rest) and l (lunch), and that satisfies the following constraints:

1. *Some activities $a \in F_t$ are not allowed to be performed at some periods t.*
2. *s covers between 3 hours and 8 hours of work activities.*
3. *If s is worked for at least 6 hours, then it includes exactly two breaks and one lunch break of 1 hour. Else, it includes only 1 break and no lunch break is planned.*
4. *If performed, the duration of an activity $a \in W$ is at least 1 hour.*
5. *A break (or lunch) is necessary between two different working activities.*
6. *Rest shifts have to be assigned only at the begining and at the end of the day.*
7. *Work activities must be inserted between breaks, lunch and rest stretches.*
8. *The maximum duration of a break is 15 minutes.*

The first condition simply consists of removing the forbidden activities F_t from the initial domain of each variable s_t: $D_t = A \setminus F_t$.

The next two regulation constraints need the definition of additional decision variables. They can then be modeled as explicit constraints as well as, implicitly, by restricting the initial domain of the variables. One way of modeling the second condition is to use one additional variable σ_a for each work activity $a \in W$, with domain $\{0, 1, \ldots, 32\}$ and representing the number of periods assigned to activity a, as well as a variable σ with domain $\{12, \ldots, 32\}$, standing for the total number of working periods. Variables σ_a and s_t may be linked by the **gcc**. In the

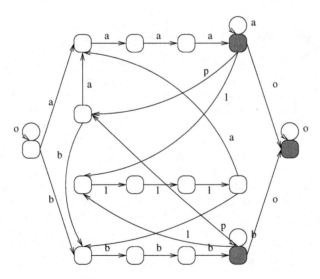

Fig. 1. An automaton for two work activities a and b. The leftmost circle represents the initial state and shaded circles correspond to accepting states

same manner, we define cardinality variables σ_p, σ_o and σ_l for the non-working activities (break, rest and lunch, respectively). To model the third condition, the domains of σ_l and σ_p are initialized to $\{0,4\}$ and $\{1,2\}$ respectively. We can also logically deduce from the whole set of conditions that any valid schedule contains a number of rest periods between 58 and 83. As redundant constraints, we can then reduce the initial domain of σ_o to $\{58,\dots,83\}$.

The last five constraints can be modeled with the help of only one **regular** constraint. Indeed, the values permitted by these constraints together for the sequence of variables $(s_1,\dots,s_T)$ can be described by a single automaton Π. Figure 1 depicts such an automaton when W contains two activities a and b.

Given automata Π, the shift scheduling problem described above can be formulated by the following Constraint Satisfaction Problem (CP):

$$\text{gcc}(< \sigma_a | a \in A >, < a \in A >, < s_1, \dots, s_T >) \tag{1}$$

$$\sigma == \sum_{a \in W} \sigma_a \tag{2}$$

$$\sigma < 24 \Rightarrow (\sigma_l == 0 \wedge \sigma_p == 1) \tag{3}$$

$$\sigma \geq 24 \Rightarrow (\sigma_l == 4 \wedge \sigma_p == 2) \tag{4}$$

$$\text{regular}(< s_1, \dots, s_T >, \Pi) \tag{5}$$

$$s_t \in A \setminus F_t, \quad \forall t = 1, \dots, T \tag{6}$$

$$\sigma_a \in \{0, \dots, 32\}, \; \forall a \in W, \; \sigma \in \{12, 32\} \tag{7}$$

$$\sigma_l \in \{0, 4\}, \; \sigma_p \in \{1, 2\}, \; \sigma_o \in \{58, \dots, 83\} \tag{8}$$

4   cost-regular Global Constraint

As indicated in Section 3.1, a **regular** constraint is specified using a deterministic finite automaton that describes the regular language to which the sequence must belong. That automaton is then unfolded into a layered directed graph where vertices of a layer correspond to states of the automaton and arcs represent variable-value pairs. This graph has the property that paths from the first layer to the last are in one-to-one correspondence with solutions of the constraint. The existence of a path through a given arc constitutes a support for the corresponding variable-value pair [7].

In the ETP, assigning a given activity at a given period has a cost. For the CP model, this translates to associating costs to variable-value pairs. For this purpose, we define **cost-regular**(X, Π, z, C) constraining X as in **regular** but also requiring that z, a bounded-domain continuous variable, represent the cost of a solution with respect to the constraint, given cost matrix C. This cost is computed as the sum of the costs of the individual arcs in the solution. [1]

---

[1] Note that we could refine the costs by associating one to every combination of variable, value, and state of the automaton.

Instead of simply maintaining paths, the filtering algorithm must from now on consider the cost of these paths. Supports do not come from just any path but rather from a path whose length falls within the domain of z. To check this efficiently, it is sufficient to compute and maintain shortest and longest paths from the first layer to every vertex and from every vertex to the last layer: if the shortest way to build a path through a given arc is larger than the upper limit of the interval for z, the arc cannot participate in a solution and can thus be removed; if the longest way to build a path through a given arc is smaller than the lower limit of that interval, the arc can again be removed. In this way, domain consistency is achieved for the variables of X. The domain of z can also be trimmed using the shortest and longest paths from the first to the last layer.

The time complexity for the initial computation of the shortest and longest paths is linear in the size of X and in the number of transitions appearing in the automaton, due to the special structure of the graph. Subsequently these paths are updated incrementally.

5   Integer Linear Formulations and Column Generation

This section presents linear programs modeling three different ETP. These linear programs are based on a well-known Dantzig formulation for this kind of problems, involving integer variables indexed by the set of valid schedules. Solving these programs yields an optimal staff timetable by selecting the best set of individual schedules and choosing how many employees (or which employees) will be assigned to each of these schedules.

The first two problems differ on the definition of the overall cost of the staff schedule. In the first one, the overall cost equals the sum of the costs of the schedules assigned to each employees, while in the second one it includes also an additional cost of overcoverage and undercoverage for each work activity on each period. Both models assume that employees are interchangeable. In other words, any schedule in S is valid for any employee. The problem is then to find how many employees will be assigned to each schedule in S.

On the contrary, the last problem takes into account individual preferences and skills, which requires to define a different set S_e of valid schedules for each employee e. Here, the problem consists of choosing one schedule in S_e for each employee e (or none if e is not used).

In this kind of formulations, the number of variables grows exponentially with the number of activities. Beyond two work activities, the set of valid schedules which is the set of solutions of the CSP presented in Section 3, becomes too large to be computed. At this point, we use a column generation procedure to solve the linear relaxation of the integer program.

The column generation algorithm is detailed in Section 5.1 for the first formulation while the corresponding pricing problem alone is described for the two other variants.

5.1   Column Generation

The first ETP problem is to minimize the sum of the costs of the schedules while covering all the workforce requirements. The following linear formulation (ETP) uses an integer variable x_s for each valid schedule $s \in \mathcal{S}$, standing for the number of employees assigned to schedule s:

$$\min \quad \sum_{s \in \mathcal{S}} c^s x_s \tag{9}$$

$$\text{s.t.} \quad \sum_{s \in \mathcal{S}} b_{at}^s x_s \geq r_{at} \qquad \forall\, a \in W, \forall\, t \in [1..T], \tag{10}$$

$$x_s \geq 0 \qquad \forall\, s \in \mathcal{S}, \tag{11}$$

$$x_s \in \mathbb{Z} \qquad \forall\, s \in \mathcal{S}. \tag{12}$$

Let $(P) : ((9)s.t.(10),(11))$ be the linear relaxation of (ETP). The dual (D) of (P) can be written as:

$$\max \quad \sum_{a \in W} \sum_{t=1}^{T} r_{at} \lambda_{at} \tag{13}$$

$$\text{s.t.} \quad \sum_{a \in W} \sum_{t=1}^{T} b_{at}^s \lambda_{at} \leq c^s \qquad \forall\, s \in \mathcal{S}, \tag{14}$$

$$\lambda_{at} \geq 0 \qquad \forall\, a \in W, \forall\, t \in [1..T]. \tag{15}$$

Column generation applied to (P) is an iterative algorithm where, at each iteration, the so-called *master problem* (P'), that is linear program (P) restricted to a subset of columns, is solved to optimality. Duality considerations allow to formulate a *pricing problem* (SP) such that: 1) the unfeasiblity of (SP) proves that the optimal solution of the master problem can be extended to an optimal solution of (P) by setting to 0 any variables x_s that are not present in the restricted master program. 2) if (SP) is feasible then its solutions correspond to columns that can improve the solution of the master problem when added at the next iteration.

More precisely, let $\mathcal{S}' \subset \mathcal{S}$ be the subset of valid schedules corresponding to the restricted set of columns of master problem (P') at a given iteration. If x, x' and λ' are optimal solutions for (P), (P') and (D'), the dual of P', respectively, then weak duality says that:

$$\sum_{a \in W} \sum_{t=1}^{T} r_{at} \lambda_{at}' = \sum_{s \in \mathcal{S}'} c^s x_s' \geq \sum_{s \in \mathcal{S}} c^s x_s.$$

Hence, x' (completed with 0) is an optimal solution of (P) if λ' is a feasible solution of (D), or in other words, if λ' satisfies all constraints (14), that is if:

$$\{s \in \mathcal{S} \mid c^s - \sum_{a \in W} \sum_{t=1}^{T} b_{at}^s \lambda_{at}' < 0\} = \emptyset.$$

At each iteration, the pricing problem (SP) is to find valid schedules $s \in \mathcal{S}$ with negative reduced costs rc'_s, where:

$$rc'_s = c^s - \sum_{a \in W} \sum_{t=1}^{T} b^s_{at} \lambda'_{at}.$$

Section 3.2 gives a CSP formulation (CP) for the problem of finding valid schedules. At each iteration, the pricing problem can be solved by adding to (CP) a negative reduced cost constraint. One way is to add this constraint to (CP) by using global constraint `element` :

$$\sum_{t=1}^{T} (c_{s_t t} - \lambda'_{s_t t}) < 0.$$

We propose a more efficient way to tackle this additional constraint by replacing in (CP) the `regular` constraint (5) by its variant detailed in Section 4:

$$\texttt{cost-regular}(< s_1, \ldots, s_T >, \Pi, z, < c_{at} - \lambda'_{at}, a \in A, t = 1, \ldots, T >). \quad (16)$$

This constraint ensures, as constraint (5), that the sequence of values taken by $< s_1, \ldots, s_T >$ belongs to the language defined by automaton Π of Section 3.2. But it also forces variable z to be equal to the sum of the costs of the variable assignments for $s_1, \ldots, s_T$, the cost of assigning variable s_t to activity $a \in A$ being set to $c_{at} - \lambda'_{at}$. In order to model the negative reduced cost constraint within the (CP) model we just need to define such an additional variable z with the appropriate domain:

$$z \in \,] -\infty, 0[. \quad (17)$$

Hence, the filtering algorithm of `cost-regular` processes simultanously the domains of variables $s_1, \ldots, s_T$ and z such that schedules with non-negative reduced cost are removed from the search space by this algorithm alone.

5.2   Overcoverage and Undercoverage

Overcoverage and undercoverage costs may be taken into account by slightly modifying the previous formulation of (ETP) in this way: for each activity a and period t, let variables $\hat{x}_{at} \in \mathbb{Z}$ and $\check{x}_{at} \in \mathbb{Z}$ represent the overcoverage and the undercoverage respectively. In other words, when N employees are assigned to activity a at period t in the timetable, then $\hat{x}_{at} = N - r_{at}$ and $\check{x}_{at} = 0$ if N is greater than the request r_{at} (overcoverage) and $\check{x}_{at} = r_{at} - N$ and $\hat{x}_{at} = 0$ in the other case (undercoverage).

The modified problem $(ETPl)$ becomes set partitioning problem:

$$\min \quad \sum_{s\in\mathcal{S}} c^s x_s + \sum_{a\in W}\sum_{t=1}^{T}(\hat{c}_{at}\hat{x}_{at} + \check{c}_{at}\check{x}_{at}) \tag{18}$$

$$\text{s.t.} \quad \sum_{s\in\mathcal{S}} b^s_{at}x_s + \check{x}_{at} - \hat{x}_{at} = r_{at} \qquad \forall\, a\in W, \forall\, t\in[1..T], \tag{19}$$

$$x_s \in \mathbb{N} \qquad\qquad\qquad \forall\, s\in\mathcal{S}, \tag{20}$$

$$\hat{x}_{at}\in\mathbb{N}, \check{x}_{at}\in\mathbb{N} \qquad\qquad \forall\, a\in W, \forall\, t\in[1..T]. \tag{21}$$

Since $(ETPl)$ simply contains additional columns (and no more constraints than ETP), the pricing problem associated to a Dantzig-Wolfe decomposition of $(ETPl)$ is identical to the pricing problem for (ETP) (denoted (SP)).

5.3   Individual Shift Scheduling

Most practical timetabling problems take into consideration the preferences and skills of the employees, for instance excluding them from specific activities or work periods. The set $\mathcal{S}_e$ of possible schedules is thus different for each employee $e\in\mathcal{E}$. Such a variant of the ETP can be modeled by the following binary linear program $(ETPe)$:

$$\min \quad \sum_{e\in\mathcal{E}}\sum_{s\in\mathcal{S}_e} c^{es}x_{es} \tag{22}$$

$$\text{s.t.} \quad \sum_{e\in\mathcal{E}}\sum_{s\in\mathcal{S}_e} b^s_{at}x_{es} \geq r_{at} \qquad \forall\, a\in W, \forall\, t\in[1..T], \tag{23}$$

$$\sum_{s\in\mathcal{S}_e} x_{es} \leq 1 \qquad\qquad \forall\, e\in\mathcal{E}, \tag{24}$$

$$x_{es}\in\{0,1\} \qquad\qquad \forall\, e\in\mathcal{E}, \forall\, s\in\mathcal{S}_e, \tag{25}$$

In this model, x_{es} is a binary variable that is equal to 1 if and only if employee $e\in\mathcal{E}$ is assigned to schedule $s\in\mathcal{S}_e$. Constraints (24) enforce that at most one schedule is assigned to each employee.

Within a column generation approach for this problem, the pricing problem (SPe) can be written[2]:

$$\{(e,s)\in\mathcal{E}\times\mathcal{S}_e \mid c^{es} - \sum_{t=1}^{T}\lambda_{s_t t} + \mu_e < 0\},$$

where (λ,μ) are the current dual values (associated to constraints (23) and (24), respectively) of the master problem at a given iteration of the column generation

---

[2] Note that constraints (25) are redundant and that in practice x is simply set to be greater or equal to 0. Otherwise, (25) requires the introduction of another dual variable.

process. (SPe) can clearly be decomposed as several problems (one for each employee), which are treated in the same way as problem (SP). The CSP to solve for each employee e includes the personal constraints for e as well as the cost-regular constraint and the cost variable z with initial domain equal to $] - \infty, -\mu_e[$.

6   Computational Results

We ran some preliminary experiments on a set of generated benchmarks. The data of these instances are based on a realistic ETP for a retail store where employees can be assigned to many sale activities and the work regulation constraints are the ones described in the example of Section 3.2. As a first step, we evaluate the efficiency of our algorithm on the first formulation of ETP presented in Section 5.1. In other words, we assume that employees are interchangeable and that the objective is to minimize the sum of the employee schedule costs while covering the demand curves on the activities.

Generated benchmarks are distributed into eight groups of instances denoted $ETP_1, \ldots, ETP_8$. Each set ETP_n contains 10 instances of the timetabling problem with parameter n indicating the number of work activities. In these instances, the demand curves can require the presence of up to 12 employees at the same period. Even for instances in group ETP_1, the set of valid schedules is too large to be pre-generated quickly. Remember that for benchmarks in ETP_1, schedules are potentially any assignment from the set T of 96 periods to the set A of 4 activities.

For this reason, we directly apply the column generation algorithm to the linear relaxation (P) of program (ETP) as described in Section 5.1. In the rest of this section, we present the details of the implementation of this algorithm, a summary of the computational results obtained on the 80 generated instances and a discussion about these results as well as future research directions for improvements.

Our program was implemented in C++ using ILOG Concert libraries (ILOG Solver 6.0 to solve the CP models and ILOG Cplex 9.0 to solve the linear relaxation), compiled using g++ 3.3 and run on a biprocessor Intel Xeon 2.8GHz under Gnu/Linux 2.6.

The column generation proceeds as follows: An initial minimal set of columns is generated in order to make feasible the master linear program (P) at the first iteration. These columns correspond to mono-activity shedules worked for only four consecutive periods (one hour). Since these schedules are not valid (they violate constraint (7)), we give them "infinite" costs in the LP. At each iteration of the column generation process, the reduced costs returned by the resolution of (P) are used to update the subproblem (SP) of finding improving columns. (SP) is formulated as the CSP of Section 3.2 with the cost constraints (16) and (17) and it is solved by a backtracking algorithm returning the 50 (or less) first computed valid schedules of negative reduced cost. These schedules are then added as new columns to the master program. The algorithm stops when the subproblem of a given iteration is unfeasible. The optimal value of the master problem at the last iteration is then a lower bound LB of the optimal value of the original problem (equal to the linear relaxation bound). To estimate the quality of LB we compute an upper bound UB by solving the integer linear program with the generated columns alone. UB is the value of the best integer solution obtained by the default branch-and-bound of Cplex ran during one hour.

Table 1. Column generation algorithm results on the generated instances ETP

Group	nb	$\Delta LB/UB$		nb iterations		nb columns		CPU in sec.			
		av.	(max)	av.	(max)	av.	(max)	av.	(max)	CPav.	(CPmax)
ETP_1	10	4.9%	(16.6%)	20	(29)	914	(1361)	1.9	(5.7)	0.1	(2.3)
ETP_2	10	5.6%	(15.6%)	51	(76)	2466	(3722)	6.1	(12.0)	0.1	(3.3)
ETP_3	10	5.5%	(9.2%)	76	(106)	3749	(5226)	16.7	(45.6)	0.2	(5.8)
ETP_4	10	4.6%	(8.7%)	137	(204)	6818	(10142)	92.9	(452.4)	0.6	(13.9)
ETP_5	10	5.4%	(12.6%)	132	(207)	6558	(10300)	108.4	(354.4)	0.7	(24.6)
ETP_6	10	5.0%	(11.0%)	203	(337)	10103	(16798)	355.6	(884.6)	1.6	(268.8)
ETP_7	9	5.6%	(7.9%)	244	(337)	12186	(16814)	793.6	(2115.1)	3.1	(130.9)
ETP_8	9	5.4%	(8.5%)	296	(548)	14776	(27377)	950.3	(2531.2)	3.0	(159.9)

Table 1 provides details of the algorithm execution on each problem set ETP_n. The first column gives for each set of 10 instances the number of instances among them that have been processed by column generation in at most one hour. The following columns give, by pair, average and maximal results on these sets of solved instances. These results are, in order: the deviation of LB to the upper bound UB, the number of iterations of the column generation process, the number of generated columns, the total computation time in seconds and the time spent to solve the subproblem (SP) at one iteration.

Computation results given here come from preliminary experiments but give some insight on how to improve the general algorithm. In order to decrease the computational times, we aim to improve the subproblem resolution since the processing time is mainly spent in the CP phases. A good branching strategy has to be found for solving the CSP. For instance, it would be interesting to implement a value ordering heuristic based on the shortest path computed by the filtering algorithm of `cost-regular`. Something similar is performed in the Branch and Price library Maestro [2].

In fact, our random generated instances seem to be really diversified in each group. While, for some of them, we hardly find schedules at each iteration, some others contain a large number of valid schedules. For these last instances, the computation of the (negative reduced cost) schedules is very quick but the number of iterations of the column generation process can then be more important. In these cases, a basic backtracking algorithm has a tendency to provide a set of solutions that are almost identical. Convergence of the column generation process can then be slower when columns added at each iteration are too similar. Many stabilization techniques have been proposed to accelerate convergence

([11]). Another simple way (suggested in [10]) is to add diversity in the set of solutions generated by CP with a multi-start search process, by introducing some randomization in the variable ordering heuristic or by implementing a dedicated diversity constraint.

7   Conclusion

This paper presented a hybrid constraint programming-linear programming solution method for employee timetabing problems. The proposed decomposition applies to several formulations of this kind of problem. By using CP to generate the permitted shift schedules, it also offers a flexible way to tackle the various work regulation constraints that arise in real world timetabling problems.

The optimization criterion on the staff scheduling is handled by LP when assigning shift schedules to employees. With a column generation approach, only "lowest cost" schedules are iteratively generated by CP. The newly introduced global constraint `cost-regular` allows to efficiently take into account the cost of the schedules within their generation process by CP.

The method has now only been implemented to compute lower bounds on generated benchmarks for a first formulation of ETP with general regulation constraints. An obvious continuation of this work is to elaborate branch-and-price algorithms based on these bounds in order to solve at optimality various realistic timetabling problems.

References

1. Benoist, T., Gaudin, E., Rottembourg, B.: Constraint programming contribution to Benders decomposition: a case study. In Proc. 8th Int. Conf. on Principles and Practice of Constraint Programming – CP'02, Springer-Verlag LNCS **2470** (2000) 603–617
2. Chabrier, A.: Maestro: A column and cut generation modeling and search framework. Technical report, ILOG SA (2002)
3. Fahle, T., Junker, U., Karish, S.E., Kohl, N., Vaaben, N., Sellmann, M.: Constraint programming based column generation for crew assignment. Journal of Heuristics **8** (2002) 59–81
4. Fahle, T., Sellmann, M.: Cost based filtering for the constrained knapsack problem. Annals of Operations Research **115** (2002) 73–93
5. Junker, U., Karish, S.E., Kohl, N., Vaaben, N., Fahle, T., Sellmann, M.: A framework for constraint programming based column generation. In Proc. 5th Int. Conf. on Principles and Practice of Constraint Programming – CP'99, Springer-Verlag LNCS **1713** (1999) 261–274
6. Pesant G.: A filtering algorithm for the stretch constraint. In Proc. 7th Int. Conf. on Principles and Practice of Constraint Programming – CP'01, Springer-Verlag LNCS **2239** (2001) 183–195
7. Pesant G.: A regular language membership constraint for finite sequences of variables. In Proc. 10th Int. Conf. on Principles and Practice of Constraint Programming – CP'04, Springer-Verlag LNCS **3258** (2004) 482–495

8. Régin J.-C.: Generalized arc consistency for global cardinality constraints. In Proc. of AAAI'96, AAAI Press/The MIT Press (1996) 209–215
9. Régin J.-C., Puget J.-F.: A filtering algorithm for global sequencing constraints. In Proc. 3rd Int. Conf. on Principles and Practice of Constraint Programming – CP'97, Springer-Verlag LNCS **1330** (1997) 32–46
10. Rousseau, L.-M., Gendreau, M., Pesant, G., Focacci, F.: Solving VRPTWs with constraint programming based column generation. Annals of Operations Research **130** (2004) 199–216
11. Rousseau L.-M.: Stabilization issues for constraint programming based column generation. In Proc. 1st Int. Conf. on Integration of AI and OR Techniques in Constraint Programming for Combinatorial Optimization Problems – CPAIOR'04, Springer-Verlag LNCS **3011** (2004) 402–408
12. Sellmann, M., Zervoudakis, K., Stamatopoulos, P., Fahle, T.: Crew assignment via constraint programming: integrating column generation and heuristic tree search. Annals of Operations Research **115** (2002) 207–225

Scheduling Social Golfers Locally

Iván Dotú[1] and Pascal Van Hentenryck[2]

[1] Departamento De Ingeniería Informática, Universidad Autónoma de Madrid
[2] Brown University, Box 1910, Providence, RI 02912

Abstract. The scheduling of social golfers has attracted significant attention in recent years because of its highly symmetrical and combinatorial nature. In particular, it has become one of the standard benchmarks for symmetry breaking in constraint programming. This paper presents a very effective, local search, algorithm for scheduling social golfers. The algorithm find the first known solutions to 11 instances and matches, or improves, state-of-the-art results from constraint programming on all but 3 instances. Moreover, most instances of the social golfers are solved within a couple of seconds. Interestingly, the algorithm does not incorporate any symmetry-breaking scheme and illustrates the nice complementarity between constraint programming and local search on this scheduling application.

1 Introduction

The social golfer problem has attracted significant interest since it was first posted on sci.op-research in May 1998. It consists of scheduling $n = g \times p$ golfers into g groups of p players every week for w weeks so that no two golfers play in the same group more than once. An instance of the social golfer is specified by a triple $g - p - w$, where g is the number of groups, p is the size of a group, and w is the number of weeks in the schedule.

The scheduling of social golfers is a highly combinatorial and symmetric problem and it is not surprising that it has generated significant attention from the constraint programming community (e.g., [5, 12, 6, 11, 10, 2, 9]). Indeed, it raises fundamentally interesting issues in modeling and symmetry breaking, and it has become one of the standard benchmarks for evaluating symmetry-breaking schemes. Recent developments (e.g., [2, 9]) approach the scheduling of social golfers using innovative, elegant, but also complex, symmetry-breaking schemes.

This paper approaches the problem from a very different angle. It proposes a local search algorithm for scheduling social golfers, whose local moves swap golfers within the same week and are guided by a tabu-search meta-heuristic. The local search algorithm matches, or improves upon, the best solutions found by constraint programming on all instances but 3. It also found the first solutions to 11 instances that were previously open for constraint programming.[1] Moreover, the local search algorithm solves

---

[1] For the statuses of the instances, see Warwick Harvey's web page at http://www.icparc.ic.ac.uk/wh/golf.

R. Barták and M. Milano (Eds.): CPAIOR 2005, LNCS 3524, pp. 155–167, 2005.
© Springer-Verlag Berlin Heidelberg 2005

almost all instances easily in a few seconds and takes about 1 minute on the remaining (harder) instances. The algorithm also features a constructive heuristic which trivially solves many instances of the form $odd - odd - w$ and provides good starting points for others.

The main contributions of this paper are as follows.

1. It shows that local search is a very effective way to schedule social golfers. It found the first solutions to 11 instances and matches, or improves upon, all instances solved by constraint programming but 3. In addition, almost all instances are solved in a few seconds, the harder ones taking about 1 minute.
2. It demonstrates that the local search algorithm uses a natural modeling and does not involve complex symmetry-breaking schemes. In fact, it does not take symmetries into account at all, leading to an algorithm which is significant simpler than constraint programming solutions, both from a conceptual and implementation standpoint.
3. The experimental results indicate a nice complementarity between constraint programming and local search, as some of the hard instances for one technology are trivially solved by the other.

The rest of the paper is organized as follows. The paper starts by describing the basic local search algorithm, including its underlying modeling, its neighborhood, its metaheuristic, and its experimental results. It then presents the constructive heuristic and reports the new experimental results when the heuristic replaces the random configurations as starting points of the algorithm. Finally, the paper discusses related work and concludes by giving some preliminary results on generalizations of the problem.

2   The Modeling

There are many possible modelings for the social golfer problem, which is one of the reasons why it is so interesting. This paper uses a modeling that associates a decision variable $x[w, g, p]$ with every position p of every group g of every week w. Given a schedule σ, i.e., an assignment of values to the decision variables, the value $\sigma(x[w, g, p])$ denotes the golfer scheduled in position p of group g in week w. There are two kinds of constraints in the social golfer.

1. A golfer plays exactly once a week;
2. Two golfers can play together (i.e., in the same group of the same week) at most once.

The first type of constraints is implicit in the algorithms presented in this paper: It is satisfied by the initial assignments and is preserved by local moves. The second set of constraints is represented explicitly. The model contains a constraint $m[a, b]$ for every pair (a, b) of golfers: Constraint $m[a, b]$ holds for an assignment σ if golfers a and b are not assigned more than once to the same group. More precisely, if $\#_\sigma(a, b)$ denotes the number of times golfers a and b meet in schedule σ, i.e.,

$$\#\{(w, g) \mid \exists p, p' : \sigma(x[w, g, p]) = a \ \& \ \sigma(x[w, g, p']) = b\},$$

constraint $m[a, b]$ holds if

$$\#_\sigma(a, b) \le 1.$$

To guide the algorithm, the model also specifies violations of the constraints. Informally speaking, the violations $v_\sigma(m[a, b])$ of a constraint $m[a, b]$ is the number of times golfers a and b are scheduled in the same group in schedule σ beyond their allowed meeting. In symbols,

$$v_\sigma(m[a, b]) = \max(0, \#_\sigma(a, b) - 1).$$

As a consequence, the social golfer problem can be modeled as the problem of finding a schedule σ minimizing the total number of violations $f(\sigma)$ where

$$f(\sigma) = \sum_{a,b \in \mathcal{G}} v_\sigma(m[a, b]).$$

and $\mathcal{G}$ is the set of $g \times p$ golfers. A schedule σ with $f(\sigma) = 0$ is a solution to the social golfer problem.

3   The Neighborhood

The neighborhood of the local search consists of swapping two golfers from different groups in the same week. The set of swaps is thus defined as

$$\mathcal{S} = \{(\langle w, g_1, p_1 \rangle, \langle w, g_2, p_2 \rangle) \mid g_1 \ne g_2\}.$$

It is more effective however to restrict attention to swaps involving at least one golfer in conflict with another golfer in the same group. This ensures that the algorithm focuses on swaps which may decrease the number of violations. More formally, a triple $\langle g, w, p \rangle$ is said to be in conflict in schedule σ, which is denoted by $v_\sigma(\langle g, w, p \rangle)$, if

$$\exists p' \in P : v_\sigma(m[\sigma(x[w, g, p]), \sigma(x[w, g, p'])]) > 1.$$

With this restriction in mind, the set of swaps $\mathcal{S}^-(\sigma)$ considered for a schedule σ becomes

$$\mathcal{S}^-(\sigma) = \{(\langle w_1, g_1, p_1 \rangle, \langle w_2, g_2, p_2 \rangle) \in \mathcal{S} \mid v_\sigma(\langle w_1, g_1, p_1 \rangle)\}.$$

4   The Tabu Component

The tabu component of the algorithm is based on three main ideas. First, the tabu list is distributed across the various weeks, which is natural since the swaps only consider golfers in the same week. The tabu component thus consists of an array $tabu$ where $tabu[w]$ represents the tabu list associated with week w. Second, for a given week w, the tabu list maintains triplet $\langle a, b, i \rangle$, where a and b are two golfers and i represents the first iteration where golfers a and b can be swapped again in week w. Observe that the tabu lists store golfers, not positions $\langle w, g, p \rangle$. Third, the tabu tenure, i.e., the time a pair of golfers (a, b) stays in the list, is dynamic: It is randomly generated in the interval $[4, 100]$. In other words, each time a pair of golfers (a, b) is swapped, a random value ρ is drawn uniformly from the interval $[4, 100]$ and the pair (a, b) is tabu for the next ρ iterations. At iteration k, swapping two golfers a and b is tabu, which is denoted by

$$tabu[w](a, b, k)$$

if the Boolean expression

$$\langle a, b, i \rangle \in tabu[w] \ \& \ i \leq k$$

holds. As a consequence, for schedule σ and iteration k, the neighborhood consists of the set of moves $\mathcal{S}^t(\sigma, k)$ defined as

$$\mathcal{S}^t(\sigma, k) = \{(t_1, t_2) \in \mathcal{S}^-(\sigma) \mid \neg tabu[w](\sigma(x[t_1]), \sigma(x[t_2]), k)\}.$$

where we abuse notations and use $x[\langle w, g, p \rangle]$ to denote $x[w, g, p]$.

Aspiration In addition to the non-tabu moves, the neighborhood also considers moves that improve the best solution found so far, i.e., the set $\mathcal{S}^*(\sigma, \sigma^*)$ defined as

$$\mathcal{S}^*(\sigma, \sigma^*) = \{(t_1, t_2) \in \mathcal{S}^-(\sigma) \mid f(\sigma[x[t_1] \leftrightarrow x[t_2]]) < f(\sigma^*)\},$$

where $\sigma[x_1 \leftrightarrow x_2]$ denotes the schedule σ where the values of variables x_1 and x_2 have been swapped and σ^* denotes the best solution found so far.

5   The Tabu-Search Algorithm

We are now ready to present the basic local search algorithm SGLS. The algorithm, depicted in Figure 1, is a tabu search with a restarting component. Lines 2-7 perform the initializations. In particular, the tabu list is initialized in lines 2-3, the initial schedule is generated randomly in line 4, while lines 6 and 7 initialize the iteration counter k, and the stability counter s. The initial configuration σ randomly schedules all golfers in the various groups for every week, satisfying the constraint that each golfer plays exactly once a week. The best schedule found so far σ^* is initialized to σ.

The core of the algorithm is given in lines 8-23. They iterate local moves for a number of iterations or until a solution is found. The local move is selected in line 9. The key idea is to select the best swaps in the neighborhood

$$\mathcal{S}^t(\sigma, k) \ \cup \ \mathcal{S}^*(\sigma, \sigma^*),$$

i.e., the non-tabu swaps and those improving the best schedule. Observe that the expression

$$f(\sigma[x[t_1] \leftrightarrow x[t_2]])$$

represents the number of violations obtained after swapping t_1 and t_2. The tabu list is updated in line 11, where $week(\langle w, g, p \rangle)$ is defined as

$$week(\langle w, g, p \rangle) = w.$$

```
1.    SGLS(W, G, P)
2.        forall w ∈ W
3.            tabu[w] ← {};
4.        σ ← random configuration;
5.        σ* ← σ;
6.        k ← 0;
7.        s ← 0;
8.        while k ≤ maxIt & f(σ) > 0 do
9.            select (t₁, t₂) ∈ Sᵗ(σ, k) ∪ S*(σ, σ*) minimizing f(σ[x[t₁] ↔ x[t₂]]);
10.           τ ← RANDOM([4, 100]);
11.           tabu[week(t₁)] ← tabu[week(t₁)] ∪ {⟨σ(x[t₁]), σ(x[t₂]), k + τ⟩};
12.           σ ← σ[x[t₁] ↔ x[t₂]];
13.           if f(σ) < f(σ*) then
14.               σ* ← σ;
15.               s ← 0;
16.           else if s > maxStable then
17.               σ ←random configuration;
18.               s ← 0;
19.               forall w ∈ W do
20.                   tabu[w] = {};
21.           else
22.               s++;
23.           k++;
```

Fig. 1. Algorithm SGLS for Scheduling Social Golfers

The new schedule is computed in line 12. Lines 13-15 update the best schedule, while lines 16-20 specify the restarting component.

The restarting component simply reinitializes the search from a random configuration whenever the best schedule found so far has not been improved upon for *maxStable* iterations. Note that the stability counter s is incremented in line 22 and reset to zero in line 15 (when a new best schedule is found) and in line 18 (when the search is restarted).

6   Experimental Results

This section reports the experimental results for the SGLS algorithm. The algorithm was implemented in C and the experiments were carried out on a 3.06GHz PC with 512MB of RAM. Algorithm SGLS was run 100 times on each instance and the results report average values for successful runs, as well as the percentage of unsuccessful runs (if any).

Tables 1 and 2 report the experimental results for SGLS. Given a number of groups g and a group size p, the tables only give the results for those instances $g - p - w$ maximizing w since they also provide solutions for $w' < w$. Table 1 reports the number of iterations (moves), while Table 2 reports the execution times. Bold entries indicate that SGLS matches the best known number of weeks for a given number of groups and a given group size. The percentage of unsuccessful runs is shown between parentheses in Table 2.

Table 1. Number of Iterations for SGLS with Maximal Number of Weeks. Bold Entries Indicate a Match with the Best Known Number of Weeks

g	size 3 w	I	size 4 w	I	size 5 w	I	size 6 w	I	size 7 w	I	size 8 w	I	size 9 w	I	size 10 w	I
6	**8**	**282254.0**	6	161530.3	6	16761.5	3	15.8	-	-	-		-	-	-	-
7	**9**	**12507.6**	7	274606.0	5	102.9	4	100.4	3	23.4	-		-		-	
8	**10**	**653.9**	8	323141.5	6	423.7	5	1044.9	4	237.5	4	153301.6	-		-	
9	**11**	**128.3**	8	84.4	6	52.7	5	55.5	4	44.8	3	27.7	3	43.9	-	
10	**13**	**45849.1**	9	100.2	7	80.8	6	110.7	5	94.6	4	61.8	3	36.1	3	53.3

Table 2. CPU Time in Seconds for SGLS with Maximal Number of Weeks. Bold Entries Indicate a Match with the Best Known Number of Weeks

g	size 3 w	T	%F	size 4 w	T	%F	size 5 w	T	size 6 w	T	size 7 w	T	size 8 w	T	size 9 w	T	size 10 w	T
6	**8**	**48.93**	6	6	47.75		**6**	**107.18**	3	0.01	-	-	-		-		-	
7	**9**	**3.06**		7	107.62	8	5	0.07	4	0.09	3	0.03	-		-		-	
8	**10**	**0.23**		8	207.77	9	6	0.37	5	1.21	4	0.39	4	360.00	-		-	
9	**11**	**0.08**		8	0.09		6	0.09	5	0.13	4	0.14	3	0.09	3	0.19	-	
10	**13**	**30.82**		9	0.16		7	0.19	6	0.34	5	0.41	4	0.33	3	0.20	3	0.39

As can be seen from the tables, Algorithm SGLS finds solutions to all the instances solved by constraint programming except 4. Moreover, almost all entries are solved in less than a second. Only a few instances are hard for the algorithm and require around 1 minute of CPU time. Interestingly, algorithm SGLS also solves 7 new instances: $9 - 4 - 9, 9 - 5 - 7, 9 - 6 - 6, 9 - 7 - 5, 9 - 8 - 4, 10 - 5 - 8$ and $10 - 9 - 4$.

It is interesting to observe that algorithm SGLS does not break symmetries and does not exploit specific properties of the solutions. This contrasts with constraint-programming solutions that are often quite sophisticated and involved. See, for instance, the recent papers [2, 9] which report the use of very interesting symmetry-breaking schemes to schedule social golfers.

7   A Constructive Heuristic

The quality of SGLS can be further improved by using a constructive heuristic to find a good starting, and restarting, configuration. The heuristic [3] trivially solves $p - p - (p + 1)$ instances when p is prime and provides good starting points (or solutions) for other instances as well. Examples of such initial configurations are given in Tables 3 and 4, which will be used to explain the intuition underlying the constructive heuristic. The heuristic simply aims at exploiting the fact that all golfers in a group for a given week must be assigned a different group in subsequent weeks. As a consequence, the heuristic attempts to distribute these golfers in different groups in subsequent weeks.

Table 4 is a simple illustration of the heuristic with 5 groups of size 5 (i.e., 25 golfers) and 6 weeks. The first week is simply the sequence 1..25. In the second week, group i consists of all golfers in position i in week 1. In particular, group 1 consists

Table 3. The initial configuration for the problem $4 - 3 - 3$

weeks	group 1	group 2	group 3	group 4
week 1	1 2 3	4 5 6	7 8 9	10 11 12
week 2	1 4 7	10 2 5	8 11 3	6 9 12
week 3	1 5 9	10 2 6	7 11 3	4 8 12

Table 4. The intial configuration for the problem $5 - 5 - 6$

weeks	group 1	group 2	group 3	group 4	group 5
week 1	1 2 3 4 5	6 7 8 9 10	11 12 13 14 15	16 17 18 19 20	21 22 23 24 25
week 2	1 6 11 16 21	2 7 12 17 22	3 8 13 18 23	4 9 14 19 24	5 10 15 20 25
week 3	1 7 13 19 25	2 8 14 20 21	3 9 15 16 22	4 10 11 17 23	5 6 12 18 24
week 4	1 8 15 17 24	2 9 11 18 25	3 10 12 19 21	4 6 13 20 22	5 7 14 16 23
week 5	1 9 12 20 23	2 10 13 16 24	3 6 14 17 25	4 7 15 18 21	5 8 11 19 22
week 6	1 10 14 18 22	2 6 15 19 23	3 7 11 20 24	4 8 12 16 25	5 9 13 17 21

of golfers $1, 6, 11, 16, 21$, group 2 is composed of golfers $2, 7, 12, 17, 22$ and so on. In other words, the groups consist of golfers in the same group position in week 1. The third week is most interesting, since it gives the intuition behind the heuristic. The key idea is to try to select golfers whose positions are j,j+1,j+2,j+3,j+4 in the first week, the addition being modulo the group size. In particular, group 1 is obtained by selecting the golfers in position i from group i in week 1, i.e., golfers $1, 7, 13, 19, 25$. Subsequent weeks are obtained in similar fashion by simply incrementing the offset. In particular, the fourth week considers sequences of positions of the form j,j+2,j+4,j+6,j+8 and its first group is $1, 8, 15, 17, 24$. Table 3 illustrates the heuristic on the 4-3-3 instance. Note that the first group in week 2 has golfers in the first position in groups 1, 2, and 3 in week 1. However, the first golfer in week 4 must still be scheduled. Hence the second group must select golfer 10, as well as golfers 2 and 5.

Figure 2 depicts the code of the constructive heuristic. The code takes the convention that the weeks are numbered from 0 to $w-1$, the groups from 0 to $g-1$, and the positions from 0 to $p - 1$, since this simplifies the algorithm. The key intuition to understand the code is to recognize that a week can be seen as a permutation of the golfers on which the group structure is superimposed. Indeed, it suffices to assign the first p positions to the first group, the second set of p positions to the second group and so on. As a consequence, the constructive heuristic only focuses on the problem of generating w permutations $P_0, \ldots, P_{w-1}$.

The top-level function is HEURISTICSCHEDULE which specifies the first week and calls function SCHEDULEWEEEK for the remaining weeks. Scheduling a week is the core of the heuristic. All weeks start with golfer 1 (line 7) and initialize the position po to 0 (line 8), the group number gr to 1 (line 9), and the offset Δ to $we - 1$. The remaining golfers are scheduled in lines 11-15.

The key operation is line 12, which selects the first *unscheduled* golfer s from group gr of week 0 (specified by P_0) starting at position $(po + \Delta)\%p$ and proceeding by viewing the group as a circular list. The next three instructions update the position po to the position of s in group gr of week 0 (line 13), increment the group to select a

1. HEURISTICSCHEDULE(w, g, p)
2. $\quad n \leftarrow g \times p;$
3. $\quad P_0 \leftarrow \langle 1, \ldots, n \rangle;$
4. $\quad$ **forall** $we \in 1..w - 1$
5. $\quad\quad P_{we} \leftarrow scheduleWeek(we, g, p, n);$

6. SCHEDULEWEEK(we, g, p, n)
7. $\quad P_{we} \leftarrow \langle 1 \rangle;$
8. $\quad po \leftarrow 0;$
9. $\quad gr \leftarrow 1;$
10. $\quad \Delta \leftarrow we - 1;$
11. $\quad$ **forall** $go \in 1..n - 1$
12. $\quad\quad s \leftarrow$ SELECT$(gr, (po + \Delta)\%p);$
13. $\quad\quad po \leftarrow$ POSITION$(s);$
14. $\quad\quad gr \leftarrow (gr + 1)\%g;$
15. $\quad\quad P_{we} \leftarrow P_{we} :: \langle s \rangle;$
16. $\quad$ **return** $P_{we};$

Fig. 2. The Constructive Heuristic for Scheduling Social Golfers

golfer from the next group, and extend the permutation by concatenating s to P_{we}. By specification of SELECT, which only selects unscheduled golfers and the fact that the heuristic selects the golfers from the groups in a round-robin fashion, the algorithm is guaranteed to generate a permutation.

8   Experimental Results Again

This section discusses the performance of algorithm SGLS-CH that enhances SGLS with the constructive heuristic to generate starting/restarting points. Although the starting point is deterministic, the algorithm still uses restarting, since the search itself is randomized, i.e., ties are broken randomly.

8.1   The $odd - odd - w$ Instances

It is known that the constructive heuristic finds solutions for $p - p - (p + 1)$ instances when p is prime. Moreover, it also provides solutions to many instances of the form $odd - odd - w$ as we now show experimentally. The results were performed up to $odd = 49$. For all (odd) prime numbers p lower than 49, the heuristic solves the instances $p - p - w$, where w is the maximal number of weeks for p groups and periods. When odd is divisible by 3, the heuristic solves instances of the form $odd - odd - 4$, when odd is divisible by 5, it solves instances of the form $odd - odd - 6$, and when odd is divisible by 7, it solves instances of the form $odd - odd - 8$. For instance, the constructive heuristic solves instance 49-49-8.

It is interesting to relate these results to mutually orthogonal latin squares. Indeed, it is known that finding a solution for instances of the form $g - g - 4$ is equivalent to the problem of finding two orthogonal latin squares of size g. Moreover, instances of the form $g - g - n$ are equivalent to the problem of finding $n - 2$ mutually orthogonal latin

Table 5. Results on the $odd - odd - w$ Instances

instances	$CH : w$	Gol:LB
3-3-w	4	4
5-5-w	6	6
7-7-w	8	8
9-9-w	4	10
11-11-w	12	12
13-13-w	14	14
15-15-w	4	6
17-17-w	18	18
19-19-w	20	20
21-21-w	4	7
23-23-w	24	24
25-25-w	6	26
27-27-w	4	28
29-29-w	30	30
31-31-w	32	32
33-33-w	4	7
35-35-w	6	7
37-37-w	38	38
39-39-w	4	6
41-41-w	42	42
43-43-w	44	44
45-45-w	4	8
47-47-w	48	48
49-49-w	8	50

squares of size g [3, 10]. Instances of the form $g - g - 4$ can be solved in polynomial time when g is odd. This provides some insight into the structure of these instances and some rationale why the constructive heuristic is able to solve many of the $odd - odd - w$ instances. Table 5 summarizes the results on the $odd - odd - w$ instances. The columns respectively specify the instances, the largest w found by the constructive heuristic, and the number of weeks w for the social golfers that corresponds to the best lower bound on the latin square as given in [4]. Rows in bold faces indicate closed instances.

It is interesting to observe that the lower bounds on the mutually orthogonal latin squares vary significantly. Indeed, the lower bound for size 17 is 16, while it is 4 for size 15. These lower bounds give some additional insights on the inherent difficulty of these instances and on the behavior of the constructive heuristic.

8.2   Hard Instances

Table 6 compares the tabu-search algorithm with and without the constructive heuristic on the hard instances from Table 2. Note that $7 - 7 - 7$ and $7 - 7 - 8$ are now trivially solved, as well as $9 - 9 - 4$ which was also open. SGLS-CH does not strictly dominates SGLS, as there are instances where it is slightly slower. However, on some instances,

Table 6. Comparison between SGLS and SGLS-CH

instances	random			new		
	I	T	%F	I	T	%F
6-3-8	282254.07	48.93	6	250572	43.84	4
6-4-6	161530.35	47.75		168000	49.66	
7-4-7	274606.00	107.18		200087	124.15	
8-4-8	323141.52	107.62	8	316639	141.91	3
8-8-4	153301.61	360.00		8380.45	19.54	
8-8-5	–	–	100	108654.00	496.82	
10-3-13	45849.00	30.82		51015.00	34.28	

Table 7. Experimental Results of SGLS-CH on the New Solved Instances

instance	I	T	%solved
7-5-6	487025.0	370.50	10
9-4-9	469156.4	402.55	100
9-5-7	4615.0	5.39	100
9-6-6	118196.7	196.52	100
9-7-5	64283.9	155.16	100
9-8-4	1061.3	2.92	100
10-4-10	548071.6	635.20	100
10-5-8	45895.4	76.80	100
10-9-4	5497.9	24.42	100

it is clearly superior (including on $8 - 8 - 5$ which can now be solved). Algorithm SGLS-CH also closes two additional open problems: $7 - 5 - 6$ and $10 - 4 - 10$. Table 7 depicts the performance of algorithm SGLS-CH on the new solved instances.

8.3   Summary of the Results

Table 8 summarizes the results of this paper. It depicts the status of maximal instances for SGLS-CH and whether the instances are hard (more than 10 seconds) or easy (less than 10 seconds). The results indicate that SGLS-CH matches or improves the best results for all but 3 instances. In addition, it produces 11 new solutions with respect to earlier results. These results are quite remarkable given the simplicity of the approach. Indeed, constraint-programming approaches to the social golfer problem are typically very involved and use elegant, but complex, symmetry-breaking techniques. Algorithm SGLS-CH, in contrast, does not include any such symmetry breaking.

It is interesting to observe the highly constrained nature of the instances for which SGLS-CH does not match the best-known results. Hence it is not surprising that constraint programming outperforms local search on these instances. Note also that Brisset and Barnier [2] proposed a very simple constraint-programming model to solve $8-4-9$ in a few seconds. So, once again, there seems to be a nice complementarity between constraint programming and local search on the social golfer problem.

Table 8. Summary of the Results for SGLS-CH with Maximal Number of Weeks. Bold entries represent a match or an improvement over existing solutions. The status is *new* (for improvement), hard (> 10 seconds), and easy (≤ 10 seconds)

#groups	size 3 w status	size 4 w status	size 5 w status	size 6 w status	size 7 w status	size 8 w status	size 9 w status	size 10 w status
6	**8 Hard**	6 Hard	**6 Hard**	**3 Easy**	- -	- -	- -	- -
7	**9 Easy**	**7 Hard**	6 **New**	**4 Easy**	**8 New**	- -	- -	- -
8	**10 Easy**	8 Hard	**6 Easy**	**5 Easy**	**4 Easy**	5 Hard	- -	- -
9	**11 Easy**	9 **New**	7 **New**	6 **New**	5 **New**	4 **New**	4 **New**	- -
10	**13 Hard**	**10 New**	**8 New**	**6 Easy**	**5 Easy**	**4 Easy**	4 **New**	**3 Easy**

Table 9. Summary of the Results for Atmost Two Meetings. Easy < 10s. Medm > 20s & < 5% unsolved. Hard < 50% unsolved. Chal > 50% unsolved

#groups	size 3 w status	size 4 w status	size 5 w status	size 6 w status	size 7 w status	size 8 w status	size 9 w status	size 10 w status
3	8 Easy	6 Easy	6 Easy	6 Easy	4 Easy	4 Easy	4 Easy	2 -
4	11 Easy	10 Easy	8 Chal	6 Easy	6 Easy	7 Chal	6 Easy	6 Easy
5	14 Hard	11 Easy	9 Easy	8 Easy	7 Easy	7 Hard	6 Easy	6 Easy
6	16 Easy	13 Easy	11 Easy	10 Easy	9 Easy	8 Easy	7 Easy	7 Easy
7	19 Easy	15 Easy	13 Easy	12 Easy	11 Medm	10 Easy	9 Easy	8 Easy
8	22 Medm	18 Medm	15 Easy	14 Hard	12 Easy	11 Easy	10 Easy	10 Medm
9	25 Chal	20 Easy	17 Easy	15 Easy	14 Easy	13 Medm	12 Easy	11 Easy
10	27 Easy	22 Easy	19 Easy	17 Easy	15 Easy	14 Easy	13 Easy	12 Easy

9 Related Work

There is a considerable body of work on scheduling social golfers in the constraint programming community. References [2, 6, 9] describe state-of-the art results using constraint programming and are excellent starting points for more references.[2] See also [10] for interesting theoretical and experimental results on the social golfer problem, as well as the description of SBDD, a general scheme for symmetry breaking. Reference [1] describes a tabu-search algorithm for scheduling social golfers, where the neighborhood consists of swapping the value of a single variable and where all constraints are explicit. The results are very far in quality and performance from those reported here.

10 Conclusion

This paper reconsidered the scheduling of social golfers, a highly combinatorial and symmetric application which has raised significant interest in the constraint programming community. It presented an effective local search algorithm which found the first

---

[2] Reference [9] contains much more general results on symmetry breaking but the scheduling of social golfers is one of the main applications in evaluating the new techniques.

solutions to 11 new instances and matched, or improved upon, all instances solved by constraint programming solutions but 3. Moreover, the local search algorithm was shown to find almost all solutions in less than a couple of seconds, the harder instances taking about 1 minute. The algorithm also features a constructive heuristic which trivially solves many instances of the form $odd - odd - w$.

It is interesting to conclude with a number of interesting observations. First, the social golfer is a problem where the properties of the instances seem to determine which approach is best positioned to solve them. In particular, hard instances for constraint programming are easy for local search and vive-versa. There are of course other applications where this also holds. What is interesting here is the simplicity of local search compared to its constraint programming counterpart and the absence of symmetry-breaking schemes in local search. Whether this observation generalizes to other, highly symmetric, problems is an interesting issue for future work. See, for instance, [7, 8] for early results along these lines.

Second, there are many interesting variations around the social golfer problem that are generating increasing interest. These variations include the possibility of golfers to meet more than once, as well as the superimposition of a referee assignment minimizing the number of referees subject to "fairness" constraints. Our preliminary results with local search on these problems, which are motivated by real-life applications, are extremely encouraging. In particular, Table 9 reports some very preliminary results on the generalizations where golfers are allowed to meet more than once. The instances are classified into easy, medium, hard, and challenging. Hard instances mean that local search may occasionally fail to find a solution in 100,000 iterations (but in less than 50% of the time), while challenging instances fail in finding a solution more than 50% of the time within the iteration limit. Once again, given a number of groups g and a group size p, the tables only give the results for those instances $g - p - w$ maximizing w since they also provide solutions for $w' < w$.

Finally, there are many connections with Latin squares that could be exploited further. It is likely that new heuristic solutions based on this connection would close additional instances and provide good starting points on others.

Acknowledgements

Thanks to Warwick Harvey and Meinolf Sellmann for pointing out relevant results on social golfers and Latin squares, and to the reviewers for several interesting comments and suggestions. This work was partially supported by NSF ITR Awards ACI-0121497.

References

1. M. Ågren. Solving the Social Golfer Using Local Search. peg.it.uu.se/ saps02/ MagnusAgren/, 2003.
2. N. Barnier and P. Brisset. Solving kirkman's schoolgirl problem in a few seconds. *Constraints*, 2005. To appear in the Special Issue on CP'02.
3. C. Colbourn and Dinitz. *The CRC Handbook of Combinatorial Design*. CRC Press, Boca Raton, FL, 1996.

4. C.J. Colbourn and J.H. Dinitz. Mutually orthogonal latin squares: A brief survey of constructions. *"Journal of Statistical Planning and Inference"*, 95:9–48, 2001.

5. T. Fahle, S. Schamberger, and M. Sellmann. Symmetry breaking. In Toby Walsh, editor, *Principles and Practice of Constraint Programmingo*, volume 2293 of *LNCS*, pages 93–107. Springer Verlag, 2001.

6. S. Prestwich. Randomized backtracing for linear pseudo-boolean constraint problems. In *Proceedings of the Fourth International Workshop on Integration of AI and OR Techniques in Constraint Programming for Combinatorial Optimisation Problems (CP-AI-OR'02)*, pages 7–19, Le Croisic, France, March 2002.

7. S.D. Prestwich. Supersymmetric Modeling for Local Search. In *Second International Workshop on Symmetry in Constraint Satisfaction Problems*, 2002.

8. S.D. Prestwich. Negative Effects of Modeling Techniques on Search Performance. *Annals of Operations Research*, 118:137–150, 2003.

9. J.F. Puget. Symmetry breaking revisited. *Constraints*, 2005. To appear in the Special Issue on CP'02.

10. M. Sellmann. *Reduction Techniques in Constraint Programming and Combinatorial Optimization*. PhD thesis, University of Paderborn, Germany, 2003.

11. Meinolf Sellmann and Warwick Harvey. Heuristic constraint propagation – Using local search for incomplete pruning and domain filtering of redundant constraints for the social golfer problem. In Narendra Jussien and François Laburthe, editors, *Proceedings of the Fourth International Workshop on Integration of AI and OR Techniques in Constraint Programming for Combinatorial Optimisation Problems (CP-AI-OR'02)*, pages 191–204, Le Croisic, France, March 2002.

12. B. Smith. Reducing Symmetry in a Combinatorial Design Problem. In *CP-AI-OR'2001*, pages 351–359, Wye College (Imperial College), Ashford, Kent UK, April 2001.

Multiconsistency and Robustness with Global Constraints

Khaled Elbassioni and Irit Katriel

Max-Planck-Institut für Informatik, Saarbrücken, Germany
{elbassio, irit}@mpi-sb.mpg.de

Abstract. We propose a natural generalization of arc-consistency, which we call multiconsistency: A value v in the domain of a variable x is k-multiconsistent with respect to a constraint C if there are at least k solutions to C in which x is assigned the value v. We present algorithms that determine which variable-value pairs are k-multiconsistent with respect to several well known global constraints. In addition, we show that finding super solutions is strictly harder than finding arbitrary solutions and suggest multiconsistency as an alternative way to search for robust solutions.

1   Introduction

A value v in the domain of a variable x is *consistent* with respect to the constraint C if there is at least one solution to the constraint in which x is assigned the value v. Identifying values which are not consistent is a fundamental task for a constraint solver; it is crucial for reducing the exponential-size search space that would otherwise need to be explored.

In this paper we generalize the notion of consistency: A value v in the domain of the variable x is k-multiconsistent with respect to a constraint C if there are at least k solutions to C in which x is assigned the value v. Intuitively, a value that appears in many solutions is a *"useful"* value. Knowing which values are useful can be helpful in several ways. For example, usefulness can be used as a heuristic while searching for a solution: While it is not guaranteed, it seems reasonable to assume that if the constraint program has a solution s, then the more useful a variable-value pair is with respect to individual constraints that are defined on it, the more likely it is to be used in s. This implies that it makes sense to regard the usefulness of the values as a heuristic that guides the search.

Another possible application is in the search for robust solutions, i.e., solutions that can be repaired if a small change occurs. In their recent paper on the topic, Hebrard et al. [5] give the example of a schedule: A robust schedule does not collapse if one job takes slightly longer to execute than planned. Rather, the schedule changes locally and the overall makespan changes little if at all. In the same paper, they define the notion of *super solutions*, which is a generalization of super models in propositional satisfiability. An (a, b)-super solution is a solution

R. Barták and M. Milano (Eds.): CPAIOR 2005, LNCS 3524, pp. 168–182, 2005.

such that if a variables lose their values, a new solution can be constructed by assigning new values to these a variables and changing the values of at most b other variables.

This is a very strong guarantee of robustness, and Hebrard et al. note that it is quite rare to have a solution for which all of the variables can be repaired. Therefore, they formulate the optimization problem of seeking the "most robust" solution, i.e., the solution that maximizes the number of repairable variables. They then study several approaches for finding super solutions, and the super MAC search algorithm that they have developed for this purpose emerged as the most promising. As for complexity, Hebrard et al. show that it is, in general, NP-hard to find an (a, b)-super solution for a constraint program, for any fixed a. They show this by proving that any constraint program P can be transformed in polynomial time into a second constraint program P' such that P has a solution iff P' has an (a, b)-super solution. Thus, finding a super solution is as hard as finding an arbitrary solution, which is NP-hard. We will show that, in fact, finding a super solution is strictly harder than finding an arbitrary solution. In particular, we will prove that it is NP-hard to determine whether an *AllDifferent* constraint has a $(1, 0)$-super solution. Finding an arbitrary solution to an *AllDifferent* constraint can be done in polynomial time [6, 11].

On the other hand, we will show that there are efficient algorithms to determine which values are k-multiconsistent with respect to *AllDifferent* and other global constraints. This information can easily be used to search for a k-multiconsistent solution, i.e., a solution that uses only k-multiconsistent assignments of values to the variables, or, if no such solution exists, a solution that maximizes the number of k-multiconsistent values. It is not guaranteed that a k-multiconsistent solution can be easily repaired if some of the variables lose their values. We are certainly not guaranteed that a local change will give a new solution, or even that another solution exists. However, with a k-multiconsistent solution, we do know that the remaining variables are assigned to values that were once considered "useful", and our purpose is to show that the computational price we need to pay for this knowledge is not very high in the case of the constraints that we consider. We therefore believe that it would be worthwhile to conduct experimental and theoretical research on the concept of multiconsistency and ways in which it can be applied.

Section 2 contains a formal definition of k-multiconsistency and other notions that will appear in the following sections. In Section 3 we show that it is NP-hard to determine whether an *AllDifferent* constraint has a $(1, 0)$-super solution. In Section 4 we describe an algorithm that computes k-multiconsistency for the *AllDifferent* constraint when the number of variables is equal to the number of values. In Section 5 we show that this basic algorithm can be generalized for the general *AllDifferent* constraint and for other global constraints. Finally, in Section 6 we list some open problems that arise from our work.

2   Multiconsistency and Preliminaries

2.1   Multiconsistency

The formal definition of multiconsistency appears below. It is a straightforward generalization of the definition of arc-consistency [1] as appears, e.g., in [1].

Definition 1. *Let C be a constraint on the variables $x_1, \ldots, x_\ell$ with respective domains $D(x_1), \ldots, D(x_\ell)$ and let $S \subseteq D(x_1) \times \ldots \times D(x_\ell)$ be the set of solutions to C. Then a variable-value pair (x_j, v_i) is k-multiconsistent with respect to C if there are at least k tuples in S in which the jth component is v_i.*

The rest of the paper deals with multiconsistency of individual values. However, for completeness, we include the definition of a multiconsistent solution.

Definition 2. *Let C be a constraint on the variables $x_1, \ldots, x_\ell$ with respective domains $D(x_1), \ldots, D(x_\ell)$. Let s be a solution to C and for all $1 \le j \le \ell$, let $s(x_j)$ be the value assigned by s to x_j. s is a k-multiconsistent solution if for every $1 \le j \le \ell$, $(x_j, s(x_j))$ is k-multiconsistent with respect to C.*

2.2   A Few Global Constraints

The global constraints that we will consider in this paper are:

- The *AllDifferent*$(x_1, \ldots, x_n)$ [8, 9, 10, 13, 15] constraint is specified on n assignment variables. A solution s assigns each variable x_i a value $s(x_i) \in D(x_i)$ such that for any $1 \le i < j \le n$, $s(x_i) \neq s(x_j)$.
- The *Global Cardinality Constraint* $GCC(x_1, \cdots, x_n, c_{v_1}, \cdots, c_{v_{n'}})$ [7, 11, 12, 14] is specified on n assignment variables $x_1, \ldots, x_n$ and n' count variables $c_{v_1}, \ldots, c_{v_{n'}}$. A solution s assigns each assignment variable x_j a value $s(x_j) \in D(x_j) \subseteq D = \{v_1, \cdots, v_{n'}\}$ and assigns each count variable c_{v_i} a value $s(c_{v_i}) \in D(c_{v_i})$ such that each value v_i is assigned to exactly $s(c_{v_i})$ assignment variables. We will assume that the domains of the count variables are intervals, each of which is specified by a lower and upper bound, i.e., $D(c_{v_i}) = [L_i, U_i]$.
- The *Same*$(X = \{x_1, \ldots, x_n\}, Z = \{z_1, \ldots, z_n\})$ [2] constraint is defined on two sets X and Z of distinct variables such that $|X| = |Z|$. A solution s assigns each variable $v \in X \cup Z$ a value $s(v) \in D(v)$ such that the multiset of values assigned to the variables of X is identical to the multiset of values assigned to the variables of Z.

2.3   Matchings

The solutions to the global constraints we consider in this paper will be represented as subsets of the edges of a graph that models the constraint. The following terms will be used:

---

[1] Some texts refer to *hyper-arc-consistency* when speaking of global constraints and reserve the term *arc-consistency* for the special case of binary constraints.

Definition 3. *Given a graph $G = (V, E)$, a subset M of E is a matching if every node is incident to at most one edge from M. It is a perfect matching if every node is incident to exactly one edge from M.*

In the case of a bipartite graph we have the following definition:

Definition 4. *Let $G = (V, E)$ be a bipartite graph with $V = X \cup Y$ such that X is the set of nodes on one side, Y is the set of nodes on the other side, and $|X| \leq |Y|$. Then a subset M of E is called an X-perfect matching if every node in X is incident to exactly one edge from M, and every node Y is incident to at most one edge from M.*

We turn to the more general case where each node of the graph has a capacity requirement that specifies how many of the edges incident to it should be included in the matching.

Definition 5. *Let $G = (V, E, C)$ be a capacitate graph, where C is a function that maps every node $v \in V$ to an interval $C(v) = [L_v, U_v]$. We call $C(v)$ the* capacity requirement *of v. A generalized matching [7] in G is a subset M of its edges such that each node $v \in V$ is incident to at least L_v and at most U_v edges in M.*

Alternating cycles and paths will appear as an important tool in our algorithms.

Definition 6. *Let G be a graph and let M be a subset of its edges. An* alternating path (cycle) *in G with respect to M is a simple path (cycle) in G where each edge belonging to M in the path (cycle) (except the last in the case of a path), is followed by an edge which is not in M, and vice versa.*

2.4   Flows

When the graph is directed and has capacities on the edges, and not on the nodes, we can view it as a flow network.

Definition 7. *Given a directed graph $\vec{G} = (V, \vec{E})$ with lower and upper capacities l_e, u_e for each arc $e \in \vec{E}$, a* feasible flow *in $\vec{G}$ is a function $f : E \to \mathbb{R}$ such that*

1. **Flow conservation:** *For each node $v \in V$,*

$$\sum_{\{u \mid (v,u) \in \vec{E}\}} f(v, u) = \sum_{\{w \mid (w,v) \in \vec{E}\}} f(w, v).$$

2. **Capacities:** *For each $e \in \vec{E}$, $l_e \leq f(e) \leq u_e$.*

An integral feasible flow *is a feasible flow such that for all $e \in \vec{E}$, $f(e)$ is an integer.*

The residual graph, defined below, appears in one of our algorithms:

Definition 8. *Given a directed graph $\vec{G} = (V, \vec{E})$ with lower and upper capacities l_e, u_e for each arc $e \in \vec{E}$ and a flow f in it, the* residual graph $\vec{G}_f$ *is defined as follows: For each $e = (u, v) \in \vec{E}$, (1) if $f(e) < u_e$ then the arc (u, v) appears in $\vec{G}_f$ with capacity $[0, u_e - f(e)]$. (2) if $f(e) > l_e$ then the arc (v, u) appears in $\vec{G}_f$ with capacity $[0, f(e) - l_e]$.*

It is not hard to show that for a directed graph $\vec{G}$, if f is a feasible flow in $\vec{G}$ and f' is a feasible flow in $\vec{G}_f$, then $f'' = f \oplus f'$ is a feasible flow in $\vec{G}$, where the operation $\oplus$ is defined as follows: If e has the same direction in $\vec{G}$ and $\vec{G}_f$ then $f(e) \oplus f'(e) = f(e) + f'(e)$ and otherwise, $f(e) \oplus f'(e) = f(e) - f'(e)$. Thus, the residual graph enables us to transform one feasible flow into another by finding a positive-weight cycle.

2.5   Enumeration Algorithms

Generally speaking, given a property $\pi : 2^U \mapsto \{0, 1\}$ defined over all subsets of a ground set U, an enumeration algorithm for π is a procedure that lists, one by one, all subsets Y of U satisfying π, i.e., for which $\pi(Y) = 1$. For example, assume that $U = E$ is the edge set of a bipartite graph $G = (V, E)$ and $\pi(Y)$ is the property that the edge set $Y \subseteq E$ is a perfect matching. Since, in general, the size of the output of an enumeration algorithm (in our case, the perfect matchings) is typically exponential in the size of the input (in our case, the size of the graph $|V| + |E|$), it is common to measure the efficiency of the algorithm in terms of the combined size of the input and output. Such an algorithm is said to be *incrementally polynomial* if, after generating a subset $\mathcal{X}$ of elements (satisfying π), the time to generate a new element is polynomial in both $|\mathcal{X}|$ and the size of the input. A stronger requirement on an enumeration algorithm is to run with *polynomial delay*, in which case, the time to generate a new element is polynomial only in the size of the input, i.e., does not depend on how many elements have been generated so far. As we shall see below, the enumeration algorithms we use are of the latter type.

3   NP-Hardness of Finding a $(1, 0)$-Super Solution to the *AllDifferent* constraint

In this section we show that it is NP-hard to determine whether an *AllDifferent* constraint has a $(1, 0)$-super solution. Since it takes $O(n^{3/2} n')$ time to determine whether it has an arbitrary solution [6, 11], this implies that finding super solutions is strictly harder than finding arbitrary solutions.

Theorem 1. *Given n variables, $x_1, \ldots, x_n$, with respective domains $D(x_1), \ldots, D(x_n)$, it is NP-hard to determine whether there exists a $(1, 0)$-super solution for the AllDifferent$(x_1, \ldots, x_n)$ constraint, even if $|D(x_i)| \le 4$ for all $1 \le i \le n$.*

Proof. We use a polynomial-time transformation from the 3SAT problem: Given a conjunctive normal form formula $\phi(y_1, \ldots, y_N) = C_1 \wedge \ldots \wedge C_m$, where each C_j is a disjunction of 3 literals in $\{y_1, \bar{y}_1, \ldots, y_N, \bar{y}_N\}$, determine whether there exists a truth assignment to $y_1, \ldots, y_N$ which satisfies all clauses of $\phi(y_1, \ldots, y_N)$.

We show that we can construct an instance of *AllDifferent* that has a $(1, 0)$-super solution iff $\phi(y_1, \ldots, y_N)$ is satisfiable. Let $n = N + m$, and $x_1, \ldots, x_n$ be n variables, the union of whose domains is $D = \{a_1, b_1, \ldots, a_N, b_N, c_1, \ldots, c_m\}$. The domains of specific variables are defined as follows. For $i = 1, \ldots, N$, let $D(x_i) = \{a_i, b_i\}$. For $j = 1, \ldots, m$, let $D(x_{j+N}) = \{c_j\} \cup \{a_i : i = 1, \ldots, N, \text{and } y_i \in C_j\} \cup \{b_i : i = 1, \ldots, N, \text{and } \bar{y}_i \in C_j\}$ (see Figure 1).

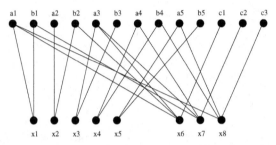

Fig. 1. The *AllDifferent* instance generated from the 3SAT formula $(y_1 \vee \bar{y}_2 \vee y_3) \wedge (\bar{y}_1 \vee y_3 \vee y_4) \wedge (y_1 \vee \bar{y}_4 \vee y_5)$

We claim that there is a $(1, 0)$-super solution to the *AllDifferent*$(x_1, \ldots, x_n)$ constraint with the specified variable domains if and only if the formula ϕ is satisfiable. Indeed, given a satisfying assignment σ for ϕ, we can construct a $(1, 0)$-super solution to our constraint as follows. For $i = 1, \ldots, N$, x_i is assigned either the value b_i or the value a_i, depending, respectively, on whether y_i is assigned *True* or *False* by σ. For $j = 1, \ldots, m$, the variable x_{j+N} is assigned the value c_j. Clearly, each variable was assigned a different value so the *AllDifferent* constraint is satisfied. Furthermore, for each variable x_i, there exists a value $v \in D(x_i)$ that has not been assigned to any other variable. This is obvious for $i = 1, \ldots, N$, and follows, for $i = N + 1, \ldots, n$ from the fact that σ is satisfying, i.e. for each clause C_j, $j = 1, \ldots, m$, there is a literal in C_j that is assigned *True* by σ. Conversely, given a $(1, 0)$-super solution to the *AllDifferent* constraint, we define a truth assignment σ to the Boolean variables $y_1, \ldots, y_N$, by setting $y_i = True$ if x_i is assigned the value b_i, and setting $y_i = False$ if x_i is assigned the value a_i. Note that, for $j = 1, \ldots, m$, each variable x_{j+N} is assigned the value c_j by the super solution since otherwise some variable x_i, $i \in \{1, \ldots, N\}$ would have all its domain values assigned to variables, contradicting the requirement of a $(1, 0)$-super solution. In particular, for each $j = 1, \ldots, m$, variable x_{j+N} must have at least one value in its domain that is also an unassigned value in the domain of some variable x_i, $i \in \{1, \ldots, N\}$. This implies that each clause C_j is satisfied by σ. $\square$

4   Multiconsistency for a Restricted Case of the *AllDifferent* Constraint

In this section we consider the following problem: Given an integer k and an *AllDifferent*$(x_1, \ldots, x_n)$ constraint where the domains of $x_1, \ldots, x_n$ are all contained in the set $\{1, \ldots, n\}$, determine which values are k-multiconsistent with respect to the constraint[2].

It is common to represent the *AllDifferent* constraint by a bipartite graph G with a node for each variable on one side, a node for each value on the other side, and an edge between the node representing the variable x_j and the node representing the value v_i iff v_i is in the domain of x_j. Then, there is a one-to-one correspondence between the solutions to the constraint and the matchings of cardinality n in G. In the restricted case that we consider in this section, a matching of cardinality n is also a perfect matching.

The k-multiconsistency problem for the restricted *AllDifferent* constraint, then, is the following: Given a bipartite graph G with n nodes on each side and m edges, determine which edges of G belong to at least k perfect matchings.

The algorithm in Figure 2 uses a recursive reformulation of an algorithm by Fukuda and Matsui that enumerates the perfect matchings in a bipartite graph [4]. For each edge e in G, the algorithm attempts to enumerate k perfect matchings that contain e. If it fails, it determines that e is not k-multiconsistent. The algorithm uses the following operations on graphs.

Definition 9. *Let $e = (u, v)$ be an edge in G. Then $G - e$ is the graph obtained by removing the edge e from G and $G \backslash e$ is the graph obtained by removing from G the nodes u and v and all edges incident to them.*

Clearly, there are k perfect matchings in $G \backslash e$ iff there are k perfect matchings in G that contain the edge e. Hence, checking whether e is k-multiconsistent is equivalent to checking whether $G \backslash e$ contains k perfect matchings.

Let $T(n, n', m)$ be the time required to find a maximum cardinality matching in a bipartite graph with n nodes on one side, n' nodes on the other, and m edges. In the $n' = n$ case, the enumeration algorithm by Fukada and Matsui needs $T(n, n, m)$ time to find the first perfect matching and then $O(n + m)$ time to generate each additional perfect matching. We get that given a single perfect matching that contains the edge e, we can check in $O(k(m + n))$ time whether e is k-multiconsistent. Note that once we have a perfect matching M, we can find, in linear time, a perfect matching M' that contains a specified edge e which is not in M: All we need is an alternating cycle that contains e. Thus, the total running time required to check k-multiconsistency for all edges is $T(n, n, m) + O(mk(m + n))$ time.

---

[2] This restriction of the *AllDifferent* constraint is also equivalent to a special case of the *Sortedness* constraint [3, 9].

Enumerating k Perfect Matchings

The basic idea of Fukuda and Matsui's enumeration algorithm is the following: First, it finds two perfect matchings M and M' in the graph. Then, it selects an edge e which belongs to one but not the other. e is used to partition the problem into two subproblems: The first is to generate all perfect matchings that contain e and the second is to generate all perfect matchings that do not contain e. Clearly, the outputs of the two subproblems are disjoint.

The procedure *NextPerfectMatchings* shown in Figure 2 implements this algorithm, with the additional upper bound k on the number of perfect matchings that should be generated. It receives a graph G, a perfect matching M in G and an integer k that indicates how many more perfect matchings should be generated. If $k > 0$, the procedure searches for an alternating cycle and generates a new perfect matching M'. Then it selects an edge $e \in M' \setminus M$ and makes two recursive calls to itself: The first receives the graph $G - e$ and the matching M. It generates all matchings in G that do not contain the edge e. The second recursive call receives the graph $G \backslash e$ and the matching $M' \setminus \{e\}$. It generates the matchings in G that contain the edge e. The procedure returns the number of matchings it has generated, which is k if it was successful and an integer smaller than k otherwise.

5   Generalizations for Other Constraints

In this section we show that the basic algorithm described in Section 4 can be generalized for the (unrestricted) *AllDifferent*, *GCC* and *Same* constraints.

5.1   *AllDifferent*

The bipartite graph representing the *AllDifferent* constraint is defined similarly to that of the restricted *AllDifferent* constraint, with one difference: Instead of n nodes on each side, there are n variable nodes and n' value nodes, with $n' \geq n$. Of course, when $n' > n$ the graph does not contain any perfect matching. A solution to the constraint now corresponds to a matching that matches all of the variable nodes, i.e., an X-perfect matching where the set of variable nodes is denoted by X.

The algorithm of Figure 2 can be modified for this case as follows: Replace all references to perfect matchings by X-perfect matchings. Algorithmically, this means that a new matching can be generated from an existing matching in one of two ways: By an alternating cycle as in the previous section, or by an alternating path from a matched value node to an unmatched value node.

5.2   *GCC*

The graph with which we represent the *GCC* constraint is a *capacitated* graph, i.e., a bipartite graph which topologically looks like the graph used for *AllDifferent*, but which has a capacity associated with each node. The capacity

Procedure *kMultiConsistency*(G,k)
    (* Initialization and tests for trivial inputs: *)
    foreach edge e in G **do** $kCons[e] \leftarrow TRUE$
    if $k \leq 0$ **then return**
    if there is no perfect matching in G **then**
        foreach edge e in G **do** $kCons[e] \leftarrow FALSE$;
        return
    end if
    $M \leftarrow$ a perfect matching in G
    $k \leftarrow k - 1$

    (* The main loop: *)
    foreach edge e in G **do**
        $M' \leftarrow$ a perfect matching in G which contains e
        $k' \leftarrow NextPerfectMatchings(G \backslash e, M' \setminus \{e\}, k)$
        if $k' < k$ **then** $kCons[e] \leftarrow FALSE$
    end for
end

Procedure *NextPerfectMatchings*(G, M, k)
    if $k \leq 0$ **then return** 0
    else if there is an alternating cycle C in G **then**
        $M' \leftarrow M \oplus C$
        $k \leftarrow k - 1$

        $e \leftarrow$ an edge from $M' \setminus M$
        (* First recursive call: perfect matchings without e *)
        $k' \leftarrow NextPerfectMatchings(G - e, M, k)$
        (* Second recursive call: perfect matchings containing e *)
        $k'' \leftarrow NextPerfectMatchings(G \backslash e, M' \setminus \{e\}, k - k')$
        return $k' + k'' + 1$
    else
        return 0
    endif
end

Fig. 2. k-multiconsistency for the restricted *AllDifferent* constraint

of a node v, denoted $C_v = [L_v, U_u]$, is an interval. With capacity $[1, 1]$ for each variable node and $[L_i, U_i]$ for the value node that corresponds to the value v_i, we get that there is a one-to-one correspondence between the generalized matchings in G and the solutions of the GCC. Note that the different generalized matchings in G do not, in general, have the same cardinality.

To modify the algorithm of Figure 2 for the GCC constraint, we generalize the $G \backslash e$ operation that Fukuda and Matsui use with uncapacitated graphs, to the case of capacitated graphs. For a capacitated graph G and an edge $e = (u, v)$ in G, $G \backslash e$ is the graph obtained by subtracting 1 from the lower and upper capacities of each of u and v. Note that reducing the upper capacity of a node to 0 is equivalent to removing the node and all edges incident to it from the graph.

We also need to generalize the manner in which the algorithm searches for M'. Given the capacitated graph $G = (V, E)$ and a generalized matching M in it, another generalized matching can be found by searching for a directed cycle in the directed graph $\vec{G} = (\vec{V}, \vec{E})$ defined as follows: $\vec{V} = V \cup \{s\}$. For each edge $e = (x, v) \in E$ between a variable node x and a value node v, $\{x, v\} \in \vec{E}$ if $e \in M$ and $\{v, x\} \in \vec{E}$ otherwise. Finally, for each value node v, $\{v, s\} \in \vec{E}$ if v is incident to more than L_v edges in M and $\{s, v\} \in \vec{E}$ if v is incident to less than U_v edges in M [7, 14].

5.3   Same

The basic algorithm can also be modified to support the *Same* constraint, but in this case the changes are more substantial.

The *Same*$(X = \{x_1, \ldots, x_n\}, Z = \{z_1, \ldots, z_n\})$ constraint [2] is modelled by a graph with three sets of nodes: One set for the variables of X (called x-nodes), a second set for the variables of Z (called z-nodes) and a third set for the values (called y-nodes). For each variable $u \in X \cup Z$ and for each value v in the domain of u, there is an edge in the graph between the node that represents u and the node that represents v. Let M be a subset of the edges and let y be a y-node. We denote by $M_X(y)$ ($M_Z(y)$) the set of x-nodes (z-nodes) adjacent to y by edges in M. An edge between an x-node (z-node) and a y-node is called *an xy-edge (a yz-edge)*. A *parity matching* in such a graph is a subset M of the edges such that every x-node or z-node is incident to exactly one edge from M and for every y-node y, $|M_X(y)| = |M_Z(y)|$ [2]. There is a one-to-one correspondence between the parity matchings and the solutions to the *Same* constraint.

In the previous cases, after removing an edge from the graph we remained with a subproblem of the same type as the original problem. However, with the *Same* constraint the situation is slightly different. Suppose that we wish to enumerate all parity matchings that contain the xy-edge $e = (x, y)$. Then the algorithm will explore the graph $G \backslash e$ for sets of edges which are *almost* parity matchings. More precisely, we are interested in subsets M such that (1) $|M_Z(y)| = |M_X(y)| + 1$, (2) $|M_X(y')| = |M_Z(y')|$ for all $y' \neq y$ and (3) every variable in $X \cup Z$ except for x is matched. Then, $M \cup \{e\}$ is a parity matching which contains e. Since the algorithm recursively removes edges from the graph, the desired difference between $|M_X(y)|$ and $|M_Z(y)|$ can change in each recursive step, and not necessarily for the same y-node every time.

To support such demands, the algorithm of Figure 3 associates an imbalance requirement $I(y)$ to each y-node y, which is equal to the desired value of $|M_Z(y)| - |M_X(y)|$. Initially, $I(y) = 0$ for all y. When the algorithm makes a recursive call, there are three cases:

1. The recursive call needs to enumerate all solutions in which an xy-edge $e = (x, y)$ is contained. Then e is removed from the graph along with all other edges incident to x, and $I(y)$ is incremented.

Table 1. Domains of the variables for our example

j	$D(x_j)$	$D(z_j)$
1	{1,2}	{2,3}
2	{3,4}	{4,5}
3	{4,5,6}	{4,5}

2. The symmetric case for a yz-edge $e = (y, z)$. e is removed from the graph along with all other edges incident to z, and $I(y)$ is decremented.
3. The recursive call needs to enumerate all solutions in which an edge $e = (v, y)$ is not contained, for some x-node (or z-node) v and y-node y. Then e is removed from the graph and $I(y)$ remains unchanged.

Definition 10. *Let $G = (X \cup Z, Y, E)$ be a bipartite graph with $|X| = |Z|$ and an integer $I(y)$ associated with every $y \in Y$. A* generalized parity matching *is a subset $M \subseteq E$ such that for all y, $|M_Z(y)| - |M_X(y)| = I(y)$ and each $v \in X \cup Z$, is incident to exactly one edge in M.*

The algorithm needs to find a generalized parity matching from scratch only once, when $I(y) = 0$ for all y. Since in this case a generalized parity matching is just a parity matching, there already exists an algorithm for this task [2], which is based on finding a flow in the following network: We direct the arcs from x-nodes to y-nodes and from y-nodes to z-nodes, and place a capacity of $[0, 1]$ on each of them. In addition, we add two nodes s and t to the graph, add an arc with capacity $[1, 1]$ from s to each x-node, an arc with capacity $[1, 1]$ from each z-node to t and an arc with capacity $[n, n]$ from t to s, where $n = |X|$. There is a one-to-one correspondence between the integral feasible flows in this network and the parity matchings in the graph. Figure 4 shows the network constructed for the following example: $|X| = |Z| = 3$, $|Y| = 6$ and the domains of the variables of $X \cup Z$ are as in Table 1. Figure 5 shows an integral feasible flow in this network.

It remains to show how, given a generalized parity matching M, we can determine in linear time whether another generalized parity matching M' exists in the graph. To do this, we show how to generalize the graph described above such that there is a one-to-one correspondence between integral feasible flows and *generalized* parity matchings. Then, we can use the standard flow theory technique of finding another integral feasible flow by searching for a cycle in the residual graph (see Section 2).

As shown in Figure 6, we add an additional node $\mathcal{I}$ and connect it to the y-nodes with arcs that enforce the imbalances: For each y such that $I(y) > 0$, we add the arc $\{\mathcal{I}, y\}$ with capacity $[I(y), I(y)]$ and for each y such that $I(y) < 0$, we add the arc $\{y, \mathcal{I}\}$ with capacity $[-I(y), -I(y)]$. Since in a feasible flow, the flow into each y-node is equal to the flow out of this y-node, the required imbalances must be respected.

Procedure *kMultiConsistencySame*(G,k)
    (* Initialization and tests for trivial inputs: *)
    foreach edge e in G **do** $kCons[e] \leftarrow TRUE$
    foreach value node y **do** $I(y) \leftarrow 0$
    if $k \leq 0$ **then return**
    if there is no parity matching in G **then**
        foreach edge e in G **do** $kCons[e] \leftarrow FALSE$;
        return
    end if
    $M \leftarrow$ a parity matching in G
    $k \leftarrow k - 1$

    (* The main loop: *)
    foreach edge $e = (v, y)$ in G **do**
        $M' \leftarrow$ a parity matching in G which contains e
        if e is an xy-edge **then** $I(y) \leftarrow 1$ **else** $I(y) \leftarrow -1$
        $k' \leftarrow NextParityMatchings(G \backslash e, I, M' \setminus \{e\}, k)$
        if $k' < k$ **then** $kCons[e] \leftarrow FALSE$
        $I(y) \leftarrow 0$
    end for
end

Procedure *NextParityMatchings*(G, I, M, k)
    if $k \leq 0$ **then return** 0
    else if there is another generalized parity matching M' in G **then**
        $k \leftarrow k - 1$

        $e = (v, y) \leftarrow$ an edge from $M' \setminus M$
        (* First recursive call: matchings without e *)
        $k' \leftarrow NextParityMatchings(G - e, I, M, k)$

        (* Second recursive call: matchings containing e *)
        (* Update $I(y)$ *)
        if e is an xy-edge **then** $I(y) \leftarrow I(y) + 1$
        else $I(y) \leftarrow I(y) - 1$
        $k'' \leftarrow NextParityMatchings(G \backslash e, I, M' \setminus \{e\}, k - k')$
        (* Restore the previous $I(y)$ *)
        if e is an xy-edge **then** $I(y) \leftarrow I(y) - 1$
        else $I(y) \leftarrow I(y) + 1$

        return $k' + k'' + 1$
    else
        return 0
    endif
end

Fig. 3. k-multiconsistency for the *Same* constraint

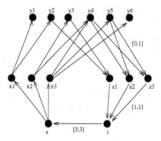

Fig. 4. The directed network for the example in Table 1

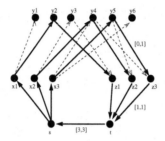

Fig. 5. A feasible flow in the graph of Figure 4

6   Discussion and Future Directions

In this paper, we have defined multiconsistency in the natural way and argued that the term corresponds to the intuitive notion of a "useful" value. Therefore, we believe that determining which values are k-multiconsistent with respect to a global constraint can be a component of reasonable heuristics for finding a solution to a constraint program, or for preferring solutions that can be expected to be more robust. In the realm of the search for robust solutions, we noted that the super solutions as defined by Hebrard et al. [5] seem to offer a better guarantee of robustness than k-multiconsistent solutions. However, we show that while it is NP-hard to determine whether an *AllDifferent* constraint has a $(1,0)$-super solution, computing k-multiconsistency for the *AllDifferent*, *GCC* and *Same* constraints can be performed in time $T(n, n', m) + O(mk(m+n))$, where $T(n, n', m)$ is the time required to find a single solution, and is upper bounded by $O(n^{3/2}n')$ for *AllDifferent* and *GCC* [6, 11] and to $O(n^2n')$ for *Same* [2]. The complexity of computing arc-consistency (which we can now call 1-multiconsistency) for these constraints is $O(T(n, n', m))$. Thus, while there is a computational cost for k-multiconsistency, for constant k the algorithms can still be considered useful. We are currently working to further reduce the complexity of these algorithms.

There are many questions that remain to be explored in the context of multiconsistency. On the theoretical level, one would hope that efficient specialized algorithms can be found for many global constraints, whether exact algorithms

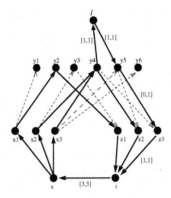

Fig. 6. Example: $I(y_1) = I(y_2) = I(y_3) = I(y_6) = 0$, $I(y_4) = -1$ and $I(y_5) = 1$. A flow in the augmented graph corresponds to a generalized parity matching

such as the ones in this paper, or faster approximation algorithms (i.e., algorithms that determine k-multiconsistency correctly for edges that participate in much fewer or much more than k solutions, but might make errors regarding edges that participate in approximately k solutions). In addition, it would be good to have a theoretical analysis that will better clarify the meaning of multiconsistency. An example of such a result could be a probabilistic analysis that correlates the robustness of the solution to a random *AllDifferent* constraint with the level of consistency of this solution, i.e., the maximal k for which this solution is k-multiconsistent. A third type of theoretical result could be the following: Given a constraint, efficiently compute a "reasonable" value of k for this constraint, where "reasonable" could mean a k such that at least $1/4$ and at most $3/4$ of the edges are k-multiconsistent. It seems desirable to determine such values, because we do not gain much information by computing k-multiconsistency with an "unreasonable" k: Most of the edges fall into the same set (consistent or inconsistent), so we cannot conclude any preferences among them.

Finally, there are questions that will need to be explored experimentally. What is the practical value of the heuristics we propose? Does a constraint solver really find a solution faster if it prefers to assign the "useful" values? Does the robustness increase in practice (even if there does not exist a theoretical guarantee?) Are there other heuristics that can be conceived and which use multiconsistency?

References

1. K.R. Apt. *Principles of Constraint Programming*. Cambridge University Press, 2003.
2. N. Beldiceanu, I. Katriel, and S. Thiel. Filtering algorithms for the Same constraint. In *CP-AI-OR 2004*, volume 3011 of *LNCS*, pages 65–79, 2004.
3. N. Bleuzen-Guernalec and A. Colmerauer. Optimal narrowing of a block of sortings in optimal time. *Constraints*, 5(1–2):85–118, 2000.

182   K. Elbassioni and I. Katriel

4. K. Fukuda and T. Matsui. Finding all the perfect matchings in bipartite graphs. *Appl. Math. Lett.*, 7(1):15–18, 1994.
5. E. Hebrard, B. Hnich, and T. Walsh. Super solutions in constraint programming. In *CP-AI-OR 2004*, volume 3011 of *LNCS*, pages 157–172. Springer-Verlag, 2004.
6. J.E. Hopcroft and R.M. Karp. An $n^{5/2}$ algorithm for maximum matching in bipartite graphs. *SIAM J. Computing*, 2(4):225–231, 1973.
7. I. Katriel and S. Thiel. Fast bound consistency for the global cardinality constraint. In *CP 2003*, volume 2833 of *LNCS*, pages 437–451, 2003.
8. A. Lopez-Ortiz, C.-G. Quimper, J. Tromp, and P. van Beek. A fast and simple algorithm for bounds consistency of the alldifferent constraint. In *Proceedings of the 18th International Joint Conference on Artificial Intelligence*, 2003.
9. K. Mehlhorn and S. Thiel. Faster Algorithms for Bound-Consistency of the Sortedness and the Alldifferent Constraint. In *CP 2000*, volume 1894 of *LNCS*, pages 306–319, 2000.
10. J.-F. Puget. A fast algorithm for the bound consistency of alldiff constraints. In *Proceedings of the fifteenth national/tenth conference on Artificial intelligence/Innovative applications of artificial intelligence*, pages 359–366, July 1998.
11. C.-G. Quimper, A. López-Ortiz, P. van Beek, and A. Golynski. Improved algorithms for the *global cardinality* constraint. In *CP 2004*, volume 3258 of *LNCS*, pages 542–556. Springer-Verlag, 2004.
12. C.-G. Quimper, P. van Beek, A. Lopez-Ortiz, A. Golynski, and S. B. Sadjad. An efficient bounds consistency algorithm for the global cardinality constraint. In *CP 2003*, pages 600–614, 2003.
13. J.-C. Régin. A filtering algorithm for constraints of difference in CSPs. In *AAAI-94*, pages 362–367, 1994.
14. J.-C. Régin. Generalized Arc-Consistency for Global Cardinality Constraint. In *AAAI 1996*, pages 209–215, 1996.
15. W.J. van Hoeve. The alldifferent constraint: A survey. In *Manuscript*, 2001.

Mixed Discrete and Continuous Algorithms for Scheduling Airborne Astronomy Observations

Jeremy Frank and Elif Kürklü*

NASA Ames Research Center, Mail Stop N269-3,
Moffett Field CA 94035-1000
{frank, ekurklu}@email.arc.nasa.gov

Abstract. We describe the problem of scheduling astronomy observations for the Stratospheric Observatory for Infrared Astronomy, an airborne telescope. The problem requires maximizing the number of requested observations scheduled subject to a mixture of discrete and continuous constraints relating the feasibility of an astronomical observation to the position and time at which the observation begins, telescope elevation limits, Special Use Airspace limitations, and available fuel. Solving the problem requires making discrete choices (e.g. selection and sequencing of observations) and continuous ones (e.g. takeoff time and setup actions for observations by repositioning the aircraft). Previously, we developed an incomplete algorithm called ForwardPlanner using a combination of AI and OR techniques including progression planning, lookahead heuristics, stochastic sampling and numerical optimization, to solve a simplified version of this problem. While initial results were promising, ForwardPlanner fails to scale when accounting for all relevant constraints. We describe a novel combination of Squeaky Wheel Optimization (SWO), an incomplete algorithm designed to solve scheduling problems, with previously devised numerical optimization methods and stochastic sampling approaches, as well as heuristics based on reformulations of the SFPP to traditional OR scheduling problems. We show that this new algorithm finds as good or better flight plans as the previous approaches, often with less computation time.

1 Introduction

The Stratospheric Observatory for Infrared Astronomy (SOFIA) is NASA's next generation airborne astronomical observatory. The facility consists of a 747-SP modified to accommodate a 2.5 meter telescope. SOFIA is expected to fly an average of 140 science flights per year over its 20 year lifetime, and will commence operations in 2005. The SOFIA telescope is mounted aft of the wings on the port side of the aircraft and is articulated through a range of 20° to 60° of elevation. The telescope has minimal lateral flexibility; thus, the aircraft must turn constantly to maintain the telescope's focus on an object during observations. A significant problem in future SOFIA operations is that of scheduling Facility Instrument (FI) flights in support of the SOFIA General Investigator (GI) program, called the SFPP (Single Flight Planning Problem). GIs are expected

---

* QSS Group, Inc.

R. Barták and M. Milano (Eds.): CPAIOR 2005, LNCS 3524, pp. 183–200, 2005.

to propose small numbers of observations, and many observations must be grouped together to make up single flights. Approximately 70 GI flight per year are expected, with 5-15 observations per flight.

Flight planning for the previous generation airborne observatory, the Kuiper Airborne Observatory (KAO), was done by hand; planners had to choose takeoff time, observations to perform, and decide on setup-actions called "dead-legs" to position the aircraft prior to observing. This task frequently required between 6-8 hours to plan one flight [1]. The scope of the flight planning problem for supporting GI observations with the anticipated flight rate for SOFIA makes the manual approach for flight planning daunting. There has been considerable success in automating observation scheduling for ground-based telescopes [1], space-based telescopes such as Hubble Space Telescope [2], Earth Observing Satellites [3] and planetary rovers [4]. However, the SOFIA flight planning problem differs from these problems in a variety of ways. Observations are feasible over large, continuous regions of space and time; observations that can't be done at the current position and time may have an infinite number of setup actions enabling them. The principal feasibility condition for observations is goverened by a nonlinear function over the solution to the equations of motion, complicating the task of finding good heuristics. Temporal constraints are implicit in these continuous constraints; bounding above approximations are hard to calculate and generally weak, making temporal constraint propagation unlikely. Finally, the expense of checking feasibility conditions impacts the speed of automated planning.

The SFPP is an intractable constrained optimization problem, containing both an exponential discrete sub-problem (selecting and ordering observations) as well as continuous choices (takeoff time and setup steps). Previously, we developed an algorithm to solve a simplified version of the SFPP, called ForwardPlanner [5, 6]. ForwardPlanner is a novel combination of AI and OR techniques, including progression planning, lookahead heuristics, biased stochastic sampling, approximations and continuous optimization methods. Initial results with ForwardPlanner on a simplified version of the SFPP were promising; however, we show in this paper that ForwardPlanner fails to scale as more and more constraints (Special Use Airspace (SUAs), runway and airway selection, high-fidelity fuel consumption on takeoff and landing, in-flight altitude changes, calculation of initial fuel load) on valid flight plans are added to the problem description. Computationally expensive lookahead search is needed to obtain good results from ForwardPlanner. Introducing approximations to reduce the costs of lookahead improves runtime, but ultimately leads to poor quality flight plans. Consequently, we seek a new approach to solving the problem.

Squeaky Wheel Optimization (SWO) [7] was originally developed for scheduling problems with an optimization objective. SWO accepts as input a permutation of tasks to schedule, and a fast procedure called a *Constructor* that treats each task in order, ultimately scheduling tasks or rejecting them. The permutation and its resulting schedule are then analyzed by a *Critic* to determine a new permutation that might schedule tasks that were previously rejected. The cycle repeats until all tasks are scheduled or for a fixed number of iterations. SWO was originally evaluated on Graph Coloring [7], and

---

[1] Anecdotal evidence provided from conversations with SOFIA staff who worked with KAO.

has since been employed for satellite observation scheduling [8] and range scheduling [9], as well as project scheduling with temporal constraints [10]. The promise of SWO for solving the SFPP is that good plans can be found using fewer expensive feasibility checks than ForwardPlanner.

The rest of the paper is organized as follows. We first formally describe the SFPP, the constraints on flight plans, and the optimization criteria used to compare valid flight plans. We then briefly describe the ForwardPlanner and discuss its problems. We then introduce Squeaky Wheel Optimization (SWO) and discuss how to apply it to the SFPP using numerical optimization methods and approximate solutions to OR problems. We show that SWO improves upon ForwardPlanner on a small set of examples. We then discuss a variety of ways to improve the performance of SWO. We describe experiments to validate the approach. Finally, we conclude and discuss future work.

2   Describing SOFIA's Choice

The input to the Single Flight Planning Problem (SFPP) consists of a set of observation requests, each consisting of the Right Ascension (RA) α and Declination (Dec) δ, observation duration, priority; a flight date; maximum fuel load; an altitude profile mapping flight time to maximum altitude; earliest takeoff time θ_l and latest landing times θ_u; the designated takeoff and landing airports (which need not be the same); predicted wind and temperature; and a list of SUAs. For a flight plan to be valid, the aircraft must take off from the takeoff airport, land at the landing airport, avoid all SUAs, and consume less than the available fuel at takeoff. The objective is to find a flight plan that maximizes the number of requested observations performed. During flight, *Flight-legs* require tracking an object for a period of time, and are only valid if the object stays within the telescope elevation limits for the requested duration. The observation must also take place in darkness (the sun must be below the horizon). *Dead-legs*, when no observations are performed, can be used to reposition the aircraft to enable flight-legs. A distinguished class of dead-legs are used to take off and return to the landing airport. Since it is intractable to find the best possible plan, we limit ourselves to searching for *good* plans that perform many observations of high priority. Solving the SFPP requires choosing a takeoff time, selecting the set of observations to service, ordering them and inserting necessary dead-legs to ensure that all flight legs are valid.

2.1   Constraints on Valid Flights

In this section we describe the constraints on valid solutions to the SFPP in more detail. The telescope is carried aboard a Boeing 747-SP aircraft. The fuel consumption of each engine depends on the aircraft weight, mach number, outside air temperature, initial altitude and final altitude. The fuel consumption constraints are represented in a lookup table provided by Boeing. The aircraft follows a pre-determined *altitude profile* that describes the maximum permitted altitude at an absolute time after takeoff. Climbs are allowed periodically to decrease fuel consumption. At the end of a leg, if the aircraft is allowed to climb, it climbs to the maximum altitude permitted by the fuel performance table or the altitude profile. The profiles used in this paper were developed assuming standard atmosphere [11]; actual atmospheric conditions and aircraft weight may force the aircraft to fly lower than the altitude profile permits. Predicted wind and temperature are used to calculate the ground track and fuel consumption. Finally, SUAs constrain the ground track of the aircraft by forcing dead-legs to reposition the aircraft. Space precludes describing the fuel consumption constraint in more detail.

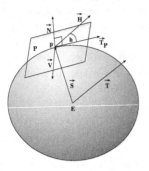

The constraints linking aircraft motion and observation feasibility are the most complex and important component of the problem, so we describe them further here. SOFIA can view objects between 20° and 60° of elevation (from the plane of flight). If an observation is scheduled, then it must be performed for the requested duration without interruption, and the object must stay within the elevation limits throughout the observation. The elevation of an object depends on the object's coordinates, the aircraft's position and the time.

Fig. 1. The Cartesian formulation of the instantaneous equations of motion of the aircraft and the elevation

Checking this constraint requires computing the aircraft's ground track throughout the course of the observation. Figure 1 shows the interaction between the object's coordinates, the aircraft's position, the time, and the telescope elevation. The Earth is modeled as an oblate spheroid $\mathbf{E}$, whose surface is defined by the equation $\frac{x^2}{a^2} + \frac{y^2}{a^2} + \frac{z^2}{c^2} = 1$ where $c < a$. Let $\mathbf{p}$ be the aircraft's current position, (latitude γ and longitude L) and θ be the (Sidereal) time that the aircraft is at $\mathbf{p}$. Let S be the vector from the center of $\mathbf{E}$ to $\mathbf{p}$. Let T be the vector to an astronomical object o at time θ, and $\mathbf{P}$ as the plane tangent to $\mathbf{E}$ at $\mathbf{p}$. Let $\hat{\mathbf{i}}, \hat{\mathbf{j}}, \hat{\mathbf{k}}$ be the unit vectors in the x, y, z directions respectively. Let N be the vector normal to $\mathbf{P}$: $N = \frac{p_x}{a^2}\hat{\mathbf{i}} + \frac{p_y}{a^2}\hat{\mathbf{j}} + \frac{p_z}{c^2}\hat{\mathbf{k}}$ (Note that S and N are generally not parallel since $\mathbf{E}$ is a spheroid.) Let T_P be the projection of T onto $\mathbf{P}$; this is the *object azimuth* at $\mathbf{p}$, and is given by $T_P = T - \frac{TN}{||N||^2}N$. Let V be the desired heading of the aircraft. The observatory must track the object inducing T, subject to the constraint that the angle between V and T_P is 270°, because the telescope points out the left-hand side of the aircraft. Let $\mathbf{R}_N(270°)$ be a rotation matrix that rotates a vector 270° around N, and v be the airspeed of the aircraft; then $V = v\mathbf{R}_N(270°)\frac{T_P}{||T_P||}$. Let H be the elevation vector with respect to $\mathbf{P}$. We also require the angle h between H and T_P obey the constraint $20° \leq h \leq 60°$ throughout an observation. Most targets are sufficiently far from Earth that we can assume $H = T + S$. From vector calculus we then get the equation for the elevation: $h = \cos^{-1}\left(\frac{HT_P}{||H||\,||T_P||}\right)$. The angle r between V_d and the object azimuth at the new position T_P is given by: $r = \cos^{-1}\left(\frac{V_d T_P}{||V_d||\,||T_P||}\right)$. Now, T is a function of o and θ; this is because the Earth rotates on its axis. The vector T traces a circle of radius $x^2 + y^2 = \frac{c^2 - d}{c^2}$, where $d = |\frac{\delta}{90°}|$ in 24 hours (see [12] for an explanation of this).

The instantaneous change in $\mathbf{p}$ as the aircraft tracks o is $\frac{d\mathbf{p}}{d\theta} = V$. Since V is a function of T, it is a function of $o, \mathbf{p}$ and θ. Solving for the ground track is necessary to compute h over the entire duration of the observation and check the elevation constraints. It is worth noting that this formulation also makes it easy to add the effect of winds by adding the appropriate vectors to V, and also correct for aircraft pitch by rotating about $V \times N$, but we omit these for brevity.

3   ForwardPlanner and Its Discontents

The first fully automated approach to solving the SFPP was ForwardPlanner [5, 6]. We originally assumed no SUAs and ignored runway and airway selection, ascent and descent, thus simplifying the fuel consumption constraint. ForwardPlanner combines progression based search, continuous numerical optimization, dispatch heuristics and stochastic sampling, resulting in an incomplete randomized algorithm. The ground track and elevation constraints are solved using a specialized 5th-order Runge-Kutta [13] with error-adaptive step sizing. ForwardPlanner evaluates the feasible observations at each phase of a flight, and selects one observation to add to the flight. When checking feasibility, rather than considering all possible setup actions, ForwardPlanner only considers the *shortest dead-leg* making an observation visible for long enough and allowing the aircraft to subsequently fly to the landing airport. If the shortest dead-leg crosses an SUA, the heading is shifted minimally left or right from the heading of the shortest dead leg until the dead leg misses all SUAs. The duration of the leg is then adjusted to ensure the object is visible for the required duration. If the resulting dead leg is longer than D (an operational limitation on the longest permissible dead-leg), then the observation is rejected. If the flight-leg following this dead-leg crosses any SUA, the observation is rejected. If the observation begins before sunset or ends after sunrise *at the local position*, the observation is rejected. (Remember, changing your position changes the time at which the sun rises or sets.) Finally, if the aircraft cannot return to the landing airport after the observation is performed, the observation is rejected. If the observation survives all of these checks, ForwardPlanner considers it is feasible.

Each feasible observation is then evaluated by first adding it to the flight plan, then heuristically adding a fixed number of additional observations. This "lookahead" is performed to estimate the best flight plan possible after adding each observation. These short extensions are evaluated using a weighted sum of the *priority* of the observations performed so far, the *efficiency* (ratio of time spent observing to total flight time) of the (incomplete) flight, the estimated time to return to the designated landing airport, and the total time spent in turns. The heuristic rank of each observation is treated as the mass of a probability distribution used to select the next observation. Thus, if we have a set of choices C and heuristic values of of these choices $v(c)$, we choose an element $c \in C$ with probability $\frac{v(c)}{\sum_{d \in C} v(d)}$. This technique is similar to Heuristic Biased Stochastic Sampling (HBSS), a technique used for scheduling ground based telescopes [1]. This means that the "best" candidate need not be selected at any stage of the process, but has the highest probability of being selected. The process of evaluating the feasible observations and adding the next observation to a flight is shown pictorally in Figure 2. ForwardPlanner is a stochastic algorithm, and can be run several times to generate better flights; the ForwardPlanner algorithm sketch is shown in Figure 3.

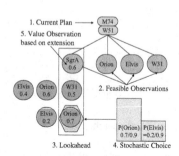

Fig. 2. ForwardPlanner's Evaluate() routine. Each feasible observation in the current plan 1) is added to the plan 2). A fixed number of observations are used to extend the plan 3). Each of these observations is evaluated individually, and the values are used to form a probability distribution; this distribution is sampled 4) to determine how to extend the flight. Once the maximum number of observations in lookahead (2 in this example) is reached, the resulting flight is used to determine how good it is to add the first observation to the current flight 5)

```
ForwardPlanner()
# F is (initially empty) current flight plan
for MaxRepeats
    Select takeoff time
    while not done
        # E is set of feasible observations
        for each unscheduled observation o
            if Feasible(o, p, θ)
                Add p to F; update p, θ
                v=Evaluate(o, F)
                Add (o, v) to E
                Remove o from F
        end for
        if E is not empty
            Use values v to select e from E
            Extend F by e; empty E
        else done
    end for
return F
end
```

Fig. 3. A sketch of the ForwardPlanner Algorithm. At each step, all feasible observations are considered as the next observation in the plan. For each feasible observation, the Evaluate() routine builds an extension of the plan to evaluate how good a flight will result. Feasible() is described in Figure 7

The principal cost of ForwardPlanner is in the lookahead phase, where many legs are constructed to test observation feasibility solely to evaluate an observation, and then are thrown away. Let N be the number of observation requests, let K be the lookahead depth, and let M be the maximum number of observations that can be in any flight plan. Each of *MaxRepeats* loops in ForwardPlanner makes $O(N^2KM)$ calls to Feasible(); a proof of this appears in [5]. It was found empirically that $K = 4$ struck a good balance between computational cost and flight plan quality [5]. ForwardPlanner was improved upon in [6] by observing that many expensive dead-leg construction steps could be eliminated. Suppose an observation is not visible at the current position and time. If we drop the condition on reaching the landing airport, an approximation of the shortest dead leg b, d (b is the heading and d is the duration) has the property that $F_1(b, d) = < f_1(b, d), f_2(b, d) > = < 0, 0 >$ where f_1 is the difference between the object azimuth and the final heading of the aircraft after flying the dead-leg defined by b, d, and f_2 is the difference between the object elevation after flying the dead-leg and

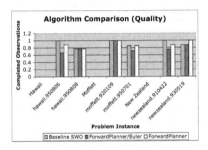
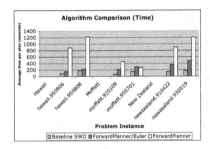

Fig. 4. Comparison of solution quality for ForwardPlanner (with and without Euler's method approximation of flight dynamics) and SWO

Fig. 5. Comparison of average CPU time for ForwardPlanner (with and without Euler's method approximation of flight dynamics) and SWO

the telescope elevation limit closest to the initial object elevation. A similar formulation exists for the shortest dead leg ensuring an observation is visible throughout a flight leg [6]. We solve for b, d using a Secant Method[2]; the final version of Feasible() used in FowardPlanner appears in Figure 7. The resulting algorithm is a novel combination of AI progression planning and stochastic sampling and OR numerical optimization techniques for solving a complex constrained optimization problem. This approach reduces the runtime of ForwardPlanner without impacting the value of the flight plans found.

Initial results on solving the simplified version of SFPP with ForwardPlanner were promising [6]. However, adding requirements to avoid SUAs, calculate initial fuel loads in the face of predicted weather, runway and airway selection, and calculating fuel consumption based on altitude changes (especially complex for takeoff and landing). made ForwardPlanner too slow. In particular, SUA evasion and fuel consumption during climb vastly increase the expense of the feasibility check. This is problematic, given that large lookahead and many samples are needed to find good quality plans. Further reductions in runtime can be accomplished by approximately calculating aircraft position after flight legs in the lookahead phase using Euler's Method instead of Runge-Kutta. Euler's Method approximates the solution to the ground track by flying a constant heading for a fixed (small) duration relative to the total observation time. The approximation we used does not adequately account for the ellipsoid Earth, wind speed and direction, change of altitude, and estimates fuel consumption based on the last calculated fuel consumption rate. Our intuition was that these approximations would permit a good, fast estimate of the value of inserting an observation. Unfortunately, the heuristic quality degrades too much and leads to poor quality plans. Figures 4 and 5 compare performance on 6 sample problems (we will discuss the SWO results later in the paper). In these experiments, ForwardPlanner was run with MaxRepeats = 20. ForwardPlanner finds good quality plans, but takes 8-20 minutes per flight generated. Employing Euler's Method

---

[2] The previous work incorrectly identified the method used as Newton's Method; since numerical derivatives are used, we actually use a Secant Method, which is the term we will use in this paper.

reduces ForwardPlanner's computation time considerably, but leads to plans with fewer scheduled observations in 4 of 6 cases. While some of the cost savings is in lookups to outside air temperature and the fuel table, as well as the switch in integration methods, the vast increase in software complexity required to correct these problems led us to search for new solutions to the problem that allow us to generate good quality flights fast.

4   Squeaky Wheel Optimization for the SFPP

SWO takes as input a permutation of tasks to schedule, and a fast procedure called a *Constructor* that treats each task in order, ultimately scheduling tasks or rejecting them. The permutation and its resulting schedule are then analyzed by a *Critic* to construct a new permutation that might schedule tasks that were previously rejected. The cycle repeats until all tasks are scheduled or for a fixed number of iterations. Figures 6 and 7 describe a *family* of SWO algorithms specialized for solving the SFPP. We discuss the features of this specialized SWO in more detail below.

The constructor assumes that the flight begins at the takeoff time, and that the permutation P imposes a precedence ordering on the observations, and attempts to construct a schedule. If an observation is not trivially visible for the requested duration, the shortest dead-leg is constructed by solving the zero finding problem. If this leg is short enough, SUAs can be avoided, and sufficient fuel remains the observation is added, otherwise it is rejected. This is identical to the procedure used in ForwardPlanner, and is shown in Figure 7. Rejecting the ith observation in P does not imply rejection of $j > i$; all observations are processed. The best flight B is the flight maximizing $\frac{s}{2} + \frac{e}{2}$, where s is the percentage of requested observations scheduled, and e is the efficiency of the flight (the ratio of time spent observing to flight time) [3]. The final flight plan is checked for SUA violations on the return leg; if there are any, the flight is rejected.

In order to modify the permutation P, a critic must both select a rejected observation r in R and decide where in P to move r. We use the flight plan F built with permutation P to decide how to modify P. Each observation in a flight plan defines a "slot" in which a new observation could be placed. Unlike SWO approaches taken in [9] and [8], we do not perform "blind" migration of jobs in the permutation that might not lead to a new flight plan. Rather, we identify where in the permutation we can move rejected observations to ensure that the resulting schedule is modified. Since we guarantee that a rejected observation will be scheduled during the next construction phase, we run the risk that some observations later in the flight might be displaced. Thus, it is important to estimate how much we "regret" moving an observation to a particular place. Critics must both be fast and produce good quality flights by moving rejected observations without displacing many scheduled observations.

To ensure rejected observations are scheduled, SWO checks the feasibility of every rejected observation at every slot in the flight. At worst, this might require $O(N^2)$ feasibility checks to determine which slots rejected observations can occupy. While each of *MaxRepeats* calls in ForwardPlanner makes $O(N^2KM)$ flight leg feasibility

---

[3] Efficiency is a secondary criteria for good quality flights.

SWO(MaxFlights,MaxRepeats)
F is current flight plan
B is best flight plan
P is a permutation of observations
R is rejected observations
for MaxRepeats
  1. *Generate permutation P*
  for MaxFlights
    2. *Select the takeoff time θ*
    # Construct flight from P
    # p is the current position of F
    for observation $o \in P$
      if Feasible(o, p, θ)
      Add p to F
      Update p, θ
      else add p to R
    end for
    Update best flight plan B
    if $R = \emptyset$ return F
    else
      3. *Modify P by analyzing F and R*
  end for
end for
if dead leg home does not violate SUA
  return B

Feasible(o, p, θ)
o is the observation
p is the current position
D is maximum dead leg duration
(b, d, z) = FindDeadLeg(o, p, θ)
b = heading, d = duration, z = SUA zone
if the dead-leg crosses any SUA zone z
  #Revise dead legs to avoid SUA
  b' is closest heading s.t. all z not crossed
  d' is new duration
  $d = d'; b = b'$
  if $d > D$ **return false**
if observation starts and ends in darkness
  if dead leg home possible following o
    return true
return false

FindDeadLeg(o, p, θ)
#e is the elevation limit o violates at p, θ
Guess dead-leg b, d; calculate r, h after dead-leg
#$f_1(b,d) = r, f_2(b,d) = e - h$
while $\langle f_1, f_2 \rangle \neq \langle 0, 0 \rangle$
$$J = \begin{pmatrix} \frac{\partial f_1}{\partial b}(b,d) & \frac{\partial f_1}{\partial d}(b,d) \\ \frac{\partial f_2}{\partial b}(b,d) & \frac{\partial f_2}{\partial d}(b,d) \end{pmatrix} \equiv \begin{pmatrix} p & q \\ r & s \end{pmatrix}$$
$|J| = ps - qr$
if $|J| < t$ **then** $|J| = t$ (preserve sign of $|J|$)
$db = \frac{qf_2 - sf_1}{|J|}$ and $dd = \frac{pf_1 - rf_2}{|J|}$
$b = b + db, d = d + dd$
Update r, h

Fig. 6. A sketch of the family of SWO-based Flight Planning Algorithm. Later sections elaborate on options for *1. Generate permutations, 2. Select the takeoff time, and 3. Modify P by analyzing F and R*

Fig. 7. The feasibility test with the Secant Method for finding dead legs. t, db, dd are tuning parameters. Derivatives are all calculated numerically. r and h are calculated as discussed in Section 2.1

checks, each such call in SWO makes $O(MaxFlights(N + N^2))$ feasibility checks. As long as *MaxFlights* $< KM$, SWO costs less per invocation than ForwardPlanner; this seems likely since, to ensure good performance of ForwardPlanner. *MaxFlights*, M and K likely scale with N. This makes SWO a good candidate for improving upon ForwardPlanner.

In SWO, there is a complex interplay between the permutation modification and takeoff time selection. It is possible to construct very bad flight plans by poor selection of the takeoff time. Also, the combination of the takeoff time, permutation and the fast scheduler implicitly schedules a subset of the observations. Finally, the fact that permutations are constantly modified allows reconsideration of the takeoff time based on the new permutation. For these reasons, this version of SWO ensures that new takeoff times can be chosen after each modification of the permutation.

4.1   Useful Concepts

In preparation for building our SWO, we introduce some useful concepts.

Time windows during which an object o at Right Ascension α and declination δ is visible at a fixed position can be constructed as follows. If the aircraft is at position $p = \gamma, L$, the earliest and latest times $\theta_{r,p}(o), \theta_{s,p}(o)$ at which the observation is visible by SOFIA at p are given by $\theta_{r,s}(o) = \cos^{-1}\left(\frac{\sin(20) - (\sin \delta)(\sin \gamma)}{(\cos \delta)(\cos \gamma)}\right) + L + \alpha$ [12]. The $\sin(20)$ term arises from the fact that SOFIA's lower elevation limit is 20°. Note that $\cos^{-1}(x)$ has 2 solutions, which provide the earliest rise time $\theta_{r,p}(o)$ and latest set time $\theta_{s,p}(o)$ of the object at this position. The time of sunset and sunrise at this position can be used to further tighten this window. There can be at most 2 feasible windows since all objects period is 24 (sidereal) hours and the aircraft stays aloft less than 10 hours. For example, an object can rise above the maximum elevation limit, then drop back into view. In our critics, by default we use the *first* feasible window. We will also use the time at which an object reaches its maximum elevation (above the local horizon), called the *transit time*. This is simply $\frac{\theta_{s,p}(o) + \theta_{r,p}(o)}{2}$.

The SFPP can be relaxed by approximating time windows for observations as described in the previous paragraph, effectively pretending that the observatory is fixed at some location. This leaves a problem in which observations have release times (earliest rise times), due dates (latest set times), occupy a unary resource (the telescope). This approximation is not bounding, because objects may rise earlier and set later at different positions than the one used to calculate the time windows. Since SOFIA has a maximum and maximum telescope elevation limit, the true feasibility windows of objects may not be convex. Additionally, objects could set then rise during the night, but usually objects are observed at times of year when they are visible all night (and thus achieve their maximum elevation sometime during the night). The resulting problem is $1|r_i; p_i; d_i| \sum w_i U_i$ according to Graham's hierarchy, a well-studied problem in AI and OR which Karp proved $\mathcal{NP}$-complete [14]. Note that p_i are generally not equal and that tasks are not interruptible. The relaxation is too crude to use directly; we will use approximate solutions of this problem in our takeoff-time selection method.

4.2   Generating Initial Permutations

We considered the following ways of generating the initial permutation:

Random selection **Uniform**: If there are N observations, one of the $N!$ permutations is chosen uniformly at random.

Sort by Earliest Start Time **Rise** at the takeoff airport: We calculate $\theta_{r,p}(o)$ as described in the previous section. The intuition behind this ordering is that flights often occupy the whole night, so beginning observations as early as possible is a good initial guess. Furthermore, this allows the largest time window to observe any object.

Sort by Latest Start Time **Set** at the takeoff airport: We calculate $\theta_{s,p}(o)$ as described in the previous section. Observing an object as late as possible may be a cheap method of ensuring enough time remains to schedule necessary dead-legs.

Sort by Transit Time **Transit** at the (landing) airport: The intuition here is that this allows observing very nearby the airport; while one object is being observed, the next object moves closer to the landing airport, allowing the aircraft to "loiter" nearby.

4.3   Generating Takeoff Time

As we previously observed, due to the complex nature of the visibility constraints, choosing a good takeoff time is important to constructing good flight plans. We considered several takeoff time methods:

Estimated flight duration **FlightDur**: If we simply assume that the aircraft will stay aloft as long as possible, we can estimate the flight duration f from the initial fuel load and flight profile. The takeoff time range is $[\theta_l, \theta_u - f]$. Since this quantity is independent of the permutation, it needs be calculated only once. However, this approach will usually overestimate the actual flight duration. Furthermore, especially in the summertime for long flights, f will exceed the duration of the night and reduce the takeoff time range to one time (roughly half an hour before sunset).

Minimum of Earliest Start Times **Min Rise**: We can calculate the minimum over all o of $\theta_{r,p}(o)$ at the takeoff airport, and "pad" this by the amount of time needed to climb to operational altitude. Since this quantity is independent of the permutation, it needs be calculated only once. Only one takeoff time is generated by this approach.

Optimize **First-Observation** in Permutation: It is clear that $\theta_{r,p}(o)$ is a bounding above approximation to the earliest time when an observation can be performed; to see why, observe that flying towards the observation makes it possible to observe it earlier. If we assume that the first observation in a permutation is meant to be observed, we can calculate the earliest time at which this observation can be performed and takeoff at that time. Binary search over takeoff times is performed to find the takeoff time leading to the earliest feasible observation time for the first observation. Only one feasible takeoff time is generated by this approach. As the first observation in the permutation can change, the takeoff time will need to be recalculated each time the permutation changes.

Approximate solution to the relaxed scheduling problem **Feas-Sched**: We use the $\theta_{r,p}(o)$ and $\theta_{s,p}(o)$ calculated at the takeoff airport to approximate the time windows for the observations and induce the relaxed scheduling problem $1|r_i; p_i; d_i| \sum w_i U_i$. Solving this problem optimally is pointless, since it is a crude approximation of the original problem. A feasible solution to the relaxed scheduling problem can be generated using the permutation as an ordering heuristic, and either greedily scheduling from the beginning or the end of the permutation. It is trivial to see that different feasible schedules, and different takeoff time ranges, can be generated by scheduling forwards or backwards; this leads to two methods, **Feas-Sched (Fwd)** and **Feas-Sched (Bkwd)**. Once a feasible solution is generated, we calculate the slack of the first feasible observation, again "padding" for the time to climb to altitude.

If a range of takeoff times is generated, we select from them uniformly at random.

4.4   Modifying the Permutation with Critics

In what follows, assume the problem instance contains N observation requests. All of our critics use the biased sampling approach described earlier to make selections. Recall that if we have a set of choices C and values of of these choices $v(c) \in C$, we choose an element $c \in C$ with probability $\frac{v(c)}{\sum_{d \in C} v(d)}$.

We explored the following five critics to modify the permutation:

1-Phase: We first determine for each rejected observation o whether it is feasible in each slot s. This test uses the feasibility test in Figure 7 assuming the aircraft begins at the position and time at the beginning of slot s. For each feasible pair (o, s) we examine the time at which the new observation o ends. Since the new observation is guaranteed to be feasible, successive observations will be delayed, both due to the duration of the new observation and its dead leg (if any). We then evaluate the rate of change of the elevation of each successive observation to find out if it would still be visible at the same position at the later time. This is obviously an approximation, since the aircraft position would change after the newly inserted observation. Furthermore, we don't consider the possibility that unscheduled observations in the permutation could be added, so it is a conservative regret estimate. Let $X_{o,s}$ be the set of observations we estimate are made infeasible by performing o in slot s. We then calculate $v(o, s)$ for the sampling probabilities as follows. If s is the first or last slot or one for which $X_{o,s} = \emptyset$, then $v(o, s) = N$. Otherwise, $v(o, s) = \left(\sum_{x \in X_{o,s}} u(x)\right)^{-1}$, where $u(x) = 0.5$ if x had a dead-leg before it, and $u(x) = 1$ if not. This penalizes choices that incur more regret, with the assumption that replaced observations with dead-legs are regretted less.

Obs-Slot: We first determine for each rejected observation o whether it is feasible in each slot s. We then randomly choose a feasible observation o from those that could go into some slot s. We calculate sampling probabilities as follows: if an observation is visible in s slots, the heuristic is $v(o) = N + 1 - s$. (Observations visible nowhere are not chosen.) We then calculate $v(o, s)$ as described above for those s in which o is feasible, and randomly choose the slot for o.

Slot-Obs: We first determine for each rejected observation o whether it is feasible in each slot s. We then randomly choose a slot in which at least one rejected observation is feasible. We calculate sampling probabilities as follows: if v observations are visible in a slot, and the problem instance contains N observations the heuristic is $v(s) = N + 1 - v$. We then calculate $v(o, s)$ as described above for those o feasible in s, and choose randomly the observation to move to s.

Time$_t$: For this critic, we use $\theta_{s,p}(o)$ and $\theta_{r,p}(o)$ at the takeoff airport, which can be calculated once and needs never be repeated. The critic first chooses a feasible observation o. We calculate sampling probabilities as follows: $v(o) = \frac{1}{\theta_{s,p}(o) - \theta_{r,p}(o)}$. We then determine which slots s are feasible for o, and calculate $v(o, s)$ as described above. Finally, we randomly choose the observation to move to s using $v(o, s)$.

Time$_f$: Calculating $\theta_{r,p}(o), \theta_{s,p}(o)$ at the takeoff airport is clearly inaccurate. We can instead calculate $\theta_{s,p}(o)$ and $\theta_{r,p}(o)$ at each slot in the current flight, but at a higher computational cost. We calculate sampling probabilities as follows: if P is the set of positions at the beginning of the slots in the flight, $v(o) = \min_{p \in P} \frac{1}{(\theta_{s,p}(o) - \theta_{r,p}(o))}$. We then choose o according to $v(o)$. We then determine which slots s are feasible for o, and calculate $v(o, s)$ as described above. Finally, we randomly choose the observation to move to s using $v(o, s)$.

As a final wrinkle, we can modify the permutation by moving k rejected objects rather than just one. The idea here is that multiple rejected observations could be re-ordered *independently* and potentially improve the flight plan using fewer construction steps. This idea was successfully employed by [8] and [9] to speed up SWO.

5   Identifying the Right SWO Features

Our approach to finding the best SWO features is to begin with a baseline algorithm: **Flight-Duration** based takeoff time range selection, **Uniform** random initial permutation, and the **Time**$_t$ critic. We will use the Wilcoxon Signed Ranked Test [15] to determine whether using one feature is superior (finds better quality flights) to the baseline SWO; we will select a small subset of promising algorithms to generate the next algorithm. In the presentation of the Wilcoxon test results, X indicates the tests leading to different values, positive z indicates an algorithm variant is likely to perform better than the baseline, while a negative z indicates an algorithm variant is likely to perform worse than baseline. Criticality measurements are typically given in ranges; criticalities of > 0.05 are not considered statistically significant.

6   Empirical Results

In this section we present empirical results for varying facets of SWO in order to find the best overall algorithm for solving the SFPP.

6.1   Sample Problems

We used as a benchmark flights previously flown on KAO, described in [5], to determine the utility of our new techniques. In Figure 8 we tabulate the number of observations, he archived flight duration, and the airport. Flights from Moffett Field, CA are denoted with an M; flights originating in Moffett and ending in Hawaii are denoted MH; flights from Hawaii are denoted H, and flights from New Zealand are denoted N. Takeoff time is between sunset and sunrise (calculated for each day and year of flight). Wind and temperature data from European Center for Medium Range Weather Forecasting [4] are used to calculate ground tracks and fuel consumption. The initial fuel load is also calculated for each flight, and is based on the altitude profile 4 from [11]. This profile conforms to realistic expectations that good observing will require an altitude of at least 39000 ft. Finally, SUAs impact flights from Moffett and Hawaii; we use data from the National Geospatial Intelligence Agency's Digital Aeronautical Flight Information File.

---

[4] www.ecmwf.int

Index	1	2	3	4	5	6	7	8	9	10	11	12	13	14	15	16
Airport	H	H	H	H	M	M	M	M	M	M	M	M	M	M	M	M
Date	8/6	8/8	8/10	8/12	1/9	1/10	1/16	6/16	6/18	6/19	6/30	7/6	8/12	8/16	4/4	4/5
# Obs	9	9	10	10	7	8	8	6	10	8	8	6	11	10	9	9

Index	17	18	19	20	21	22	23	24	25	26	27	28	29	30	31	32
Airport	M	M	M	M	M	M	M	M	M	M	M	M	M	M	M	M
Date	4/6	4/11	4/12	4/14	4/19	5/4	5/8	7/1	7/6	8/2	8/22	8/24	8/26	8/29	9/1	9/19
# Obs	10	8	8	8	10	10	6	7	4	6	9	8	11	10	8	7

Index	33	34	35	36	37	38	39	40	41	42	43	44	45	46	47
Airport	M	M	M	M	M	M	M	MH	MH	MH	N	N	N	N	N
Date	9/20	9/21	9/23	9/26	9/28	9/29	10/4	6/21	7/12	8/4	11/25	4/22	5/11	5/15	5/19
Obs	7	3	10	8	8	8	4	8	7	7	10	8	8	8	8

Fig. 8. Characteristics of Single Day Instances

The priorities of all observations are identical, and all observations could be scheduled for the KAO flights. While SOFIA's performance characteristics differ from KAO and its elevation limits are different, we found ForwardPlanner was able to schedule all observations for most of the tests we constructed [6]. Thus, the principal goal is to find an efficient flight with all of the observations scheduled. The maximum dead-leg duration D was set to 4 hours. For the dead-leg search using Secant Method we used a step cutoff of 150 and error tolerance $t = 10^{-6}$. The step parameters used in the Secant Method were: $s_1 = 0.01°$ and $s_2 = 60$ seconds. When CPU times are reported, these experiments were run on a Sun Workstation with dual 600 MHz CPUs and 2048 Mb memory. Unless otherwise stated, *MaxFlights* = 20 and *MaxRepeats* = 10.

6.2   Choosing Takeoff Times

The results of varying the takeoff time selection while holding all other aspects of the baseline SWO algorithm the same are shown in Figure 9. In this figure we present the Wilcoxon Ranked Sign test output for the best percentage of the observations found by SWO. Recall that we compare each new SWO variant to the baseline SWO described in the previous section according to the quality of the flights. In what follows, our "best" SWO variants are those "most likely to exceed the quality of the baseline SWO".

Takeoff Range	X	Z	Crit.
Min Rise	17	-2.218	[0.01,0.025]
First Observation	18	-2.057	[0.01,0.025]
Feas-Sched (Fwd)	12	1.313	>0.05
Feas-Sched (Bkwd)	16	1.822	[0.025,0.05]

Fig. 9. Wilcoxon Ranked Sign Test results comparing SWO Takeoff Time variants to SWO Baseline

Backwards scheduling to produce a relaxed feasible schedule **Feas-Sched(Bkwd)** did best. The least "informed" approach, **Min Rise**, performs worst. Optimizing the takeoff time range of the first observation also did not perform well. Both of these approaches perform worse than the baseline SWO, which uses **Flight-Duration**.

Curiously, **Feas-Sched(Fwd)** did not perform as well as **Feas-Sched(Bkwd)**. It is possible that scheduling backwards produces a larger takeoff time range, thereby increasing flexibility, but more work is needed to understand this result.

6.3   Generating Initial Permutations

For this series of tests, we tested the **Feas-Sched (Bkwd)** variant of takeoff time selection with the different initial permutation methods. The results of varying the permutation selection while using the baseline critic are shown in Figure 10. Notice that **Uniform** is our baseline permutation method, and thus the first line of Figure 10 repeats the last line from table 9.

Permutation	X	Z	Crit.
Uniform	16	1.822	[0.025,0.05]
Rise	14	2.055	[0.01,0.025]
Set	14	1.428	> 0.05
Transit	14	1.490	> 0.05

Fig. 10. Wilcoxon Ranked Sign Test results comparing SWO Initial Permutation ordering variants to SWO Baseline

Previous work indicates that "informed" initial permutations improve the performance of SWO when compared to random permutations. We find this to be the case as well; **Rise** coupled with the **Feas-Sched (Bkwd)** performs best when compared to the baseline SWO. Surprisingly, **Uniform** performs second best, but is not as good as **Rise**.

6.4   Modifying Permutations

For this series of tests, we tested the **Feas-Sched (Bkwd)** takeoff time generation method and **Rise** initial permutation generation method with each critic method. In each case, only one rejected observation was moved per critic application. The results of varying the critics are shown in Figure 11. Notice that **Time**$_t$ is our baseline critic method, and thus the first line of Figure 11 repeats the second line from table 10.

As expected, **1-Phase** is quite good. Also as expected, we see that **Time**$_t$ is not as good as **Time**$_f$. Somewhat surprisingly, though, **Time**$_t$ and **Time**$_f$ are superior to **Obs-Feas** and **Slot-Feas**, even though the former do not correctly identify the feasible observation-slot combinations, while the latter do not. This suggests that even crude estimates of time are important when building the critics, and demonstrates that simply using slot counts is not good enough.

Our final critic experiments use **Feas-Sched (Bkwd)** takeoff time generation, **Rise** based initial permutation selection, and **1-Phase** critic. In this experiment we vary the number of rejected observations that are moved. The regret values are still used to sample, and are renormalized between samples. The number of observations is moderately low, so we limited ourselves to experiments moving 2, 3 or all rejected observations. As we see, we don't always benefit from increasing the number of rejected observations that are moved; moving 2or 3 rejects is worse than moving 1, but moving all rejects is clearly better than moving 1.

Critic	X	Z	Crit.
Time$_f$	16	2.210	[0.01 0.025]
Time$_t$	14	2.055	[0.01,0.025]
Obs-Feas	14	1.710	[0.025 0.05]
Slot-Feas	16	1.641	>0.05
1-Phase	14	2.338	[0.005 0.01]

Rejects	N	Z	Crit.
1	14	2.338	[0.005, 0.01]
2	13	2.148	[0.01, 0.025]
3	13	2.253	[0.01, 0.025]
all	15	2.541	≈ 0.005

Fig. 11. Wilcoxon Ranked Sign Test results comparing SWO Critic variants to SWO Baseline

Fig. 12. Wilcoxon Ranked Sign Test results comparing critics moving variable numbers of rejected observations to SWO Baseline

6.5 The Best Algorithms

First, we revisit Figures 4 and 5. The baseline SWO generates plans of as good or better quality as ForwardPlanner. It runs at a fraction of the time of ForwardPlanner without the Euler's Method approximation speedup, and often is faster than ForwardPlanner with Euler's Method. The results show that, for these 6 problems, the baseline SWO is capable of producing quality plans.

We next compare the CPU performance of the SWO algorithms. In order to make sense of this analysis, it is important to note that SWO terminates if all observations are scheduled. We compare algorithm performance in Figure 13 using the mean and standard deviation in CPU times for all 20 runs of the different algorithms; CPU times are given in seconds. We also reproduce the Wilcoxon signed rank test results comparing the quality of the flights of each SWO version to the SWO baseline. Overall, adding features that further improve the quality of flights leads to roughly a factor of two increase in CPU time. The takeoff time selection method imposes a significant computational burden on SWO, as can be seen by the increase in the mean CPU time. While the critics also impose a computational burden on SWO, we actually see a *reduction* in CPU time compared to those methods without the intelligent critics; this is likely due to the early termination of SWO when all observations are scheduled.

Analyzing the CPU time on a case by case basis, we find that our worst-case performance hit is roughly a factor of 10 increase in CPU time between the baseline SWO and

Name	Baseline	T/O	Perm.	Critic	Swaps
Takeoff Range	**FlightDur**	**Feas-Sched (Fwd)**	⇒	⇒	⇒
Permutation	**Uniform**	⇒	**Rise**	⇒	⇒
Critic	**Rise$_t$**	⇒	⇒	**1-Phase**	⇒
Swaps	1	1	1	1	all
Mean	63.728	113.071	187.612	166.486	145.501
Sdev	29.976	77.623	144.755	108.985	86.427
X	-	16	14	14	15
Z	-	1.823	2.055	2.338	2.541
Crit	-	[0.025,0.05]	[0.01,0.025]	[0.005,0.01]	≈ 0.005

Fig. 13. Comparison of mean and variance of SWO CPU times for all "incremental best" SWO variants identifying best SWO features

the best SWO, which is moderately high. Howerver, the vast majority of the time the CPU time hit is under a factor of 2. The resulting SWO algorithms deliver significantly better quality flights than ForwardPlanner with Euler's Approximation, at roughly comparable run times.

7   Conclusions and Future Work

We described the SFPP, a difficult mixed discrete and continuous constrained optimization problem. ForwardPlanner, an initially promising approach mixing techniques from AI and OR, ultimately fails to scale for the SFPP. We have described the application of SWO to the SFPP problem. As with our previous approach, ForwardPlanner, the resulting algorithm combines AI and OR techniques to solve a difficult constrained optimization scheduling problem. Our results indicate that SWO is a powerful technique that delivers higher quality flight plans in less time than ForwardPlanner, our previous approach to the SFPP. The quality of flights found by SWO can be increased even further, at a reasonably increase in CPU time.

SWO utilizes numerous techniques in a novel combination to solve the SFPP. The combination of relaxations and continuous optimization method used in ForwardPlanner to reduce the infinite space of setup actions lead to an efficient constructor for our SWO algorithm. We also show that relaxations of the SFPP lead to traditional OR problems, and employ heuristic solutions to these problems in our SWO approach to good effect. In particular, the takeoff time selection method based on greedy solutions to $1|r_i; p_i; d_i| \sum w_i U_I$ proved to be an important component of the best quality SWO algorithm. The use of critics that guarantee each step of SWO produces a change in schedule is a novel contribution that we believe is an important component of our algorithm. Finally, we verify two conclusions from previous work in SWO. First, informed permutation construction techniques improve SWO performance over random permutation generation. Second, swapping many rejected observations per critic application pays off well in terms of both the quality of solutions and speed of SWO. These lessons may serve others working on complex constrained optimization problems with mixes of discrete and continuous variables.

There is considerably work left to do on the SFPP. Our experiments assumed all observations were of equal value; it is easy to generalize our SWO to handle variable priority, but empirical studies are needed to ensure SWO finds high quality flights. Our benchmark included problems for which it was always possible to schedule all observations. SWO can be modified for problems where this is impossible. Ongoing work shows SWO works well even when this is not the case; again, further tests are required to ensure good performance. In particular, CPU times will likely increase when early termination is no longer likely. Additionally, for each observation, minimizing average line-of-sight water vapor is an important objective. Initial results with SWO show promise, but more work is needed. Finally, the SFPP also requires that we build series of flights rather than just a single flight. Preliminary flight series testing indicates that SWO is a promising technique for building flight series, but the basic algorithm requires some modifications to ensure good performance.

Acknowledgments

We would like to thank European Center for Medium Range Weather Forecasting for the use of the climatology data, Michael A. K. Gross for his ongoing assistance in this project, and Tien Ba Dinh for prototyping SWO for the SFPP. This work was funded by the SOFIA Projects Office and by the NASA Intelligent Systems Program.

References

1. Bresina, J.: Heuristic-biased stochastic sampling. In: Proceedings of the 13th National Conference on Artificial Intelligence. (1996)
2. Johnston, M., Miller, G.: Spike: Intelligent scheduling of the hubble space telescope. In Zweben, M., Fox, M., eds.: Intelligent Scheduling. Morgan Kaufmann Publishers (1994)
3. Potter, W., Gasch, J.: A photo album of earth: Scheduling landsat 7 mission daily activities. In: Proceedings of the International Symposium Space Mission Operations and Ground Data Systems. (1998)
4. Smith, D.: Choosing objectives in over-subscription planning. Proceedings of the 14th International Conference on Automated Planning and Scheduling (2004)
5. Frank, J., Kürklü, E.: Sofia's choice: Scheduling observations for an airborne observatory. In: Proceedings of the 13th International Conference on Automated Planning and Scheduling. (2003)
6. Frank, J., Gross, M.A.K., Kürklü, E.: Sofia's choice: An ai approach to scheduling airborne astronomy observations. In: Proceedings of the 16th Conference on Innovative Applications of Artificial Intelligence. (2004)
7. Joslin, D., Clements, D.: Squeaky wheel optimization. Journal of Artificial Intelligence Research **10** (1999) 353 – 373
8. Globus, A., Crawford, J., Lohn, J., Pryor, A.: A comparison of techniques for scheduling earth observing satellites. In: Proceedings of the 16th Conference on the Innovative Applications of Artificial Intelligence. (2004)
9. Barbalescu, L., Whitley, D., Howe, A.: Leap before you look: An effective strategy in an oversubscribed scheduling problem. In: Proceedings of the 19th National Conference on Artificial Intelligence. (2004)
10. Smith, T., Pyle, J.: An effective algorithm for project scheduling with arbitrary temporal constraints. In: Proceedings of the 19th National Conference on Artificial Intelligence. (2004)
11. Becklin, E., Horn, J.: High-latitude observations on sofia. Publications of the Astronomical Society of the Pacific **113** (2001)
12. Meeus, J.: Astronomical Algorithms. Willmann-Bell, Inc. (1991)
13. Cash, J.R., Karp, A.H.: A variable order runge-kutta method for initial value problems with rapidly varying right hand sides. ACM Transactions on Mathematical Software **16** (1990) 201–222
14. Brücker, P.: Scheduling Algorithms. Springer (1998)
15. Lindgren, B.: Statistical Theory. Macmillian (1976)

Shorter Path Constraints for the Resource Constrained Shortest Path Problem

Thorsten Gellermann[1], Meinolf Sellmann[2], and Robert Wright[3]

[1] University of Paderborn, Computer Science,
Fuerstenallee 11, 33098 Paderborn
[2] Brown University, Computer Science,
115 Waterman St, Providence, RI 02912
[3] Air Force Research Lab, Inform. Directorate,
525 Brooks Road, Rome, NY 13441

Abstract. Recently, new cost-based filtering algorithms for shorter-path constraints have been developed. However, so far only the theoretical properties of shorter-path constraint filtering have been studied. We provide the first extensive experimental evaluation of the new algorithms in the context of the resource constrained shortest path problem. We show how reasoning about path-substructures in combination with CP-based Lagrangian relaxation can help to improve significantly over previously developed problem-tailored filtering algorithms and investigate the impact of required-edge detection, undirected versus directed filtering, and the choice of the algorithm optimizing the Lagrangian dual.

1 Introduction

Path constraints play a key role in many applications. Examples range from airline crew scheduling [8, 14] and vehicle routing [20] to the traveling salesman [2] and the resource constrained shortest path problem [1, 4, 6, 7, 12]. Of special interest in the context of combinatorial optimization are path constraints that incorporate the objective function. Shorter path constraints do exactly that by stating that a set of binary variables that are semantically linked to the edges of a graph must form a path from some designated source to a designated sink, whereby the length of this path must not exceed a given threshold value. Unfortunately, in [17] it was shown that achieving generalized arc-consistency for shorter path constraints is NP-hard. Consequently, filtering algorithms were developed that achieve weaker, so called *relaxed consistency* in the same time that it takes to solve the shortest path problem itself. This work was purely theoretical, though. Therefore, we consider it of interest to evaluate the performance of these filtering algorithms in practical experiments.

For this purpose, we focus on the resource constrained shortest path problem (RC-SPP) that consists in finding a shortest route from some given source to a designated sink such that some given resources that are consumed while traversing the edges are not exhausted. While the RCSPP is of interest in itself, for example in the context of traffic guiding systems and route planners for cars and trucks, the problem also evolves as a natural subproblem in the context of even more complex problems like vehicle routing [20].

R. Barták and M. Milano (Eds.): CPAIOR 2005, LNCS 3524, pp. 201–216, 2005.
© Springer-Verlag Berlin Heidelberg 2005

Based on the new filtering algorithms presented in [17], we provide an evaluation of relaxed consistency for shorter path constraints in the context of the resource constrained shortest problem. In the following section, we briefly review the filtering algorithms developed in [17]. In Section 3 we define the resource constrained shortest path problem formally and present a filtering approach that considerably outperforms previous filtering algorithms for this problem, as we will then see in Section 4.

2   Relaxed Consistency for Shorter Path Constraints

Before we review the filtering algorithms that we are going to use for the RCSPP later, let us start out by defining what shorter path constraints are. In words, they express our wish to search for paths in a (directed or undirected) graph such that the length is smaller than some given threshold value. Formally, we define:

Definition 1. *Denote with $G = (V, E, c)$ a weighted (directed or undirected) graph with $||c||_\infty \in O(\mathrm{poly}(|E|, |V|))^1$, and let $h \in \mathbb{N}$.*

- *A sequence of nodes $P = (i_1, \ldots, i_h) \in V^h$ with $(i_f, i_{f+1}) \in E$ for all $1 \leq f < h$ is called a path from i_1 to i_h in G.*
- *A path P is called* simple *iff P visits every node at most once. For all $i, j \in V$, denote with $\pi(i, j)$ the set of all simple paths from i to j.*
- *For all paths P, nodes $i \in V$ and edges $(i, j) \in E$, we write $i \in P$ or $(i, j) \in P$ iff P visits node i or the edge (i, j), respectively. For a set of nodes or edges S, we write $S \subseteq P$, iff $s \in P$ for all $s \in S$. Correspondingly, we write $P \subseteq S$ iff $s \in S$ for all $s \in P$.*
- *The cost of a path $P = (i_1, \ldots, i_h)$ is defined as $\mathrm{cost}(P) := \sum_{1 \leq j < h} c_{i_j i_{j+1}}$. Accordingly, for any set $S \subseteq E$ we define $\mathrm{cost}(S) := \sum_{(i,j) \in S} c_{ij}$.*

Definition 2. *Let $G = (V, E, c)$ denote a (directed or undirected) graph with $n = |V|$ and $m = |E|$, a designated source $v_1 \in V$ and sink $v_n \in V$, and arc costs $c_{ij} \in \mathbb{Z}$. Further, assume we are given binary variables $X_1, \ldots, X_m$, and an objective bound $B \in \mathbb{Z}$.*

- *A constraint $SPC(X_1, \ldots, X_m, G, v_1, v_n, B)$ that is true, iff*
  1. *the set $\{e_i \mid X_i = 1\} \subseteq E$ determines a simple path in the graph G from the source v_1 to the sink v_n, and*
  2. *the cost of the path defined by the instantiation of X is lower than B*
  is called a shorter path constraint.
- *We call every simple path in G from source to sink with costs less than B* admissible.
- *A path P is called a* k-simple path *in G iff for all $j \in V$ the path P visits j at most k times. Note that a 1-simple path is a simple path in G.*
- *Given a shorter path constraint, a k-simple path P from v_1 to v_n is called a k-admissible path iff $\mathrm{cost}(P) < B$.*

---

[1] This is the common *similarity assumption* that states that the largest cost is bounded by some polynomial in $|E|$ and $|V|$.

As mentioned before, it can be shown that achieving generalized arc consistency for shorter path constraints is an NP-hard task. Therefore, in [17] filtering algorithms for shorter path constraints on arbitrary undirected and directed graphs were developed that achieve *relaxed consistency*. With that term we denote filtering algorithms for minimization constraints [9] that only guarantee that those variable assignments are identified that would cause a bound rather than the optimal solution in the current subtree itself to exceed the current best known upper bound.

The formal relaxations considered in [17] are very technical and do not give a particularly useful insight into the task of filtering shorter path constraints. Therefore, we do not to repeat them here but just outline the filtering algorithms that we will use later.

1. On both directed and undirected graphs, the filtering algorithm starts with two shortest path computations once from the source and the other starting at the sink node whereby, in the directed case, the computation is performed in the reverse graph.
2. As a result, we get the shortest path value from source to sink. If this value exceeds the objective bound B, the constraint fails.
3. Otherwise, as a byproduct of the shortest path computations we get the shortest path distances from the source and to the sink of every node for free. We use this information to identify those nodes and arcs of the graph for which the shortest 2-simple path that visits them is above the threshold B. For the nodes, we get this value by adding the shortest path distance from the source and that to the sink, for edges, we add the weight of the edge to that value.

2.1   Exploiting Bridges in Undirected Graphs

After having shrunk the graph in step 3, as a last step of our filtering algorithm we try to identify those edges that must be visited by all paths having a length below the given threshold. This step will be different for undirected and directed graphs. In the undirected case, there exists a simple exact classification of the edges that must be visited. In [17], it was shown that the edges to be required are exactly the bridges[2] in the reduced graph that fall onto the shortest path:

4a. We compute the set of bridges in the reduced graph. The bridges that are also on the shortest path from source to sink must be visited by all admissible paths, and we mark them as required.

On top of this last step of the filtering procedure that was proposed in [17], we add one more idea: We observe that bridges that are not on the shortest path cannot be visited by any simple path from source to sink. Therefore, we can remove those bridges and the entire part of the graph behind them as well:

5a. Remove all bridges from the graph that are not on the shortest path.

---

[2] A bridge is an edge whose removal disconnects the graph.

2.2   Required Arcs in Directed Graphs

Unfortunately, we do not know a similar classification of required arcs as we have it for undirected graphs where the edges to be required are exactly the bridges in the reduced graph that we get after step 3. The algorithm in [17] tries to bound the shortest path distance when having to detour around an arc on the shortest path. When implementing this algorithm, we realized that actually we do not need to compute this bound after having reduced the graph in step 3 of the algorithm. While preserving the same filtering effectiveness of the original algorithm, we can save the overhead of using a heap data structure, because it is completely sufficient to know whether such a detour still exists; since the arc that we use in our detour has not been deleted, we know already that the value of the detour will not exceed the given path-length threshold.

In order to state the last step of our filtering algorithm for directed graphs, we briefly repeat some of the terminology introduced in [17]. Let $T \subseteq E$ denote a shortest-path

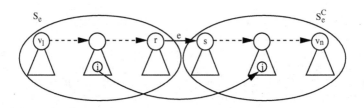

Fig. 1. The figure schematically shows a shortest-path tree T rooted at v_1. Solid lines denote arcs in G, dashed lines mark parts of the shortest path $P(v_1, v_n)$ from v_1 to v_n. The triangles symbolize shortest-path subtrees. For an edge $e = (r, s) \in P(v_1, v_n)$, the nodes in V are partitioned into two non-empty sets S_e and S_e^C. If e is removed from the graph, the shortest path from v_1 to v_n must visit an edge $(i, j) \in (S_e \times S_e^C) \setminus T$

tree in G rooted at v_1. Without loss of generality, we may assume that every node in the graph $G = (V, E)$ can be reached from the source node v_1. Obviously, when $e \in E$ is removed from T, the nodes in V are partitioned into two sets: the set $v_1 \in S_e \subset V$ of nodes that are still connected with v_1 in $T \setminus \{e\}$, and the complement of S_e in V, S_e^C (see Fig. 1). Using these naming conventions, the last step of our filtering algorithm for directed graphs reads:

4b. Denote with E^R the reduced arc set after step 3. For all arcs $e = (r, s)$ on the shortest path from v_1 to v_n, check whether there exists any other arcs $e \neq f = (i, j) \in (S_e \times S_e^C) \cap E^R$. If not, then e must be visited by all paths from source to sink and it is therefore required.

Note that this last step can be implemented with the help of a simple set data structure for an asymptotic cost of $O(m + n)$.

3   A Filtering Approach for Resource Constrained Shortest Paths

In order to evaluate the filtering algorithms for directed and undirected graphs as described in the previous section, we apply them in the context of resource constrained shortest path:

Definition 3. *Given a (directed or undirected) graph $G = (V, E)$, $n = |V|$, $m = |E|$, with $R + 1$ edge-weight functions $l^k : E \rightarrow \mathbb{N}$, $0 \leq k \leq R$, R resource limits $L^1, ..L^R$, and two designated source- and sink-nodes $v_1, v_n \in V$, the* resource constrained shortest path problem *(RCSPP) consists in the computation of a path $P \subseteq E$ such that $\sum_{e \in P} l_e^k \leq L^k$ for all $1 \leq k \leq R$ and $\sum_{e \in P} l_e^0$ is minimal.*

When we denote the best known solution value found with B and set $L^0 := B$, any RCSPP-instance can be modeled as a conjunction of $R + 1$ shorter path constraints $SPC(X_1, \ldots, X_m, (V, E, l^k), v_1, v_n, L^k)$ for $0 \leq k \leq R$.

Of course, we could use these constraints to perform an ordinary tree search. However, for the RCSPP it was found that tree search approaches perform rather poorly. Instead, to solve the RCSPP, state-of-the-art solvers compute lower and upper bounds on the problem first and then close the duality gap. The latter task is carried out by an enumeration procedure such as dynamic programming [15] or labeling approaches [6]. The tightening of the initial problem is vital for an effective gap closing procedure and is therefore essential for the overall performance and the practical success of the entire approach.

Following this framework, we assume that an initial upper bound B has been computed before the filtering phase that will, as a byproduct, also provide a lower bound on the problem. Instead of having the shorter path constraints communicate via variable domains only, we use the CP-based Lagrangian relaxation framework as published in [16, 18, 19]. Precisely, we relax all linear constraints $\sum_{1 \leq i \leq m} l_i^k X_i \leq L^k$, $1 \leq k \leq R$, and penalize their violation in the objective function. Given any vector of Lagrangian multipliers $0 \leq \lambda \in \mathbb{Q}^R$, we consider the constraint

$$SPC(X_1, \ldots, X_m, (V, E, l^0 + \sum_{1 \leq k \leq R} \lambda_k l^k), v_1, v_n, L^0 + \sum_{1 \leq k \leq R} \lambda_k L^k).$$

As usual, the question arises how to compute good Lagrangian multipliers that will yield a good lower bound on the problem and allow us to filter effectively. In general, we can use any subgradient, bundle, or volume algorithm for this purpose [3, 5, 10, 11]. Since most benchmark sets for the RCSPP contain only one resource (i.e., $R = 1$), we use a specialized algorithm for the optimization of the Lagrangian dual with only one multiplier.

3.1   Maximizing One-Parameter Piecewise Linear Concave Functions

A schematic view on the Lagrangian dual for RCSPP-instances with $R = 1$ is given in Figure 2. Assume that we know an interval $[A, B]$ in which the function (let us denote it with f) must take its maximum.

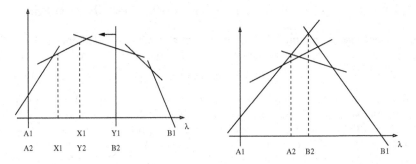

Fig. 2. Maximizing a piece-wise linear concave function by interval partitioning (left) or by cutting planes (right)

Interval Partitioning One way to find the function's maximum in the given interval is to trisect the interval by introducing two interior points $A < X < Y < B$. When we evaluate the function at X and Y and find that $f(X) > f(Y)$ ($f(X) < f(Y)$), due to its concaveness we can deduce that f must take its maximum in the interval $[A, Y]$ ($[X, B]$). Thus, we have found a smaller interval in which the function must take its maximum. We can repeat this process until the width of the interval has become small enough. Of course, in every iteration we could choose new interior points in the current interval. However, in order to save some evaluations of the function f we should try to reuse one of the former inner points. If we partition the current interval according to the golden section, we know that the interval length will decrease geometrically and that one inner point can always be reused, which means that we need to perform only one evaluation of f in each iteration. The procedure is sketched graphically in Figure 2.

For the optimization of the Lagrangian dual this means that we can ϵ-approximate the best Lagrangian multiplier λ in $O(\log \frac{L}{\epsilon})$ iterations, whereby L denotes the width of the initial interval. Each iteration involves the solution of only one Lagrangian subproblem. In the context of the RCSPP, the subproblem is a shortest-path problem. Moreover, assuming that there exists a path that obeys the resource restriction (i.e., when a primal solution exists at all and consequently the dual is not unbounded), it is easy to show that the optimal Lagrangian multiplier cannot be greater than $n||l^0||_\infty$. Consequently, we can solve the Lagrangian dual in time $O(\log \frac{n||l^0||_\infty}{\epsilon}(m + n \log n))$.

With this method, the proposed filtering algorithm for the RCSPP with one resource works as follows: We choose an initial interval $[0, n||l^0||_\infty]$. Denote with L the length of the current interval with left end-point l. Then, we solve the Lagrangian shortest-path subproblem for interior points $X = l + \frac{\sqrt{5}-1}{2}L$ and $Y = l + \frac{\sqrt{5}+1}{2}L$. While solving the shortest-path problem, we also apply our filtering algorithm as described in the previous section. Depending on the point that achieves a larger shortest-path value, we cut off either the right or the left part of the interval and proceed by solving one more Lagrangian dual and filtering in each successive iteration of the algorithm.

Cutting Planes Another way of computing the function's maximum in the given interval is to use a cutting plane algorithm [13]. Clearly, if the function has a negative slope on the left end point of the interval, then this is the point where the function takes its maximum in the given interval. Analogous reasoning holds for the right end point of the interval if the slope there is positive. Now, if the left end point has a positive slope and the right end point a negative one, then the two lines intersect for some point in the middle of the interval. We evaluate the function at that point and check whether the slope is positive or negative (if it is horizontal then we are obviously done). If it is positive (negative), then this point becomes the new left (right) end point of our search interval, and we continue until the computed inner point does not change anymore.

With this method, the proposed filtering algorithm for the RCSPP with one resource works as follows: We choose an initial interval $[0, n||l^0||_\infty]$. We solve the Lagrangian subproblem at the two end points and perform cost-based filtering. Then we repeatedly intersect the slopes at the end points of our interval to determine inner points for which we process the Lagrangian subproblem again. Depending on whether the solution to the subproblem exceeds the resource limit or not, the corresponding Lagrange multiplier becomes the new left or right end point of our search interval (see Figure 2).

Note that our filtering routine is actually changing the problem while we are solving the Lagrangian dual. In general, this is problematic since really both algorithms designed for maximizing concave functions over convex polytopes are not designed to cope with changes in the problem during the optimization. Instead, one could just mark the changes to be made. However, we found that both our algorithms for maximizing the Lagrangian dual were very robust and yielded good results even when incorporating the changes "on the fly". As a matter of fact, this was very beneficial since the successive calls to the shortest path algorithm become cheaper and cheaper, since the graph size reduces considerably during filtering, as we will see in the following section.

4   Numerical Results

We have outlined the shorter-path filtering algorithms and described how CP-based Lagrangian relaxation can be applied for two-shorter-path-constraint problems. The latter correspond to resource constrained shortest path problems (RCSPPs). We have chosen to base our experimentation on the RCSPP for various reasons. Of course, when evaluating the practical efficiency of shorter-path constraint filtering, we would like to eliminate all possible side effects caused by other constraints of the problem under consideration. Therefore, the purest evaluation would be to consider the shortest path problem. However, since all filtering algorithms for shorter-path constraints are actually based on efficient shortest-path algorithms, this is not a feasible choice.

Note also that the application of filtering algorithms usually only makes sense for NP-hard problems. So the natural idea is to consider a problem that consists in the conjunction of two shorter-path constraints, which corresponds to the search of improving solutions for the RCSPP with one resource. This problem is NP-hard and filtering methods for it have been studied a long time before the idea of constraint programming was developed [1, 4]. Therefore, it is of particular interest to investigate how CP performs in comparison with those problem-tailored filtering algorithms.

This being said, it is important to note here that we do not aim at providing a complete state-of-the-art algorithm for the RCSPP itself. Our goal is instead to evaluate the practical performance of shorter-path constraint filtering, and the RCSPP appears as a very reasonable benchmark for such an evaluation. There exist very efficient algorithms for the optimization of the RCSPP [1, 4, 6, 7, 12]. Most of them incorporate a filtering component, but it could be interweaved with the specific algorithm. Note also that an upper bound is required to perform filtering. Now, in order to avoid that we are actually measuring the performance of an upper bounding procedure and not the quality of shorter-path filtering, we do not provide a primal heuristic for the RCSPP. Instead, we base our experimentation on upper bounds of predefined and controlled accuracy, so that we are able to evaluate the performance of the existing and the new filtering algorithms when the quality of the primal heuristic varies.

Thus, when interpreting the following experimental results, keep in mind that shorter-path filtering is just one component in an RCSPP solver, and that we do not provide a complete solver for this problem here. Especially, we do not provide algorithms for the computation of good upper bounds.

4.1   Overview of Experiments and Benchmark Sets Used

In our experiments, we run tests to determine under which parameters our algorithms, that combine shorter-path filtering with CP-based Lagrangian relaxation (SPFCP), perform best. The performance of the algorithms is measured by the number of edges filtered and the CPU time taken. We seek to answer the following questions. Is there any advantage towards using the undirected version (marked by SPFCP-U) over the directed version (SPFCP-D) of our filtering method? Does using required-arc (-RE) and bridge detection (-BD) as part of the filtering have any benefit? And, which method for optimizing the Lagrangian dual is better, interval partitioning (-IP) or cutting plane (-CP)? Finally, we add a comparison of the SPFCP algorithm with two existing filtering algorithms when used for the RCSPP.

For the optimization of the RCSPP after the initial filtering phase, we use our own implementation of a standard RCSPP label setting algorithm (LSA) or our implementation of the RCSPP algorithm by Mehlhorn and Ziegelmann (MZ) [15]. The experiments measure CPU time in seconds and were performed on an Intel Pentium 4 2.5GHz, 1Gb RAM machine running Red Hat Linux 9. The filtering programs, LSA, and MZ were compiled using gcc version 3.2.2 with the optimizing flag.

We use the RCSPP benchmark files provided by Mehlhorn and Ziegelmann [15][3]. All input graphs specify a designated source and a sink, edge cost and resource, and a resource limit. We use two variants of the benchmark files: the original directed files and converted undirected versions. The latter were generated by viewing the arcs as undirected and flipping a coin in case of multi-edges. Files that were generated in that way are marked with an extra '*'. Note that an undirected graph can be viewed as a bi-directed graph where resource and cost coefficient for all edges are the same in both directions. This interpretation allows us to use the directed version of our filtering algorithm on this benchmark set as well, so that we can compare the undirected and the directed filtering variants on this benchmark set. Table 1 shows information on the size of the graphs as well as the time needed to solve them using MZ and LSA. The following is a description of the types of RCSPP problems the input graphs represent.

---

[3] Data files are available at http://www.mpi-sb.mpg.de/~mark/cnop/.

Table 1. The table shows the initial number of edges in the undirected and directed versions of the test files and the time needed by LSA and MZ to solve them. A '-' indicates a solver was unable to compute a solution due to exhaustive memory consumption

	Austria S	Austria B	Scotland S	Scotland B	Road S	Road B
Undirected	46160	165584	65024	252432	50826	171536
LSA	7.589	252.590	7.784	-	0.651	3.472
MZ	0.492	2.186	0.891	12.378	0.791	3.273
Directed	46160	165584	65024	252432	50826	171536
LSA	6.489	257.361	7.581	-	0.651	3.461
MZ	0.474	2.339	2.914	11.053	0.792	3.283
	Curve 1	Curve 2	Curve 3	Curve 4	Curve 5	Curve 6
Undirected	19890	39580	99890	199580	199890	399580
LSA	5.607	18.603	155.749	-	-	-
MZ	0.055	1.436	2.438	0.596	0.797	-
Directed	9945	19790	49945	99790	99945	199790
LSA	3.537	12.286	29.939	-	-	-
MZ	0.039	1.128	2.078	0.394	0.594	183.034

Digital elevation models (DEM): These graphs are grid graphs representative of elevation data over areas of Austria and Scotland. The problem is to find the path with the minimum total height difference while satisfying a constraint on distance.

Road graphs: This benchmark set contains US road graphs. Edges in these graphs are weighted by distance and congestion. The problem is to find the route that takes minimal time while satisfying constraints on fuel consumption.

Curve approximation: In some applications, such as computer graphics programs, it is necessary to represent infinitely detailed curves with less complex functions. In this benchmark set, curves are estimated by many straight lines/edges joined at breakpoints/nodes which lay on the original curve. It is desirable to reduce the number of breakpoints used to estimate the curve while satisfying a constraint on the amount of error introduced. Modeled as an RCSPP, solutions to these instances minimize the number of sampling points when approximating a curve by a piecewise linear function.

4.2   Undirected and Directed SPFCP

In Section 2, we proposed two implementations of the SPFCP algorithm, one that filters on directed graphs and one that filters on undirected graphs. We explained how the undirected version has the advantage of being able to reason via the detection of bridges. We now want to compare the two variants by using the bi-directed benchmark set. We varied the upper bound on the objective from optimal to +5% optimal to examine how the performance of the SPFCP algorithms degrade. Table 2 shows the results of the comparison using both the directed and undirected versions of SPFCP with required-arc and bridge detection used.

Table 2. The table shows the number of remaining edges and the CPU-time in seconds taken to filter the bi-directed graphs using both the directed and undirected versions of the SPFCP algorithm with bridge and required-arc detection. We vary the quality of the upper bounds between optimal and 5%. The Lagrangian dual is optimized using the cutting plane algorithm

Graph	Optimal				+5%			
	SPFCP-D-RE		SPFCP-U-BD		SPFCP-D-RE		SPFCP-U-BD	
	# Edges	Time	# Edges	Time	# Edges	Time	# Edges	Time
Austria Small*	213	0.367	426	0.153	3667	0.443	6882	0.260
Austria Big*	436	1.650	872	1.017	8903	1.860	14742	1.243
Scotland Small*	652	0.547	1304	0.287	6879	0.613	11258	0.347
Scotland Big*	494	2.793	988	1.927	24155	3.170	30584	2.377
Road Small*	899	0.610	1596	0.320	1559	0.627	2180	0.340
Road Big*	1278	1.807	2476	0.997	2755	1.870	3904	1.000
Curve 1*	301	0.107	602	0.040	13555	0.127	15896	0.053
Curve 2*	300	0.193	600	0.063	15824	0.247	20138	0.110
Curve 3*	811	0.660	1622	0.287	99865	0.837	99886	0.430
Curve 4*	810	1.150	1620	0.423	188321	1.293	191684	0.577
Curve 5*	2018	1.380	4036	0.647	199890	1.820	199890	0.990
Curve 6*	2091	2.387	4182	0.953	392448	3.243	393974	1.697

When comparing the raw numbers, the directed version is capable of filtering more edges than the undirected version on the same graphs. However, on most of the tests where the algorithms were given an optimal bound on the objective the filtered graphs from the directed algorithm has exactly half as many edges as the graphs from the undirected algorithm. This is because the undirected algorithm must meet the constraint of leaving a bi-directed graph after filtering whereas the directed version does not. So, the filtered graphs from both algorithms when given an optimal bound on the objective are relatively the same, the directed version just additionally filters out return edges on the shortest path. The undirected version runs faster though, by 53% on average when given an optimal upper bound. When the value of the upper bound is 5% above the optimal value, the directed version filters on average 20% more edges than the undirected. However, the undirected version is still 45% faster on average.

In general, the time taken by the SPFCP algorithms to perform the filtering increases as the quality of the upper bound decreases. This phenomenon can easily be explained in that successive iterations of the filtering algorithm during the optimization of the Lagrangian dual require more time when previous iterations were not as effective at removing edges. While the undirected version works twice as fast as the directed variant of the SPFCP, the directed version is more effective and more general since it can filter both directed and undirected graphs.

4.3   Required-Arc and Bridge Detection

Next, we would like to investigate what the benefit of identifying edges that must be visited by any improving path is. Note that this aspect was one of the main contribu-

Table 3. The table shows the number of remaining edges and the CPU-time, in seconds, taken to filter the bi-directed road graphs using both the directed and undirected versions of the SPFCP algorithm with and without bridge and required-arc detection. The Lagrangian dual is optimized using the interval partitioning algorithm

Graph	Algo	Optimal		+1%		+3%		+5%	
		# Edges	Time	# Edges	Time	# Edges	Time	# Edges	Time
Road Small SPFCP-	-D	1181	1.457	1314	1.460	1597	1.487	2294	1.657
	-D-RE	899	1.570	965	1.560	1171	1.563	1813	1.770
	-U	1718	1.357	1812	1.360	2046	1.383	2640	1.583
	-U-BD	1596	1.137	1658	1.143	1852	1.153	2332	1.323
Road Big SPFCP-	-D	1414	2.097	1493	2.093	1717	2.127	2100	2.127
	-D-RE	1278	2.217	1307	2.217	1385	2.243	1672	2.223
	-U	2528	1.417	2584	1.423	2724	1.427	3178	1.460
	-U-BD	2476	1.360	2506	1.360	2574	1.373	2980	1.407

tions in [17]. We found that, in the DEM and curve approximation graphs the required-arc and bridge detection algorithms were ineffective. This is caused by the structure of these graphs that have many alternate optimal routes. However, on the road graph test files required-arc and bridge detection turned out to be quite effective and also caused the filtering of more edges than just using the SPFCP algorithm without the detection of required arcs. Table 3 shows the results for running both the undirected and directed versions of the SPFCP algorithm on the bi-directed road graphs with and without required-arc and bridge detection.

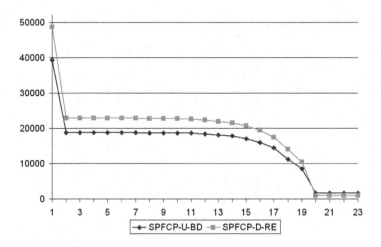

Fig. 3. This figure shows the remaining size of the Road Small* instance after each iteration of the interval partitioning algorithm

The test results show how required-arc and bridge detection improve the SPFCP algorithm's ability to filter edges on all of the road graph test files. They also show that, as the value of the initial upper bound on the objective deviates from optimality required-arc and bridge detection becomes more valuable. In the case of using the undirected SPFCP on the Road Small* test file, bridge detection filters 7% more edges with an optimal upper bound and 13% more with an upper bound of +5% from optimal. Generally, SPFCP-U-BD takes less time to complete than SPFCP-D-RE and SPFCP-U. This can be attributed to the bridge detection being most effective in the early iterations of the filtering algorithm and is illustrated in Figure 3.

4.4   Interval Partitioning Versus Cutting Plane

In the following experiments, we compare the performance of SPFCP-D while using the two algorithms for closing the duality gap, cutting plane and interval partitioning. Table 4 summarizes our results. Using the cutting plane algorithm improves the speed of the SPFCP filtering algorithm dramatically over using interval partitioning: SPFCP-D-CP is 63% faster on average when given an optimal upper bound and 65% faster on average when given an upper bound of +5% from optimal. In the optimality proof, both methods filtered roughly the same amount of edges, while interval partitioning is slightly more effective than the cutting plane algorithm.

The faster computation times and the slightly diminished effectivity of the cutting plane algorithm are explained by the fact that the method is able to close the duality gap in far fewer iterations, which can be seen by comparing Figures 4 and 5. We can see clearly how the cutting plane algorithm considers more meaningful Lagrangian

Table 4. The table shows the number of remaining edges and the CPU-time, in seconds, taken by SPFCP-D-RE using the cutting plane and interval partitioning algorithms for solving the Lagrangian dual. The quality of the upper bound was varied from optimal to +5% from optimal

Graph	interval partitioning				cutting plane			
	Optimal		+5%		Optimal		+5%	
	# Edges	Time	# Edges	Time	# Edges	Time	# Edges	Time
Austria S	229	1.293	3150	1.750	231	0.367	3341	0.430
Austria B	410	7.623	7097	8.160	410	1.607	7323	1.767
Scotland S	304	2.040	3864	2.317	263	0.537	4737	0.607
Scotland B	494	14.113	17220	14.970	494	2.770	21578	3.113
Road S	899	1.563	1813	1.767	899	0.607	1559	0.620
Road B	1278	2.217	1672	2.223	1278	1.793	2755	1.813
Curve 1	306	0.217	7948	0.300	301	0.067	7948	0.090
Curve 2	300	0.270	10067	0.403	300	0.107	10069	0.170
Curve 3	836	1.090	49943	1.797	811	0.447	49943	0.600
Curve 4	803	1.643	95842	2.673	810	0.710	95842	0.847
Curve 5	2018	2.587	99945	4.067	2018	0.953	99945	1.310
Curve 6	2034	3.953	196987	5.947	2091	1.553	196987	2.263

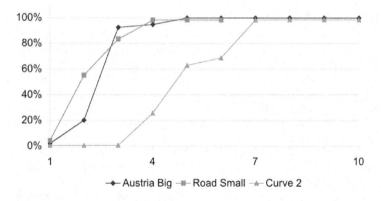

Fig. 4. This figure shows the percentage of edges filtered by SPFCP-D at each iteration of the cutting plane algorithm used to solve the Lagrangian relaxation

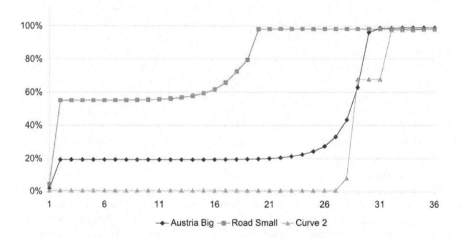

Fig. 5. This figure shows the percentage of edges filtered by SPFCP-D at each iteration of the interval partitioning algorithm used to solve the Lagrangian relaxation

multipliers much earlier in the search, which results in a much quicker computation of the lower bound as well as more filtering at earlier stages of the optimization. Also, it needs less iterations close to the optimal multipliers where the interval partitioning algorithm considers quite a few very near optimal multipliers before the desired approximation quality is achieved. The cutting plane needs just one iteration once that it is close enough to the optimum. We believe it is for that reason that the algorithm is slightly less effective in its filtering abilities. Still, we prefer the cutting plane algorithm over interval partitioning since it is able to filter almost as many edges in a fraction of the time.

4.5   Filtering for the RCSPP

In this section we compare the performance of the SPFCP algorithm against previously developed filtering algorithms for the RCSPP. Particularly, we compare against the algorithm from Aneja et al. (AN) and the algorithm from Beasley and Christofides (BC) [1, 4]. Both AN and BC only remove edges from the graph without detecting those edges that must be visited by all improving paths. AN considers the pure shorter-path constraints on the objective and the resource only, without integrating them in a Lagrangian fashion. BC performs filtering for the optimal Lagrangian multiplier. It is important to note here that suboptimal Lagrangian multipliers can have stronger filtering abilities than the optimal ones [16]. Therefore, the idea of CP-based Lagrangian relaxation makes sense, i.e. it is a reasonable approach to filter even during the optimization of the Lagrangian dual and not just for optimal multipliers only.

Table 5 shows the results for experiments using the directed input graphs and SPFCP-D-BD versus AN and BC. Comparing BC and AN first, we find that BC filters much better, but also takes significantly more time to do so. This is not surprising, since BC needs to solve the Lagrangian dual whereas AN works by just four shortest-path computations. We observe that SPFCP can increase the filtering effectiveness further (by 40% on average) while using less computation time than BC but still about twice as much as AN. The fact that SPFCP runs faster than BC has to be attributed to the algorithm's ability to filter out most of the edges in the first few iterations of solving the Lagrangian dual, thereby reducing the graph size and making successive iterations quicker. While SPFCP filtering works slower than AN, from the solution times of the RCSPP algorithms LSA and MZ on the filtered graphs we see that the additional effort is very worthwhile in the context of the RCSPP. Since this problem, though NP-hard, is still relatively easy to solve, we conjecture that the improved filtering power of shorter-path filtering in combination with CP-based Lagrangian relaxation will probably pay off even more in the context of more complex problems that incorporate shorter-path constraints.

5   Conclusions

We provided an experimental evaluation of shorter-path filtering by applying it to the resource constrained shortest path problem. We have compared the undirected and the directed versions of shorter-path filtering and found that the undirected version, where applicable, works about twice as fast while the directed version is more effective and enjoys wider applicability. Regarding the identification of edges that must be visited by all improving routes, we found that this ability is of use only in rare special cases where no alternative improving paths exist.

Further, we have seen that, in the context of CP-based Lagrangian relaxation, the choice of the algorithm solving the Lagrangian dual can have a significant impact on the overall performance of the filtering algorithm. For one-parameter relaxations, we have found that a method based on cutting planes can be much more efficient than an interval partitioning algorithm.

Finally, our experiments showed that, even for this comparably simple problem, an increase in filtering power can yield to significant performance improvements and that shorter-path constraint filtering outperforms previously developed filtering algorithms for the RCSPP.

Table 5. The table shows the number of remaining edges after filtering wrt to an optimal lower bound plus one (so that the solvers still need to compute the optimum after filtering), the CPU-time in seconds taken for the filtering on directed graphs, and the CPU-time taken by the RCSSP solvers LSA and MZ to find an optimal solution for the filtered graphs. A '-' indicates that a solver was unable to find a solution due to exhaustive memory consumption

Graph	AN				BC				SPFCP D-IP				SPFCP D-CP			
	Filter		Solver		Filter		Solver		Filter		Solver		Filter		Solver	
	Edges	Time	LSA	MZ	Edges	Time	LSA	MZ	Edges	Time	LSA	MZ	Edges	Time	LSA	MZ
Austria S	28480	.287	5.113	.338	245	.620	.002	.039	229	1.293	.002	.040	231	.367	.002	.040
Austria B	131647	1.190	258.596	1.786	410	3.143	.007	.049	410	7.623	.007	.052	410	1.607	.007	.049
Scotland S	43667	.423	7.221	2.683	906	.883	.020	.064	304	2.040	.005	.044	263	.537	.003	.040
Scotland B	209546	1.933	-	10.34	1287	4.823	.033	.084	494	14.113	.008	.052	494	2.770	.009	.050
Road S	21446	.430	.288	.318	1309	1.010	.011	.054	899	1.563	.009	.053	899	.607	.009	.053
Road B	109711	1.770	1.222	1.818	1641	4.247	.019	.072	1278	2.217	.017	.065	1278	1.793	.018	.063
Curve 1	9945	.040	3.693	.038	9945	.067	3.581	.038	306	.217	.004	.001	301	.067	.004	.001
Curve 2	19679	.073	12.225	1.143	305	.133	.004	.040	300	.270	.004	.040	300	.107	.004	.040
Curve 3	49945	.223	30.175	2.078	826	.450	.009	.053	836	1.09	.009	.004	811	.447	.009	.004
Curve 4	99790	.387	-	.394	99790	.653	-	.393	803	1.643	.009	.004	810	.710	.010	.002
Curve 5	99945	.470	-	.592	2018	.940	.049	.018	2018	2.587	.037	.014	2018	.953	.036	.018
Curve 6	199790	.810	-	190.62	199790	1.513	-	190.105	2034	3.953	.036	.006	2091	1.553	.038	.031

References

1. Y. Aneja, V. Aggarwal, K. Nair. Shortest chain subject to side conditions. *Networks*, 13:295-302, 1983.
2. D. Applegate, R. Bixby, V. Chvátal, and W. Cook. On the solution of Traveling Salesman Problems. *Doc. Math. J. DMV*, Extra Volume ICM III, pp.645–656, 1998.
3. F. Barahona and R. Anbil. The Volume Algorithm: producing primal solutions with a subgradient algorithm. *Mathematical Programming*, 87:385–399, 2000.
4. J. Beasley, N. Christofides. An Algorithm for the Resource Constrained Shortest Path Problem. *Networks*, 19:379-394, 1989.
5. H. Crowder. Computational improvements for subgradient optimization. *Symposia Mathematica*, XIX:357–372, 1976.
6. J. Desrosiers, Y. Dumas, M. Solomon, F. Soumis. Time Constrained Routing and Scheduling. *Handbook in Operations Research and Management Science 8: Network Routing*, North-Holland, 8:35–139, 1995.
7. I. Dumitrescu, N. Boland. The weight-constrained shortest path problem: preprocessing, scaling and dynamic programming algorithms with numerical comparisons. *International Symposium on Mathematical Programming (ISMP)*, 2000.
8. T. Fahle, U. Junker, S.E. Karisch, N. Kohl, M. Sellmann, B. Vaaben. Constraint programming based column generation for crew assignment. *Journal of Heuristics*, 8(1):59-81, 2002.
9. F. Focacci, A. Lodi, M. Milano. Cost-Based Domain Filtering. *Principles and Practice of Constraint Programming (CP)* Springer LNCS 1713:189–203, 1999.
10. A. Frangioni. A Bundle type Dual-ascent Approach to Linear Multi-Commodity Min Cost Flow Problems. *Technical Report*, Dipartimento di Informatica, Universita di Pisa, TR-96-01, 1996.
11. A. Frangioni. Dual Ascent Methods and Multicommodity Flow Problems. *Doctoral Thesis*, Dipartimento di Informatica, Universita di Pisa, TD-97-05, 1997.
12. G. Handler, I. Zang. A Dual Algorithm for the Restricted Shortest Path Problem. *Networks*, 10:293-310, 1980.
13. J.E. Kelley. The Cutting Plane Method for Solving Convex Programs. *Journal of the SIAM*, 8:703–712, 1960.
14. A. Makri and D. Klabjan. A New Pricing Scheme for Airline Crew Scheduling. *INFORMS Journal on Computing*, 16:56–67, 2004.
15. K. Mehlhorn, M. Ziegelmann. Resource Constrained Shortest Paths. *Proc. 8th European Symposium on Algorithms (ESA)*, Springer LNCS 1879:326-337, 2000.
16. M. Sellmann. Theoretical Foundations of CP-Based Lagrangian Relaxation. *Principles and Practice of Constraint Programming (CP)*, Spirnger LNCS 3258:634–647, 2004.
17. M. Sellmann. Cost-Based Filtering for Shorter Path Constraints. *Principles and Practice of Constraint Programming (CP)*, Springer LNCS 2833:679–693, 2003.
18. M. Sellmann and T.Fahle. Coupling Variable Fixing Algorithms for the Automatic Recording Problem. *Annual European Symposium on Algorithms (ESA)*, Springer LNCS 2161: 134–145, 2001.
19. M. Sellmann and T. Fahle. Constraint Programming Based Lagrangian Relaxation for the Automatic Recording Problem. *Annals of Operations Research*, 118:17–33, 2003.
20. P. Toth and D. Vigo. The Vehicle Routing Problem. *Monographs on Discrete Mathematics and Applications*, SIAM, Philadelphia, 2001.

Improving the Cooperation Between the Master Problem and the Subproblem in Constraint Programming Based Column Generation

Bernard Gendron[1,2], Hocine Lebbah[2,3], and Gilles Pesant[2,3]

[1] Département d'informatique, et de recherche opérationnelle,
Université de Montréal, C.P. 6128, succ. Centre-ville,
Montreal, Quebec H3C 3J7
`bernard@crt.umontreal.ca`
[2] Centre de recherche sur les transports, Université de Montréal,
C.P. 6128, succ. Centre-ville, Montreal, Quebec H3C 3J7
[3] Département de génie informatique, École Polytechnique de Montréal,
C.P. 6079, succ. Centre-ville,
Montréal, Québec H3C 3A7

Abstract. Constraint programming (CP) based column generation uses CP to solve the pricing subproblem. We consider a set partitioning formulation with a huge number of variables, each of which can be generated by solving a CP subproblem. We propose two customized search strategies to solve the CP subproblem, which aim to improve the coordination between the master problem and the subproblem. Specifically, these two strategies attempt to generate more promising columns for the master problem in order to counter the effect of slow convergence and the difficulty of reaching integer solutions. The first strategy uses the dual variables to direct the search towards columns that drive the relaxed master problem faster to optimality. The second strategy exploits the structure of the constraints in the master problem to generate columns that help to reach integer solutions more quickly. We use a physician scheduling problem to test the strategies.

1 Introduction

Constraint programming (CP) based column generation uses CP to solve the pricing subproblem in a column generation algorithm. First introduced by Junker et al. [10], it has been subsequently used for various applications such as airline crew scheduling [6, 17] and vehicle routing [15]. In these applications, the master problem corresponds to a set partitioning model and the pricing subproblem is formulated as a constrained shortest path problem. Although dynamic programming is generally used to solve constrained shortest path problems, some complex constraints can be difficult to handle within this framework. In this context, CP allows more flexibility and can extend the scope of applicability of column generation algorithms. It is also worthwhile to note that not all

R. Barták and M. Milano (Eds.): CPAIOR 2005, LNCS 3524, pp. 217–227, 2005.

practical applications give rise to constrained shortest path subproblems. For example, in a constrained cutting-stock problem, the pricing subproblem is a constrained knapsack problem. Again, in this context, CP can be useful since the subproblems can include complex constraints in addition to the knapsack structure [7].

Column generation for linear programming (LP) [4] and integer programming (IP) [8,9] dates back to the 60s. During the last 20 years or so, it has been used extensively to solve IP problems with a huge number of variables. In this context, the LP relaxation is solved by column generation and integer solutions are obtained by branching. When column generation is repeated at each node of the branch-and-bound tree, we obtain the so-called branch-and-price algorithm [1,5,18,19]. As pointed out by many authors, the main difficulty when solving these huge IPs by column generation is to achieve the right balance between two different objectives: 1) try to solve the LP relaxation of the master problem, and 2) try to obtain an integer solution to the master problem. In branch-and-price algorithms, the second objective is attained by clever branching rules, while the first one is achieved through the coordination between the master problem and the pricing subproblem provided by the dual variables associated to the constraints of the LP relaxation.

The goal of this paper is to show that, in the context of CP based column generation, these objectives can also be reached by selecting an appropriate branching scheme in the pricing subproblem, e.g., by devising customized selection strategies in a standard CP based tree search algorithm. These selection strategies will guide the search towards "good" columns and thus help the master problem to reach LP relaxation optimality, as well as integrality. Note that using such customized selection strategies in solving the pricing subproblem makes the CP based column generation framework even more attractive.

The paper is organized as follows. First, we present the formulation of the master problem: we consider a set partitioning formulation with different restrictions on the subsets. Since many applications found in the literature fall into this category (many others are also set partitioning models, but with the same restrictions on the subsets), we will use this framework to develop our selection strategies. The next section discusses how to model the pricing subproblem and describes two selection strategies: the first one uses the dual variables to direct the search towards columns that drive the LP master problem to optimality; the second strategy exploits the structure of the constraints in the master problem to generate columns that help to reach integer solutions. We illustrate these strategies on a difficult personnel scheduling problem, dealing with the planning of work schedules for physicians in the emergency room of a major hospital. Section 4 presents the particular CP model used in this case. In Section 5, we present computational results that illustrate the behavior of the search strategies in the context of a price-and-branch implementation (column generation to solve the LP, followed by branch-and-bound on the limited set of variables obtained after column generation). We conclude by summarizing our results and by discussing extensions to our work.

2   The Master Problem

Suppose we have a set of shifts K to assign to a set of employees I. Each shift must be assigned to exactly one employee. There are several constraints that limit the number of feasible schedules for each employee (see Section 4 for an illustrative example). To model this problem, we introduce a set P_i of feasible schedules for each employee $i \in I$. Each feasible schedule $p \in P_i$ is made of a number of shifts; we let δ_k^p equal 1, if shift k belongs to schedule p, and 0 otherwise. Assuming that there is a cost c_i^p for assigning schedule p to employee i, we obtain one of the most common forms of huge IP models amenable to solution by column generation, the set partitioning formulation with different restrictions on the subsets [1]:

$$\min \sum_{i \in I} \sum_{p \in P_i} c_i^p y_i^p \tag{1}$$

$$\sum_{i \in I} \sum_{p \in P_i} \delta_k^p y_i^p = 1, \quad k \in K, \tag{2}$$

$$\sum_{p \in P_i} y_i^p = 1, \quad i \in I, \tag{3}$$

$$y_i^p \in \{0, 1\}, \quad i \in I, p \in P_i. \tag{4}$$

The partitioning constraints, (2), simply state that each shift must be assigned to exactly one employee. Constraints (3), often called convexity constraints, assure that each employee gets one schedule, when combined with the integrality constraints (4). This model is the column generation formulation of a classical scheduling problem, where each shift must be assigned to exactly one employee and there are several constraints for each employee. In some applications, multiple employees can be assigned to the same shift, in which case the partitioning constraints (2) are simply rewritten with a right-hand side corresponding to the number of employees assigned to that shift. This model arises in many applications, such as airline crew scheduling [6, 17] and the classical generalized assignment problem [16] (the pricing subproblem in this case is a collection of knapsack problems). When the pricing subproblems are identical for all employees, we can aggregate the y_i^p variables and replace them by $y^p = \sum_{i \in I} y_i^p$. The convexity constraints are then often removed because it is common in many applications to have $|I|$ not fixed. Examples of this type of master model arise in vehicle routing [15], as well as in the classical cutting-stock problem [8, 9].

Since the number of variables in this set partitioning formulation is huge, they are generated dynamically by solving the LP relaxation of a restricted model (with only a subset of the variables) and by collecting the values of the dual variables associated to the constraints of the problem. These dual values are then transfered to the pricing subproblem that will try to find at least one variable with negative reduced cost (such variables have not yet been generated, since at optimality of the restricted LP relaxation, all variables have non negative reduced costs). In our case, if we denote by λ_k and μ_i the dual variables corresponding to constraints (2) and (3), respectively, the reduced cost for variable y_i^p is given by

$$\bar{c}_i^p = c_i^p - \sum_{k \in K} \delta_k^p \lambda_k - \mu_i. \tag{5}$$

The pricing subproblem decomposes into one problem for each employee, defined by the constraints corresponding to that employee. The objective of that employee subproblem is to find the solution with the least reduced cost. In practice, we do not want to solve this subproblem optimally; it is generally enough to identify a small number of negative reduced cost solutions. Once these variables have been identified, they are added to the restricted LP relaxation and we proceed with another iteration, until the LP relaxation is solved (to prove optimality, we need an exact algorithm for solving the pricing subproblem) or a maximum number of iterations has been reached. Typically, the LP relaxation of such huge set partitioning formulations are very difficult to solve as they exhibit degeneracy and slow convergence (see [11] and [14] for recent contributions on improving the solution of the LP relaxation in column generation algorithms).

Once the LP relaxation is solved (in an exact or heuristic way) an integer solution can be found by branching. One alternative is to perform branch-and-bound using the set of columns obtained after solving the LP relaxation, but this is a heuristic method, as some non generated columns might appear in an optimal solution. If the column generation scheme is repeated at each node of the search tree, we obtain a branch-and-price algorithm, which is an exact method, provided the pricing subproblem can be solved to optimality.

3   Search Strategies for the Pricing Subproblem

Recall that the pricing subproblem decomposes by employee. In each employee subproblem, we will assume that the schedule can be decomposed by day because no more than one shift can be assigned to the employee on any given day. This allows us to define one variable x_j for each day $j \in J$ in the employee subproblem; thus we have a vector of variables $X = (x_1, x_2, ..., x_n)$, where n is the number of days in the planning horizon. The domain of each variable x_j, $D(x_j)$, is defined as the set of shifts required on day j, plus a dummy shift that represents the outcome that the employee does not work on day j. One can take this dummy shift into account in the master problem by adding partitioning constraints for each day corresponding to the dummy shift, with a right-hand side equal to (number of employees - number of shifts required on that day). In this way, dual values associated to these dummy shift constraints can be taken into account when solving the pricing subproblem.

Another alternative for modeling the employee subproblem is to use a set variable representing the shifts that can be assigned over the whole planning horizon [6]. In this setting, a constraint such as "no more than one shift per day" would be modeled using a *shift graph*, where each node corresponds to a shift on a given day and each arc connects two possible consecutive shifts; thus, in this graph, there would be no arc connecting two shifts on the same day. With our formulation, these constraints are implicitly taken into account. Moreover, we will use them to define global constraints specialized to the physician scheduling problem that we consider in our study (see Section 4). Note that adapting our search strategies to the model with set variables is a straightforward task.

Often, the dominant terms in the reduced costs, (5), are the dual values corresponding to the λ_k variables (since feasibility in the master is an issue). Moreover, the costs associated to the assignment of a particular shift do not differ significantly by employee (this is the case in the application presented in Section 4). In this situation, by solving the employee subproblems independently with the same search strategy, we would end up with many schedules that are similar from one employee to another. This implies that reaching an integer solution that satisfies the partitioning constraints (2) is a difficult issue because several columns for many employees would share the same shifts.

The first search strategy, based on the values of the dual variables, attempts to drive the search towards solutions with negative reduced costs. The objective of this *Dual strategy* is thus to speed up the resolution of the LP relaxation of the master problem. A similar strategy, called Lowest Reduced First, has been proposed by Fahle et al. [6]. In this strategy, the shift with the smallest reduced cost over the whole planning horizon is selected and assigned to the employee. In our approach, we aim to introduce more variability in the selection strategy from one employee to another to avoid generating similar schedules. We achieve this objective by scanning the days and, for each day, by selecting the shift with the lth largest dual value λ_k (which is the dominant term in the reduced cost), where l is randomly chosen, with a bias towards the largest values. By choosing different seed values for the random number generator, we obtain variability in the schedules generated for different employees.

The second strategy tries to speed up the search for integer solutions in the branch-and-bound (-and-price) phase; we call it the *Master strategy*, since the idea is to take into account the partitioning constraints (2) of the master problem in the selection process. We simply store N_{sj}, the number of times each shift s on any given day j has been assigned to another employee when solving the current pricing subproblem. When solving an employee subproblem, we scan the days, and for each day j, we choose the first shift s (in arbitrary order) for which N_{sj} is less than the right-hand side of the partitioning constraint corresponding to shift s (note that this definition allows for right-hand side values larger than 1). If there is no such shift, we choose a shift arbitrarily. In addition, to avoid generating similar schedules from one column generation iteration to the next, we change the order for solving the employee subproblems. Note that, instead of choosing the shifts in arbitrary order, we could have biased the selection towards the shifts with the largest dual values, as in the dual strategy. However, we first wanted to examine the effects of each strategy independently.

We will present computational results of these two strategies on a physician scheduling problem, to be described in the next section. Before examining the particular constraints of that problem, we comment on how we have modeled the negative reduced cost constraint. As in [6], we have used a simple version where the constraint is used only for pruning and not for domain reduction. As shown by these authors, this version of the reduced cost constraint is clearly inferior, especially for large-scale instances, to the shortest path constraint they developed, which performs domain reduction in an efficient way. Since we focus on the search strategies in the pricing subproblems and their impact on solving the master problem, we expect that our conclusions will hold as well if we use shortest path constraints instead of the simple version of the reduced cost constraints.

4   Illustrative Example: A Physician Scheduling Problem

As illustrative example of our approach, we use a simplified version of the physician scheduling problem described in [2] (a similar problem, also modeled with CP, can be found in [3]). In this problem, a group of physicians must be assigned to a predefined set of shifts in order to satisfy the requirements of an emergency room of a major hospital. The schedules are planned for the next n days; typically, the planning horizon varies between two weeks and six months.

Several types of rules must be satisfied in order to obtain acceptable working schedules. First, there are a few compulsory rules, e.g., rules that must absolutely be enforced. Demand rules are the most basic in this category. They define how many physicians should work at different periods of a day and which responsibilities are attached to particular shifts. Each day is divided into three periods of eight hours: day, evening and night. Three physicians (two on weekends or holidays) work during day and evening shifts, including one exclusively in charge of traumas ("heavy" emergencies). "Trauma" shifts are considered heavier than "regular" shifts (which mostly involve the treatment of "light" cases and patients in stabilized condition). At night, there is only one night shift, the physician assuming the responsibilities of trauma and regular shifts. Three days per week, one physician works a four-hour shift, the "follow-up" shift, when he receives by appointment patients that have recently been treated at the emergency room. Other compulsory rules include: vacations, days-off, or particular shifts requested by the physicians, and the basic ergonomic rule: "there must be at least 16 hours between the end of one shift and the beginning of another one".

The demand rules are implicitly taken into account in the definition of our variables. The other rules dealing with preassigned shifts are easily modeled, as well as the basic ergonomic rule, which is formulated as follows:

$$x_j = s \Rightarrow x_{j+1} \neq t, \quad j \in \{1, 2, ..., n-1\}, s \in S_j, t \in F_{sj}, \qquad (6)$$

where S_j is the set of required shifts on day j and F_{sj} is the set of forbidden shifts once shift s is assigned on day j (all forbidden shifts are located on day $j+1$). Another way to model this constraint is by using the shift graph described in Section 3.

In addition to these basic constraints, the physicians have their own requirements regarding the number of consecutive night shifts they accept to work: some prefer to work three consecutive nights, while others do not want to work more than one night at a time, and some others accept any number of consecutive nights, as soon as it never exceeds three. These specific requirements can be easily modeled using the **stretch** constraint [12]. If we denote by $S = \{1, 2, \ldots, m\}$ the set of shifts that can be assigned on any given day (e.g., $S = \cup_{j \in J} S_j$),

$$\textbf{stretch}(X, S, L^-, L^+) \tag{7}$$

ensures that, in any instanciation of the variables $X = (x_1, x_2, \ldots, x_n)$, the length of any sequence of consecutive days assigned to shift $s \in S$ is between l_s^- and l_s^+, where $L^- = (l_1^-, l_2^-, \ldots, l_m^-)$ and $L^+ = (l_1^+, l_2^+, \ldots, l_m^+)$. The stretch constraint also serves to limit to four (in some cases, five) the number of consecutive shifts of any type.

In addition, there are constraints on the minimum and maximum number of hours each physician can work every week. These constraints are modeled using the two global constraints **distribute** [13] and **ext**. The constraint **distribute**(C, S, X) ensures that each value $s \in S$ appears exactly c_s times in X, where $C = (c_1, c_2, \ldots, c_m)$. The constraint **ext**$(M, \Gamma)$ lists all the admissible d-tuples of variable $M = (m_1, m_2, \ldots, m_d)$. Let

- H, an m-dimensional vector such that H_s equals the number of hours worked during shift s (there are only three possible values: 0, for the dummy shift, 4, for the follow-up shift, and 8, for all other shifts);
- $Y = (y_1, y_2, \ldots, y_n)$, an n-dimensional vector of variables defined as $y_j = H_{x_j}$;
- $M = (m_0, m_4, m_8)$ a 3-dimensional vector of variables representing the number of times 0-hour, 4-hour and 8-hour shifts appear in Y.

The constraint on the minimum $(minH)$ and maximum $(maxH)$ number of working hours every week can then be written as:

$$Y = H_X \wedge \textbf{distribute}(M, \{0, 4, 8\}, Y) \wedge \textbf{ext}(M, \Gamma), \tag{8}$$

where the set of admissible triples Γ is given by $\Gamma = \{(m_0, m_4, m_8) \in [0, \overline{m_0}] \times [0, \overline{m_4}] \times [0, \overline{m_8}] \mid minH \leq 4m_4 + 8m_8 \leq maxH, m_0 + m_4 + m_8 = 7\}$, with $\overline{m_4} = \lfloor maxH/4 \rfloor$, $\overline{m_8} = \lfloor maxH/8 \rfloor$ and $\overline{m_0} = 7 - (\overline{m_4} + \overline{m_8})$.

Many other rules exist in the application described in [2], but our simplified version contains only the constraints described so far. This choice allows to capture enough of the complexity of the problem, and to illustrate as well the flexibility of the modeling approach.

The objective function is defined so as to ensure that different types of shifts are fairly distributed among physicians. In our simplified version, we try to achieve a fair distribution of two types of antagonist shifts: days versus evenings, and regular versus trauma. Positive and negative deviations with respect to an ideal ratio are penalized, thus defining the cost of a schedule over the whole planning horizon.

5 Computational Results

The objective of our computational experiments is to compare the two search strategies, Dual and Master, when embedded within a simple column generation algorithm that proceeds in three phases:

- *Initialization*: During this phase, an initial set of columns is generated. For this purpose, we use a CP engine to solve a global problem containing all the constraints of the subproblem, as well as the partitioning constraints; this method ensures that at least one globally feasible set of schedules is generated (a similar approach is presented in [17]).
- *Column generation*: This phase is the CP based column generation method, with either the Dual or the Master search strategy being used to solve the subproblem; the search is stopped for each employee subproblem when one solution with negative reduced cost is obtained. We stop this phase after a predetermined number of iterations (40 in our tests).
- *Branch-and-bound*: We invoke a branch-and-bound algorithm that solves the formulation obtained at the end of the column generation phase. This allows to evaluate if the columns generated during the column generation phase are sufficient to improve the initial integer solution.

The overall method is programmed with ILOG Concert (version 1.1). ILOG Solver (version 5.1) is used for solving the subproblems and the global problem in the initialization phase. ILOG CPLEX (version 7.1) solves the restricted LP relaxations in the column generation phase, as well as the restricted IP model in the branch-and-bound phase. The default parameters are used for all software packages.

We tested the approach on an instance of the physician scheduling problem with 18 physicians over a period of two weeks. Table 1 compares the results obtained with the two search strategies, Dual and Master, as well as the default search strategy implemented in Solver. For each strategy, five values are displayed: $Z(INIT)$, the objective function value of the integer solution obtained after the initialization phase; $Z(LPCG)$, the objective function value of the best solution to the restricted LP relaxation of the master problem obtained at the end of the column generation phase (recall that we stop this phase after 40 iterations); $Z(IPBB)$, the value of the best integer solution obtained at the end of the branch-and-bound phase; Avg.CPU/iter, the average CPU time in seconds spent per iteration during the column generation phase; Avg.Fail/iter, the av-

Table 1. Comparing the Default, the Dual and the Master Search Strategies

Strategy	$Z(INIT)$	$Z(LPCG)$	$Z(IPBB)$	Avg.CPU/iter	Avg.Fail/iter
Default	28816	28816	28816	3.68	200.41
Dual	28816	12028	28816	1.84	0.12
Master	28816	13481	13952	2.28	37.27

erage number of failures per iteration when solving the employee subproblems during the column generation phase.

These results indicate that both the Dual and Master strategies improve significantly upon the default strategy. The Dual strategy quickly identifies negative reduced cost solutions, as shown by the low number of failures and the modest CPU time required. This strategy also exhibits the lowest objective value of $Z(LPCG)$, which indicates that it is the most effective for solving the LP relaxation. However, with the set of columns obtained after the column generation phase, the branch-and-bound algorithm is unable to identify an improving integer solution. By contrast, the Master strategy is less effective at solving the LP relaxation, but identifies a significantly better integer solution during the branch-and-bound phase.

Figure 1 presents, for the three search strategies, the evolution of the objective value during the column generation phase. The default strategy generates negative reduced cost solutions at every iteration but these added columns do not contribute to change the objective value. The Dual strategy is unable to improve the objective for the first 25 iterations, but then exhibits a sudden drop and a constant improvement at every iteration. These observations are consistent with the common knowledge in the column generation literature that the dual information is relatively poor during the first iterations. The Master strategy improves the objective value very early and then continues to make progress, although it is outperformed by the dual strategy in the last iterations. These results suggest that a hybrid method combining the Dual and Master strategies would be indicated to take advantage of the relative merits of both approaches.

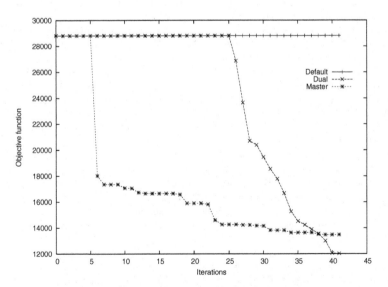

Fig. 1. Comparing the Strategies During Column Generation

6   Conclusion

In this paper, we have presented and compared two search strategies for solving
the pricing subproblem in CP based column generation. The dual strategy aims
to accelerate the solution of the LP relaxation of the master problem, while the
master strategy drives the search towards integer solutions. We have presented
computational results on an instance of a physician scheduling problem. These
results show that these two strategies are promising and suggest that combining
them might be even more effective. Further tests on other classes of problems
would confirm the interest of such customized search strategies in CP based
column generation. In addition, several other comparisons might be instructive:

- A comparison of the search strategies using the shortest path-based negative
  reduced cost constraint [6] rather than the simple pruning version used here;
- A comparison of the relative gains in terms of CPU time and branch-and-
  bound tree size obtained by sophisticated master problem branching rules
  (instead of the CPLEX default strategy used in our experiments) versus
  subproblem search strategies;
- An investigation of whether the subproblem search strategies can add to the
  gains achieved by more sophisticated master problem branching rules in a
  branch-and-price algorithm.

References

1. Barnhart, C., Johnson, E.L., Nemhauser G.L., Savelsbergh, M.W.P., Vance P.H.:
   Branch-and-Price: Column Generation for Solving Huge Integer Programs. Operations Research **46** (1998) 316–329
2. Beaulieu, H., Ferland, J.A., Gendron, B., Michelon, P.: A Mathematical Programming Approach to Scheduling Physicians in the Emergency Room. Health Care
   Management Science **3** (2000) 193–200
3. Bourdais, S., Galinier, P., Pesant, G.: HIBISCUS: A Constraint Programming Application to Staff Scheduling in Health Care. Principles and Practice of Constraint
   Programming, Proceedings of CP'03, Lecture Notes in Computer Science **2833**
   (2003) 153–167
4. Dantzig, G.B., Wolfe, P.: Decomposition Principle for Linear Programs. Operations
   Research **8** (1960) 101–111
5. Desrosiers, J., Dumas, Y., Solomon, M.M., Soumis, F.: Time Constrained Routing
   and Scheduling. Handbooks in Operations Research and Management Science **8**:
   Network Routing, edited by Ball, M.O., Magnanti, T.L., Monma, C.L., Nemhauser,
   G.L. (1995) 35–139 (North-Holland, Amsterdam)
6. Fahle, T., Junker, U., Karish, S.E., Kohl, N., Vaaben, N., Sellmann, M.: Constraint Programming Based Column Generation for Crew Assignment. Journal of
   Heuristics **8** (2002) 59–81
7. Fahle, T., Sellmann, M.: Cost Based Filtering for the Constrained Knapsack Problem. Annals of Operations Research **115** (2002) 73–93
8. Gilmore, P.C., Gomory, R.E.: A Linear Programming Approach to the Cutting
   Stock Problem. Operations Research **9** (1961) 849–859

9. Gilmore, P.C., Gomory, R.E.: A Linear Programming Approach to the Cutting Stock Problem: Part II. Operations Research **11** (1963) 863–888

10. Junker, U., Karish, S.E., Kohl, N., Vaaben, N., Fahle, T., Sellmann, M.: A Framework for Constraint Programming Based Column Generation. Principles and Practice of Constraint Programming, Proceedings of CP'99, Lecture Notes in Computer Science **1713** (1999) 261–274

11. Lübbecke, M.E., Desrosiers, J.: Selected Topics in Column Generation. Technical Report G-2002-64, GERAD, Montréal (2002)

12. Pesant, G.: A Filtering Algorithm for the Stretch Constraint. Principles and Practice of Constraint Programming, Proceedings of CP'01, Lecture Notes in Computer Science **2239** (2001) 183–195

13. Régin, J.C.: Generalized Arc Consistency for Global Cardinality Constraints. Proceedings of AAAI-96 (1996) 209–215 (AAAI Press/MIT Press)

14. Rousseau, L.-M., Gendreau, M., Feillet, D.: Interior Point Stabilization for Column Generation. Publication CRT-2003-39, Centre de recherche sur les transports, Université de Montréal (2003)

15. Rousseau, L.-M., Gendreau, M., Pesant, G., Focacci, F.: Solving VRPTWs with Constraint Programming Based Column Generation. Annals of Operations Research **130** (2004) 199–216

16. Savelsbergh, M.W.P.: A Branch-and-Price Algorithm for the Generalized Assignment Problem. Operations Research **45** (1997) 831-841

17. Sellmann, M., Zervoudakis, K., Stamatopoulos, P., Fahle, T.: Crew Assignment via Constraint Programming: Integrating Column Generation and Heuristic Tree Search. Annals of Operations Research **115** (2002) 207–225

18. Vanderbeck, F.: On Dantzig-Wolfe Decomposition in Integer Programming and Ways to Perform Branching in a Branch-and-Price Algorithm. Operations Research **48** 111–128

19. Vanderbeck, F., Wolsey, L.A.: An Exact Algorithm for IP Column Generation. Operations Research Letters **19** 151–159

Group Construction for Airline Cabin Crew: Comparing Constraint Programming with Branch and Price

Jesper Hansen[1] and Tomas Lidén[2]

Carmen Systems
[1] Købmagergade 53, DK-1150 København K, Denmark
[2] Maria Bangata 6, SE-118 63 Stockholm, Sweden
{jesper.hansen, tomas.liden}@carmensystems.com

Abstract. Producing work schedules for airline crew normally results in individually different schedules. Some airlines do however want to give the same schedule to groups of people. The construction of such groups must respect certain rules, provide a good matching of certain factors and fit well into the normal process of producing anonymous trips starting and ending at the home base and assigning these to the crew. In this paper we present an application, implemented and delivered to a large European airline, which addresses these needs. The problem is challenging to solve for certain cases. Hence two different approaches have been applied, one using constraint programming and the other using column generation. These two methods are described and compared – along with computational results.

1 Introduction

Constructing work schedules for airline crew is typically divided into a crew pairing problem and a crew rostering problem. In the pairing problem anonymous pairings, or trips starting and ending at the home base, are constructed from the flight legs such that the crew need of each flight is covered. Following the pairing construction, the pairings are assigned to individual crew together with other activities such as ground duties, reserve duties and off-duty blocks, to form *rosters*. For more information on crew pairing and rostering we refer to the surveys of Andersson et. al. [1] and Kohl and Karisch [7].

Some airlines want to give the same work schedule/roster to a group of cabin crew members (purser and cabin attendants). This is to ease planning, increase the robustness of the schedule and for social reasons. Constructing such groups can be done before the rostering step. Then, a representative person from each group (normally the purser) is assigned a roster by the crew rostering system, which is then copied to the rest of the group members. If the group has been poorly formed, it will not be possible to copy all pairings. For example, if one crew member has a pre-assigned duty (e.g. course), a pairing touching that day cannot be assigned. Such "drop-out" pairings must be resolved manually.

R. Barták and M. Milano (Eds.): CPAIOR 2005, LNCS 3524, pp. 228–242, 2005.
© Springer-Verlag Berlin Heidelberg 2005

The construction of crew groups should therefore be solved so that the number of problems to handle in successive steps is minimized (both for the current scheduling period and future ones). Thus we want to achieve "homogenous" groups.

An application solving the crew-grouping problem has been developed and delivered to the Spanish airline Iberia as an addition to the Carmen Crew Rostering system. It has been used in production since spring 2002. Each problem instance solved includes 300-1200 crew and results in 30-200 groups of 3-13 persons each.

There are several approaches to formulating and solving this problem. The first delivered version used pure Constraint Programming (CP) [2]. During spring 2004 a second version was delivered based on Column Generation (CG) where the generation makes use of CP techniques (enumeration with some simple look-ahead domain reduction). The idea of combining CG and CP is not new (see e.g. Junker et. al. [6]) and many papers have been dedicated to comparing Integer Programming and CP methods for various applications. Grönkvist [4, 5] for instance, compares the methods on the Aircraft Scheduling Problem.

The initial reasons for choosing CP were the uncertainty of the problem formulation and constraints involved. Further the initial descriptions indicated a highly constrained problem including non-linear constraints and objectives. The need for modeling flexibility and a restricted budget were also important factors.

During the elaboration and development of the CP version the problem proved to be less constrained than anticipated and turned out to be more of an optimization problem than a feasibility problem. The idea of a mathematical programming formulation started to evolve and was investigated in a master's thesis work [3]. The results were good, however not all issues were considered. Hence it was decided to develop a CG version. The normal cases (January-November) to be introduced in the following were easily solved, but the "December" problem for the medium and large fleets proved to be surprisingly hard to solve. Thus the development of several advanced features was needed, such as stabilized column generation [8], and branch and price using constraint branching [9]. All these features along with a careful tuning of parameters have been necessary in order to achieve the goal of finding better solutions within similar or shorter execution times.

After giving a problem description, the pure CP model is described followed by the CG model. In sections 4 and 5 the solution procedures are outlined. In section 6 we describe the special considerations for the month of December. Computational results are presented in section 7 along with discussions on the suitability of the different methods, properties of the problem and alternative approaches. We end with a conclusion in section 8.

2   Problem Description

The problem is basically to assign crew members to groups such that the best possible matching is achieved minimizing the exceptions to handle when assigning schedules to the groups. The number of crew per group depends on the aircraft type. The cost factors in the matching problem are:

- **Preferred days off:** Match crew with the same or overlapping wishes for days off. The matching is done towards the purser if she has any preferred days off, otherwise towards the crew with maximum number of preferred days off.
- **Pre-assignments:** Match crew with the same or overlapping pre-assignments (courses etc). This matching is also done towards the purser if she has any pre-assignments, otherwise towards the crew with the most pre-assignments.
- **Historic values:** Match crew with excess or deficit for selected fairness values (e.g. flights to certain destinations, overtime etc).
- **Work reduction:** Match crew with the same work reduction together. Specifically it should be avoided to mix fulltime and reduction crew in the same group.

For the month of December there are special considerations described in section 6. In the normal problem (January - November), the minimum number of groups shall be constructed. The total cost for the constructed groups, which we define later, shall be minimized.

A number of rules are considered:

- **Buddy Bids**. The crew members give buddy bids saying that two or more people want to fly together. These buddy bids are "closed", meaning that one person can only participate in one bid. The buddy bids must be respected. Crew in buddy bids must be placed in a group while other crew can supplement groups if necessary.
- **Pursers**. There must be exactly one purser in each group.
- **Incompatibles**. There can also be crew forbidden combinations, which must be avoided.
- **Granted Days Off**. The number of granted days off is limited and certain patterns of granted days off are not allowed: Before and after a sequence of two or more consecutive granted days off there must be a stretch of at least 3 working days. Allowed patterns where × is a granted day off are:

    ×000×× ××000×

  Not allowed patterns are:

    ×00×× ×0×× ××00× ××0×

- **Inexperienced/transitions**. Limits can be set on the number of inexperienced crew per group, or crew in a transition phase from one type of aircraft to another.
- **Work reduction**. Cabin attendants must have less or the same work reduction as the purser.

3 Model Formulation

For modeling purposes we consider all crew to be in a so-called *subgroup*. A subgroup can consist either of all crew in a buddy bid or a single crew in no buddy bid. A crew member belongs to only one subgroup. Further since a group must have exactly one purser, each subgroup with a purser is associated with exactly one group. Let S be the set of subgroups. Further S is partitioned into two parts: S^B containing subgroups that must be in a group (buddies) and S^C containing subgroups that do not necessarily have to be in a group (single crew). Let F denote the set of fairness factors.

3.1 The CP Model

We first present the normal model and then in section 6.1 present the additions necessary for handling the special considerations in December.

Each subgroup i has the following basic properties, calculated from its set of crew:

N_i	Number of crew in the subgroup.
D_i	A set of preferred days off.
D_i^G	A set of granted days off, which is a subset of D_i.
P_i	A set of pre-assigned days.
H_{fi}^L, H_{fi}^H	Lowest and highest historic value (integer) for fairness factor $f \in F$.
$N_{fi}^-, N_{fi}^0, N_{fi}^+$	Number of crew with negative, zero and positive historic value for fairness factor $f \in F$.
N_i^I, N_i^T	Number of inexperienced crew and crew in transition phase
R_i	Work reduction factor (takes values 5, 3, 2, 1 or 0, where 5 is half-time and 0 is full-time)

Let S_j denote the set of subgroups for group j, with the initial domain S. Thus the following properties can be calculated for the group:

$$n_j = \sum_{i \in S_j} N_i$$

$$d_j = \bigcup_{i \in S_j} D_i \qquad\qquad d_j^G = \bigcup_{i \in S_j} D_i^G \qquad\qquad p_j = \bigcup_{i \in S_j} P_i$$

$$h_{jf}^L = \min_{i \in S_j} \left(H_{fi}^L \right) \qquad\qquad h_{jf}^H = \max_{i \in S_j} \left(H_{fi}^H \right)$$

$$n_{jf}^- = \sum_{i \in S_j} N_{fi}^- \qquad\qquad n_{jf}^0 = \sum_{i \in S_j} N_{fi}^0 \qquad\qquad n_{jf}^+ = \sum_{i \in S_j} N_{fi}^+$$

$$n_j^I = \sum_{i \in S_j} N_i^I \qquad\qquad n_j^T = \sum_{i \in S_j} N_i^T$$

$$r_j^L = \min_{i \in S_j} \left(R_i \right) \qquad\qquad r_j^H = \max_{i \in S_j} \left(R_i \right)$$

Let N_j, D_j, etc denote the properties for the purser subgroup associated with group j. Then the cost c_j for group j is calculated, using weight factors W^D, W^P, W^{HS}, W^{HR}, W^R and cost factors as follows:

$$c_j = W^D c_j^D + W^P c_j^P + \sum_{f \in F} \left(W_f^{HS} c_{jf}^{HS} + W_f^{HR} c_{jf}^{HR} \right) + W^R c_j^R \tag{1}$$

- **Days Off**. Limit the number of non-matching days off compared to crew with max number (use purser as max if she has any days off).

$$c_j^D = \begin{cases} \left| |d_j| - |D_j| \right| & \text{if } |D_j| > 0 \\ |d_j| - \max_{i \in S_j} |D_i| & \text{if } |D_j| = 0 \end{cases} \tag{2}$$

- **Pre-Assignments**. Same as for days off.

$$c_j^P = \begin{cases} \left| |P_j| - |P_j| \right| & \text{if } |P_j| > 0 \\ \left| |P_j| - \max_{i \in S_j} |P_i| \right| & \text{if } |P_j| = 0 \end{cases} \tag{3}$$

- **Historic Values – Sign**. Try to get all crew on excess or deficit.

$$c_{fj}^{HS} = \min\left(n_{fj}^-, n_{fj}^+\right) \tag{4}$$

- **Historic Values – Range**. Minimize the span in historic values.

$$c_{jf}^{HR} = h_{jf}^H - h_{jf}^L \tag{5}$$

- **Reduction**. 1) Don't mix full-time and reduction. 2) Avoid spread in reduction.

$$c_j^R = \begin{cases} 3 + r_j^H - r_j^L & \text{if } r_j^H > 0 \text{ and } r_j^L = 0 \\ r_j^H - r_j^L & \text{if } r_j^H = r_j^L = 0 \text{ or } r_j^L > 0 \end{cases} \tag{6}$$

The first row in (6) is used when both full-time crew and crew with reduction are placed in the same group. Since this is considered especially bad, an extra penalty of 3 is added. The other row is used when all crew are full-time or have reduction.

The formulation of these costs has been designed in close collaboration with the customer and reflects what they consider as a desirable group construction. However, this formulation has several numerical problems, since factors (2) and (3) are non-additive (the cost does not increase monotonically when adding subgroups). This makes it difficult to calculate a good lower bound for the cost during the construction of a group and hence the possibility of pruning the search tree. The applied lower bound calculation for days off is given below.

$$\lfloor c_j^D \rfloor = \begin{cases} \left| |d_j| - |D_j| \right| & \text{if } |D_j| > 0 \\ 0 & \text{if } |D_j| = 0 \end{cases} \tag{2.1}$$

Let S_k^I denote the set of subgroups that should not be put in the same subgroup according to incompatibility k. Then the following constraints must hold for each group j:

- **Incompatibilities**

$$\left| S_j \cap S_k^I \right| \leq 1 \qquad \forall k \tag{7}$$

- **Limit number of inexperienced crew and crew in Transition**

$$n_j^I \leq L^I, \ n_j^T \leq L^T, \ n_j^{I+T} \leq L^{I+T} \tag{8}$$

- **Limit number of granted days off**

$$d_j^G \leq L^G \tag{9}$$

- **No illegal patterns for granted days off**

$$NoIllegalPattern\left(d_j^G\right) \qquad (10)$$

- **Reduction**. Same or less work reduction as the purser. Let $S^{CR} \subseteq S^C$ denote the set of single crew with reduction. Single crew with reduction cannot join a purser without reduction.

$$\begin{cases} r_j^H \leq R_j & \text{if } R_j \neq 0 \\ S_j \cap S^{CR} \equiv \varnothing & \text{if } R_j = 0 \end{cases} \qquad (11)$$

Of the above constraints, (9) has the most limiting effect.

During the construction of a group the above constraints will reduce the domain of S_j. This domain reduction is however rather weak, mainly due to the non-additive property of factors (2) and (3) – these values will therefore be bounded late in the construction. Moreover, since all constraints apply per group they will not propagate any domain reduction to other groups.

When solving a pure CP version of the problem, we also need the following global constraints:

- **Each subgroup is used only once**

$$\bigcap_j S_j \equiv \varnothing \qquad (12)$$

- **All crew in buddy bids must be assigned to a group**

$$S^B \subseteq \bigcup_j S_j \qquad (13)$$

This CP model does not take advantage of the fact that several subgroups may have the same characteristics that permits us to strengthen (12) and (13).

3.2 The CG Model

Let G be the set of all feasible groups that can be constructed from the subgroups. Now we can formulate the problem as the following Set Partitioning/Set Packing problem:

$$\min \sum_{j \in G} \left(M + c_j\right)x_j \qquad (14)$$

s.t.

$$\sum_{j \in G} a_{ij}x_j = b_i, \quad \forall i \in S^B$$

$$\sum_{j \in G} a_{ij}x_j \leq b_i, \quad \forall i \in S^C$$

$$x_j \in \{0,1\}, \quad j \in G$$

where c_j is the cost of group j and M is a large negative number if we are maximizing the number of groups, large positive if we are minimizing and 0 if we are only focusing on costs. Further, a_{ij} is 1 if subgroup i is in group j, and 0 otherwise. When several subgroups have similar characteristics it is possible to aggregate the model so that the b_i's represent the number of similar subgroups of type i. Each subgroup type can then participate in the same group more than once and some subgroups can be used more than once resulting in the following limitations:

$$0 \leq a_{ij} \leq b_i \text{ and } 0 \leq x_j \leq \left\lfloor \min_{i \in S} \left(b_i / a_{ij} \right) \right\rfloor \tag{15}$$

The size of the set G is potentially exponential in the number of subgroups. The model is hence solved for a restricted number of groups where the values of the dual variables of the LP-relaxed model are used for finding new groups that correspond to columns with negative reduced cost. When no group exists with negative reduced cost the LP-relaxation is optimal. The problem of finding these groups is very similar to the problem formulated in section 3.1. Here we are searching for one feasible group at a time resulting in a new column in the model. The optimal solution to the LP-relaxation is however very unlikely to be integer. To get optimal integer solutions, the column generation needs to be embedded in a Branch & Bound procedure where columns can be generated in every node of the Branch & Bound tree. This procedure is referred to as Branch & Price.

4 Solving the CP Model

To solve the pure CP model, a customized constraint $NoIllegalPattern\left(d_j^G\right)$ was implemented. The filtering algorithm applies the following propagations whenever the domain of d_j^G changes.

Remove impossible values at start or end of a sequence (0 means not in domain, ? means possible in domain, × means required in domain. Affected values are underlined.)	0?×× → 00×× ??0×× → 000×× ××?0 → ××00 ××0?? → ××000
Add required values at start or end of a sequence	×?×× → ×××× ×??×× → ××××× ××?× → ×××× ××??× → ×××××

We then apply the following three-step approach:

1. Initial solution plus refinement. The search strategy selects a group (variable S_j) and then finds a suitable subgroup (value) to assign to that group. This continues in a depth-first search. Having found an initial solution, the refinement process adds additional constraints that improve the quality of the solution by setting limitations on the max values of c_j^D (days off), c_j^P (pre-assignments), c_{fj}^{HS} (historic values, sign) and then resolving the problem. This approach has been used to reduce the size of the problem and increase the domain reduction.
2. Global search for cheaper solutions. The solution found in step 1 is further improved by adding a constraint that forces the total cost to decrease for each new

solution and then search for better solutions. The same search strategy is used as in step 1. When the cost decreases slowly, some additional constraints are added that forces very expensive groups to get a lower cost (and hence be modified). Due to the very large search tree this step cannot be completed within reasonable time. Thus early decisions made in the search strategy are not likely to be altered.

3. Local search for cheaper solution. In this final step, groups are picked in pairs from the current solution and all possible solutions for these two groups with their used subgroups are calculated to see if a cheaper solution can be found. All pairs of groups are treated in this way. This step is repeated until no better solution can be found.

It should be noted that the pure CP model does not assume minimizing/maximizing the number of groups. Instead this is built into the search heuristic by first assigning buddy bids (large before small) and then filling up the groups to the desired size. If the minimum number of groups is wanted, no more groups are constructed. If the maximum number of groups is wanted, the algorithm continues constructing groups until the available groups or subgroups are exhausted. This method does not guarantee to produce the minimum possible number of groups, but will come reasonably close.

The order in which groups and subgroups are picked and assigned is crucial, since that will set the quality of the solution after step 1 and 2. The ordering is a trade-off between greediness (to find high-quality/low-cost solutions) and feasibility (to actually find solutions). Since the domain reduction is rather poor (see the end of section 3.1) we get a very large search tree. Therefore a very greedy approach can easily result in not finding feasible solutions within a reasonable computation time.

The search heuristic gives special consideration to granted days off, since that is crucial for finding feasible solutions (due to constraint (9) and (10)). Apart from that, it looks at the domain of the cost (c_j) and the lower bound of the additional cost when adding different subgroups. If a good solution has been found after step 2, the local search will only make minor adjustments, but if the solution has poor quality after step 2, the last step will repeat many times and can therefore be very time-consuming (although it is surprisingly effective in decreasing the cost).

Several drawbacks can be noted regarding the local search in step 3: It only uses subgroups that are already in the solution (not unused subgroups); it cannot achieve modifications that include swapping subgroups between three or more groups; it cannot escape from local minima.

5 Solving the CG Model

To solve the CG model we apply the following three-step approach:

1. Set-up an initial problem matrix with a) columns from a simple initial construction heuristic, which ensures the existence of at least one IP-feasible solution and gives an upper bound, b) columns from an initial generation.
2. Solve the LP-relaxation and perform pricing (column generation) to generate better columns based on the dual values until no better columns can be constructed.
3. Perform branch & price until no better solution can be found or the time limit is reached. When the time limit is reached a final IP is solved consisting of all obtained columns and no variables fixed.

5.1 The Pricing Problem

A constraint handling procedure is implemented for testing legality of a group. With this procedure a subgroup infeasibility table is set-up: For each subgroup we have a set of subgroups for which the subgroup is pair-wise infeasible. During the construction of groups this infeasibility table is used to reduce the domain of the problem. The constraint handling is used to verify the legality of each group before and after adding a subgroup to it.

The generation of columns is done by considering one group at a time and assigning subgroups from a list of feasible subgroups in a depth first search procedure. The list of subgroups to consider is randomized (to increase search coverage) and sorted according to dual value (positive before zero before negative). During the assignment of subgroups two pruning techniques are applied:

- **Cost cuts:** When assigning a subgroup with dual value ≤ 0 we know that the sum of duals cannot increase (due to the sorting applied). Then we calculate a lower bound for the group cost, which is also a lower bound on the reduced cost. When the lower bound becomes positive, we stop exploring the current node and backtrack.

- **Pattern cuts:** The constraint $NoIllegalPattern\left(d_j^G\right)$ can be used to check whether any specific values must be added to d_j^G in order to maintain legality. For example: if $d_j^G = \{5,8,9,10\}$ we know that 6 and 7 must be added – otherwise the constraint will not hold. In such situations we loop over the remaining subgroups to see if these needed additions can be found. If not, there is no need to explore the current node any further and we backtrack. During the looping over remaining subgroups we can also remove all subgroups that are infeasible with the currently assigned ones according to the infeasibility table and any subgroups that would result in breaking any constraints. If this results in an empty list of remaining subgroups we backtrack.

Apart from the above no special domain reductions are used. We have investigated the use of no-goods, forward-checking etc, but the computational effort is not worthwhile. Solving the CP model in a CP framework as in [3] could also have been used, but the domain reduction did not seem to be sufficient for introducing the extra overhead.

Only columns that will improve the LP solution (according to the reduced cost) will be used. In addition the number of backtracks in each node is limited and the generation reverts to the top node as soon as a valid column has been found. Thus we get a form of backtrack-bounded search that will get a broad coverage of the search tree instead of going deep into the search tree. This will improve the quality of the dual information. A limited time of a few seconds is spent on constructing columns for each group. There is hence no guarantee of finding a column with negative reduced cost even if such exists and we cannot prove that no column exists with negative reduced cost. To speed up the generation, we stop generating for a group, if no columns were generated for the group in the previous iteration. When no groups remain, we restart generating for all groups again. This greatly reduces the wasted time of unsuccessful search for columns.

The above handling relies heavily on the dual values from the previous LP solution. For some problem instances the dual values can fluctuate substantially, resulting in larger uncertainty and more iterations needed to converge. To remedy such cases, stabilized column generation due to du Merle et. al. [8] has been implemented. An alternative to this approach is to use an interior point or subgradient solver resulting in more well-behaved duals.

As can be seen from the above, the column generation does not have any problem specific considerations.

5.2 Branch and Price

Branching is performed if no columns were generated in the last pricing loop or the reduced cost of the best column is above a certain threshold. Note that this only applies if the search for columns is done for all groups.

Constraint branching according to Ryan and Foster [9] is applied for pairs of subgroups that have the partitioning constraint =1. The interpretation in this context is that either two subgroups must be in the same group or they must not be in the same group. When branching on the packing constraints ≤1, at least one of the constraints must be a partitioning constraint. Otherwise we cannot force them to be together since only one of them has to be used.

When finding the best branching candidate (a pair of subgroups) and its fractional sum f, the following branching scheme is applied:

Find f closest to 1. If f is larger than some threshold value (parameter setting) then apply the 1-branch, which bans columns with only one of the two subgroups, and continue branching. The 0-branch is not investigated upon backtrack.

If f is smaller than the threshold value then pricing is performed both in the 1-branch and the 0-branch. Continue branching, first in the branch that has the best LP objective and in the other when backtracking. When applying the 0-branch columns are banned which include both subgroups.

When no more branching candidates are found, we search for an improving IP solution and backtrack.

To fathom a node in the tree the lower bound of the node must be equal or larger than the best upper bound found so far. The LP bound is however only a valid lower bound if no column with a negative reduced cost exists. To establish this fact would require a complete enumeration of the pricing problem, which is generally not possible. Since we have no ambition of solving the problem to optimality by completely enumerating the branching tree, we can cut off a branch, which could contain an improving solution by assuming that the LP objective is a valid lower bound. So, if the LP objective in a node is worse than the best found upper bound, that node is fathomed and we backtrack.

The branching tree is only explored to a limited depth. If the max depth or time limit is reached we call the IP solver to search for a better IP solution and backtrack to the top of the tree. At the end of the search we finally call the IP solver to search for an improving solution on all generated columns.

After a branching decision the pricing procedure must ensure that columns are generated which follows earlier made branching decisions. For a specific group and corresponding purser subgroup we can identify branching decisions, which force a set of other subgroups to be together with this subgroup. These can be added a priori. Given the remaining branching decisions a number of clusters of subgroups can be identified, which are forced to be together. During the pricing search procedure, clusters are added before individual subgroups not in clusters. When adding both clusters and individual subgroups, we check that this does not violate any branching decision that two subgroups must not be together. This procedure guarantees that all branching decisions are fulfilled.

The constraint branching scheme does not apply for constraints where $b_i>1$, $i \in S$. In that case one could use the general branching scheme of Vanderbeck and Wolsey [10]. Both the branching and the pricing procedure would however be even more complex and has hence not been implemented.

6 Special Considerations in December

We have in this section collected all problem specifics for the month of December. In December the following changes apply compared to the normal case:

- The number of groups shall be maximized.
- There is special consideration to pre-assignments – any pre-assignments on New Years Day (called *day32*) must be very well matched.
- Due to the strong focus on day32, buddy bids containing both persons with and without day32 will be split in two parts. If possible, perfect matching of day32 is wanted. If not, it is preferred that split bids are rejoined.

6.1 Modeling the December Problem

First we introduce an additional basic property per subgroup

$\qquad B_i \qquad\qquad$ Other subgroup (the one without day32) from split buddy bid

and then calculate the following parameters per group: $n_j^{32} = \sum\limits_{\substack{i \in S_j \\ 32 \in P_i}} N_i$, $b_j = \bigcup\limits_{i \in S_j} B_i$

The cost function is modified so that an additional element is added to c_j:

$$c_j = [\text{as given in (1)}] + W^{P32} c_j^{P32} \qquad\qquad (16)$$

To avoid combining crew with and without day32, and achieve rejoining of buddies from split bids, the cost factor and lower bound for pre-assigned day32 are calculated as follows:

$$c_j^{P32} = \begin{cases} 0 & \text{if } n_j^{32} = 0 \\ n_j^{0,not\_rejoined} + \dfrac{n_j^{0,rejoined}}{n_j} & \text{if } n_j^{32} > 0 \end{cases} \qquad\qquad (17)$$

$$\text{where} \quad n_j^{0,not\_rejoined} = n_j - n_j^{32} - n_j^{0,rejoined}$$

$$n_j^{0,rejoined} = \sum\limits_{i \in (S_j \cap b_j)} N_i$$

$$\left\lfloor c_j^{P32} \right\rfloor = \begin{cases} 0 & \text{if } n_j^{32} = 0 \\ \dfrac{n_j - n_j^{32}}{n_j} & \text{if } n_j^{32} > 0 \end{cases} \qquad (17.1)$$

6.2 Solving the CG Model in December

The search heuristic is modified for this problem, taking the day32 factor into consideration. The crucial thing is to decide whether to go for rejoining buddy bids or not, however the demand for always obtaining a feasible solution restricts the amount of greediness.

6.3 The Pricing Problem for December

A few tricks are applied in the pricing phase to make it work better for the December problem. The sorting of subgroups during generation is slightly refined, and as soon as a possible rejoining of buddy bids is found, that rejoining is tried as well.

7 Computational Results

The application has been implemented in C++. The CP version uses ILOG Solver 5.2 as the constraint solver library, while the CG version uses XPress-MP 2003F as the optimization engine (with Coin/Osi as the programming API).

The system has been run for all fleet types used at Iberia – the below table show some basic data.

Fleet name	Group size	No. of crew	No. of subgroups	Produced no. of groups
MD87	3	~350-500	~180-230	~100-150
A320	4	~700-900	~310-450	~150-200
B757	5	~300-400	~150-200	~40-80
A340	10	~1000-1200	~350-420	~90-110
B747	13	~400-500	~40-200	~30-40

In the following table computational results are presented for a series of instances. For the CG version three execution times are reported: time for first solution, best solution and total execution time (a time limit of 60 minutes for the normal instances and 180 minutes for the large December instances (A340 and B747) has been used).

Some comments concerning the results:

- CG is generally able to find better solutions than CP. It minimizes the number of groups if possible and it yields better matching of day32 in December.
- Instances for small groups (group size <= 5) are generally solved faster with CG.
- Instances for large groups (group size > 5) require substantial amount of time and are generally slower with CG.

Num groups Total cost Σc_j Execution time (min)[1]	CP	CG	CG-CP
MD87	104 985,278 3.7	104 870,356 0.1/0.1/0.1	-11.7 % -3.6/-3.6/-3.6
A320	150 1,966,488 14.2	150 1,643,148 0.2/0.2/0.2	-16.4% -14.0/-14.0/-14.0
B757	38 437,446 1.5	35 425,913 0.1/4.3/4.3	-3 groups -2.6% -1.4/+2.8/+2.8
A340	82 521,000 9.1	73 502,000 13.6/61.1/61.1	-9 groups -3.6% +4.5/+52.0/+52.0
B747	30 476,335 2.2	26 445,554 0.8/30.1/30.1	-4 groups -6.5% -1.4/+27.9/+27.9
MD87, December	120 148,966 3.9	120 113,985 0.2/0.2/30.1	-23.5% -3.7/-3.7/+26.2
A320, December	198 3,309,710 32.0	198 3,042,610 2.4/55.0/55.0	-8.1% -29.6/+23.0/+23.0
B757, December	68 249,204 2.4	68 173,245 1.1/15.2/15.3	-30.5% -1.3/+12.8/+12.9
A340, December	110 560,935 67.6	110 470,909 51.0/70.5/181.1	-16.0% -16.6/+2.9/+113.5
B747, December	31 285,880 1.3	31 241,046 1.3/129.9/129.9	-15.7% -0.0/+129/+129
B747, December with no crew excess	39 909,892 4.6	39 1,076,933 76.9/181.5/181.5	+18.4 +72.3/+177/+177

- For cases where there is no or very little excess of crew for group construction (last example), CP finds better solutions more quickly than CG.
- A340 is the hardest since it both has a large number of subgroups and has a large group size, which gives a large combinatorial complexity.

---

[1] The results have been produced on the following platforms: SunOS 2.7 on a SparcServer 10000 with 16 UltraSparcII, 250MHz each, (CP) and Linux on a PentiumIII, 1300 MHz. The SPECint2000 figures are about 130 for the Sparc and 600 for the Pentium. Thus the times in the CP column are a result of dividing the actual CP times by 4.6.

- To achieve good matching for day32 for December, the tuning of parameters is crucial. The application has parameters for setting tolerance, min/max number of columns to generate per group and iteration, branching depth, threshold, max number of backtracks per generation node etc.

Another difference is that the CP version will improve the solution over the execution time and stop when it fails to do so. The CG version however will find a very good solution early, which can be slightly improved, but it will take very long time to do so – see Fig.. Therefore the user has the possibility to stop the execution whenever an acceptable solution has been found.

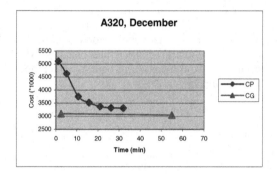

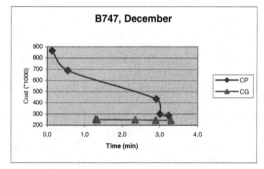

Fig. 1. Solution progress

8 Conclusion

First it should be noted that the numerical properties of the cost factors and constraints are a major cause for the difficulties of this problem. Reformulating some of these factors would probably be beneficial but such a redesign together with the customer was not possible to perform. Also, it is not obvious that alternative formulations exist that fulfills the needs of the customer.

There are some additional properties that make this problem hard to solve even with decomposition: It is a Set Partitioning/Packing Problem where there is no natural ordering (like connection time in crew pairing/rostering and distance in vehicle routing). Instead there are multiple factors that should be matched well, and these factors have no correlation between subgroups.

When comparing CP and CG for this problem it is not easily said that one is better than the other. The expressiveness and ease of modeling was beneficial for the development of the CP version. The CG version does however perform better – after the inclusion of some rather sophisticated techniques. Further enhancements could of course be applied to the CP version. For example, it could consider subgroups with similar properties (which reduce both domains and symmetries), taking advantage of the mirrored formulation/channeling constraints (letting the subgroup have a variable for which group to belong to) and using a meta-heuristic like Tabu search instead of the simple local search adopted.

On the other hand the CG formulation is more general, since it only has one problem specific consideration (day32) built into the solving methods while the CP search heuristic is much more tailored to the specific problem properties. Thus the CG version should be more stable to future additions and changes, which was also a reason for choosing that method. And there are of course further enhancements that could be considered to the CG version.

References

1. Andersson, E., Housos, E., Kohl, N., Wedelin, D.: *Crew pairing optimization*, in: OR in airline industry, Eds. Yu, G., Kluwer Acad. Publ., 1997.
2. Carmen Consulting: *Carmen Crew Grouping Module*, (Product sheet), http://www.carmenconsulting.com, 2002.
3. A. Ekenbäck: *Optimal Crew Groups – A column generation heuristic for a combinatorial optimization problem*, Royal Institute of Technology, Sweden, (Master's Thesis), 2002.
4. M. Grönkvist: *Using Constraint Propagation to Accelerate Column Generation in Aircraft Scheduling*, In proceedings of CP-AI-OR'03, 2003.
5. M. Grönkvist: *A Constraint Programming Model for Tail Assignment*, In proceedings of CP-AI-OR'04, 2004.
6. U. Junker, S.E. Karisch, N. Kohl, B. Vaaben, T. Fahle and M. Sellmann: *A Framework for Constraint Based Column Generation*, In proceedings of CP'99, 1999.
7. N. Kohl and S. E. Karisch: Airline Crew Rostering: Problem types, modeling, and optimization, Annals of Operations Research 127, 223—257, 2004.
8. O. du Merle, D. Villeneuve, J. Desrosiers and P. Hansen: *Stabilized Column Generation*, Discrete Mathematics 194, 229—237, 1999.
9. D. M. Ryan and B. A. Foster: *An integer programming approach to scheduling*, In A. Wren, editor, Computer Scheduling of Public Transport Urban Passenger Vehicle and Crew Scheduling, 269—280, North-Holland, 1981.
10. F. Vanderbeck and L. A. Wolsey: *An exact algorithm for IP column generation*, Operations Research Letters, 19, 1996.

A Search-Infer-and-Relax Framework for Integrating Solution Methods

J.N. Hooker

Tepper School of Business,
Carnegie Mellon University,
Pittsburgh, USA
john@hooker.tepper.cmu.edu

Abstract. We present an algorithmic framework for integrating solution methods that is based on search, inference, and relaxation and their interactions. We show that the following are special cases: branch and cut, CP domain splitting with propagation, popular global optimization methods, DPL methods for SAT with conflict clauses, Benders decomposition and other nogood-based methods, partial-order dynamic backtracking, various local search metaheuristics, and GRASPs (greedy randomized adaptive search procedures). The framework allows elements of different solution methods to be combined at will, resulting in a variety of integrated methods. These include continuous relaxations for global constraints, the linking of integer and constraint programming via Benders decomposition, constraint propagation in global optimization, relaxation bounds in local search and GRASPs, and many others.

1 Introduction

The constraint programming and optimization communities have developed a wide variety of effective methods for solving combinatorial optimization problems. Yet they are described in different literatures using different terminology and implemented in a growing collection of different solvers. Recent advances in hybrid methods show how to integrate algorithmic ideas from several sources, but hybrid methods themselves are multiplying, since there are so many ways to hybridize. Practical application would be much more effective if a single solver could bring a wide variety of methods under one roof, not only to allow the user to select the best one, but to allow the integration of techniques from different methods.

We suggest that the goal of integration should be addressed at a fundamental and conceptual level rather than postponing it to the software design stage. The growing repertory of combinatorial optimization methods should be interpreted as special cases of a single solution method that can be adjusted to exploit the structure of a given problem. This overarching method would then dictate the architecture of a general-purpose solver.

One approach, some elements of which are proposed in [4, 10, 11, 12, 13, 15, 17], is to view solution methods as instances of a *search-infer-and-relax algo-*

R. Barták and M. Milano (Eds.): CPAIOR 2005, LNCS 3524, pp. 243–257, 2005.

rithm. The search phase enumerates restrictions of the problem, perhaps by branching, neighborhood search, or creation of subproblems. Inference may take the form of cutting planes, filtering, or nogood generation. Relaxation provides bounds on the optimal value that can reduce the search space.

We show in this paper that a wide variety of solution methods have this structure, including branch and cut, standard CP methods, popular global optimization methods, DPL methods for the propositional satisfiability problem, generalizations of Benders decomposition and other varieties of nogood-based search, partial-order dynamic backtracking and related methods, local search metaheuristics, and GRASPs (greedy randomized adaptive search procedures).

However, it is one thing to observe in a general way that solution algorithms tend to have a search-infer-and-relax structure, and another thing to demonstrate it in precise algorithmic terms. While such methods as branch and cut or standard CP methods readily fit into this framework, it is less obvious how to treat some of the other methods. The main contribution of this paper, relative to previous work, is to extend the range of solution methods that can be viewed as having common structure, while trying to make their commonality more precise.

In particular, we extend the analysis to "heuristic" methods, such as local search and GRASPs. Although one can distinguish exact from inexact methods, this distinction need not imply a fundamental distinction of the algorithmic approach. We view them as special cases of the same search strategy, adjusted in some cases to be exhaustive and in other cases to be inexhaustive.

Some aspects of the integration scheme described here are implemented in the solution and modeling system SIMPL [1], which combines integer and constraint programming but has not yet been extended to other methods.

2   The Basic Ideas

- *Search* is an enumeration of problem restrictions, each of which is obtained by adding constraints to the problem. The motivation for examining problem restrictions is that they may be easier to solve than the original. In branching search, for example, the problem restrictions correspond to nodes of the search tree. In Benders decomposition and its generalizations, the restrictions are subproblems. In local search, each neighborhood is the feasible set of a problem restriction.
- *Inference* derives valid constraints that were only implicit in the constraint set. They can rule out infeasible or suboptimal restrictions that would otherwise be solved. Popular forms of inference include the identification of valid inequalities in integer programming, the generation of nogoods (such as Benders cuts and conflict clauses), and domain filtering in constraint programming.
- *Relaxation,* like restriction, is motivated by the desire to solve a problem that is easier than the original. Solution of a relaxation may provide an optimal solution of the original problem, but more often it provides a bound on the optimal value. Popular forms of relaxation include the constraint

store in constraint programming, continuous relaxations of 0-1 inequalities or global constraints, and the master problem in Benders decomposition and its generalizations.

The interaction of these elements is key to problem solving.

- *Search and inference.* Inference reduces the number of restrictions that must be enumerated in the search. For instance, domain filtering reduces branching by eliminating values on which one must branch. Conversely, restricting the problem can make inference more effective. Branching on variables, for example, reduces domains and triggers further domain reduction through propagation.
- *Search and relaxation.* Relaxation provides valuable information for directing the search. For instance: the solution of a continuous relaxation suggests how to branch (perhaps on a variable with a fractional value); the solution of a master problem can define the next subproblem (in Benders-like methods); and the result of a neighborhood search can provide the center of the next neighborhood to be searched. Conversely, problem restriction during search can yield a tighter relaxation, perhaps one whose optimal solution is feasible in the original problem. Relaxation and restriction also interact in a bounding mechanism that is used by branch-and-relax methods but has much wider application. If the relaxation of a restriction has an optimal value that is no better than that of the best solution found so far, then the restriction need not be solved.
- *Inference and relaxation.* The solution of a relaxation can help identify useful inferences, such as separating cuts in integer programming. Conversely, inference can generate constraints that strengthen the relaxation, as cutting planes strengthen a continuous relaxation.

Inference and relaxation are most effective when they exploit problem structure. For instance, specialized cutting planes or domain filtering methods can be developed for constraints or subsets of constraints that have special characteristics. Arguably the success of combinatorial optimization relies on the identification of structure, and the problem formulation should indicate to the solver where the structure lies.

3   Overview of the Solution Method

For the purposes of this paper, an optimization problem P can be written

$$\min\ f(x)$$
$$S(x)$$
$$x \in D$$

where $f(x)$ is a real-valued function of variable x and D is the *domain* of x. The function $f(x)$ is to be minimized subject to a set $S(x)$ of constraints, each of which is either satisfied or violated by any given $x \in D$. Generally x is a vector $(x_1, \ldots, x_n)$ and D a Cartesian product $D_1 \times \cdots \times D_n$, where each $x_j \in D_j$.

Any $x \in D$ is a *solution* of P. A *feasible* solution is one that satisfies all the constraints in $S(x)$, and the *feasible set* of P is the set of feasible solutions. A feasible solution x^* is *optimal* if $f(x^*) \leq f(x)$ for all feasible x. An *infeasible* problem is one with no feasible solution and is said to have optimal value ∞.

3.1   Search

Search is carried out by solving a series of problem restrictions $P_1, P_2, \ldots, P_m$ of P and picking the best *candidate* solution. The search is *complete* if the feasible set of P is equal to the union of the feasible sets of $P_1, \ldots, P_m$. In incomplete search the restrictions may not be solved to optimality.

The most basic kind of search simply enumerates elements of the domain D and selects the best feasible solution. This is can be viewed as a search over problem restrictions P_k, each of which is defined by fixing x to a particular value. It is generally more practical, however, to define restrictions by branching, constraint-directed search, or local search.

3.2   Inference

Search can often be accelerated by inference, that is, by inferring new constraints from the constraint set of each P_k. The new constraints are then added to P_k. Constraints that can be inferred from P alone are added to P_k and all subsequent restrictions.

Inference procedures are typically applied to individual constraints or small highly-structured groups of constraints rather than the entire problem. As a result, implications of the entire constraint set may be missed.

One can partially address this problem by propagating constraints through a constraint store S. When inferences are drawn from constraint C, they are actually drawn from $\{C\} \cup S$. Processing each constraint enlarges S and thereby strengthens the implications that can be derived from the next constraint. Propagation of this sort is practical only if the constraint store contains elementary constraints that all of the inference algorithms can accommodate. Constraint programming solvers typically store in-domain constraints, and mixed integer solvers store linear inequalities.

3.3   Relaxation

Relaxation is often used when the subproblems P_k are themselves hard to solve. A relaxation R_k of each P_k is created by dropping some constraints in such a way as to make R_k easier than P_k. For instance, one might form a continuous relaxation by allowing integer-valued variable to take any real value.

The optimal value v of the relaxation R_k is a lower bound on the optimal value of P_k. If v is greater than or equal to the value of the best candidate solution found so far, then there is no need to solve P_k, since its optimal value can be no better than v.

Let $v_{\mathrm{UB}} = \infty$ and $S = \{P_0\}$. Perform **Branch**.
The optimal value of P_0 is v_{UB}.

Procedure **Branch**.
    If S is nonempty then
        Select a problem restriction $P \in S$ and remove P from S.
        If P is too hard to solve then
            Add restrictions $P_1, \ldots, P_m$ of P to S and perform Branch.
        Else
            Let v be the optimal value of P and let $v_{\mathrm{UB}} = \min\{v, v_{\mathrm{UB}}\}$.

Fig. 1. *Generic branching algorithm for solving a minimization problem P_0. Set S contains the problem restrictions so far generated but not yet attempted, and v_{UB} is the best solution value obtained so far*

The relaxation, like the constraint store, must contain fairly simple constraints, but for a different reason: they must allow easy optimal solution of the relaxed problem. In traditional optimization methods, these are generally linear inequalities in continuous variables, or perhaps nonlinear inequalities that define a convex feasible set.

4   Branching Search

Branching search uses a recursive divide-and-conquer strategy. If the original problem P is too hard to solve as given, the branching algorithm creates a series of restrictions $P_1, \ldots, P_m$ and tries to solve them. In other words, it *branches* on P. If a restriction P_k is too hard to solve, it attacks P_k in a similar manner by branching on P_k. The most popular branching mechanism is to branch on a variable x_j. The domain of x_j is partitioned into two or more disjoint subsets, and restrictions are created by successively restricting x_j to each of these subsets.

Branching continues until no restriction so far created is left unsolved. If the procedure is to terminate, problems must become easy enough to solve as they are increasingly restricted. For instance, if the variable domains are finite, then branching on variables will eventually reduce the domains to singletons. Figure 1 displays a generic branching algorithm.

4.1   Branch and Infer

Inference may be combined with branching by inferring new constraints for each P_k before P_k is solved. When inference takes the form of domain filtering, for example, some of the variable domains are reduced in size. When one branches on variables, this tends to reduce the size of the branching tree because the domains more rapidly become singletons. *Constraint programming solvers* are typically built on a branch-and-infer framework.

Let $v_{\mathrm{UB}} = \infty$ and $S = \{P_0\}$. Perform **Branch**.
The optimal value of P is v_{UB}.

Procedure **Branch**.
    If S is nonempty then
        Select a problem restriction $P \in S$ and remove P from S.
        Repeat as desired:
            Add inferred constraints to P.
            Let v_{R} be the optimal value of a relaxation R of P.
        If $v_{\mathrm{R}} < v_{\mathrm{UB}}$ then
            If R's optimal solution is feasible for P then let $v_{\mathrm{UB}} = \min\{v_{\mathrm{R}}, v_{\mathrm{UB}}\}$.
            Else add restrictions $P_1, \ldots, P_m$ of P to S and perform **Branch**.

Fig. 2. *Generic branching algorithm, with inference and relaxation, for solving a mini-mization problem P_0. The repeat loop is typically executed only once, but it may be executed several times, perhaps until no more constraints can be inferred or R becomes infeasible. The inference of constraints can be guided by the solution of previous relaxations*

4.2   Branch and Relax

Relaxation can also combined with branching in a process that is known in the operations research community as *branch and bound*, and in the constraint pro-gramming community as *branch and relax*. One solves the relaxation R_k of each restriction, rather than P_k itself. If the solution of R_k is feasible in P_k, it is opti-mal for P_k and becomes a candidate solution. Otherwise the algorithm branches on P_k. To ensure termination, the branching mechanism must be designed so that R_k's solution will in fact be feasible for P_k if one descends deeply enough into the search tree.

Branch-and-relax also uses the bounding mechanism described earlier. If the optimal value of R_k is greater than or equal to the value of the best candidate solution found so far, then there is no point in solving P_k and no need to branch on P_k.

The addition of inferred constraints to P_k can result in a tighter bound when one solves its relaxation R_k. This is the idea behind *branch-and-cut* methods, which add cutting planes to the constraint set at some or all of the nodes. Conversely, the solution of R_k can provide guidance for generating further con-straints, as for instance when *separating cuts* are used. A generic branching algorithm with inference and relaxation appears in Fig. 2.

It is straightforward to combine elements of constraint programming and in-teger programming in this framework. Domain filtering can be applied to integer inequalities as well as global constraints at each node of the search tree, and tight relaxations can be devised for global constraints as well as specially-structured inequalities.

4.3   Continuous Global Optimization

A continuous optimization problem may have a large number of locally optimal solutions and can therefore be viewed as a combinatorial problem. The most popular and effective global solvers use a branch-and-relax approach that combines relaxation with constraint propagation [21, 22, 24]. Since the variable domains are continuous intervals, the solver branches on a variable by splitting an interval into two or more intervals. This sort of branching divides continuous space into increasingly smaller "boxes" until a global solution can be isolated in a very small box.

Two types of propagation are commonly used: bounds propagation, and propagation based on Lagrange multipliers. Bounds propagation is similar to that used in constraint programming solvers. Lagrange multipliers obtained by solving a linear relaxation of the problem provide a second type of propagation. If a constraint $ax \leq \alpha$ has Lagrange multiplier λ, v is the optimal value of the relaxation, and L is a lower bound on the optimal value of the original problem, then the inequality

$$ax \geq \alpha - \frac{v - L}{\lambda}$$

can be deduced and propagated. Reduced-cost-based variable fixing is a special case.

Linear relaxations can often be created for nonlinear constraints by "factoring" the functions involved into more elementary functions for which linear relaxations are known [24].

5 Constraint-Directed Search

A ever-present issue when searching over problem restrictions is the choice of which restrictions to consider, and in what order. Branching search addresses this issue in a general way by letting problem difficulty guide the search. If a given restriction is too hard to solve, it is split into problems that are more highly restricted, and otherwise one moves on to the next restriction, thus determining the sequence of restrictions in a recursive fashion.

Another general approach is to create the next restriction on the basis of lessons learned from solving past restrictions. This suggests defining the current restriction by adding a constraint that excludes previous solutions, as well as some additional solutions that one can determine in advance would be no better. Such a constraint is often called a *nogood*.

Restrictions defined in this manner may be hard to solve, but one can solve a more tractable relaxation of each restriction rather than the restriction itself. The only requirement is that the relaxation contain the nogoods generated so far. The nogoods should therefore be chosen in such a way that they do not make the relaxation hard to solve.

More precisely, the search proceeds by creating a sequence of restrictions $P_0, P_1, \ldots, P_m$ of P, where $P_0 = P$ and each P_k is formed by adding a nogood N_{k-1} to P_{k-1}. It solves a corresponding series of relaxations $R_0, R_1, \ldots, R_m$ of P to obtain solutions $x^0, x^1, \ldots, x^m$. Each relaxation R_k contains the nogoods $N_0, \ldots, N_{k-1}$ in addition to the constraints in R_0.

Step k of the search begins by obtaining a solution x^k of R_k. If x^k is infeasible in P, a nogood N_k is designed to exclude x^k and possibly some other solutions

Let $v_{\mathrm{UB}} = \infty$, and let R be a relaxation of P.
Perform **Search**.
The optimal value of P is v_{UB}.

Procedure **Search**.
    If R is feasible then
        Select a feasible solution $x = s(R)$ of R.
        If x is feasible in P then
            Let $v_{UB} = \min\{v_{UB}, f(x)\}$.
            Define a nogood N that excludes x and possibly other solutions x'
                with $f(x') \geq f(x)$.
      Else
            Define a nogood N that excludes x and possibly other solutions
                that are infeasible in P.
        Add N to R and perform **Search**.

Fig. 3. *Generic constraint-directed search algorithm for solving a minimization problem P with objective function $f(x)$, where s is the selection function. R is the relaxation of the current problem restriction*

that are infeasible for similar reasons. If x^k is feasible in P, it may or may not be optimal, and it is recorded as a candidate for an optimal solution. A nogood N_k is designed to exclude x^k and perhaps other solutions whose objective function values are no better than that of x^k. The search continues until R_k becomes infeasible, indicating that the solution space has been exhausted. An generic algorithm appears in Fig. 3.

The search is exhaustive because the infeasibility of the final relaxation R_m implies the infeasibility of P_m. Thus any feasible solution x of P that is not enumerated in the search is infeasible in P_m. This is possible only if x has been excluded by a nogood, which means x is no better than some solution already found.

If the domains are finite, the search will terminate. Each relaxation excludes a solution that was not excluded by a previous relaxation, and there are finitely many solutions. If there are infinite domains, more care must be exercised to ensure a finite search and an optimal solution.

Interestingly, there is no need to solve the relaxations R_k to optimality. It is enough to find a feasible solution, if one exists. This is because no solution is excluded in the course of the search unless it is infeasible, or an equally good or better solution has been found.

There is normally a good deal of freedom in how to select a feasible solution x^k of R_k, and a constraint-directed search is partly characterized by its *selection function*; that is, by the way it selects a feasible solution $s(R_k)$ for a given R_k. Certain selection functions can make subsequent R_k's easier to solve, a theme that is explored further below.

We briefly examine three mechanisms for generating nogoods: constraint-directed branching, partial-order dynamic backtracking, and logic-based Benders decomposition.

5.1   Constraint-Directed Branching

Constraint-directed branching stems from the observation that branching on variables is a special case of constraint-directed search. The leaf nodes of the branching tree correspond to problem restrictions, which are defined in part by nogoods that exclude previous leaf nodes. The nogoods take the form of "conflict clauses" that contain information about why the search backtracked at previous leaf nodes. Well-chosen conflict clauses can permit the search to prune large portions of the enumeration tree.

Search algorithms of this sort are widely used in artificial intelligence and constraint programming. Conflict clauses have played a particularly important role in fast algorithms for the propositional satisfiability problem (SAT), such as Chaff [20].

Branching can be understood as constraint-directed search in the following way. We branch on variables in a fixed order $x_1, \ldots x_n$. The original problem P corresponds to the first leaf node of the branching tree, and its relaxation R contains only the domain constraints of P. The branching process reaches the first leaf node by fixing $(x_1, \ldots, x_n)$ to certain values $(v_1, \ldots, v_n)$, thereby creating P_1. If the search backtracks due to infeasibility, typically only some of the variables are actually responsible for the infeasibility, let us say the variables $\{x_j \mid j \in J\}$. A nogood or *conflict clause* N is constructed to avoid this partial assignment in the future:

$$\bigvee_{j \in J} (x_j \neq v_j) \tag{1}$$

If a feasible solution with value z is found at the leaf node, then a subset of variables $\{x_j \mid j \in J\}$ is identified such that $f(x) \geq z$ whenever $x_j = v_j$ for $j \in J$. A nogood N of the form (1) is created.

Each of the subsequent leaf nodes corresponds to a relaxation R_k, which contains the nogoods generated so far. A feasible solution $s(R_k)$ of R_k is now selected to define the next solution to be enumerated. A key property of constraint-based branching is that the selection function s is easy to compute. The solution $s(R_k)$ sequentially assigns $x_1, x_2, \ldots$ the values to which they are fixed at the current leaf node, until such an assignment violates a nogood in R_k. At this point the unassigned variables are sequentially assigned any value that, together with the assignments already made, violates none of the nogoods in R_k. Constraint-based search does not actually construct a search tree, but the values to which x_j is fixed at the current leaf node are encoded in the nogoods: if one or more nogoods in R_k contain the disjunct $x_j \neq v_j$, then x_j is currently fixed to v_j (all disjuncts containing x_j will exclude the same value v_j).

It is shown in [12] that this procedure finds a feasible solution of R_k without backtracking, if one exists, provided the nogoods are processed by *parallel resolution* before computing $s(R_k)$. Consider a set $S = \{C_i \mid i \in I\}$ where each C_i has the form

$$\bigvee_{j \in J_i} (x_j \neq v_j) \vee (x_p \neq v_{pi})$$

and where p is larger than all the indices in $J = \bigcup_{i \in I} J_i$. If $\{v_{pi} \mid i \in I\}$ is equal to the domain of x_p, then S has the parallel resolvent

$$\bigvee_{j \in J} (x_j \neq v_j)$$

Thus parallel resolution always resolves on the *last* variable x_p in the clauses resolved. (In constraint-directed branching, J_i is the same for all $i \in I$, but this need not be true in general to derive a parallel resolvent.) Parallel resolution is applied to R_k by deriving a parallel resolvent from a subset of nogoods in R_k, deleting from R_k all nogoods dominated by the resolvent, adding the resolvent to R_k, and repeating the process until no parallel resolvent can be derived. In the context of constraint-based branching, parallel resolution requires linear time and space.

The Davis-Putnam-Loveland (DPL) algorithm for SAT is a special case of constraint-directed branching in which the *unit clause rule* is applied during the computation of $s(R_k)$. The SAT problem is to determine whether a set of logical clauses is satisfiable, where each clause is a disjunction of literals (x_j or $\neg x_j$). The unit clause rule requires that whenever x_j (or $\neg x_j$) occurs as a unit clause (a clause with a single literal), x_j is fixed to true (respectively, false) and the literal $\neg x_j$ (respectively, x_j) is eliminated from every clause in which it occurs. The procedure is repeated until no further variables can be fixed. During the computation of $s(R_k)$, the unit clause rule is applied after each x_j is assigned a value, and subsequent assignments must be consistent with any values fixed by the rule.

An infeasible assignment $(x_1, \ldots, x_n) = (v_1, \ldots, v_n)$ for the SAT problem is one that violates one or more clauses. A conflict clause (1) is obtained by identifying a subset of variables $\{x_j \mid j \in J\}$ for which the assignments $x_j = v_j$ for $j \in J$ violate at least one clause. Thus if setting $(x_1, x_2) = (\text{true}, \text{false})$ violates a clause, the conflict clause is $\neg x_1 \vee x_2$.

5.2   Partial-Order Dynamic Backtracking

Partial-order dynamic backtracking (PODB) is a generalization of branching with conflict clauses [3, 8, 9]. In a conventional branching search, one backtracks from a given node by de-assigning the assigned variables in a certain order. In PODB, this complete ordering of the assigned variables is replaced by a partial ordering. Thus the search cannot be conceived as a tree search, but it remains exhaustive while allowing a greater degree of freedom in how solutions are enumerated.

PODB can be viewed as constraint-based search in which the selection function $s(R_k)$ is computed in a slightly different way than in constraint-based branching. In constraint based branching, there is a complete ordering on the variables, and $s(R_k)$ is computed by assigning values to variables in this order. In PODB, this ordering is replaced by a partial ordering.

The partial ordering is defined as follows. Initially, no variable precedes another in the ordering. At any later point in the algorithm, the partial ordering is defined by the fact that some variable x_j in each nogood N of R_k has been designated as *last* in N. Every other variable in N is *penultimate* and precedes x_j in the partial ordering. The ordering is updated whenever a new nogood is created. Any variable x_j in the new nogood can be chosen as last, provided the choice is consistent with the current partial ordering. Thus if x_k is a penultimate variable in the nogood, x_j must not precede x_k in the current partial ordering.

The nogoods are processed by parallel resolution exactly as in constraint-based branching, which as shown in [], again consumes linear time and space. The selection function $s(R_k)$ is computed as follows. It first assigns values to variables x_j that are penultimate in some nogood. As before, it assigns x_j the value v_j if the disjunct $x_j \neq v_j$ occurs in a nogood (all penultimate disjuncts containing x_j exclude the same value v_j). The remaining variables are assigned values as in constraint-based branching, but in any desired order.

5.3 Logic-Based Benders Decomposition

Benders decomposition [2, 7] is a constraint-directed search that enumerates possible assignments to a *subset* of the variables, which might be called the *search variables*. Each possible assignment defines a *subproblem* of finding the optimal values of the remaining variables, given the values of the search variables. Solution of the subproblem produces a nogood that excludes the search variable assignment just tried, perhaps along with other assignments that can be no better. Since the subproblems are restrictions of the original problem, a Benders algorithm can be viewed as enumerating problem restrictions.

Benders is applied to a problem P of the form

$$\min \ f(x)$$
$$C(x, y) \tag{2}$$
$$x \in D_x, \ y \in D_y$$

where x contains the search variables and y the subproblem variables. $C(x, y)$ is a constraint set that contains variables x, y. To simplify exposition we assume that the objective function depends only on x. The more general case is analyzed in [12, 16].

In the constraint-directed search algorithm, each problem restriction (subproblem) P_k is obtained from P by fixing x to the solution x^{k-1} of the previous relaxation R_{k-1}. P_k is therefore the feasibility problem

$$C(x^{k-1}, y)$$
$$y \in D_y \tag{3}$$

where $C(x^{k-1}, y)$ is the constraint set that remains when x is fixed to x^{k-1} in $C(x, y)$.

Unlike many constraint-directed methods, a Benders method obtains nogoods by solving the restriction P_k. If P_k has a feasible solution y^k, then $(x, y) = (x^{k-1}, y^k)$ is optimal in P, and the search terminates. Otherwise a nogood or

Benders cut $N_k(x)$ is generated. $N_k(x)$ must exclude x^{k-1}, perhaps along with other values of x that are infeasible for similar reasons.

In classical Benders decomposition, the subproblem is a continuous linear or nonlinear programming problem, and $N_k(x)$ obtained from Lagrange multipliers associated with the constraints of the subproblem. In a more general setting, $N_k(x)$ is based on an analysis of how infeasibility of the subproblem is proved when $x = x^{k-1}$. The same proof may be valid when x takes other values, and these are precisely the values that violate $N_k(x)$. The result is a "logic-based" form of Benders.

The Benders cut $N_k(x)$ is added to the previous relaxation R_{k-1} to obtain the current relaxation or *master problem* R_k:

$$
\begin{aligned}
&\min f(x) \\
&N_i(x), \quad i = 0, \ldots, k-1 \\
&x \in D_x
\end{aligned}
\tag{4}
$$

If the master problem is infeasible, then P is infeasible, and the search terminates. Otherwise we select any optimal solution $s(R_k)$ of R_k and denote it x^k, and the algorithm proceeds to the next step. A Benders method therefore requires solution of both P_k and R_k, the former to obtain nogoods, and the latter to obtain P_{k+1}.

Logic-based Benders decomposition can be combined with constraint programming in various ways [5, 6, 12, 14, 16, 18, 19, 26]. One type of integration is to solve the subproblem by constraint programming (since it is naturally suited to generate Benders cuts) and the master problem R_k by another. Planning and scheduling problems, for example, have been solved by applying integer programming to R_k (task allocation) and constraint programming to P_k (scheduling) [12, 14, 19, 26]. This approach has produced some of the largest computational speedups available from integrated methods, outperforming conventional solvers by several orders of magnitude.

6   Heuristic Methods

Local search methods solve a problem by solving it repeatedly over small subsets of the solution space, each of which is a "neighborhood" of the previous solution. Since each neighborhood is the feasible set of a problem restriction, local search can be viewed as a search-infer-and-relax method.

In fact, it is useful to conceive local search as belonging to a family of *local-search-and-relax* algorithms that resemble branch-and-relax algorithms but are inexhaustive (Fig. 4). A number of other heuristic methods, such as GRASPs, belong to the same family. The analogy with branch and relax suggests how inference and relaxation may be incorporated into heuristic methods.

The generic local-search-and-relax algorithm of Fig. 4 "branches" on a problem restriction P_k by creating a further restriction P_{k+1}. For the time being, only one branch is created. Branching continues in this fashion until a restriction is

Let $v_{\mathrm{UB}} = \infty$ and $S = \{P_0\}$.
Perform **LocalSearch**.
The best solution found for P_0 has value v_{UB}.

Procedure **LocalSearch**.
  If S is nonempty then
      Select a restriction $P =$ from S.
      If P is too hard to solve then
          Let v_{R} be the optimal value of a relaxation of P.
          If $v_{\mathrm{R}} < v_{\mathrm{UB}}$ then
              Add a restriction of P to S.
              Perform **LocalSearch**.
          Else remove P from S.
      Else
          Let v be the value of P's solution and $v_{\mathrm{UB}} = \min\{v, v_{\mathrm{UB}}\}$.
          Remove P from S.

Fig. 4. *Generic local-search-and-relax algorithm for solving a minimization problem P_0*

created that is easy enough to solve, whereupon the algorithm returns to a previous restriction (perhaps chosen randomly) and resumes branching. There is also a bounding mechanism that is parallel to that of branch-and-relax algorithms.

Local search and GRASPs are special cases of this generic algorithm in which each restriction P_k is specified by setting one or more variables. If all the variables $x = (x_1, \ldots, x_n)$ are set to values $v = (v_1, \ldots, v_n)$, P_k's feasible set is a neighborhood of v. P_k is easily solved by searching the neighborhood. If only some of the variables $(x_1, \ldots, x_k)$ are set to $(v_1, \ldots, v_k)$, P_k is regarded as too hard to solve.

A pure local search algorithm, such as simulated annealing or tabu search, branches on the original problem P_0 by setting all the variables at once to $v = (v_1, \ldots, v_n)$. The resulting restriction P_1 is solved by searching a neighborhood of v. Supposing P_1's solution is v', the search backtracks to P_0 and branches again by setting $x = v'$. Thus in pure local search, the search tree is never more than one level deep.

In simulated annealing, P_k is "solved" by randomly selecting one or more elements of the neighborhood until one of them, say v', is accepted. A solution v' is accepted with probability 1 if it is better than the currently best solution, and with probability p is it is no better. The probability p may drop (reflecting a lower "temperature") as the search proceeds.

In tabu search, P_k is solved by a complete search of the neighborhood, whereupon the best solution becomes v'. In this case the neighborhood of v' excludes solutions currently on the tabu list.

Each iteration of a GRASP has two phases, the first of which constructs a solution in a greedy fashion, and the second of which uses this solution as a starting point for a local search [25]. In the constructive phase, the search branches by setting variables one at a time. At the original problem P_0 it branches by setting one variable, say x_1, to a value v_1 chosen in a randomized greedy fashion. It then branches again by setting x_2, and so forth. The resulting restrictions $P_1, P_2, \ldots$ are regarded as too hard to solve until all the variables x are set to some value v. When this occurs, a solution v' of P is found by searching a neighborhood of v, and the algorithm moves into the local search phase. It backtracks directly to P_0 and branches by setting $x = v'$ in one step. Local search continues as long as desired, whereupon the search returns to the constructive phase.

A GRASP provides the opportunity to use the bounding mechanism of the local-search-and-relax algorithm, a possibility already pointed out by Prestwich [23]. If a relaxation of P_k has an optimal value that is no better than that of the incumbent solution, then there is nothing to be gained by branching on P_k.

References

1. Aron, I., J. N. Hooker, and T. H. Yunes, SIMPL: A system for integrating optimization techniques, in J.-C. Régin and M. Rueher, eds., *Conference on Integration of AI and OR Techniques in Constraint Programming for Combinatorial Optimization Problems (CPAIOR 2004), Lecture Notes in Computer Science* **3011** (2004) 21–36.
2. Benders, J. F., Partitioning procedures for solving mixed-variables programming problems, *Numerische Mathematik* **4** (1962) 238–252.
3. Bliek, C. T., Generalizing partial order and dynamic backtracking, *Proceedings of AAAI* (AAAI Press, 1998) 319–325.
4. Bockmayr, A., and T. Kasper, Branch and infer: A unifying framework for integer and finite domain constraint programming, *INFORMS Journal on Computing* **10** (1998) 287–300.
5. Cambazard, H., P.-E. Hladik, A.-M. Déplanche, N. Jussien, and Y. Trinquet, Decomposition and learning for a hard real time task allocation algorithm, in M. Wallace, ed., *Principles and Practice of Constraint Programming (CP 2004), Lecture Notes in Computer Science* **3258**, Springer (2004).
6. Eremin, A., and M. Wallace, Hybrid Benders decomposition algorithms in constraint logic programming, in T. Walsh, ed., *Principles and Practice of Constraint Programming (CP 2001), Lecture Notes in Computer Science* **2239**, Springer (2001).
7. Geoffrion, A. M., Generalized Benders decomposition, *Journal of Optimization Theory and Applications* **10** (1972) 237–260.
8. Ginsberg, M. L., Dynamic backtracking, *Journal of Artificial Intelli- gence Research* **1** (1993) 25–46.
9. Ginsberg, M. L., and D. A. McAllester, GSAT and dynamic backtrack- ing, *Second Workshop on Principles and Practice of Constraint Pro- gramming (CP1994)* (1994) 216–225.
10. Hooker, J. N., Logic-based methods for optimization, in A. Borning, ed., *Principles and Practice of Constraint Programming, Lecture Notes in Computer Science* **874** (1994) 336–349.
11. Hooker, J. N., Constraint satisfaction methods for generating valid cuts, in D. L. Woodruff, ed., *Advances in Computational and Stochasic Optimization, Logic Programming and Heuristic Search*, Kluwer (Dordrecht, 1997) 1–30.
12. Hooker, J. N., *Logic-Based Methods for Optimization: Combining Optimization and Constraint Satisfaction*, John Wiley & Sons (New York, 2000).

13. Hooker, J. N, A framework for integrating solution methods, in H. K. Bhargava and Mong Ye, eds., *Computational Modeling and Problem Solving in the Networked World* (Proceedings of ICS2003), Kluwer (2003) 3–30.

14. Hooker, J. N., A hybrid method for planning and scheduling, in M. Wallace, ed., *Principles and Practice of Constraint Programming (CP 2004), Lecture Notes in Computer Science* **3258**, Springer (2004).

15. Hooker, J. N., and M. A. Osorio, Mixed logical/linear programming, *Discrete Applied Mathematics* **96-97** (1999) 395–442.

16. Hooker, J. N., and G. Ottosson, Logic-based Benders decomposition, *Mathematical Programming* **96** (2003) 33–60.

17. Hooker, J. N., G. Ottosson, E. Thorsteinsson, and Hak-Jin Kim, A scheme for unifying optimization and constraint satisfaction methods, *Knowledge Engineering Review* **15** (2000) 11–30.

18. Hooker, J. N., and Hong Yan, Logic circuit verification by Benders decomposition, in V. Saraswat and P. Van Hentenryck, eds., *Principles and Practice of Constraint Programming: The Newport Papers (CP95)*, MIT Press (Cambridge, MA, 1995) 267–288.

19. Jain, V., and I. E. Grossmann, Algorithms for hybrid MILP/CP models for a class of optimization problems, *INFORMS Journal on Computing* **13** (2001) 258–276.

20. Moskewicz, M. W., C. F. Madigan, Ying Zhao, Lintao Zhang, and S. Malik, Chaff: Engineering an efficient SAT solver, *Proceedings of the 38th Design Automation Conference (DAC'01)* (2001) 530–535.

21. Neumaier, A., Complete search in continuous global optimization and constraint satisfaction, in A. Iserles, ed., *Acta Numerica 2004* (vol. 13), Cambridge University Press (2004).

22. Pinter, J. D., *Applied Global Optimization: Using Integrated Modeling and Solver Environments*, CRC Press, forthcoming.

23. Prestwich, S., Exploiting relaxation in local search, *First International Workshop on Local Search Techniques in Constraint Satisfaction*, 2004.

24. Sahinidis, N. V., and M. Tawarmalani, *Convexification and Global Optimization in Continuous and Mixed-Integer Nonlinear Programming*, Kluwer Academic Publishers (2003).

25. Silva, J. P. M., and K. A. Sakallah, GRASP–A search algorithm for propositional satisfiability, *IEEE Transactions on Computers* **48** (1999) 506–521.

26. Thorsteinsson, E. S., Branch-and-Check: A hybrid framework integrating mixed integer programming and constraint logic programming, T. Walsh, ed., *Principles and Practice of Constraint Programming (CP 2001), Lecture Notes in Computer Science* **2239**, Springer (2001) 16–30.

Combining Arc-Consistency and Dual Lagrangean Relaxation for Filtering CSPs*

Mohand Ou Idir Khemmoudj, Hachemi Bennaceur, and Anass Nagih

LIPN-CNRS UMR 7030 Av J-B. Clément,
93430 Villetaneuse France
{MohandOuIdir.Khemmoudj, Hachemi.Bennaceur,
Anass.Nagih}@lipn.univ-paris13.fr

Abstract. This paper presents a CSPs filtering method combining arc-consistency and dual Lagrangean relaxation techniques. First, we model the constraint satisfaction problem as a 0/1 linear integer program (IP); then, the consistency of a value is defined as an optimization problem on which a dual Lagrangean relaxation is defined. While solving the dual Lagrangean relaxation, values inconsistencies may be detected (dual Lagrangean inconsistent values); the constraint propagation of this inconsistency can be performed by arc-consistency. After having made the CSP arc-consistent, the process iteratively selects values of variables which may be dual Lagrangean inconsistent. Computational experiments performed over randomly generated problems show the advantages of the hybrid filtering technique combining arc-consistency and dual Lagrangean relaxation.

Keywords: Arc-Consistency, Lagrangean Relaxation, Subgradient Algorithm.

1 Introduction

Constraint Programming (CP) and Integer Linear Programming (ILP) are two approaches to model and solve combinatorial optimization problems. From a modeling point of view, constraint programming is in general preferable to Integer Linear Programming. However, from a solving time point of view, the two approaches can claim different success, but none of them can claim to be universally the best.

Several works have discussed the synergy between the two approaches, leading to the conclusion that using them together can be quite beneficial [5] [6] [12] [13].

In ILP, the most used algorithm is the Branch and Bound one. It is a complete tree based search algorithm in which the branching is intertwined with a relaxation, which sterilizes nodes for which the relaxation is infeasible or which lead to solutions of worst value than the best known. Lagrangean relaxation [11]

* This work is supported in part by the French Electricity Board (EDF).

R. Barták and M. Milano (Eds.): CPAIOR 2005, LNCS 3524, pp. 258–272, 2005.

is one of the most widely used relaxations. It consists in dualizing a subset of the constraints and defining a dual problem which is then solved by a convergent iterative method, like the subgradient algorithm [4].

Like ILP, CP relies on branching to enumerate regions of the search space. While ILP solves a relaxed problem at each node, CP uses deductive methods to reduce the amount of choices to explore. This process is called constraint propagation, domain filtering, pruning or consistency technique [15] [18].

It is now recognized that integrating propagation and relaxation techniques do yield to substantial results. One issue to do this is the development of optimization constraints [9] [17].

Some recent works [19] [8] [2] have discussed how optimization constraints can strengthen their propagation abilities by the way of Lagrangean relaxation.

Other works propose mathematical programming models for CSPs. A quadratic formulation has been presented in [1] and used to define the weighted arc-consistency for MAX-CSP. In [14] a 0-1 programming model has been proposed and used to solve efficiently some instances of the frequency assignment problem.

In this work, we are interested in the binary Constraint Satisfaction Problem. We propose a new 0-1 formulation for binary CSPs on which we study the combination of arc-consistency technique and Lagrangean relaxation. We propose a technique combining arc-consistency and dual Lagrangean relaxation to reduce efficiently CSPs. First, arc-consistency is achieved on the CSP, then some arc-consistent values are selected. For each selected value we check its consistency by solving an associated dual Lagrangean relaxation. If the solving of the dual problems detect inconsistent values, then constraint propagation of these inconsistencies is performed by arc-consistency. Computational experiments performed over randomly generated problems show the advantages of the hybrid filtering technique.

The remainder of this paper is organized as follows. In section 2 we recall some notions of CSPs and we describe the Lagrangean relaxation of integer programming problems. Section 3 describes the proposed new 0-1 formulation for CSPs. In section 4, we describe how the Lagrangean relaxation technique can be exploited for filtering CSPs; the presented computational experiments show that this technique can be of a great interest when it is judiciously exploited. Section 5 presents the hybrid technique to reduce CSPs and the computational experiments performed over randomly generated problems. Finally, section 6 concludes.

2 Preliminaries

In this section, we recall some notions of CSPs and we describe Lagrangean relaxation of integer programming problems.

2.1 Constraint Satisfaction Problem

Constraint Satisfaction Problems (CSPs) involve the assignment of values to variables which are subject to a set of constraints. Formally, a binary CSP is defined by a quadruplet (X,D,C,R) where:

- X is a set of n variables $\{X_1, X_2, \ldots, X_n\}$;
- D is a set of n domains $\{D_1, D_2, \ldots, D_n\}$ where each D_i is a set of d_i possible values for X_i;
- C is a set of m binary constraints where each constraint C_{ij} involving variables X_i and X_j ($i \neq j$) is defined by its relation R_{ij};
- R is a set of m relations, where R_{ij} is a subset of the Cartesian product $D_i \times D_j$.
  The predicate $R_{ij}(k, l)$ holds iff the pair (v_k, v_l) belongs to R_{ij}.
  We will restrict our study to problems which verify : $(v_k, v_l) \in R_{ij}$ $\Leftrightarrow$ $(v_l, v_k) \in R_{ji}$.

A solution of the CSP is a total assignment which satisfies each constraint.
The constraint graph represents the variables and the constraints of the CSP within a network, where each variable is represented by a vertex and each constraint by an edge.

- A value $v_k \in D_i$ is consistent iff there exists a solution such that $X_i = v_k$.
- A CSP is consistent iff it has a solution.

We recall that:

1. A domain D_i is arc-consistent iff $\forall v_k \in D_i$, v_k is arc-consistent, ie: $\forall X_j \in X$ such that $C_{ij} \in C$, there exists $v_l \in D_j$ such that $R_{ij}(k, l)$.

2. A binary CSP is arc-consistent iff $\forall D_i \in D$, $D_i \neq \emptyset$ and D_i is arc-consistent.

2.2 Lagragean Relaxation

Consider the following integer programming problem:

$$(IP) \begin{cases} \max & cx \\ \text{subject to } Ax \leq a \\ & Bx \leq b \\ & x \in S \subset \mathbb{N}^n \end{cases}$$

where c and x are vectors of order $1 \times n$, a is a vector of order $1 \times m$, b is a vector of order $1 \times p$, A is a matrix of order $m \times n$ and B is a matrix of order $p \times n$; $V(IP)$ will denote its value.

It is well known that the efficiency of a Branch-and-Bound scheme to solve (IP) depends on the quality of the bounds used to evaluate $V(IP)$.

To define a better bound (i.e. a smaller value) than the value provided by the linear relaxation, one can use Lagrangean relaxation which consists in dualizing a subset of constraints [11].

Suppose that the problem (IP) becomes (relatively) easy to solve if we remove complicating constraints $Ax \leq a$, the associated Lagrangean function is classically defined as follows, for a given $u \in \mathbb{R}_+^m$:

$$l(x, u) = cx + u(a - Ax) = ua + (c - uA)x.$$

The corresponding Lagrangean relaxation problem is given by:

$$(LR(u)) \begin{cases} \max ua + (c - uA)x \\ \text{s.t.} \quad Bx \leq b \\ \qquad x \in S \subset \mathbb{N}^n \end{cases}$$

and the Lagrangean dual problem is defined by

$$(D) \begin{cases} \min LR(u) \\ \text{s.t.} \ u \in \mathbb{R}_+^m. \end{cases}$$

$V(D)$, the value of (D), is an upper bound for (IP) which may be approximately computed, by any convergent method solving nondifferentiable optimization problems, like subgradient algorithm.

The aim of the subgradient algorithm is to provide in an iterative way the vector u of multipliers for which the value $V(LR(u))$ is as close as possible to the optimal value of (IP). Given an initial Lagrangean multiplier vector u^0, a sequence $(u^k)_{k \in \mathbb{N}}$ is generated. More precisely, at iteration k:

− The subgradient s^k is given by

$$s^k = a - Ax^k$$

where x^k is an optimal solution of $(LR(u^k))$.
− The new step direction is given by

$$d^k = (1 - \mu)s^k + \mu d^{k-1}$$

The direction (d^k) at iteration k is a convex combination of the subgradient s^k and the previous direction d^{k-1} ($= 0$ if $k = 1$) with μ is a real parameter chosen in $[0,1[$ (CFM method [4]).
− The new Lagrangean multiplier vector is defined by

$$u_i^{k+1} = \max\left\{0, u_i^k - t_k d_i^k\right\}, i = 1, ..., m,$$

where the step size $t_k > 0$ is given by the commonly used formula

$$t_k = \rho_k \frac{v^k - \underline{v}}{||d^k||^2}.$$

The value of the step size parameter ρ_k is a scalar satisfying $0 < \rho_k \leq 2$; $\underline{v}$ is the lower bound on $V(IP)$ available at iteration k, whereas $v^k = V(LR(u^k))$ is the value of the Lagrangean function at iteration k.

3   Modeling

In this section, we describe a new 0-1 formulation for CSPs. First, we introduce for each variable X_i, d_i binary variables x_i^k, $k = 1, ..., d_i$ (where d_i is the size of D_i), such that

$$x_i^k = \begin{cases} 1 \text{ if } X_i = v_k \quad (v_k \in D_i) \\ 0 \text{ otherwise.} \end{cases}$$

We will denote by x the vector of components x_i^k, $x = \left(x_i^k\right)_{i=\overline{1,n}}^{k=\overline{1,d_i}}$.

It is well known that any incompatible pair of values (v_k, v_l) of two variables (X_i, X_j) can be written as a 0-1 linear inequality as follows:

$$x_i^k + x_j^l \leq 1.$$

The expression of each incompatible pair of values may require a huge number of inequalities. To reduce this number, let us denote, for each variable X_i and each value $v_k \in D_i$, by a_i^k the number of variables X_j $(j \neq i)$ that contain in their domain at least one value which is incompatible with v_k,

$$a_i^k = \sum_{j \neq i : \exists v_l \in D_j \wedge \neg R_{ij}(k,l)} 1.$$

Then, we can write:

$$a_i^k x_i^k + \sum_{j \neq i} \sum_{v_l \in D_j, \neg R_{ij}(k,l)} x_j^l \leq a_i^k. \tag{1}$$

This means that if X_i, is set to v_k, then a variable X_j will not be set to each value v_l such that $\neg R_{ij}(k, l)$.

Remark 1. *The inequality (1) is violated if and only if $x_i^k = 1$ and there is at least one binary variable x_j^l, $(j \neq i)$ such that $\neg R_{ij}(k, l)$ and $x_j^l = 1$.*

Thus, if we associate one inequality of type (1) to each couple (X_i, v_k), then we express all the incompatible pairs of values.

Remark 2. *The number of inequalities of type (1) required is bounded by $n.d$, where n is the number of variables and d is the size of the largest domain.*

To express the fact that each variable X_i must take exactly one value, we state:

$$\sum_{v_k \in D_i} x_i^k = 1 \tag{2}$$

Now, we consider the following 0-1 integer problem:

$$(IP) \begin{cases} A_i^k x \leq a_i^k \ i = \overline{1,n}, \ k = \overline{1,d_i} \\ x \in S \end{cases}$$

such that:

– $A_i^k x \leq a_i^k$ is a simple expression of the inequality of type (1) associated to the value v_k of the variable X_i;

– S is the set of solutions of the system $\begin{cases} \sum\limits_{v_k \in D_i} x_i^k = 1 \; i = \overline{1,n} \\ x_i^k \in \{0,1\} \quad i = \overline{1,n}, \; v_k \in D_i \end{cases}$

To simplify the presentation, we will denote the previous system (IP) by:

$$(IP) \begin{cases} Ax \leq a \\ x \in S \end{cases}$$

where $Ax \leq a$ is the system of the inequalities $A_i^k x \leq a_i^k : i = \overline{1,n}, \; k = \overline{1,d_i}$.

Theorem 1. *The CSP is satisfiable iff the 0-1 integer problem (IP) has a solution.*

Proof. We only need to show that to each solution of the CSP corresponds a solution of the system (IP) and conversely. Let $I = (v_1, v_2, \ldots, v_n)$ be a solution of the CSP and $\bar{x} = \left(\bar{x}_i^k\right)_{i=\overline{1,n}}^{k=\overline{1,d_i}}$ be a vector
    were $\bar{x}_i^{v_i} = 1$ and $\bar{x}_i^k = 0, \forall v_k \in D_i - \{v_i\}$.
    $\bar{x}$ is a solution of the system S because, by construction, its components are all 0 or 1 and for each $i(1 \leq i \leq n)$, only the component $\bar{x}_i^{v_i}$ has the value 1, the values of the other components are 0. Let us now consider the constraint of type (1) associated to $\bar{x}_i^k (i = \overline{1,n}, k = \overline{1,d_i})$. This constraint is violated iff $\bar{x}_i^k = 1$, i.e. $k = v_i$, and there is at least one component $\bar{x}_j^l$ $(j \neq i)$ of $\bar{x}$, such that $\neg R_{ij}(k,l)$ and $\bar{x}_j^l = 1$, i.e. $l = v_j$. However, since I is a solution of the CSP, the couple (v_i, v_j) belongs to R_{ij}. This means that the constraint is satisfied. Hence, $\bar{x}$ is a solution of (IP).
    Conversely, let $\bar{x}$ be a solution of (IP) and consider the instanciation $I = (v_1, v_2, \ldots, v_n)$ of the CSP such that $\bar{x}_i^{v_i} = 1$. Since $\bar{x}$ is a solution of (IP), the couple $(v_i, v_j)(1 \leq i, j \leq n, i \neq j)$ belongs to R_{ij}, otherwise the constraints of type (1) associated to $\bar{x}_i^{v_i}$ and $\bar{x}_j^{v_j}$ would not be satisfied. Hence, I is a solution of the CSP. $\square$

Remark 3. *In (IP), every constraint of the CSP is expressed twice. For instance, if we have $\neg R_{ij}(k,l)$, then the constraint of type (1) associated to the value v_k of X_i will express that $\neg R_{ij}(k,l)$ and the constraint of type (1) associated to the value v_l of X_j will express that $\neg R_{ji}(l,k)$. Such a redundancy can be avoided by ordering the variables of the CSP.*

Our model is linear and requires a smaller number of constraints than the classical model whose constraints are of form $x_i^k + x_j^l \leq 1$. This may lead to reduce the resolution processing time.
    Thanks to this formulation, some well known techniques of relaxations can be exploited in the CSP context.

4   Lagrangean Relaxation as Filtering Technique

In this section, we will show how the previous (IP) model can be exploited for filtering CSPs.

Consider the following system :

$$(IP_i^k) \begin{cases} \max \; x_i^k \\ s.t. \qquad Ax \leq a \\ \qquad\quad A_i^k x \leq a_i^k \\ \qquad\quad x \in S \end{cases}$$

It is an optimization problem focusing on an individual assignment.

Remark 4. *The constraint $A_i^k \leq a_i^k$ in (IP_i^k) is a redundant constraint. It is introduced to simplify the presentation.*

Theorem 2. *A value v_k of a variable X_i is consistent iff the system (IP_i^k) has a solution with value 1.*

Proof. To each solution of the CSP in which the variable X_i takes the value v_k corresponds a solution of (IP_i^k) in which the binary variable x_i^k takes the value 1. Hence, the value v_k of a variable X_i is consistent iff the system (IP_i^k) has a solution with value 1.   □

The exact resolution of (IP_i^k) enables to answer the question : is there a solution to the CSP in which the variable X_i takes the value v_k? This problem is known to be NP-complete. Therefore, one can only solve a relaxation of (IP_i^k). Any relaxation allows to define an upper bound on the value of (IP_i^k). Thus, if this bound is lower than 1, one can conclude that the value v_k of X_i is inconsistent.

In this paper, we consider the Lagrangean relaxation which consists of dualizing the constraints $Ax \leq a$ of (IP_i^k). It is motivated by its similarities with the relaxation considered in Constraint Programming (individual consideration of the constraints).

Let us denote by u the vector of Lagrangean multipliers associated to the relaxed constraints. Then, the Lagrangean relaxed problem is the following:

$$(LR_i^k(u)) \begin{cases} \max \; ua + c(u).x = ua + \sum_{j=0}^{n} \sum_{v_l \in D_j} c_j^l(u).x_j^l \\ s.t. \quad A_i^k \leq a_i^k \\ \qquad\quad x \in S \end{cases}$$

where $c(u).x = x_i^k - uAx$.

The Lagrangean relaxed problem is easy to solve, since the constraint associated to x_i^k is the only remaining constraint.

Remark 5. *A is a square matrix of order less or equal to n.d, where n is the number of variables of the CSP and d is the size of the largest domain. Thus, the complexity of computing the vector $c(u)$ is at worst$O(n^2 d^2)$.*

The Lagrangean dual problem is defined by

$$(D_i^k) : \min_u V(LR_i^k(u))$$

where $V(LR_i^k(u))$ is the optimal value of $LR_i^k(u)$.

4.1   Lagrangean Relaxation Versus Arc-Consistency

Consider the vector of multipliers with zero components u^0. For this special case of multipliers, the relaxed problem is the following:

$$(LR_i^k(u^0)) \begin{cases} \max\ x_i^k \\ s.t. \qquad A_i^k \le a_i^k \\ \qquad x \in S \end{cases}$$

The system $(LR_i^k(u^0))$ is reduced to the single type (1) constraint associated to the value v_k of X_i; then, it is clear that it has a solution with value 1 iff there exists an instanciation in which $X_i = v_k$ and $X_j \ne v_l$ if $\neg R_{i,j}(k,l), \forall j \ne i$; in other words, iff the value v_k of X_i is viable. Thus, we can conclude that checking the viability of the value v_k of X_i is equivalent to solving the problem $(LR_i^k(u^0))$. This means that the arc-consistency uses, in some extent, the same relaxation to check a value viability.

To achieve arc-consistency, several algorithms are proposed and the AC-6 [3] algorithm is one of the most commonly used. The principle of the AC-6 algorithm consists in proving that all values of the CSP are viable: it checks, for each value, one support per constraint, looking for another one only when the current support is removed from the domain. In other words, it ensures that for any not yet removed value v_k of each variable X_i, the relaxed problem $(LR_i^k(u^0))$ has a solution with value 1 (the value v_k of X_i is viable). The vector u^o is not necessarily the optimal solution for the dual problem (D_i^k). An approximate resolution of (D_i^k) can lead to the removal of the value v_k from D_i, if this value is proved to be inconsistent.

The value of $LR_i^k(u^0)$ is less than or equal to 1. This means that the difference between the optimal value of the dual Lagrangean and the optimal value of the 0-1 model is bounded by 1.

4.2   Dual Problem Resolution

The subgradient algorithm implemented to solve the dual problem (D_i^k) has the following structure.

procedure1(in i, k, μ, K, N; **out** result)

1. $it = 0; \rho = 2; d = 0; \% \ d_j^l = 0, j = \overline{1,n}, j = \overline{1,d_j}.$
2. **initialize**(it, u, d, ρ); % multiplier vector initializing.

3. **repeat**

   3.1. $it = it + 1$;

   3.2. **procedure2**$(i, k, u, x, result)$; % *resolution of the relaxed problem* $LR_i^k(u)$.

   3.3. $s = a - Ax$; % *subgradient vector computing.*

   3.4. $d = \mu.d + (1 - \mu).s$; % *direction vector computing.*

   3.5. $t = \rho \, \dfrac{result}{||d||^2}$; % *step size computing* $(\underline{v} = 0)$.

   3.6. $u_j^l = max\{0, u_j^l - t.d_j^l\}, j = \overline{1, n}, v_l \in D_j$; % *muliplier vector updating.*

   3.7. **update**(ρ); % *step size parameter updating.*

   until $(result < 1)$ **or** $(it > N)$.

The **update** procedure consists in dividing the step size parameter ρ per 2 if the subgrdient algorithm carry out K iterations without decreasing the value of the dual problem. The lower bound used to compute the step size is fixed to 0 ($\underline{v} = 0$ is a lower bound of (IP_i^k)). The **procedure2** is described below:

procedure2(**in** i, k, u; **out** $x, result$)

1. $c = -Au, c_i^k = c_i^k + 1$; % *cost vector computing* $\left(c = (c_j^l)_{j=\overline{1,n}}^{l=\overline{1,d_j}}\right)$.

2. $result = u.a$; % *fixed term computing.*

3. $X_i = v_k; result = result + c_i^k$; % *variable* X_i *assigning.*

4. $\forall j \neq i$, % *searching of the best support of the* v_k *value of* X_i *in* D_j;

   4.1. $y = -M$ % *M is a big integer;*

   4.2. $\forall v_l \in D_j : R_{ij}(k, l)$, **if** $c_j^l > y$ **then** $\{y = c_j^l; X_j = v_l;\}$

   4.3. $result = result + y$.

Since we are interested in consistency of the value v_k of the variable X_i, the procedure **procedure2** only considers the solutions of $LR_i^k(u)$ in which the variable X_i is set to v_k. The other solutions of (IP_i^k), if they exist, have a null value for x_i^k ($x_i^k = 0$).

If the value v_k of the variable X_i has a single support v_l in D_j, the variable X_j always takes the value v_l in the solutions returned by **procedure2**. This implies that the component s_j^l of the subgradient is always negative or null ($a_j^l - A_j^l x \leq 0$) and $u_j^l(it + 1) \geq u_j^l(it)$, where $u_j^l(it)$ is the component u_j^l of the multiplier vector u at iteration it of the subgradient algorithm (**procedure1**). Then, it is more probable that the value of u_j^l will be strictly positive at the end of the last iteration of **procedure1**. For this reason, u_j^l is initialized (procedure **initialize**) to 1 if v_l is a single support of the value v_k of the variable X_i and to 0 otherwise.

Remark 6. *The complexity of the procedure* **procedure2** *is* $O(n^2 d^2)$ *and the complexity of the procedure* **procedure1** *is* $O(Nn^2 d^2)$, *where N is the number of iterations allowed.*

4.3   Experimental Evaluation (AC Versus Lag)

Achieving arc-consistency on CSP is performed within two phases: the checking phase and the constraint propagation one. In order to study its performances, the Lagrangean relaxation is integrated into the checking phase: for each value v_k of each variable X_i we solve the corresponding dual problem (D_i^k). If this resolution shows that v_k is inconsistent, then the value v_k is removed from the domain D_i. We denote this process by Lag and we denote by AC the filtering process using AC-6 algorithm.

Our experiments have been carried out on a set of randomly generated problems. The generator used involves four parameters: the number n of variables, the common size d of all domains, the number c of constraints and the number p of the forbidden value pairs in a given constraint.

We have fixed $n = 16$, $d = 8$, $p = 32$ and we have varied c from 24 to 120 (complete graph) per step of 12. For each value of c we tested 100 different problems. The computed results are the average of the percentage of filtered values as well as the average time spent in filtering a value. [1]

Fig.1 gives a comparison between the Lag and the AC techniques. The results of Lag are obtained by giving to the parameters μ, K and N of the procedure **procedure1** the values 0.6, 5 and 20, respectively. The time taken to filter a value by Lag is not presented for $c < 60$. It is of 81, 31 and 13 milliseconds for c equal to 24, 32 and 48 respectively.

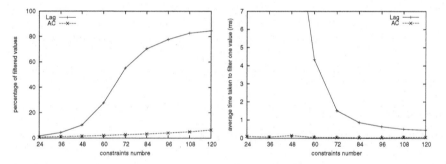

Fig. 1. AC and Lag comparison on randomly generated CSPs with $n = 16, d = 8, p = 32, 24 \leq c \leq 120$

We note that Lag allows us to remove more inconsistent values than AC. Moreover, the power of Lag increases when the constraint graph density grows. However, Lag is unable to dismount the inconsistency of any problem. This is due to the fact that the propagation of the values deletion is not performed.

---

[1] The time spent to filter a value is obtained by dividing the total running time per the number of filtered values.

When the constraint graph density is weak, the average process time necessary for Lag to detect an inconsistent value is too large compared to that of AC. This is due to the following reasons :

1. the complexity $(O(n^2d^2))$ of a single iteration of the subgradient algorithm is higher than that of the whole AC-6 algorithm $(O(md^2))$, where m is the number of constraints of the CSP;
2. the most of the values are consistent and the subgradient algorithm consumes the allowed number of iterations: it looks for an instantiation with a small number of violated constraints[2].

The average time spent by Lag to remove a value decreases when the constraint graph density grows and becomes interesting. This can be explained by the fact that when the constraint graph density is high, the majority of values are inconsistent. In this case, Lag removes much more values than AC and consumes often few iterations, since it is stopped when the inconsistency is demonstrated.

These results convinced us that this Lagrangean relaxation technique can be of a great interest if it is well exploited. Its performances can be ameliorated by:

– reducing the number of dual problems that are solved;
– combining it with deductive methods (constraint propagation techniques).

5   Combining Arc-Consistency and Lagrangean Relaxation

In this section we present a technique combining arc-consistency and Lagrangean relaxation to filter CSPs. It keeps the good features of AC (i.e. changing the domains of variables) and increases the pruning power of AC by solving the dual problems defined on some selected values of variables. The algorithm consists in the following procedure.

procedure3(in μ, K, N**)**

1. **ACfiltering();** % *filtering the problem by AC-6.*
2. $E \leftarrow X$; % *E is initialized by the set of CSP variables.*
3. **repeat**
    3.1 **domdeg**(E, i); % *A variable* X_i *is selected from E.*
    3.2 $E \leftarrow E - \{X_i\}$; % X_i *is marked as considered.*
    3.3 **if** $D_i \neq \emptyset$ **then repeat**
        3.2.1 **suspectvalue**(i, k); % *A value* v_k *is selected from* D_i.
        3.2.2 **procedure1**$(i, k, \mu, K, N, result)$; % D_i^k *solving*
        3.2.3 **if** $result < 1$ **then**
            $\{ D_i \leftarrow D_i - \{v_k\}$; **ACpropagate**$(i, k)$;$\}$ % *propagation phase*
        until $result \geq 1$ **or** $D_i = \emptyset$;
    until $E = \emptyset$;

---

[2] This is the aim of the subgradient algorithm.

After making the CSP arc-consistent (**ACfiltering** procedure), a variable X_i is selected (**domdeg** procedure) and a suspect[3] value v_k is selected from its domain (**suspectvalue** procedure), then the Dual problem D_i^k is solved. If this resolution shows that this value is inconsistent, it is removed. The propagation of this deletion is performed by arc-consistency (**ACpropagate** procedure) and the same process is repeated for another suspect value. When the suspect value is not deleted by the dual problem resolution, a not yet considered variable is selected and the process is repeated. The algorithm stops when all the variables of the CSP have been considered. We denote this technique by AC+Lag.

We use the minimum domain maximum degree heuristic to select a variable (**domdeg** procedure) [7] [10] and the suspect value heuristic (**suspectvalue** procedure) to select a value for this variable.

5.1   Experimental Evaluation (AC Versus AC+Lag)

We have tested the performance of the combining method on different classes of randomly generated problems (100 problems per class). For each class, we give a comparison between the AC+Lag and the AC techniques. The computed results are the average of the percentage of filtered values and the average time spent in filtering a value. The results of AC+Lag are obtained by giving to the parameters μ, K and N of the procedure **procedure3** the values 0.6, 5 and 10 respectively.

Fig. 2 gives the comparison between the AC+Lag and AC techniques applied to the same problems as those of the previous section ($n = 16, d = 8, p = 32, 24 \leq c \leq 120$).

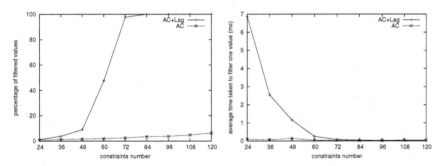

Fig. 2. AC and AC+Lag comparison on randomly generated CSPs with $n = 16, d = 8, p = 32, 24 \leq c \leq 120$

We can see that the pruning power of AC+Lag is increased compared to that of Lag (see **Fig. 1** and **Fig. 2**). AC+Lag dismounts the inconsistency for many

---

[3] The suspect value is determined by an heuristic method which selects first the values that have the minimum number of supports with another variable.

problems, even if the allowed number of iterations to solve the dual problem in AC+Lag is twice less than those of Lag.

When the constraint graph density is not under-constrained ($c > 48$), AC+Lag removes much more values than AC and dismounts the inconsistency of many problems. In this case, the average time taken to remove a value by AC+Lag is almost equal to that of AC. This can be explained by the fact that most of the values are removed during the propagation phase.

We have also tested this technique on the randomly generated problems with two other tightness of constraints : $p = 26$ and $p = 38$. The other parameters are unchanged.

Fig. 3 and **Fig. 4** summarize the results on the comparison between AC+Lag and AC. We can see that AC+Lag always removes more values than AC. However, for the under-constrained problems, the average time spent to remove a value by AC+Lag is large compared to that of AC.

Then for these experimental results, we notice that our method (combining AC and Lag) is quite beneficial for the not under-constrained CSPs. Naturally, the experiments can be extended to other classes of CSPs, in particular the non-random CSP benchmarks.

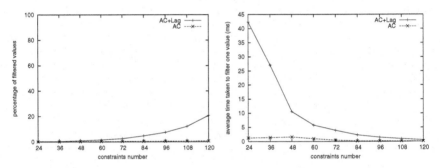

Fig. 3. AC and AC+Lag comparison on randomly generated CSPs with $n = 16, d = 8, p = 26, 24 \leq c \leq 120$

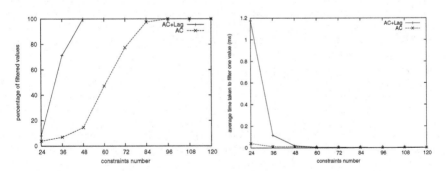

Fig. 4. AC and AC+Lag comparison on randomly generated CSPs with $n = 16, d = 8, p = 38, 24 \leq c \leq 120$

6   Conclusion

In this paper, a new formulation is proposed for CSPs. Thanks to this formulation, we show that Lagrangian relaxation and arc-consistency are two techniques which can cooperate very well.

Our experimental results show that the Lagrangean relaxation is a powerful technique when it is judiciously exploited. However its weakness is the worst case complexity of the dual problem resolution.

To keep the good features of Lagrangean relaxation and work out its weakness, a technique combining arc-consistency and Lagrangean relaxation is proposed. The experimental results show that this technique is very interesting on the not under-constrained CSPs. However, under-constrained CSPs are easy to solve. In this case, the technique combining arc-consistency and Lagrangean relaxation can be good if the CSP is provided with an objective function to optimize, i.e. when we have to solve a Constraint Satisfaction and Optimization Problems (CSOPs). One can solve only one dual problem after each branching and constraints propagation. This can lead to eliminate the exploration of uninteresting branches. This also provides interesting Lagrange multipliers useful in defining new heuristics to select the first branch to explore.

The performance of the hybrid technique can be also ameliorated by:

- searching another relaxation;
- reinforcing the model by valid inequalities;
- solving dual problems in more precise way, using other techniques such as column generation [16] or Bundle method [20].

The presented formulation can be extended to represent Valued Constraint Satisfaction Problems (VCSPs).

Naturally, since we have showed that combining arc-consistency and Lagrangean relaxation technique is very interesting on the not under-constrained CSPs and since the formulation can be extended to represent VCSPs, the topic of our future research is to study this cooperation within the VCSPs framework.

References

1. Affane, M.S., Bennaceur, H.: A Weighted Arc Consistency Technique for MAX-CSP. ECAI (1998) 209-213.
2. Benoist, T., Gaudin, E., Rottembourg, B.: Constraint Programming Contribution to Benders Decomposition: A Case Study. CP (2002) 603-617.
3. Bessière, C., Cordier, M.: Arc-consistency and Arc-consistency Again. Artificial Intelligence **65(1)** (1994) 179-190.
4. Camerini, P.M., Fratta, L., Maffioli, F.: On improving relaxation methods by modified gradient techniques. Mathematical programming Study **3** (1975) 26-34.
5. Darby-Dowman, K., Little, J.: Properties of some combinatorial optimization problems and their effect on the performance of integer programming and constraint logic programming. INFORMS Journal on Computing. **10(3)** (1998) 276-286.

6. Darby-Dowman, K., Little, J., Mitra, G., Zaffalon, M.: Constraint Logic Programming and Integer Programming Approaches and their Collaboration in Solving an Assignment Scheduling Problem. Constraints, An International Journal. **1:3** (1997) 245-264.
7. Dechter, R., Meiri, I.: Experimental evaluation of preprocessing algorithms for constraint satisfaction problems. Artificial Intelligence. **68** (1994) 211-241.
8. Focacci, F., Lodi, A., Milano, M.: Cost-Based Domain Filtering, CP (1999) 189-203.
9. Focacci, F., Lodi A., Milano, M.: Optimization-Oriented Global Constraints. Constraints **7(3-4)** (2002) 351-365.
10. Frost, D., Dechter, R.: In search of the best constraint satisfaction search. In Proceedings AAAI'94, Seatlle WA. (1994) 301-306.
11. Geoffrion, A.M.: The Lagrangean Relaxation for Integer Programming. Mathematical Programming. **2** (1974) 82-114.
12. Hooker, J.N., Osorio, M.A.: Mixed Logical/Linear Programming. Discrete Applied Mathematics. **96-97(1-3)** (1999) 395-442.
13. Hooker, J. N., Ottosson, G., E. S. Thornsteinsson, Hak-Jin Kim,: A scheme for unifying optimization and constraint satisfaction methods, Knowledge Engineering Review **15** (2000) 11-30.
14. Koster, A.M.C.:Frequency Assignment Problem, Models and Algorithms. Proefschrift Universiteit Maastricht. (1999).
15. Mackworth, A.: Consistency in networks of relations. Artificial Intelligence. **8** (1977) 99-118.
16. Maculan, N., Passini, M.M., Brito, J.A.M., Loiseau, I.: Column-Generation in Integer Linear Programming. RAIRO - Operations Research. **37** (2003) 67-83.
17. Milano, M., Ottosson, G., Refalo, P., Erlendur, S.: The Benefits of Global Constraints for the Integration of Constraint Programming and Integer Programming. Thorsteinsson, Proceedings of the Seventeenth National Conference on Artificial Intelligence, AAAI (2000).
18. Mohr, R., Henderson, T.: Arc and Path consistency revisited. Artificial Intelligence. **28** (1986) 225-233.
19. Sellmann, M., Theoretical Foundations of CP-based Lagrangian Relaxation, CP, Springer LNCS **3258**, (2004) 634-647.
20. Zhao, X., Luh, P.B.: New bundle methods for solving Lagrangian relaxation dual problems. Journal of Optimization Theory and Applications. **113(2)** (2002) 373-397.

Symmetry Breaking and Local Search Spaces

Steven Prestwich* and Andrea Roli[†]

* Cork Constraint Computation Centre, Department of Computer Science,
University College Cork, Ireland
s.prestwich@cs.ucc.ie
[†] Dipartimento di Scienze, Università degli Studi "G.D'Annunzio", Pescara, Italy
a.roli@unich.it

Abstract. The effects of combining search and modelling techniques can be complex and unpredictable, so guidelines are very important for the design and development of effective and robust solvers and models. A recently observed phenomenon is the negative effect of symmetry breaking constraints on local search performance. The reasons for this are poorly understood, and we attempt to shed light on the phenomenon by testing three conjectures: that the constraints create deep new local optima; that they can reduce the relative size of the basins of attraction of global optima; and that complex local search heuristics reduce their negative effects.

1 Introduction

Symmetry-breaking has proved to be very effective when combined with complete solvers [3, 16]. This can be explained by observing that symmetry-breaking constraints considerably reduce the search space. Nevertheless, the use of symmetry-breaking constraints (hereinafter referred to as SB constraints) seem to have the opposite effect on local search-based solvers, despite the search space reduction. In [12, 13] some examples of this phenomenon are reported. When the problem is modeled with SB constraints, the search cost[1] is higher than the one corresponding to the model with symmetries.

The reasons for this phenomenon are poorly understood. Improving our understanding may aid both the modelling process and the design of future local search algorithms; in an effort to achieve this, we pose and test some conjectures. In Sec.2, we test the conjecture that SB constraints create deep new local optima, by first providing a formal example and then by showing (indirectly) empirical evidence of this phenomenon. In Secs.3 and 4, we reinforce the conjecture by exhaustively analysing small instances. Furthermore, we extend the previous conjecture by investigating more complex characteristics of the search space and we show that, in most cases, SB constraints reduce global optima reachability. We also observe that these negative effects can be tempered by using complex search strategies.

---

[1] Measured as runtime or number of variable assignments.

R. Barták and M. Milano (Eds.): CPAIOR 2005, LNCS 3524, pp. 273–287, 2005.
© Springer-Verlag Berlin Heidelberg 2005

2   SB Constraints and Local Minima

SB constraints may generate new local minima, which might also have a negative effect on local search performance. We explain the intuition behind this idea using Boolean satisfiability (SAT). The SAT problem is to determine whether a Boolean expression has a set of satisfying truth assignments. The problems are usually expressed in *conjunctive normal form*: a conjunction of clauses $\bigwedge_i C_i$ where each clause C is a disjunction of literals $\bigvee l_j$ and each literal l is either a Boolean variable x or its negation $\bar{x}$. A Boolean variable can be assigned either T (true) or F (false). A satisfying assignment has at least one true literal in each clause. Now consider the following SAT problem:

$$\bar{a} \vee b \quad \bar{a} \vee c \quad a \vee \bar{b} \quad a \vee \bar{c}$$

which is the formula $(a \leftrightarrow b) \wedge (a \leftrightarrow c)$ in conjunctive normal form. There are two solutions: [a=T, b=T, c=T] and [a=F, b=F, c=F]. Suppose that a problem modeler realises that every solution to the problem has a symmetrical solution in which all truth values are negated. Then a simple way to break symmetry is to fix the value of any variable by adding a clause such as a to the model. Denote the first model by M and the model with symmetry breaking by M_s. Now suppose we apply a local search algorithm such as GSAT [19] to the problem. GSAT starts by making a random truth assignment to all variables, then *flipping* truth assignments (changing an assignment T to F or vice-versa) to try to reduce the number of violated clauses. In model M the state [a=F, b=F, c=F] is a solution, but in M_s the added clause a is violated. Moreover, any flip leads to a state in which *two* clauses are violated: flipping a to T removes the unit clause violation but causes the first two binary clauses to be violated; flipping b [c] to T preserves the unit clause violation and also violates the third [fourth] binary clause. In other words this state has been transformed from a solution to a local minimum. (The unit clause does not preclude this as a random first state, nor does it necessarily prevent a randomized local search algorithm from reaching this state.) In contrast M has no local minima: any non-solution state contains either two T or two F assignments, so a single flip leads to a solution (respectively TTT or FFF).

GSAT (and other algorithms) will actually escape this local minimum because it makes a "best" flip even when that flip increases the number of violations, but examples can be constructed with deeper local minima. The point is that a new local minimum has been created, and local minima degrade local search performance. We can also construct examples in which propagating the SB constraints through the model still leaves a model containing new local minima. Deep local minima require a greater level of *noise* (the probability of making a move that increases the objective function being minimized) in the search algorithm, so the creation of new local minima might be indirectly detected by analysing the performance as a function of noise. We look for this phenomenon in vertex colouring problems, encoded as SAT problems. SAT is a useful form because there are a variety of publicly-available local search algorithms.

2.1   Vertex Colouring as SAT

A graph $G = (V, E)$ consists of a set V of vertices and a set E of edges between vertices. Two vertices connected by an edge are said to be *adjacent*. The aim is to assign a colour to each vertex in such a way that no two adjacent vertices have the same colour. The task of finding a k-colouring for a given graph can be modeled as SAT by defining a Boolean variable for each vertex-colour combination, and adding clauses to ensure that (i) each vertex has at least one colour, (ii) no vertex has more than one colour, and (iii) no two adjacent vertices take the same colour. Colouring problems have a symmetry: the colours in any solution can be permuted. A simple but effective way of partially breaking this symmetry, used in an implementation of the DSATUR colouring algorithm [7], is to find a clique and assign a different colour to each of its vertices before search begins. Using a k-clique breaks $k!$ permutation symmetries. DSATUR uses a polynomial-time greedy algorithm to find a clique, preferring to spend time on colouring. We aim to test the effects of SB so we use a competitive clique algorithm described in [14].

We use the UBCSAT system [21] as a source of local search algorithms and experiment with three of them. Firstly the SKC (Selman-Kautz-Cohen) variant of the Walksat algorithm described in [18], which has become something of a standard algorithm in the SAT community. Its *random walk* approach was a significant advance over previous local search algorithms for SAT. However, advances in local search heuristics have been made since the invention of SKC. The Novelty and R-Novelty heuristics [6] use more sophisticated criteria for selecting variables to be flipped. These were later elaborated to the Novelty+ and R-Novelty+ variants [4] which have an extra noise parameter to avoid stagnation, and perform very well on many problems. The second algorithm we use is Novelty+. The third is SAPS (Scaling And Probabilistic Smoothing) [5] which incorporates further techniques: an efficient method for dynamically changing clause weights, an idea first used in [9] to escape local minima and since used in several algorithms; and subgradient optimization [17] inspired by Operations Research methods.

2.2   Experiments

We applied SKC to benchmark graphs from a recent graph colouring symposium. [2] On several graphs SKC moved more or less directly to a solution (using fewer steps than the problem has Boolean variables), which is quite surprising as these are considered to be non-trivial colouring problems. On such problems SB constraints often *improved* performance, counter to expectations. A typical example is shown in Fig. 1. Best performance is obtained with high noise, showing that a simple random walk algorithm finds the problem trivial. Adding SB constraints seems to give the algorithm a head start, transforming a trivial problem into an even more trivial one. Some other graphs gave similar results, including the mulsol.i.n graphs.

The timetabling graph school1 is nontrivial but not very hard for SKC, taking about twice as many flips as there are Boolean variables. But adding SB constraints makes the problem at least 4 orders of magnitude harder in most SKC runs, (though performance is better with frequent random restarts). Novelty+ also finds this problem extremely hard with SB constraints. SAPS finds the problem trivial without SB constraints, solving it in fewer flips than there are Boolean variables. With SB constraints it takes several times longer but is far more robust than SKC and Novelty+.

---

[2] http://mat.gsia.cmu.edu/COLOR04

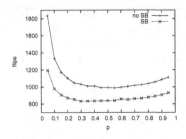

Fig. 1. SKC results for the anna graph

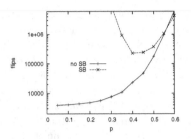

Fig. 2. SKC results for the flat graph

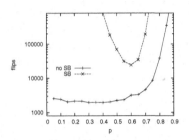

Fig. 3. Novelty+ results for the flat graph

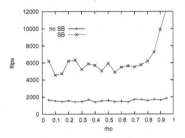

Fig. 4. SAPS results for the flat graph

It turns out that many colouring benchmarks are either too trivial to show the effects we are looking for (like anna) or too intractable to study in detail (like school1). We therefore created our own graphs using J. Culberson's graph generator. [3] We chose a flat graph, which is a randomly generated graph containing a hidden colouring. We chose an edge density of 0.5, flatness 0, 100 vertices, with a hidden 10-colouring, and found an 8-clique. The results are shown in Fig. 2 (each data point is the median over 100 runs) and confirm the effect we are looking for. Without SB constraints the problem is non-trivial, taking several times more flips than there are Boolean variables. Best results are obtained with low noise. With SB constraints the problem is about 2 orders of magnitude harder using optimal noise, which is higher than without SB constraints. In case this result is an artefact of SKC's heuristics we repeated the experiment with Novelty+, shown in Figure 3. This algorithm is often more efficient than SKC but the same pattern emerges. We believe that this indicates the increased ruggedness of the search space, caused by new local minima.

Next we tried SAPS on the same graph, which has 3 parameters besides the usual noise parameter p. SAPS performance is reported to be robust with respect to the default values of p and two of the other parameters, but less so for the smoothing parameter ρ. The greater the value of ρ the more rapidly recent history is forgotten, and the less likely the algorithm is to escape from a local minimum. Thus ρ can be viewed as a form of inverse noise parameter. Fig. 4 shows the results for the flat graph, this time varying ρ. Without SB constraints the performance is independent of ρ. With SB constraints

---

[3] http://web.cs.ualberta.ca/~joe/Coloring/index.html

performance is worse, especially for high ρ values, but it is much more robust than SKC or Novelty+. This is still consistent with our conjecture that SB constraints add new local minima. It also suggests that complex local search, with heuristics designed to escape local minima, are less affected by SB constraints — though they are still adversely affected.

The above examples show that SB constraints can have a huge impact on local search performance. These results are clearer and more extreme than those in [12, 13]. We conjectured that the cause is the creation of new, deep local minima, and our results support this conjecture but do not provide direct evidence. Ideally, we should analyse the search spaces of colouring problems, for example to count the number of local optima with and without SB constraints. Unfortunately, the search spaces are too large for an exhaustive enumeration, but in the following we analyse the search space of a hard optimization problem suitable for this kind of study. Before this, we formally define the search space explored by local search.

3   The Search Graph and Its Main Characteristics

The number of local minima can be taken as an indicator of the *ruggedness* of the search space explored by local search. In turn, it is usually recognized that the more rugged a search space is, the poorer is local search performance. Nevertheless, this parameterization of the search space might be sometimes insufficiently explicative or predictive. In this section, we provide a simple model of the search space which not only considers local and global optima, but also their *basins of attraction*, i.e., the set of states from which the optima can be reached.

The local search process can be viewed as an exploration of a landscape aimed at finding an optimal solution, or a *good* solution, i.e., a solution with a quality above a given threshold.[4] We define the search space explored by a local search algorithm as a *search graph*. The topological properties of such a graph are defined upon the neighborhood structure, that generate the *neighborhood graph*.

3.1   Neighborhood and Search Graphs

A *Neighborhood Graph* (NG), is defined by a triple: $\mathcal{L} = (S, \mathcal{N}, f)$, where:

- S is the set of feasible states;
- $\mathcal{N}$ is the neighborhood function $\mathcal{N} : S \rightarrow 2^S$ that defines the neighborhood structure, by assigning to every $s \in S$ a set of states $\mathcal{N}(s) \subseteq S$.
- f is the objective function $f: S \rightarrow \mathbb{R}^+$

The neighborhood graph can be interpreted as a graph (see Fig. 5) in which nodes are states (labeled with their objective value) and arcs represent the neighborhood relation

---

[4] For the rest of this paper, we will suppose, without loss of generality, that the goal of the search is to find an optimal solution. Indeed, the same conclusions we will draw can be extended to a set including also good solutions.

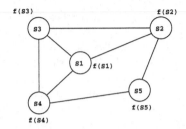

Fig. 5. Example of undirected graph representing a neighborhood graph (fitness landscape). Each node is associated with a solution s_i and its corresponding objective value $f(s_i)$. Arcs represent transition between states by means of φ. Undirected arcs correspond to symmetric neighborhood structure

between states. The neighborhood function $\mathcal{N}$ implicitly defines an *operator* φ which takes a state s_1 and transforms it into another state $s_2 \in \mathcal{N}(s_1)$. Conversely, given an operator φ, it is possible to define a neighborhood of a variable $s_1 \in S$:

$$\mathcal{N}_\varphi(s_1) = \{s_2 \in S \setminus \{s_1\} \mid s_2 \text{ can be obtained by one application of } \varphi \text{ on } s_1\}$$

In most cases, the operator is *symmetric*: if s_1 is a neighbor of s_2 then s_2 is a neighbor of s_1. In a graph representation (like the one depicted in Fig. 5) undirected arcs represent symmetric neighborhood structures. A desirable property of the neighborhood structure is to allow a path from every pair of nodes (i.e., the neighborhood is strongly optimally connected) or at least from any node to an optimum (i.e., the neighborhood is weakly optimally connected). Nevertheless, there are some exceptions of effective neighborhood structures which do not enjoy this property [10].

The exploration process of local search methods can be seen as the evolution in (discrete) time of a discrete dynamical system [1]. The algorithm starts from an initial state and describes a trajectory in the state space, that is defined by the neighborhood graph. The system dynamics depends on the strategy used; simple algorithms generate a trajectory composed of two parts: a *transient* phase followed by an *attractor* (a fixed point, a cycle or a complex attractor). Algorithms with advanced strategies generate more complex trajectories which can not be subdivided in those two phases.

It is useful to define the search as a walk on the neighborhood graph. In general, the choice of the next state is a function of the search history (the sequence of the previously visited states) and the iteration step. Formally: $s(t + 1) = \phi(\langle s(0), s(1), \ldots, s(t)\rangle, t)$ where the function ϕ is defined on the basis of the search strategy. ϕ could also depend on some parameters and can be either deterministic or stochastic.

For instance, let us consider a deterministic version of the Iterative Improvement local search. The trajectory starts from a point s_0, exhaustively explores its neighborhood, picks the neighboring state s' with minimal objective function value[5] and, if s' is better than s_0, it moves from s_0 to s'. Then this process is repeated, until a minimum $\hat{s}$ (either local or global) is found. The trajectory does not move further and we say that

---

[5] Ties are broken by enforcing a lexicographic order of states.

the system has reached a fixed point ($\hat{s}$). The set of points from which $\hat{s}$ can be reached is the basin of attraction of $\hat{s}$. Note that, especially in the case of deterministic local search algorithms, not every initial state is guaranteed to reach a global optimum.

Once we have introduced also the search strategy, the edges of the neighborhood graph can be oriented and labeled with transition probabilities (whenever it is possible to evaluate them). This will lead to the definition of concepts such as basins of attraction, state reachability and graph navigation. In the following, this resulting graph will be referred to as the *search graph*.

3.2   Basins of Attraction

The concept of *basin of attraction* (BOA) has been introduced in the context of dynamical systems, in which it is defined referring to an *attractor*. Concerning our model of local search, we will use the concept of basin of attraction of any node of the search graph. For the purposes of this paper we only consider the case of deterministic systems, even if it is possible to extend the definition to stochastic ones.[6]

Definition Given a deterministic algorithm $\mathcal{A}$, the basin of attraction $\mathcal{B}(\mathcal{A}|s)$ of a point s, is defined as the set of states that, taken as initial states, give origin to trajectories that include point s. The cardinality of a basin of attraction represents its size (in this context, we always deal with finite spaces).

Given the set S_{opt} of the global optima, the union of the BOA of global optima $I_{opt} = \bigcup_{i \in S_{opt}} \mathcal{B}(\mathcal{A}|i)$ represents the set of desirable initial states of the search. Indeed, a search starting from $s \in I_{opt}$ will eventually find an optimal solution. Since it is usually not possible to construct an initial solution that is guaranteed to be in I_{opt}, the ratio $rGBOA = |I_{opt}|/|S|$ can be taken as an indicator of the probability to find an optimal solution. In the extreme case, if we start from a random solution, the probability of finding a global optimum is exactly $|I_{opt}|/|S|$. Therefore, the higher this ratio, the higher the probability of success of the algorithm.

Given a local search algorithm $\mathcal{A}$, the topology and structure of the search graph determine the effectiveness of $\mathcal{A}$. In particular, the reachability of optimal solutions is the key issue. Therefore, the characteristics of the BOA of optimal solutions are of dramatic importance.

4   SB Constraints and Basins of Attraction

We now analyse the search space of an optimization problem to examine whether the number of local minima is indeed increased. We also test another conjecture: that SB constraints harm local search by reducing $rGBOA$ defined on the basis of a simple iterative improvement local search. In fact, even the most complex local search algorithms incorporate a greedy heuristic which is the one that characterizes iterative improvement.

---

[6] In this case, not only probabilistic basins of attraction are of interest, but also the probability a given state can be reached. Some studies in this direction are subject of ongoing work.

Therefore, if SB constraints reduce *rGBOA*, then the more a local search is similar to iterative improvement, the more it should be affected by SB constraints. Furthermore, we should also observe that local search algorithms equipped with complex exploration strategies are less affected by SB constraints. In this work, we aim at experimentally verifying this conjecture. By relating local search performance with *rGBOA* we consider a more general case than only counting local optima.[7]

We test our conjectures on the LABS optimization problem, which is to find an assignment to binary variables such that an energy function defined upon them is minimized. Given n binary variables $x_1, \ldots, x_n$, which can assume a value in $\{-1, +1\}$, we define the k-th correlation coefficient of a complete variable assignment $s = \{(x_1 = d_1), \ldots, (x_n = d_n)\}$ with $d_i \in \{-1, +1\}, i = 1, \ldots, n$, as $C_k(s) = \sum_{i=1}^{n-k} x_i \, x_{i+k}$, $k = 1, \ldots, n-1$ and the total function to be minimized is $E(s) = \sum_{k=1}^{n-1} C_k^2(s)$.

4.1   Analysis of the Space

We exhaustively explored the search space of LABS, for n ranging from 6 up to 18.[8] As neighborhood function we chose the one defined upon unitary Hamming distance, that is the most used one for problems defined over binary variables. (We emphasize that this choice determines the fundamental topological properties of the search space.)

In the model with SB constraints, only a subset of symmetric solutions has been cut, by enforcing constraints on the three left-most and right-most variables [8]. In these experiments the SB constraints are *enforced* instead of used to modify the objective function. In the SAT examples SB constraints were handled like any other clauses: when they were violated they increased the objective function being minimized (the number of violated clauses). In our LABS experiments the SB constraints are never violated so the local search space is smaller. Our experiments are therefore also a test of whether the observed negative effects still occur when SB constraints are used to restrict the search space.

It is first interesting to study how the neighborhood graph changes upon the application of SB constraints. In M, the neighborhood graph induced by single variable flips is a hypercube in which each node is connected to n other nodes. This graph has a constant degree equal to n. The neighborhood graph associated to M_s is characterized by a node degree frequency that varies in a small range, around a mean value slightly smaller than n (see an example in Fig. 6). The topological characteristics of this graph are not affecting the search, since the reachability of nodes is not significantly perturbed. Therefore, we can exclude that SB constraints in LABS affect local search by perturbing the topological properties of the neighborhood graph.

We consider now the features of the search graph, which, in general, can be algorithm-dependent. (This is the case for basins of attraction, while local and global optima only depend on the objective function and the neighborhood.) The search space characteristics of interest are the number of feasible states, the number of global and local optima and the value *rGBOA*. These values are reported in Tab. 1. We first observe that the

---

[7] Indeed, *rGBOA* can be decreased even if the density of local optima is not increased.

[8] The size limit is due to the exhaustiveness of the analysis.

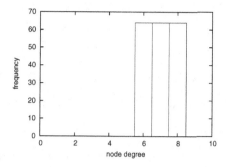

Fig. 6. Node degree frequency of the neighborhood graph in the case of the model with SB constraints ($n = 10$)

Table 1. Search space characteristics of LABS instances

n	feasible states		global optima		local optima		global BOA	
	no SB	with SB	no SB	with SB	no SB	with SB	no SB	with SB
6	64	12	28	5	0	0	1.0	1.0
7	128	24	4	1	24	5	0.40625	0.33333
8	256	48	16	3	8	2	0.86328	0.77083
9	512	96	24	4	84	16	0.42969	0.37500
10	1024	192	40	7	128	29	0.54590	0.45833
11	2048	384	4	1	240	52	0.03906	0.04427
12	4096	768	16	3	264	61	0.07544	0.06901
13	8192	1536	4	1	496	111	0.01831	0.01953
14	16384	3072	72	11	664	177	0.21240	0.15202
15	32768	6144	8	2	1384	326	0.01956	0.01742
16	65536	12288	32	8	1320	332	0.05037	0.04972
17	131072	24576	44	9	3092	721	0.05531	0.04073
18	262144	49152	16	2	5796	1372	0.02321	0.01068

search space reduction yielded by applying SB constraints is 5.33, independently of the instance size, while the global optima ratio is on average 5.19 (std.dev. is 1.22) and the local optima ratio is 4.36 (std.dev. is 0.40). Therefore, the number of local optima is reduced by a factor which is less than the search space reduction. An interesting perspective of the search space can be given by plotting the global (resp. local) optima density, i.e., the ratio of the number of global (resp. local) optima to the search space size. The density of optima is plotted in Figs. 7 and 8. These plots show that the density of local optima is always higher in the model M_s, proving that SB constraints increase the search space ruggedness. By enumerating the whole search space, we also observed that *new* local optima are created in M_s: some feasible states in M_s are local optima in M_s, but not in M. Note also that the density of global optima decreases exponentially with n. (The relation between number of global optima and n can be fitted with a good approximation by a line in a semi-logarithmic plot.) On the other side, the density of

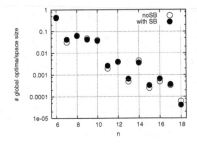

Fig. 7. Ratio of global optima w.r.t. search space size for $n = 6, \ldots, 18$, in a semi-log plot

Fig. 8. Ratio of local optima w.r.t. search space size for $n = 6, \ldots, 18$

local optima decreases much more slowly for the highest values of n. This provides an explanation for LABS being particularly difficult for local search.

Finally, we consider the basins of attraction of global optima. The basins of attraction are defined with respect to deterministic iterative improvement. We note that in all the cases, except for $n = 11$ and $n = 13$, $rGBOA(M_s) < rGBOA(M)$.

4.2 Experiments

We attacked LABS (with $n = 6, \ldots, 18$) with four different local search algorithms: Best improvement with randomly broken ties (BI), First improvement with random order among neighbors (FI), Simulated annealing (SA) and Tabu search (TS).[9] From the perspective of search space exploration, the algorithms chosen exhibit a varying explorative attitude, starting from the lowest of BI to the highest of TS, while all keeping a 'greedy' character.[10] We run each algorithm on the original model and on the model with SB constraints. The algorithms are stopped after $10n$ non-improving moves. This termination condition enables us to compare the algorithms on the basis of the best solution they returned once a steady state is reached. (In the literature of metaheuristics, this state is also commonly called *stagnation*.) Tab.2 gives a synoptic view of the algorithm performance in term of success ratio (out of 1000 runs).

A graphic comparison of the performance of each algorithm on the two problem models is given in Figs. 9, 10, 11, 12, in which we plotted the difference of solved instances (perc.) against n, i.e., $\Delta_\% = 100 \times (\text{solved(noSB)} - \text{solved(SB)})/1000$. Note that the performance on M dominates the one on M_s in all but the TS case.

The correlation between number of successes and $rGBOA$ is particularly interesting. From the plots in Figs. 13, 14, 15 and 16, we observe that for BI and FI the number of successes is proportional to the size of the global optima basin of attraction. In the case

---

[9] The initial temperature in SA has been set after a simple trial-and-test procedure. The tabu tenure in TS is randomly restarted each iteration in a range between 1 and $n/2$, in the spirit of robust tabu search [20].

[10] A deep discussion on this topic, involving also intensification and diversification, can be found in [2].

Table 2. Synopsis of the number of solved instances (out of 1000 runs) of the four local search algorithms on the original model and the model with SB constraints

n	BI		FI		SA		TS	
	no SB	with SB	no SB	with SB	no SB	with SB	no SB	with SB
6	1000	1000	1000	1000	1000	1000	1000	1000
7	417	324	530	484	1000	999	1000	1000
8	875	751	825	728	1000	960	1000	1000
9	438	342	266	267	913	834	1000	1000
10	561	426	995	720	996	920	1000	1000
11	42	41	47	30	101	136	528	928
12	77	63	39	37	308	318	835	895
13	16	17	3	5	105	111	283	251
14	200	125	202	142	857	731	992	1000
15	25	24	22	20	157	179	363	599
16	66	52	29	39	336	307	819	919
17	60	43	71	45	508	388	909	916
18	25	6	35	8	270	131	678	412

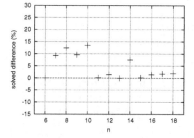

Fig. 9. Best improvement. Percentage of the difference between solved instances in the original model and solved instances in the model with SB constraints

Fig. 10. First improvement. Percentage of the difference between solved instances in the original model and solved instances in the model with SB constraints

of SA and TS, while the correlation is still observable, we note that the performance remains quite high even for low values of *rGBOA*, especially in the case of TS. The value of *rGBOA* have been measured on the basis of deterministic best improvement, therefore it is not surprising that both BI and FI show a proportional relation between successes and fraction of states that make the search converging to a global optimum. SA performs a more effective search space exploration than iterative improvement procedures, and even more TS, therefore the number of successes they achieve is much higher than that of BI and FI. It is interesting to note that the performance of both SA and TS starts to degrade (quite abruptly) when the normalized size of the global optima basin of attraction approaches a threshold value. Moreover, for TS this value is smaller than for SA, in other words, the more sophisticated an exploration strategy is, the lower the value of *rGBOA* at which the performance starts to be strongly affected.

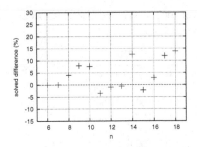

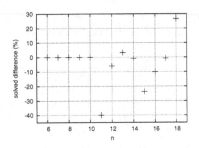

Fig. 11. Simulated annealing. Percentage of the difference between solved instances in the original model and solved instances in the model with SB constraints

Fig. 12. Tabu search. Percentage of the difference between solved instances in the original model and solved instances in the model with SB constraints

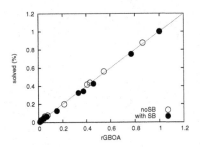

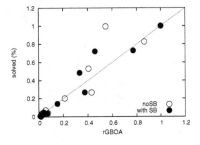

Fig. 13. Best improvement. Solved instances (perc.) plotted against the size of the global optima BOA

Fig. 14. First improvement. Solved instances (perc.) plotted against the size of the global optima BOA

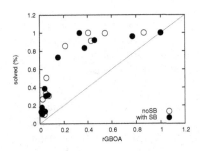

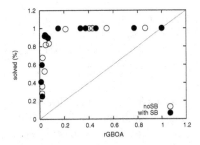

Fig. 15. Simulated annealing. Solved instances (perc.) plotted against the size of the global optima BOA

Fig. 16. Tabu search. Solved instances (perc.) plotted against the size of the global optima BOA

4.3   Discussion

The available data are still not sufficient to draw strong conclusions, but we have experimental results to support our conjectures. First of all, we have directly observed that new local optima are created and their density is higher in M_s. This is one of the reasons why *rGBOA* is reduced in the model with SB constraints. Another important observation is that local search performance is strongly affected by the size of the global optima basin of attraction. This relation is in the form of a positive correlation (i.e., the smaller the BOA, the lower the performance) and it is well approximated by a linear relation in the case of simple local search algorithms (BI and FI), while it is nonlinear in the case of more complex search strategies (SA and TS). The nonlinearity of this relation plays a big role when we compare the performance of local search algorithms (in terms of success ratio). In fact, large differences in *rGBOA* imply large deviations of the performance. But on the other side, when the difference is quite small, other factors come into play.

It is important to note that these effects occur even when SB constraints are used to *restrict* the search space. In the SAT experiments the search spaces with and without SB constraints were the same size: the SB constraints merely modified the objective function (number of violated constraints). It was therefore possible for the search to approach an excluded solution that has become a local minimum. But in the LABS experiments this was not possible: the search never strayed into the excluded subspaces, yet the SB constraints still had negative effects on local search performance. This is perhaps more surprising than the SAT results.

5   Conclusion

We showed, both empirically and by analysis, that SB constraints have two distinct negative effects on a local search space: they increase the relative size of local optima, and they reduce the relative size of global basins of attraction. These effects were observed using two different symmetry breaking methods: using SB constraints to modify an objective function, and using them to restrict the search space. The effects can be extremely strong, slowing down local search performance by several orders of magnitude. However, the effects are reduced by using more complex local search algorithms with heuristics for escaping local minima and with diversification strategies.

We believe that more research into both modelling and local search is necessary. Several lines of research are possible. Firstly, it has been suggested that, because SB constraints can have such a strong effect on local search, performance might be greatly improved by *adding* symmetry [13, 15]. We have so far found only one application of this idea (Golomb rulers [13]) but the technique seems promising. Secondly, there is a startling difference in performance between local search algorithms with and without mechanisms for escaping local minima, and our problems with SB constraints provide a useful test for such mechanisms. By focusing on such problems we might discover even better mechanisms, such as effective weighting procedures for SB constraints. (These mechanisms may include also the use of SB constraints in a TS fashion and the iterative relaxation of some of them during the search.) Thirdly, we still lack a general model (or, at least, general criteria) to both explain the phenomenon and guide the modelling phase. For example, it would be important to know if there are specific classes of SB constraints which induce a search space that is particularly unfavourable for exploration by local search. Fourthly, the effects another symmetry breaking technique have yet to be explored: *reformulation*, that is formulating a new model of the problem that does not contain the symmetry. Negative effects here would be even more surprising, and have important implications for problems modeling.

Finally, it should be noted that these negative effects are caused by symmetry *breaking*, but symmetry can be exploited in other ways to improve local search performance [11].

Acknowledgment

The authors thank the anonymous reviewers for useful comments and suggestions. This work has received support from Science Foundation Ireland under Grant 00/PI.1/C075.

References

1. Y. Bar–Yam. *Dynamics of Complex Systems*. Studies in nonlinearity. Addison–Wesley, 1997.
2. C. Blum and A. Roli. Metaheuristics in combinatorial optimization: Overview and conceptual comparison. *ACM Computing Surveys*, 35(3):268–308, September 2003.
3. James Crawford, Matthew L. Ginsberg, Eugene Luck, and Amitabha Roy. Symmetry-breaking predicates for search problems. In *KR'96: Principles of Knowledge Representation and Reasoning*, pages 148–159. Morgan Kaufmann, San Francisco, California, 1996.
4. H. H. Hoos. On the run-time behaviour of stochastic local search algorithms. In *Sixteenth National Conference on Artificial Intelligence*, pages 661–666. AAAI Press, 1999.
5. F. Hutter, D. A. D. Tompkins, and H. H. Hoos. Scaling and probabilistic smoothing: Efficient dynamic local search for sat. In *Eighth International Conference on Principles and Practice of Constraint Programming*, volume 2470 of *Lecture Notes in Computer Science*, pages 233–248. Springer, 2002.
6. D. A. McAllester, B. Selman, and H. A. Kautz. Evidence for invariants in local search. In *Fourteenth National Conference on Artificial Intelligence and Ninth Innovative Applications of Artificial Intelligence Conference*, pages 321–326. AAAI Press / MIT Press, 1997.
7. A. Mehrotra and M. A. Trick. A column generation approach to graph colouring. *INFORMS Journal on Computing*, 8:344–354, 1996.
8. S. Mertens. Exhaustive search for low-autocorrelation binary sequeneces. *J. Phys. A*, 29:L473–L481, 1996.
9. P. Morris. The breakout method for escaping from local minima. In *Eleventh National Conference on Artificial Intelligence*, pages 40–45. AAAI Press / MIT Press, 1993.
10. E. Nowicki and C. Smutnicki. A fast taboo search algorithm for the job-shop problem. *Management Science*, 42(2):797–813, 1996.
11. A. Petcu and B. Faltings. Applying interchangeability techniques to the distributed breakout algorithm. In *Proceedings of the International Joint Conference on Artificial Intelligence*, 2003.
12. S. Prestwich. First-solution search with symmetry breaking and implied constraints. In *Proc. of CP'01 Workshop on Modelling and Problem Formulation*, 2001.
13. S. Prestwich. Supersymmetric modeling for local search. In *Proc. of SymCon'02 Workshop on Symmetry and Constraint Satisfaction Problems*, 2002.
14. S. D. Prestwich. Combining the scalability of local search with the pruning techniques of systematic search. *Annals of Operations Research*, 115:51–72, 2002.
15. S. D. Prestwich. Negative effects of modeling techniques on search performance. *Annals of Operations Research*, 118:137–150, 2003.

16. J. F. Puget. Symmetry breaking revisited. In *Principle and Practice of Constraint Programming – CP02*, volume 2470 of *Lecture Notes in Computer Science*, pages 446–451. Springer-Verlag, 2002.

17. D. Schuurmans, F. Southey, and R. C. Holte. The exponentiated subgradient algorithm for heuristic boolean programming. In *Seventeenth International Joint Conference on Artificial Intelligence*, pages 334–341. Morgan Kaufmann Publishers, 2001.

18. B. Selman, H. A. Kautz, and B. Cohen. Noise strategies for improving local search. In *Twelfth National Conference on Artificial Intelligence*, pages 337–343. AAAI Press, 1994.

19. B. Selman, H. Levesque, and D. Mitchell. A new method for solving hard satisfiability problems. In *Proceedings of the Tenth National Conference on Artificial Intelligence*, pages 440–446. MIT Press, 1992.

20. E. Taillard. Robust taboo search for the quadratic assignment problem. *Parallel Computing*, 17:443–455, 1991.

21. D. A. D. Tompkins and H. H. Hoos. Ubcsat: An implementation and experimentation environment for sls algorithms for sat and max-sat. In *Seventh International Conference on Theory and Applications of Satisfiability Testing*, pages 37–46, 2004.

Combination of Among and Cardinality Constraints

Jean-Charles Régin

Computing and Information Science, Cornell University,
Ithaca NY 14850 USA
jcregin@cs.cornell.edu

Abstract. A cardinality constraint imposes that each value of a set V must be taken a certain number of times by a set of variables X, whereas an among constraint imposes that a certain number of variables of a set X must take a value in the set V.

This paper studies several combinations of among constraints and several conjunctions of among constraints and cardinality constraints. Some filtering algorithms are proposed and they are characterized when it is possible. Moreover, a weak form of Singleton arc consistency is considered. At last, it is shown how the global sequencing constraint and the global minimum distance constraint can be easily modeled by some conjunctions of cardinality and among constraints. Some results are also given for the global minimum distance constraint. They show that our study outperforms the existing constraint in ILOG Solver.

1   Introduction

Cardinality and among constraints are common to almost real-life problems. For instance, they are present in car sequencing (only some cars of a sequence can take a given option), radio frequency allocation problems (only one node of a pair of adjacent nodes can take a frequency in a set), rostering problems... The resolution of these applications can be improved if we are able to better combine these constraints.

This paper proposes to study in detail the among constraint and the combination of among constraints. We prove that the general problem of the combination of among constraints is NP-Complete. Thus, we propose to study some specific combinations which are tractable and for which we give a filtering algorithm establishing arc consistency.

Then, we consider the conjunction of cardinality constraints and among constraints and give arc consistency algorithms for two types of conjunction that are useful in practice. We also propose an original algorithm which can be viewed as a weak form of Singleton arc consistency.

At last, we will show how to model the global sequencing constraint and the global minimum distance constraint by some conjunctions of cardinality and among constraints. We also give some results for the global minimum distance constraint that outperform the existing constraint in ILOG Solver.

R. Barták and M. Milano (Eds.): CPAIOR 2005, LNCS 3524, pp. 288–303, 2005.

2   Preliminaries

2.1   Graph Theory

These definitions are based on books of [1], [2], and [3].

A **directed graph** or **digraph** $G = (X, U)$ consists of a **node set** X and an **arc set** U, where every arc (u, v) is an ordered pair of distinct nodes. We will denote by $X(G)$ the node set of G and by $U(G)$ the arc set of G. A **path** from node v_1 to node v_k in G is a list of nodes $[v_1, ..., v_k]$ such that (v_i, v_{i+1}) is an arc for $i \in [1..k-1]$. An undirected graph is **connected** if there is a path between every pair of nodes. The maximal connected subgraphs of G are its **connected components**. A directed graph is **strongly connected** if there is a path between every pair of nodes. The maximal strongly connected subgraphs of G are its strongly connected components.

Let G be a graph for which each arc (i, j) is associated with two integers l_{ij} and u_{ij}, respectively called the **lower bound capacity** and the **upper bound capacity** of the arc. A **flow** in G is a function f satisfying the following two conditions[1] :

- For any arc (i, j), f_{ij} represents the amount of some commodity that can "flow" through the arc. Such a flow is permitted only in the indicated direction of the arc, i.e., from i to j. For convenience, we assume $f_{ij} = 0$ if $(i, j) \notin U(G)$.
- A **conservation law** is observed at each node: $\forall j \in X(G) : \sum_i f_{ij} = \sum_k f_{jk}$.
  A **feasible flow** is a flow in G that satisfies the **capacity constraint**, that is, such that $\forall (i, j) \in U(G) \; l_{ij} \leq f_{ij} \leq u_{ij}$.

Definition 1. *The **residual graph** for a given flow f, denoted by $R(f)$, is the digraph with the same node set as in G. The arc set of $R(f)$ is defined as follows: $\forall (i, j) \in U(G)$:*

- $f_{ij} < u_{ij} \Leftrightarrow (i, j) \in U(R(f))$ *and upper bound capacity* $r_{ij} = u_{ij} - f_{ij}$.
- $f_{ij} > l_{ij} \Leftrightarrow (j, i) \in U(R(f))$ *and upper bound capacity* $r_{ji} = f_{ij} - l_{ij}$.

All the lower bound capacities are equal to 0.

2.2   Constraint Programming

A finite **constraint network** $\mathcal{N}$ is defined as a set of n **variables** $X = \{x_1, ..., x_n\}$, a set of current **domains** $\mathcal{D} = \{D(x_1), ..., D(x_n)\}$ where $D(x_i)$ is the finite set of possible **values** for variable x_i, and a set $\mathcal{C}$ of **constraints** between variables. $\mathcal{D}_0 = \{D_0(x_1), ..., D_0(x_n)\}$ to represent the set of

---

[1] Without loss of generality (see p.45 and p.297 in [3]), and to overcome notation difficulties, we will consider that if (i, j) is an arc of G then (j, i) is not an arc of G, and that all boundaries of capacities are nonnegative integers.

initial domains of $\mathcal{N}$. Indeed, we consider that any constraint network $\mathcal{N}$ can be associated with an initial domain $\mathcal{D}_0$ (containing $\mathcal{D}$), on which constraint definitions were stated.

A **constraint** C on the ordered set of variables $X(C) = (x_{i_1}, \ldots, x_{i_r})$ is a subset $T(C)$ of the Cartesian product $D_0(x_{i_1}) \times \cdots \times D_0(x_{i_r})$ that specifies the **allowed** combinations of values for the variables $x_1, \ldots, x_r$. An element of $D_0(x_1) \times \cdots \times D_0(x_r)$ is called a **tuple on** $X(C)$. $\tau[x]$ denotes the value of x in the tuple τ.

Let C be a constraint. A tuple τ on $X(C)$ is **valid** if $\forall x \in X(C), \tau[x] \in D(x)$. C is **consistent** iff there exists a tuple τ of $T(C)$ which is valid. A value $a \in D(x)$ is **consistent with** C iff $x \notin X(C)$ or there exists a valid tuple τ of $T(C)$ with $a = \tau[x]$. A constraint is **arc consistent** iff $\forall x_i \in X(C), D(x_i) \neq \varnothing$ and $\forall a \in D(x_i)$, a is consistent with C.

An instantiation of all variables that satisfies all the constraints is called a solution of a CN. Constraint Programming (CP) proposes to search for a solution by associating with each constraint a filtering algorithm that removes some values of variables that cannot belong to any solution. These filtering algorithms are repeatedly called until no new deduction can be made. Then, CP uses a search procedure (like a backtracking algorithm) where filtering algorithms are systematically applied when the domain of a variable is modified.

We will use the following notations:

- $\overline{x}$ (resp. $\underline{x}$) denotes the maximum (resp. minimum) value of $D(x)$.
- $D(X)$ denotes the union of domains of variables of X (i.e. $D(X) = \cup_{x_i \in X} D(x_i)$).
- $\#(a, \tau)$ is the number of occurrences of the value a in the tuple τ.
- $\#(a, X)$ is the number of variables of X such that $a \in D(x)$.

2.3   Element Constraint

The element constraint has been introduced in [4]. It defines a functional link between two variables. We propose a definition which is convenient for our purpose.

Definition 2. *Let f be a function[2] from a set S_1 to a set S_2. An **element constraint** C is a binary constraint defined on two variables x and y and associated with f and such that*
$T(C) = \{\tau \text{ s.t. } \tau \text{ is a tuple on } \{x, y\} \text{ and } \tau[y] = f(\tau[x])\}$
It is denoted by $element(y, f, x)$.

*We will say that a variable y is **created by an element constraint** element (y, f, x) if y is defined with a domain equal to $\cup_{a \in D(x)} f(a)$[3] and if the element constraint is added to the problem.*

---

[2] In mathematics, a function is a relation, such that each element of a set is associated with a unique element of another set (possibly the same).

[3] This means that the domain of x is not altered by the propagation after the definition of y and the addition of the element constraint.

Note that it is easy to maintain arc consistency for an element constraint because it is a functional constraint. This operation can be done in $O(d)$, where d is the size of the largest domain of x and y [5].

2.4   Cardinality Constraints

The Global Cardinality Constraints (GCC) has been proposed by [6]. It constraints the number of times every value is taken to be in an interval. In this initial definition, the intervals are statically given by their lower and upper bounds. Then, it has been proposed by [7] and [8] to deal with variables instead of intervals. This version is more convenient for our purpose:

Definition 3. *A* **global cardinality constraint involving cardinality variables** *defined on a set of variables X and a set of cardinality variables K and associated with a set of values V is a constraint C in which each value $a \in V$ is associated with the cardinality variable $K[a]$ and*
$T(C) = \{\tau \; s.t. \; \tau \text{ is a tuple on } X(C) \text{ and } \forall a \in V : K[a] = \#(a, \tau)\}$
It is denoted by $gcc(X, V, K)$.

This constraint has been called cardVar-GCC, but we think that there is no reason to differentiate it from a GCC because there is no ambiguity to differentiate the parameters, this is why we will use the same name.

A GCC C is consistent iff there is a flow in an directed graph $N(C)$ called the value network of C [6]:

Definition 4. *Given $C = gcc(X, V, K)$ a GCC; the* **value network** *of C is the directed bipartite graph $N(C)$ in which each arc is associated with a lower and an upper bound. The node set of $N(C)$ is defined by:*

- *the set of variables X called the variable set of $N(C)$;*
- *the set $D(X) \cup V$ called the value set of $N(C)$;*
- *a node s called the source and a node t called the sink.*

The arc set of $N(C)$ is defined as follows:

- *there is an arc from a variable x to a value a of $(D(X) \cup V)$ if and only if $a \in D(x)$. For every arc (x, a) we have $l_{xa} = 0$ and $u_{xa} = 1$;*
- *there is an arc from s to every variable $x \in X$. For every arc (s, x) we have $l_{sx} = u_{sx} = 1$.*
- *there is an arc from each value $a \in D(X) \cup V$ to the sink t. If $a \in V$ then $l_{at} = \underline{K}[a]$ and $u_{at} = \overline{K}[a]$ else $l_{at} = 0$ and $u_{at} = |X|$.*
- *there is an arc from t to s with $l_{ts} = u_{ts} = |X|$.*

Proposition 1. *[6] Let C be a GCC. Then,*

- *C is consistent if and only if there is a feasible flow in $N(C)$.*
- *Let f be a feasible flow in $N(C)$. A value a of a variable $x \in X$ is not consistent with C if and only if $f_{xa} = 0$ and a and x do not belong to the same strongly connected component in $R(f)$.*

The strongly connected component can be identified in $O(n + m)$ for a graph having n nodes and m arcs [9], thus arc consistency for the variables of X and for a GCC can be established with the same complexity. With the previous definition of $N(C)$ we have $n = |X| + |D(X) \cup V|$ and $m = (\sum_{x \in X} |D(x)|) + n + 1$. Note that we can merge all values of $D(X)$ that does not belong to V into a single value representing the fact that a variable can be assigned to a value which is not in V. This information can be easily maintained and in this case we have $m = (\sum_{x \in X} |D(x) \cap V| + 1) + n + 1$ which less than the previous value.

Arc consistency for the variables of K can be much more difficult to compute as shown by [8]:

Proposition 2. [8] *If the domain of each variable of K is a range of integers then arc consistency for the variables of K can be established in $O(nm + n^{2.66})$, else the problem is NP-Complete.*

However, a simple filtering algorithm based on constraints addition can be associated with the variables of K [7]:

Proposition 3. *Let $C = gcc(X, V, K)$ be a GCC, and f be any feasible flow in $N(C)$. Then, we have:*

- $\forall a_i \in V \ K[i] \leq \#(a_i, X)$
- $\sum_{a_i \in V} K[i] \leq |X|$ *and if $D(X) \subseteq V$ then $\sum_{a_i \in V} K[i] = |X|$*
- *for every connected component CC of $GV(X)$ we have:*
  if $vals(CC) \subseteq V$ then $\sum_{a_i \in vals(CC))} K[i] = |vars(CC)|$,
  else $\sum_{a_i \in vals(CC))} K[i] \leq |vars(CC)|$,
  where $vals(CC)$ denotes the values of V belonging to CC and $vars$ denotes the variables of X belonging to CC and $GV(X)$ is the value graph of X that is $GV(X) = (X, \cup_{x_i \in X} D(x_i), E)$ where $(x, a) \in E$ iff $a \in D(x)$.

Bound consistency of a sum constraint involving p variables can be established in $O(p)$. The strongly connected components of the residual graph are computed to establish arc consistency of the variables of X, thus bound consistency of the constraints of Proposition 3 can be implemented in $O(|K|)$.

3   The Among Constraint

Definition 5. *An **among constraint** defined on a set of variables X and a cardinality variable k and associated with a set of value V is a constraint C such that*
$$T(C) = \{\tau \ s.t. \ \tau \ is \ a \ tuple \ on \ X(C) \ and \ k = \sum_{a \in V} \#(a, \tau)\}$$
It is denoted by $among(X, V, k)$.

It is straightforward to design a filtering algorithm establishing arc consistency for this constraint. For instance, we can associate with each variable x of X a $(0,1)$ variable x_V defined as follows: $x_V = 1$ if and only if $x = a$ with $a \in V$. Then the constraint can be rewritten $\sum x_V = k$.

Definition 6. *Let $C_1 = among(X_1, V_1, k_1)$ and $C_2 = among(X_2, V_2, k_2)$ be two among constraints. If $X_1 \cap X_2 = \varnothing$ we will say that C_1 and C_2 are **variable disjoint**. If $V_1 \cap V_2 = \varnothing$ we will say that C_1 and C_2 are **value disjoint**.*

We propose to study whether it is possible to design some efficient filtering algorithms associated with a conjunction of among constraints. Three possible relations between among constraints:

1. the among constraints are variable disjoint.
2. the among constraints are value disjoint.
3. none of the previous property is satisfied.

Variable disjoint among constraints are totally independent and therefore it is trivial to study their conjunction.

3.1   Value Disjoint Among Constraints

Consider $\mathcal{A} = \{A_1, A_2, ..., A_n\}$ a set of n among constraints that are pairwise value disjoint where every A_i is equal to $among(X_i, V_i, k_i)$. This set of constraints can be efficiently combined by transforming the conjunction into another conjunction for which arc consistency can be efficiently established. This transformation requires to define new variables from the initial variables involved in the among constraints.

First, since all the V_i sets are disjoint, we can define the following function ndx:

Definition 7. *For any value a if there exists v_i such that $a \in V_i$, then $ndx(a) = i$, else $ndx(a) = -1$.*

Then, we associate every variable x_i involved in an among constraint with a more complex function denoted by f_i^{Ind}:

Definition 8. *Let Ind be the triplet $(\mathcal{A}, U, \alpha)$ where $\mathcal{A}$ is a set of value disjoint among constraints, each of them defined on a subset of X and associated with a subset of V; $U = \{u_1, u_2, ..., u_n\}$ is a set of pairwise distinct values with $U \cap (V \cup D(X)) = \varnothing$, and α is a value s.t. $\alpha \notin (U \cup V \cup D(X))$. For each variable $x_i \in X$ we define function f_i^{Ind} as follows: $\forall a \in D(x)$ with $k = ndx(a)$ if $k \neq -1$ and $x \in X_k$ then $f_i^{Ind}(a) = u_k$ else $f_i^{Ind}(a) = \alpha$.*

Now, for each variable $x_i \in X$ a variable y_i is created by the element constraint $element(y_i, f_i^{Ind}, x_i)$. Let Y be the set of these newly created variables.

Example. Consider seven variables $x_1, x_2, ..., x_7$, each having a domain equal to $[0..7]$ and 3 among constraints $A_1 = among(\{x_1, x_2, x_3, x_4\}, \{0,1\}, k_1)$, $A_2 = among(\{x_2, x_4, x_5, x_6\}, \{2,3\}, k_2)$, $A_3 = among(\{x_3, x_4, x_6, x_7\}, \{4,5\}, k_3)$. We have $\mathcal{A} = \{A_1, A_2, A_3\}$ and we define $Ind = (\mathcal{A}, \{u_1, u_2, u_3\}, \alpha)$. Then, we obtain $D(y_1) = \{u_1, \alpha\}$, $D(y_2) = \{u_1, u_2, \alpha\}$, $D(y_3) = \{u_1, u_3, \alpha\}$, $D(y_4) = \{u_1, u_2, u_3, \alpha\}$, $D(y_5) = \{u_2, \alpha\}$, $D(y_6) = \{u_2, u_3, \alpha\}$, $D(y_7) = \{u_3, \alpha\}$. For instance, we have $D(y_2) = \{u_1, u_2, \alpha\}$ because x_2 belongs to $X(A_1)$ and $X(A_2)$ and so the values of $\{0,1\}$ of x_2 correspond to the value u_1 of y_2 and the values of $\{2,3\}$ of x_2 correspond to the value u_2 of y_2 and the other values of x_2 are associated with the value α of y_2.

All the variables created by element constraints take their values from the set $\{u_1, u_2, ..., u_n\} \cup \{\alpha\}$, then by constraining the number of times these values are taken, we constrain at the same time the number of times any value of a set V_i is taken, and due to the definition of the variables created by element constraints we count only the variables of X_i that take a value in V_i.

The following proposition formally shows the link between a conjunction of among constraints and only one GCC:

Proposition 4. *The establishment of the arc consistency for the conjunction of value disjoints among constraints constraints $\{A_1, A_2, ..., A_n\}$ is equivalent to establishing arc consistency for the constraint network containing the element constraints $\{element(y_i, f_i^{Ind}, x_i), x_i \in X\}$ and the GCC: $gcc(Y, U, \{k_1, k_2, ..., k_n\})$.*

Proof. We can establish arc consistency for the conjunction of the constraints $\{element(y_i, f_i^{Ind}, x_i), x_i \in X\}$ and $gcc(Y, U, \{k_1, k_2, ..., k_n\})$ by establishing arc consistency of the constraint network (CN) consisting of these constraints, because the constraint graph associated with this constraint network is an hypergraph without any cycle and whose every pair of edges have at most one node (i.e. variable) in common. Thus, arc consistency for this CN is equivalent to arc consistency for the constraint equals to the conjunction of all the constraints in the network. Moreover, from any solution of the CN defined by $\{A_1, A_2, ..., A_n\}$ we can build a solution of the CN define by the element constraints and the GCC, and conversely. Therefore the proposition holds. $\square$

3.2   General Conjunction

Proposition 5. *Finding a tuple on the variables of X involved in among constraints is an NP-Complete problem in general.*

Proof. This problem is obviously in NP (easy polynomial certificate). We transform the NP-Complete problem TRIPARTITE MATCHING (see [10]) to this problem. TRIPARTITE MATCHING is:
Instance: Three sets B, G and H each containing n elements and a ternary relation $T \subseteq B \times G \times H$. *Question:* find a set of n triples in T, no two of which have a component in common.
We define a set X of n variables, each having a domain equal to $[1..|T|]$. For every pair $\{t_i, t_j\}$ of elements of T having a component in common we define the among constraint: $among(X, \{i, j\}, \{0, 1\})$. This constraint ensures that at most one of the element of $\{t_i, t_j\}$ can be assigned to a variable of X. This model exactly solves TRIPARTITE MATCHING. $\square$

However, it is possible to define some links between the cardinality variables of two among constraints.

Proposition 6. *Let $A_1 = among(X_1, V_1, k_1)$ and $A_2 = among(X_2, V_2, k_2)$ be two among constraints such that $X_1 \cap X_2 \neq \varnothing$ and $V_1 \cap V_2 \neq \varnothing$. Then, we have:*

$$k_1 = k_{(X_1 \cap X_2) \to (V_1 \cap V_2)} + k_{(X_1 \cap X_2) \to (V_1 - V_2)} + k_{(X_1 - X_2) \to V_1}$$
$$k_2 = k_{(X_1 \cap X_2) \to (V_1 \cap V_2)} + k_{(X_1 \cap X_2) \to (V_2 - V_1)} + k_{(X_2 - X_1) \to V_2}$$

where: $k_{Y \to W}$ is the number of times the values of W are taken by the variables of Y.

The proof of this proposition is straightforward. The sum constraints introduced by this proposition can be easily added to the constraint network and then the filtering algorithms associated them reduce the domain of the cardinality variables. This idea is more general and easier to understand than the algorithm proposed by [11] to combine sequences.

4   Integration of Some Among Constraints into Cardinality Constraints

We have seen that under some conditions it is possible to establish arc consistency for some conjunctions of among constraints. In this section we show that the same kind of result can be obtained by adding a GCC to some conjunctions of among constraints.

Definition 9. *Let X be a set of variables and V be a set of values.*

- *an X-among constraint is an among constraint defined on the set X of variables and on another variable q, that is of the form $among(X, W, q)$.*
- *a V-among constraint is an among constraint associated with the set of value V.*

4.1   Cardinality Constraint and Value Disjoint X-Among Constraints

We propose a filtering algorithm establishing arc consistency for the variables of X for a conjunction of a GCC and a set of value disjoint among constraints defined on the same set of variables X.

The efficient algorithm of the GCC is based on the flow theory and uses a specific network. The X-among constraints only introduce new constraints on the cardinality variables of the GCC. Since the among constraints are pairwise value disjoint there is no problem to take into account these new constraints: a slight modification of the value network is sufficient.

Definition 10. *Given $G = gcc(X, V, K)$ and $\mathcal{A} = \{A_1, A_2, ..., A_n\}$ a set of value disjoint among constraint such that $A_i = among(X, V_i, k_i)$; the **value network** of $C = (G \cup \mathcal{A})$ is the directed bipartite graph $N(C)$ obtained from $N(G)$ (the bipartite network associated with G), as follows:*
For each set of values V_i of an among constraint:

- *a new node w_i is defined*
- *for each value a in V_i, the arc (a, t) is replaced by the arc (a, w_i) which has the same lower and upper bounds as (a, t)*
- *an arc (w_i, t) with $l_{w_i t} = \underline{k}_i$ and $u_{w_i t} = \overline{k}_i$ is added.*

Then we immediately have a proposition similar to Prop.1:

Proposition 7. *Given* $G = gcc(X, V, K)$, $\mathcal{A} = \{A_1, A_2, ..., A_n\}$ *a set of value disjoint X-among constraints,* $C = G \cup \mathcal{A}$ *the conjunction of* G *and* $\mathcal{A}$, *and* $N(C)$ *be the bipartite value network associated with this conjunction. Then,*

- *C is consistent if and only if there is a feasible flow in $N(C)$.*
- *let f be a feasible flow in $N(C)$. A value a of a variable $x \in X$ is not consistent with C if and only if $f_{xa} = 0$ and a and x do not belong to the same strongly connected component in $R(f)$.*

The previous proposition is dedicated to the variables of X. We can obtain a filtering algorithm for the cardinality variables by adding for each among constraint $A_i = among(X, V_i, k_i)$ the constraint $\sum_{a \in V_i} K[a] = k_i$.

Note that it is possible to take into account some among constraints defined on a superset Y of X. In this case, we can transform the problem into an equivalent one for which all constraints are defined on the very same set. This transformation uses function f_i^{Ind} (See Def.8.) For every variable y_i of $Y - X$, a variable z_i is created by the element constraint $element(z_i, f_i^{Ind}, y_i)$. Let Z be the set of newly created variables. Then, the initial GCC is replaced by the $gcc(X \cup Z, V, K)$ and each among constraint A_i is replaced by the constraint $among(X \cup Z, (V_i \cap V) \cup \{u_i\}, k_i)$.

4.2   Cardinality Constraint and Variable Disjoint V-Among Constraints

We propose a filtering algorithm establishing arc consistency for the variables of X for a conjunction of a GCC and a set of variable disjoint among constraints associated with the same set of values V. We will assume that the GCC is also associated with V and that the V-among constraints are defined on subset of variables of X.

This conjunction of constraints can be efficiently taken into account by transforming it into another conjunction for which arc consistency can be efficiently established. This transformation has been proposed by [11] and requires the definition of new variables called abstract variables. In this section, we give a simpler version of this transformation.

Definition 11. *Let Red be the pair* $(\mathcal{A}, U)$ *where* $\mathcal{A}$ *is a set of n variable disjoint among constraints, each of them defined on a subset of X and associated with V; $U = \{u_1, u_2, ..., u_n\}$ is a set of pairwise distinct values with $U \cap (V \cup D(X)) = \emptyset$. For each variable $x_i \in X$ we define function f_i^{Red} as follows:* $\forall a \in D(x_i)$ *if* $a \in V$ *then* $f_i^{Red}(a) = a$ *else* $f_i^{Red}(a) = u_k$, *where k is the index of the among constraint involving x_i.*

Now, for each variable $x_i \in X$ a variable y_i is created by the element constraint $element(y_i, f_i^{Red}, x_i)$. Let Y be the set of these newly created variables. The following proposition is a reformulation of the proposition given in [11]:

Proposition 8. *Given* $G = gcc(X, V, K)$ *and* $\mathcal{A} = \{A_1, A_2, ..., A_n\}$ *a set of n V-among constraints that are pairwise variable disjoint. For each among constraint $A_i = among(X_i, V, k_i)$, we define the variable $K_A[u_i] = |X_i| - k_i$. In* *addition, we denote by $\overline{X}$ the set of variables of X that do not belong to any X_i. Then, we have:*

The establishment of the arc consistency for the conjunction of variable disjoint V-among constraints $\{A_1, A_2, ..., A_n\}$ and a $gcc(X, V, K)$ is equivalent to establishing arc consistency for the constraint network containing the element constraints $\{element(y_i, f_i^{Red}, x_i), x_i \in X\}$ and the GCC: $gcc(\overline{X} \cup Y, V \cup U, K \cup K_A)$.

It is possible to take into account some among constraints defined on $Y \supseteq X$ by applying a transformation similar to the one of the previous section. We only need to change the definition of function f_i^{Ind} as follows: $\forall x_i \in \cup_{i=1..n} X_i, \forall a \in D(x_i)$ if $a \notin V$ then $f_i^{Ind}(a) = \alpha$ else $f_i^{Ind}(a) = u_k$ where k is the index of the among constraint containing x_i.

5   Stronger Filtering Algorithm

In this section we propose an efficient algorithm to study some of the consequences of the instantiation of a variable for a GCC combined with some among constraints. In general, the conjunction of some among constraint is an NP-Complete problem. However, we have shown in the previous sections that under some conditions, we can efficiently establish arc consistency for the conjunction of a GCC and some among constraints.

Therefore, in practice, a set of GCCs and a set of among constraints will be modeled by a set of such conjunctions of constraints in addition to some among constraints and some GCCs. The conjunction of all constraints is managed by the propagation mechanism. It is sometimes worthwhile to try to deduce more information by using techniques like Singleton Arc Consistency, shaving or probing. The common idea of these methods is to instantiate some variables and to trigger the propagation mechanism after such an instantiation while expecting that a failure will occur. In this case, indeed, we know that the instantiation does not lead to any solution and so we can remove the value that was assigned from the domain of the selected variable. Unfortunately, these methods have a cost which is often too high in practice and prevent us from using them, at least if we consider all the possible instantiations. In fact, if we have n variables and d values in the domains of the variables, then nd instantiations will have to be considered and possibly several times because after a modification the constraint network has changed. Thus, some methods propose to consider only a subset of the variables and/or a subset of values and/or a subset of constraints.

In this section we will consider a problem containing $C = gcc(X, V, K)$ a GCC and some other constraints mainly dealing with the cardinality variables of the GCC. Our goal is to perform a stronger level of consistency, that is to prune more the domains of the variables of X, but we would like to avoid to be too much systematic. We aim to study the consequences for the cardinality variables of all possible instantiations of the variables of X.

A possible algorithm consists of successively trying all the possible instantiations of the variables of X and then to use the most powerful filtering algorithm associated with a GCC involving cardinality variables. However, this method requires to call nd times an algorithm in $O(nm + n^{2.66})$ (see Prop. 2) which certainly prevent us from using it.

Another possibility is to use a weaker filtering algorithm for the cardinality variables. For instance, the algorithm based on Proposition 3. The advantage of this algorithm is that it has the same complexity as the establishment of the arc consistency for the GCC (i.e. $O(m)$). Thus, we will have an new algorithm in $O(ndm)$ if we try each possible instantiation once.

In this section we present another algorithm with an $O(dm)$ time complexity, which is much more acceptable in practice.

The filtering algorithm of a GCC is based on the concepts of Hall variable set and Hall value set:

Definition 12. *Let $gcc(X, V, K)$ be a GCC.*

- *An **Hall variable set** is a set of variables $A \subseteq X$ such that*
$$|A| = \sum_{a \in D(A)} \overline{K}[a]$$
- *An **Hall value set** is a set of values $V \subseteq D(X)$ such that*
$$\sum_{v \in V} \underline{K}[v] = |vars(V)|, \text{ where } vars(V) \text{ is the set of variables having a value}$$
of V in their domain.

Proposition 9. *Let $C = gcc(X, V, K)$ be a GCC.*

- *if A is an Hall variable set then every value (x, a) with $x \in (X - A)$ and $a \in D(A)$ is not consistent with C.*
- *if V is an Hall value set then every value (x, a) with $x \in vars(V)$ and $a \notin V$ is not consistent with C.*
- *if a value (x, a) is not consistent with C then one of the previous property can prove that (x, a) is not consistent with C.*

The last property of the previous proposition proved that the application of the two first properties is sufficient to remove all the values that are not consistent with C. We have the straightforward proposition:

Proposition 10. *Let $C = gcc(X, V, K)$ be a GCC, f be a feasible flow in $N(C)$, A be an Hall variable set and V be an Hall value set.*

- (1.a) *$\forall a \in D(A)\ K[a] = \overline{K}[a]$.*
- (1.b) *the variables of A belong to strongly connected components of $R(f)$ that do not contain t.*
- (2.a) *$\forall v \in V\ K[v] = \underline{K}[v]$.*
- (2.b) *the variables of $vars(V)$ belong to strongly connected components of $R(f)$ that do not contain t.*

We propose to add to the problem some new among constraints or in some cases immediate deletions of values. For this purpose, we introduce the concept of pseudo Hall set:

Definition 13. • *A **pseudo Hall variable set** is a set of variables $A \subseteq X$ such that $|A| = (\sum\limits_{a \in D(A)} \overline{K}[a]) - 1$*

  • *A **pseudo Hall value set** is a set of values $V \subseteq D(X)$ such that $\sum\limits_{v \in V} \underline{K}[v] = |vars(V)| - 1$*

Then, suppose that A is a pseudo Hall variable set. If a variable of $X - A$ is instantiated with a value of $D(A)$ then the set A will become an Hall variable set and by Prop.10 the cardinality variables associated with the values of $D(A)$ can be set to their maximum value. So, the instantiation of one variable may imply the instantiation of some other variables that can have a huge impact on the problem. Similarly, suppose that V is a pseudo Hall value set. If a variable of $vars(V)$ is instantiated to a value which is not in V then V will become an Hall value set and by Prop.10 the cardinality variables associated with the values of V can be set to their minimum value. Once again, the instantiation of one variable may imply the instantiation of some other variables.

Pseudo Hall sets can be identified by removing values from the domains of some variables. Once a pseudo Hall set is identified we know that by instantiating some variables we can also instantiated some cardinality variables. Then, we propose to instantiate these cardinality variables and to trigger the propagation. If a failure occurs then we know that the creation of this Hall set is not possible and we introduce a constraint preventing its creation. More precisely, if we identify a pseudo Hall variable set A with $b \in D(A)$ and if the problem has no solution when b is taken by $y \notin A$ then we can introduce the constraint ensuring that at least one variable of A will take the value b. Similarly, if we identify V a pseudo Hall value set with a set of variables $Y \in vars(V)$ and if the problem has no solution when the variables of Y are instantiated to a value $b \notin V$ then we can introduce the constraint ensuring that at least one variable of Y must take a value of V.

In order to identify some pseudo Hall sets we propose to remove in turn each value. When a value is removed the variables instantiated to it are also removed. Therefore the current flow of the GCC is still a feasible flow in the new residual graph and we can establish arc consistency for the GCC in $O(m)$. This procedure will compute new strongly connected component in $R(f)$ and from Prop.10 we can identify some Hall sets from them. Then, we need to identify among these Hall sets which ones are pseudo Hall sets in the original GCC. This step is necessary because a newly created Hall sets can be independent from any pseudo Hall set. Note that it is useless to consider values belonging to the domain of variables of an Hall set.

Algorithm 1 contains an implementation of the procedure we have described and a specific procedure when the GCC is an alldiff constraint. In fact, this algorithm is complex in general, but it can be simplified when the GCC is an alldiff constraint (like for the allMinDistance constraint), because there is no Hall value set when all the lower bound capacities are equal to 0. In addition,

Algorithm 1:

```
StrongerFilteringAlgorithm(C, f)
    Let Scc(t) be the strongly connected component containing t in R(f)
    for each value a ∈ Scc(t) do
        Q(f) ← R(f)
        remove from Q(f) the value a and the set of variables Y = {y s.t. f_ya = 1}
        compute the strongly connected components of Q(f)
        for each strongly connected component S with t ∉ S do
            if vars(S) is a pseudo Hall variable set of C then
                instantiate the cardinality variables of S to their maximal value
                trigger the propagation
                if a failure occurs then
                    add the constraint among(vars(S), {a}, [1..|vars(S)|])

            if vals(S) is a pseudo Hall value set of C then
                instantiate the cardinality variables of S to their minimal value
                trigger the propagation
                if a failure occurs then
                    add the constraint among(Y, vals(S), [1..|Y|])

StrongerConsistencyWithAlldiff(C, f)
    Let Scc(t) be the strongly connected component containing t in R(f)
    for each value a ∈ Scc(t) do
        Q(f) ← R(f)
        remove from Q(f) the value a and the variable y with f_ya = 1
        compute the strongly connected components of Q(f)
        for each strongly connected component S with t ∉ S do
            if vars(S) is a pseudo Hall variable set of C then
                instantiate the cardinality variables of S to 1
                trigger the propagation
                if a failure occurs then
                    remove b from the domain of the variables of X − vars(S)
```

instead of adding an among constraint we can directly remove the value a from the domain of the variables that are not in the pseudo Hall variable set.

An example of this algorithm is presented in next section.

6   Application to the Global Minimum Distance Constraint

This constraint has been proposed by [12] and is mentioned in [13, 14, 15]. A global minimum distance constraint defined on X, a set of variables, states that for any pair of variables x and y of X the constraint $|x - y| \geq k$ must be satisfied.

Definition 14. *A **global minimum distance constraint** is a constraint C associated with an integer k such that*
$$T(C) = \{\tau \text{ s.t. } \tau \text{ is a tuple of } X(C) \text{ and } \forall a_i, a_j \in \tau : |a_i - a_j| \geq k\}$$

This constraint is present in frequency allocation problems.

A filtering algorithm has been proposed for this constraint [12]. Note that there is a strong relation between this constraint and the sequence constraint. A $1/q$ sequence constraint constrained two variables assigned to the same value to be separated by at least $q - 1$ variables, in regard to the variable ordering. Here we want to select the values taken by a set of variables such that are all pairs of values are at least k units apart.

This constraint is simply a conjunction of X-among constraints. For each value $a \in D(X)$ we define the among constraint $among(X, [a..a + k], [0, 1])$.

Then the global minimum distance constraint is equivalent to the conjunction of these X-among constraints. If we define a GCC C stating that each value of $D(X)$ has to be taken at most once by a variable of X (in other words an alldiff constraint) then we can model the global minimum constraint by using any several conjunctions of C and a set of value disjoint X-among constraints. Of course the model also uses the general conjunction of among constraints that we have proposed. The model will be equivalent to the global minimum distance constraint provided that each X-among constraint belongs to at least one conjunction. Note that it is possible to use in the conjunction less constrained among constraints for instance the constraint $among(X, [a..a + k], [0, 1])$ can be replaced by any among constraint $among(X, V, [0, 1])$ where $V \subseteq [a..a + k]$. This can be useful to ensure that every value of $D(X)$ is covered by an among constraint in the conjunction with C.

This model is powerful. For instance, consider a global minimum distance constraint involving 3 variables x, y and z with $D(x) = D(y) = \{1, 2, 3\}$, $D(z) = \{0, 1, 2, 3, 4, 5\}$ and a minimal distance equals to 2. The constructive disjunction will obtain the new domains: $D(x) = D(y) = \{1, 3\}$ and $D(z) = \{0, 1, 3, 4, 5\}$. This constraint is equivalent to the among constraints: $among(X, \{0, 1\}, \{0, 1\})$, $among(X, \{1, 2\}, \{0, 1\})$, $among(X, \{2, 3\}, \{0, 1\})$, $among(X, \{3, 4\}, \{0, 1\})$, $among(X, \{4, 5\}, \{0, 1\})$. Thus, we propose to model the constraint by the conjunction of an alldiff constraint on X and the among constraints: $among(X, \{0, 1\}, \{0, 1\})$, $among(X, \{2, 3\}, \{0, 1\})$, $among(X, \{4, 5\}, \{0, 1\})$, and by the conjunction of an alldiff constraint on X and the among constraints: $among(X, \{0\}, \{0, 1\})$, $among(X, \{1, 2\}, \{0, 1\})$, $among(X, \{3, 4\}, \{0, 1\})$, $among(X, \{5\}, \{0, 1\})$. The first conjunction will deduce that z cannot be assigned neither to $\{0, 1\}$ nor to $\{2, 3\}$. In addition, the propagation between the among constraints leads to $D(x) = D(y) = \{1, 3\}$, and $D(y) = \{5\}$.

Moreover, if the domains are $D(x) = \{0, 4\}$, $D(y) = \{1, 3\}$ and $D(z) = \{2, 3, 4, 5\}$, then only the stronger filtering algorithm we have presented is able to deduce that the only one solution is $x = 0$, $y = 3$ and $z = 5$, and so it outperforms the allMinDistance constraint of ILOG Solver.

We have also tested our algorithm on some problems for instance the Radio Link Frequency Allocation Problem. In order to build some global minimum distance constraints some cliques with a good distance value have been identified. Then, with the filtering algorithm that we have proposed we have seen dramatic improvement for some instances and mainly without specific strategies for selecting the next variable and the next value. For instance, Problem 11 is solved in 1s instead of 150s.

7   Application to the Global Sequencing Constraint

These constraints arise in many real-life problems such as car sequencing and rostering problems where a lot of min/max constraints have to be verified for each period of q consecutive time units. Sequencing constraints are useful for expressing regulations such as:

- each sequence of 7 days must contain at least 2 days off.
- A worker cannot work more than 3 night shifts every 8 days.

A global sequencing constraint is a gcc for which for each sequence S_i of q consecutive variables of X, the number of variables of S_i instantiated to any value $v_i \in V \subseteq D(C)$ must be in an interval $[min, max]$.

Definition 15. *[11] A* **global sequencing constraint** *is a constraint C associated with three positive integers min, max, q and a subset of values $V \subseteq D(C)$ in which each value $a_i \in D(C)$ is associated with two positive integers l_i and u_i and*
$$T(C) = \{ t \text{ such that } t \text{ is a tuple of } X(C)$$
$$\text{and } \forall a_i \in D(C) : l_i \leq \#(a_i, t) \leq u_i$$
$$\text{and for each sequence } S \text{ of } q \text{ consecutive}$$
$$\text{variables: } min \leq \textstyle\sum_{v_i \in V} \#(v_i, t, S) \leq max\}$$
It is noted $gsc(X(C), V, min, max, q, l, u)$, where $l = \{l_i, i = 1..|V|\}$ and $u = \{u_i, i = 1..|V|\}$.

This constraint is simply a conjunction of V-among constraints.
For each variable $x_i \in X$ we define the among constraint $among(\{x_i, ..., x_{i+q}\}, V, [min, max])$. Then the global sequencing constraint is equivalent to the conjunction of these X-among constraints and the GCC $C = gcc(X, V, K)$, where $K[i] = [l_i, u_i]$. We can model the global sequencing constraint by using several conjunctions of C with a set of variable disjoint V-among constraints. Of course the model also uses the general conjunction of among constraints that we have proposed. The model will be equivalent to the global sequencing constraint provided that each V-among constraint belongs to at least one conjunction. Note that it is possible to use in the conjunction less constrained among constraints for instance the constraint $among(\{x_i, ..., x_{i+q}\}, V, [min, max])$ can be replaced by any among constraint $among(Y, V, [min, max])$ where $Y \subseteq \{x_i, ..., x_{i+q}\}$. This can be useful to ensure that every variable of X is covered by an among constraint in the conjunction with C.

8   Conclusion

In this paper we have studied several combinations of among constraints and several conjunctions of among constraints and cardinality constraints. For each considered combination we have proposed an efficient filtering algorithm establishing arc consistency when it was possible. We have also shown that in general the combination of among constraints is an NP-Complete problem. In addition, we have proposed an original algorithm which can be viewed as a weak form of Singleton arc consistency. At last, we have proposed to model the global sequencing constraint and the global minimum distance constraint by conjunctions of cardinality and among constraints. We have also given some results for the global minimum distance constraint that outperform the existing constraint in ILOG Solver.

References

1. Berge, C.: Graphe et Hypergraphes. Dunod, Paris (1970)
2. Tarjan, R.: Data Structures and Network Algorithms. CBMS-NSF Regional Conference Series in Applied Mathematics (1983)
3. Ahuja, R., Magnanti, T., Orlin, J.: Network Flows. Prentice Hall (1993)
4. Van Hentenryck, P., Carillon, J.P.: Generality Versus Specificity: An Experience with AI and OR Techniques. In: Proceedings of AAAI-88. (1988)
5. Van Hentenryck, P., Deville, Y., Teng, C.: A generic arc-consistency algorithm and its specializations. Artificial Intelligence **57** (1992) 291–321
6. Régin, J.C.: Generalized arc consistency for global cardinality constraint. In: Proceedings AAAI-96, Portland, Oregon (1996) 209–215
7. Régin, J.C., Gomes, C.: The cardinality matrix constraint. In: CP'04, Toronto , Canada (2004) 572–587
8. Quimper, C.G., López-Ortiz, A., van Beek, P., Golynski, A.: Improved algorithms for the global cardinality constraint. In: Proceedings CP'04, Toronto, Canada (2004) 542–556
9. Tarjan, R.: Depth-first search and linear graph algorithms. SIAM Journal of Computing **1** (1972) 146–160
10. Papadimitriou, C.H.: Computational complexity. Addison Wesley (1994)
11. Régin, J.C., Puget, J.F.: A filtering algorithm for global sequencing constraints. In: CP97, proceedings Third International Conference on Principles and Practice of Constraint Programming. (1997) 32–46
12. Régin, J.C.: The global minimum distance constraint. Technical report, ILOG (1997)
13. ILOG: ILOG Solver 4.4 User's manual. ILOG S.A. (1999)
14. Régin, J.C.: Global Constraints and Filtering Algorithms. In: Constraints and Integer Programming combined. M. Milano ed, Kluwer (2003)
15. Régin, J.C.: Modélisation et Contraintes globales en programmation par contraintes. Habilitation à diriger des Recherches, Université de Nice-Sophia Antipolis (2004)

On the Tractability of Smooth Constraint Satisfaction Problems

T.K. Satish Kumar

Knowledge Systems Laboratory,
Stanford University,
tksk@ksl.stanford.edu

Abstract. We identify a property of constraints called *smoothness*, and present an extremely simple randomized algorithm for solving smooth constraints. The complexity of the algorithm is much less than the lower bound for establishing path-consistency, and because smoothness is shown to be identical to connected row-convexity (CRC) for the case of binary constraints, the time and space complexity of solving CRC constraints is improved. Central to our algorithm is the relationship of smooth constraints to random walks on directed graphs. We also provide simple deterministic algorithms to test for the smoothness of a given CSP under given domain orderings of the variables. Finally, we show that some other known tractable constraint languages, like the set of implicational constraints, and the set of binary integer linear constraints, are special cases of smooth constraints, and can therefore be solved much more efficiently than the traditional time and space complexities attached with them.

1   Introduction

While the task of solving constraint satisfaction problems (CSPs), in general, is NP-hard, much work has been done on identifying tractable subclasses. Broadly, these subclasses have resulted from restrictions imposed on: (1) the *topology* of the associated constraint network (see [3]), (2) the structure of the *constraints* themselves (see [10], [7], [5] and [1]), or (3) a combination of both (see [4]). While the notions of minimum induced-width, adaptive consistency and hypergraph acyclicity play a key role in characterizing the complexity of solving a given CSP by looking only at the topology of its associated constraint network (see [3]), row-convexity is an important property that, among others, has been identified in the context of exploiting the structure of the constraints themselves (see [10]). Row-convex constraints generalize other types of constraints like monotone and functional constraints (see [11]), and together with relational path-consistency, ensure the global consistency of a constraint network.

More specifically, if a binary constraint network is path-consistent, and all of the binary relations can be made row-convex by finding suitable domain orderings for the variables, then the network is globally consistent—enabling us to find a solution in a backtrack-free manner. Although row-convexity can be

R. Barták and M. Milano (Eds.): CPAIOR 2005, LNCS 3524, pp. 304–319, 2005.

tested and exploited in a given binary CSP, row-convex relations do not constitute a tractable language since further conditions are necessary to ensure that
the additional constraints resulting from enforcing path-consistency also remain
row-convex. Connected row-convexity (CRC) is a slightly different property of
binary constraints that, contrary to row-convexity, ensures the closure over composition, intersection and transposition, the basic operations of path-consistency
algorithms—hence making the language of CRC constraints tractable (see [5]).
Other tractable constraint languages include implicational constraints (see [2]),
max-closed constraints (see [7]), binary integer linear constraints, etc.

ALGORITHM: PATH-CONSISTENCY
INPUT: A binary constraint network
$\langle \mathcal{X}, \mathcal{D}, \mathcal{C} \rangle$.
OUTPUT: A path-consistent network.
(1) Repeat until no constraint is changed:

(a) For $k = 1, 2 \ldots N$:
  (i) For $i, j = 1, 2 \ldots N$:
    (A) $R_{ij} = R_{ij} \cap$
      $\Pi_{ij}(R_{ik} \bowtie D_k \bowtie R_{kj})$.
END ALGORITHM

Fig. 1. Shows the basic algorithm for enforcing *path-consistency* in a binary constraint
network. Here, Π indicates the *projection* operation, and $\bowtie$ indicates the *join* operation
(similar to that in database theory)

In this paper, we identify a property of constraints called *smoothness*, and
present an extremely simple randomized algorithm for solving smooth constraints.
The complexity of the algorithm is much less than the lower bound for establishing path-consistency, and because smoothness is shown to be identical to
CRC for the case of binary constraints, the time and space complexity of solving CRC constraints is improved. Central to our algorithm is the relationship
of smooth constraints to random walks on directed graphs. We also provide
simple deterministic algorithms to test for the smoothness of a given CSP under
given domain orderings of the variables. Finally, we show that some other known
tractable constraint languages, like the set of implicational constraints, and the
set of binary integer linear constraints, are special cases of smooth constraints,
and can therefore be solved much more efficiently than the traditional time and
space complexities attached with them.

2   Preliminaries and Background

In this section, we will set up some preliminary definitions, notation and other
background results that will be used (or alluded to) in the rest of the paper.

A CSP is defined by a triplet $\langle \mathcal{X}, \mathcal{D}, \mathcal{C} \rangle$, where $\mathcal{X} = \{X_1, X_2 \ldots X_N\}$ is a set
of *variables*, and $C = \{C_1, C_2 \ldots C_M\}$ is a set of *constraints* between subsets of
them. Each variable X_i is associated with a discrete-valued *domain* $D_i \in \mathcal{D}$, and
each constraint C_i is a pair $\langle S_i, R_i \rangle$ defined on a subset of variables $S_i \subseteq \mathcal{X}$,
called the *scope* of C_i. $R_i \subseteq D_{S_i}$ ($D_{S_i} = \times_{X_j \in S_i} D_j$) denotes all compatible

tuples of D_{S_i} allowed by the constraint. A *solution* to a CSP is an assignment of values to all the variables from their respective domains such that all the constraints are satisfied.

A network of binary constraints is *path-consistent* if and only if for all variables X_i, X_j and X_k, and for every instantiation of X_i and X_j that satisfies the direct relation R_{ij}, there exists an instantiation of X_k such that R_{ik} and R_{kj} are also satisfied. Conceptually, path-consistency enforcing algorithms work by iteratively "tightening" the binary constraints as shown in Figure 1. The best known algorithm that implements this procedure exploiting low-level consistency maintenance is presented in [9], and has a running time complexity of $O(N^3 K^3)$ (K is the size of the largest domain). This algorithm is optimal, since even verifying path-consistency has the same lower bound.

When binary relations are represented as matrices, path-consistency algorithms employ the three basic operations of composition, intersection and transposition. The (0,1)-matrix representation of a relation R_{ij} (denoted $M_{R_{ij}}$) between variables X_i and X_j consists of $|D_i|$ rows and $|D_j|$ columns when orderings on the domains of X_i and X_j are imposed. The '1's and '0's in the matrix respectively indicate *allowed* and *disallowed* tuples.[1]

A binary relation R_{ij} represented as a (0,1)-matrix, is *row-convex* if and only if, in each row, all of the '1's are consecutive. If there exists an ordering of the domains of $X_1, X_2 \ldots X_N$ in a path-consistent network of binary constraints, such that all the relations can be made row-convex, then the network is globally consistent. A globally consistent network has the property that a solution can be found in a backtrack-free manner. A (0,1)-matrix is *connected row-convex* if, after removing empty rows and columns, it is row-convex and *connected* (i.e. the positions of the '1's in any two consecutive rows intersect, or are consecutive). A binary relation R_{ij} constitutes a CRC constraint if both $M_{R_{ij}}$ and $M_{R_{ij}}^T$ are connected row-convex. Contrary to row-convex constraints, CRC constraints are closed under composition, intersection and transposition—hence establishing that path-consistency over CRC constraints is sufficient to ensure global consistency (see [5]). An instantiation of the generic path-consistency algorithm, that further exploits the structure of CRC constraints, has a running time complexity of $O(N^3 K^2)$ and a space complexity of $O(N^2 K)$ (see [5]).

Other tractable constraint languages include the set of implicational constraints, the set of max-closed constraints, the set of binary integer linear constraints, etc. In [8], an algebraic theory of *maximal* tractable constraint languages is developed, with the complexity of constraint languages being characterized in terms of solutions to *indicator* problems.

In this paper, we will revisit implicational constraints and binary integer linear constraints. The tractability of the former is reported in [2], and the latter is discussed as an example in [10]. An *implicational* relation, by definition, is either complete, a permutation, or a two-fan. For any $A \subseteq D_i$, $B \subseteq D_j$, $A \times B$ is called

---

[1] An extension of this representation mechanism to non-binary constraints is also straightforward.

complete; for any *bijection* $\pi : A \to B$, $\{\langle a, \pi(a) \rangle : a \in A\}$ is called a *permutation*; and for any values $x \in A$ and $y \in B$, $\{x \times B\} \cup \{A \times y\}$ is called a *two-fan*. Any CSP instance over implicational constraints can be solved by first establishing path-consistency, and then searching for a solution in a backtrack-free manner. A *binary integer linear* constraint is a linear constraint between 2 variables, each of whose domains are some finite subsets of the integers. Such constraints can be solved by enforcing path-consistency, and repeated "compaction" (removal of empty rows and columns) of the intermediate matrices to retain row-convexity.

3   Random Walks and Expected Arrival Times

In this section, we will provide a quick overview of random walks, and the theoretical properties attached with them. Figure 2(A) shows an undirected graph with weights on edges. A *random walk* on such a graph involves starting at a particular node, and at any stage, randomly moving to one of the neighboring positions of the current position. The probability with which we move to a specific neighbor of the current node is proportional to the *weight* on the edge that leads to that neighbor. One of the properties associated with such random walks on undirected graphs is that if we denote the expected time of arrival at some node (say L) starting at a particular node (say R) by $T(R, L)$, then $T(R, L) + T(L, R)$ is $O(m\mathcal{H}(L, R))$. Here, m is the number of edges, and $\mathcal{H}(L, R)$ is the "resistance" between L and R, when the weights on edges are interpreted as electrical resistance values (see [6]).

Figure 2(B) shows a particular case of the one in Figure 2(A), in which the nodes in the graph are connected in a linear fashion, and the edges are unweighted—i.e. the probabilities of moving to the left or to the right from a particular node are equal (except at the end-points). In this scenario, it is easy to note that by symmetry, $T(L, R) = T(R, L)$. Further, using the property of random walks stated above, if there are n nodes in the graph, then both $T(L, R)$ and $T(R, L)$ are $O(n^2)$.

Figure 2(C) shows a slightly modified version of that in Figure 2(B), where the graph is directed, although it is still linear. Moreover, there are weights

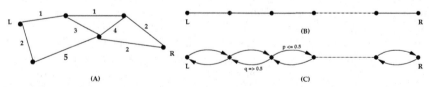

Fig. 2. Shows three scenarios in which random walks are performed. In an undirected graph (weighted as in (A), or unweighted as in (B)), for any two nodes L and R, $T(R, L) + T(L, R)$ is related to the "resistance" between them. In case (C) (when $p \leq q$ at every node), $T(R, L)$ is less than that in (B) because of an increased "attraction" towards L at every node

associated with edges which are interpreted as probabilities in the random walk; and the weight on $\langle s, s_{left}\rangle$ is, in general, not equal to that on $\langle s, s_{right}\rangle$. Here, s is some node in the graph, and s_{left} and s_{right} are respectively the nodes occurring immediately to the left and right of it. However, we are guaranteed that the probability of moving to the left at any node is greater than that of moving to the right (i.e. $p \leq q$). Given this scenario, it is easy to see that the expected time of arrival at the left end point (L), starting at the right end point (R), is also $O(n^2)$ (if there are n nodes in all). Informally, this is because at every node, there is an increased "attraction" to the left compared to that in Figure 2(B); and the expected arrival time can only be less than that in the latter.

4   The Tractability of Smooth Binary Constraints

In this section, we will define the notions of smooth binary constraints and smooth binary CSPs. We will generalize these notions to non-binary constraints in the next section.

Definition 1: Given an ordering on the domains of variables X_i and X_j—i.e. if D_i is ordered as $\langle d_{(i,1)}, d_{(i,2)} \ldots d_{(i,|D_i|)}\rangle$ and D_j as $\langle d_{(j,1)}, d_{(j,2)} \ldots d_{(j,|D_j|)}\rangle$, the *ordered distance* between two entries $\langle X_i = d_{(i,k_{i_1})}, X_j = d_{(j,k_{j_1})}\rangle$ and $\langle X_i = d_{(i,k_{i_2})}, X_j = d_{(j,k_{j_2})}\rangle$, in the (0,1)-matrix representing the binary constraint between X_i and X_j, is defined to be equivalent to the manhattan distance between the two entries—i.e. it is equal to $|\mathcal{RN}(d_{(i,k_{i_1})}, D_i) - \mathcal{RN}(d_{(i,k_{i_2})}, D_i)| + |\mathcal{RN}(d_{(j,k_{j_1})}, D_j) - \mathcal{RN}(d_{(j,k_{j_2})}, D_j)|$. Here, $\mathcal{RN}(d, D)$ denotes the rank of the element d in D (using the chosen ordering of elements in D).

Note that the ordered distance depends on the ordering chosen for the values in the domains of X_i and X_j, and could potentially be different for different orderings. Definition 4 gives a more general definition of the ordered distance between two complete assignments of values to all the variables.

Definition 2: A binary constraint between X_i and X_j (under an ordering of their domains) is said to be *smooth* if the following property is true of its 2D (0,1)-matrix representation: "At every '0', there exist two directions such that with respect to every other '1' in the matrix, moving along at least one of these directions (by 1 unit) decreases the ordered distance to it".

Figure 3 shows two examples to illustrate the notion of a smooth binary constraint. One of these is a smooth binary constraint, and the other is not. In case (A), every '0' is marked with two directions—indicating that no matter which '1' we have in mind (encircled in figure), moving along at least one of these directions reduces the ordered distance between the '0' and the '1'. In other words, if we randomly choose to move in one of these directions, we decrease the ordered distance between the '1' and the '0' by at least 1 with a probability at least 0.5, and increase it by at most 1 with a probability at most 0.5. In case (B), no such pair of directions exist for one of the '0's (marked with a cross), and it is therefore not smooth. Note that a binary constraint could be smooth under one ordering of the domains of X_i and X_j, but not so under another ordering.

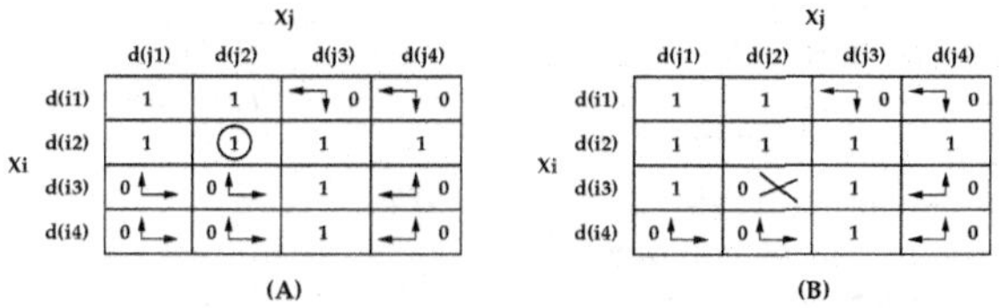

Fig. 3. Shows two examples to illustrate the notion of smooth constraints. (A) is a smooth constraint, and (B) is not

Definition 3: A *smooth binary CSP* is one in which all the constraints are smooth (under chosen domain orderings for all the variables).

We will now present algorithms for: (1) recognizing a smooth binary constraint, (2) finding the required pair of directions for each '0' in a smooth binary constraint, and (3) solving a smooth binary CSP efficiently (in randomized polynomial time).

4.1 Validating Smooth Binary Constraints

In this subsection, we will present a simple deterministic algorithm for checking whether a given binary constraint between variables X_i and X_j (under a chosen ordering of their domains) is smooth or not. We will assume that the domain values of X_i constitute the rows, and that the domain values of X_j constitute the columns. Figure 5 shows the procedure for verifying the smoothness of the given binary constraint, while Figure 6 acts as a running example. (Figure 4 is a subroutine used by the procedure in Figure 5.)

Since there are only $C_2^4 = 6$ pairs of directions possible for a binary constraint, the idea is to choose these pairs one at a time, and potentially eliminate their possibility for every '0'. Figure 4 ('ELIM-DRCTNS') shows the procedure for eliminating a specified pair of directions $\{e_1, e_2\}$ ($e_1, e_2 \in \{\mathcal{LF}, \mathcal{RT}, \mathcal{DN}, \mathcal{UP}\}$). (We will use $\mathcal{LF}$, $\mathcal{RT}$, $\mathcal{DN}$ and $\mathcal{UP}$ to respectively indicate the directions *left* (decreasing the rank of the value of X_j), *right* (increasing the rank of the value of X_j), *down* (increasing the rank of the value of X_i), and *up* (decreasing the rank of the value of X_i).) As an example, consider eliminating the directions $\{\mathcal{LF}, \mathcal{DN}\}$ as shown in Figure 6(F). The idea is to pick the right-most '1' in every row (see step (4) in Figure 4), and mark the infeasibility of the directions $\{\mathcal{LF}, \mathcal{DN}\}$ for every '0' that occurs in a rectangle (its edges inclusive) with its top-right edge rooted at that '1' (see steps (5) and (6) in Figure 4). This is because for any position in this rectangle, moving along both these directions increases the ordered distance to the '1'. The elimination of all other pairs of directions are symmetric to this case (except the more straightforward cases of (B) and (C) in Figure 6).

Lemma 1: Algorithms 'CHECK-SMOOTH' and 'ELIM-DRCTNS' are correct.

Proof: Let us first prove the correctness of 'ELIM-DRCTNS'. Given a certain pair of directions $\{e_1, e_2\}$, we first consider the case when e_1 and e_2 are opposite directions (steps (1) and (2) in Figure 4, and cases (B) and (C) respectively in

ALGORITHM: ELIM-DRCTNS

INPUT: A 2D (0,1)-matrix C representing a constraint over variables X_I (domain values constituting rows) and X_J (domain values constituting columns) with chosen domain orderings; and a pair of directions $e_1, e_2 \in \{\mathcal{LF}, \mathcal{RT}, \mathcal{DN}, \mathcal{UP}\}$.

RESULT: Eliminates the pair of directions $\{e_1, e_2\}$ for certain '0's (in accordance with the property required for smoothness).

(1) If $\{e_1, e_2\} = \{\mathcal{LF}, \mathcal{RT}\}$:
  (a) For $1 \leq j \leq |D_J|$:
    (A) If $\exists i$ s.t. $1 \leq i \leq |D_I|$ and $C[i][j] = 1$:
      (i) Eliminate $\{e_1, e_2\}$ for all '0's in that column.
(2) If $\{e_1, e_2\} = \{\mathcal{DN}, \mathcal{UP}\}$:
  (a) For $1 \leq i \leq |D_I|$:
    (A) If $\exists j$ s.t. $1 \leq j \leq |D_J|$ and $C[i][j] = 1$:
      (i) Eliminate $\{e_1, e_2\}$ for all '0's in that row.
(3) If neither (1) nor (2), let $e_1 \in \{\mathcal{LF}, \mathcal{RT}\}$ and $e_2 \in \{\mathcal{DN}, \mathcal{UP}\}$.
(4) For every row $C[i][\ldots]$ ($1 \leq i \leq |D_I|$):
  (a) Mark the right-most (left-most)

'1' if e_1 is $\mathcal{LF}$ ($\mathcal{RT}$).
  (b) Let the column index of this '1' be r_i.
(5) If $e_2 = \mathcal{UP}$:
  (a) Let $R_{|D_I|} = r_{|D_I|}$.
  (b) For $i = |D_I| - 1, |D_I| - 2 \ldots 1$:
    (A) Let $R_i = \max(\min)\{R_{i+1}, r_i\}$ if $e_1 = \mathcal{LF}$ ($\mathcal{RT}$).
    (B) For $1 \leq j \leq R_i$ ($R_i \leq j \leq |D_J|$) if $e_1 = \mathcal{LF}$ ($\mathcal{RT}$):
      (i) Eliminate the directions $\{e_1, e_2\}$ for $C[i][j]$ (if it is a '0').
(6) If $e_2 = \mathcal{DN}$:
  (a) Let $R_1 = r_1$.
  (b) For $i = 2, 3 \ldots |D_I|$:
    (A) Let $R_i = \max(\min)\{R_{i-1}, r_i\}$ if $e_1 = \mathcal{LF}$ ($\mathcal{RT}$).
    (B) For $1 \leq j \leq R_i$ ($R_i \leq j \leq |D_J|$) if $e_1 = \mathcal{LF}$ ($\mathcal{RT}$):
      (i) Eliminate the directions $\{e_1, e_2\}$ for $C[i][j]$ (if it is a '0').
(7) Eliminate the directions $\{e_1, e_2\}$ for $C[i][j]$ (if it is a '0') if:
  (a) $(e_1 = \mathcal{LF} \vee e_2 = \mathcal{LF}) \wedge j = 1$.
  (b) $(e_1 = \mathcal{RT} \vee e_2 = \mathcal{RT}) \wedge j = |D_J|$.
  (c) $(e_1 = \mathcal{DN} \vee e_2 = \mathcal{DN}) \wedge i = |D_I|$.
  (d) $(e_1 = \mathcal{UP} \vee e_2 = \mathcal{UP}) \wedge i = 1$.
END ALGORITHM

Fig. 4. A simple deterministic algorithm for potentially eliminating a given pair of directions $\{e_1, e_2\}$ at all possible '0's in a 2D matrix representing a binary constraint

ALGORITHM: CHECK-SMOOTH

INPUT: A 2D (0,1)-matrix C representing a constraint over the variables X_i (domain values constituting the rows) and X_j (domain values constituting the columns) with chosen domain orderings.

OUTPUT: 'yes' if the constraint is smooth, and 'no' otherwise.

(1) For every $e_1, e_2 \in \{\mathcal{LF}, \mathcal{RT}, \mathcal{DN}, \mathcal{UP}\}$ and $e_1 \neq e_2$:
  (A) ELIM-DRCTNS $(C, \{e_1, e_2\})$.
(2a) If there is a '0' with all pairs of directions eliminated:
  (A) RETURN: 'no'.
(2b) Else: RETURN: 'yes'.
END ALGORITHM

Fig. 5. A simple algorithm to check the smoothness of a given 2D (0,1)-matrix representing a binary constraint

Figure 6). Consider case (B) (case (C) is exactly symmetric), and consider a '0'. At least one of moving to the left or to the right decreases the ordered distance between the '0' and any other '1', except in the case when the '1' happens to be in the same column as the '0'. Thus, in step (1) of Figure 4, we eliminate all '0's that occur in non-empty columns. Now consider case (3), where one of the directions (say e_1) is always in $\{\mathcal{LF}, \mathcal{RT}\}$, and the other (say e_2) is always in $\{\mathcal{DN}, \mathcal{UP}\}$. For any '1' in the matrix, we should conceptually eliminate all '0's falling within or on the edges of a rectangle defined by that '1' as one of its corners—with the direction vectors indicating which of these corners it is. This is because for any position in this rectangle, moving along both these directions increases the ordered distance to the '1'. As an example, in Figure 6(D), for each '1', we have to eliminate all '0's falling within the rectangle defined with this '1' being its bottom-left corner. Computationally, however, we need not have to do this for all possible '1's because the rectangles defined by some of them are subsumed by the others (see Figure 7). It therefore suffices for us to consider the right-most or left-most '1' in every row, depending on whether e_1 is $\mathcal{LF}$ or $\mathcal{RT}$ respectively (see steps (4), (5) and (6) in Figure 4). Finally, step (7) in Figure 4 rules out all '0's that lie on the edges of the matrix, and for which moving along e_1 or e_2 is not possible. The correctness of 'CHECK-SMOOTH' then follows

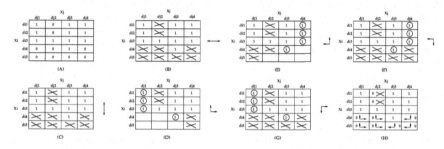

Fig. 6. Illustrates the working of the algorithms in Figure 4 and Figure 5. (A) shows the original constraint. (B)-(G) show the elimination of the 6 pairs of directions possible (each case annotated appropriately with the directions). (H) shows that there exists a '0' for which all 6 pairs of directions have been eliminated (shown by a cross)—because of which we can conclude that the constraint is not smooth

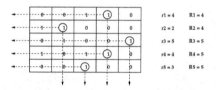

Fig. 7. Illustrates the elimination of the directions $\{\mathcal{LF}, \mathcal{DN}\}$ for a 2D matrix. The figure shows the overlapping nature of the rectangles defined for each '1'. It is this observation that allows us to look at every entry in the matrix only a constant number of times. In particular, the r_i and R_i values computed for every row (shown in figure) are employed to exploit this property (see Figure 4)

immediately from that of 'ELIM-DRCTNS' because there are only 6 pairs of directions possible for a binary constraint.

Lemma 2: The running time complexity of the procedure 'CHECK-SMOOTH' is $O(|D_i||D_j|)$.

Proof: Procedure 'CHECK-SMOOTH' makes 6 calls to the procedure 'ELIM-DRCTNS'. All the steps (cases) in the latter procedure examine every entry in the matrix at most twice. This is true even for steps (5) and (6) (see Figure 4 and Figure 7). Here, a single pass through all the entries determines all the left-most or right-most '1's in every row—from which the R_is can be easily computed. Further, in just one more pass through all the entries, all the qualifying '0's can be appropriately marked. Since there are $|D_i||D_j|$ entries in the matrix, the truth of the Lemma is established.

4.2   Solving Smooth Binary CSPs

Figure 8 provides a simple randomized algorithm for solving smooth binary CSPs. The idea is to start with an initial random assignment to all the variables from their respective domains, and use the violated constraints in every iteration to guide the search for the true assignment A^* (if it exists). In particular, in every iteration, a violated constraint is chosen, and the rank of the assignment of one of the participating variables is either increased or decreased. Since we know that the true assignment A^* satisfies all constraints, and therefore the chosen one too, randomly moving along one of the directions associated with the '0' corresponding to the current assignment A, will reduce the ordered distance to A^* with a probability ≥ 0.5. Much like the random walk in Figure 2(C), therefore, we can bound the convergence time to A^* by a quantity that is only quadratic in the maximum ordered distance between any two complete assignments.

Definition 4: Let $A_1 = \langle X_1 = d_{(1,i_1)}, X_2 = d_{(2,i_2)} \ldots X_N = d_{(N,i_N)} \rangle$ and $A_2 = \langle X_1 = d_{(1,j_1)}, X_2 = d_{(2,j_2)} \ldots X_N = d_{(N,j_N)} \rangle$ be two different complete assignments. Given orderings on the domain values of all the variables, the *ordered*

ALGORITHM: SOLVE-SMTH-BIN-CSP
INPUT: A smooth binary CSP over N variables $\{X_1, X_2 \ldots X_N\}$.
OUTPUT: A solution to the CSP.
  (1) Let the ordered domain of variable X_i be $D_i = \langle d_{(i,1)}, d_{(i,2)} \ldots d_{(i,|D_i|)} \rangle$.
  (2) Start with an initial random assignment I to all the variables.
  (3) While the current assignment A violates some constraint $C(X_i, X_j)$:
    (a) Let $d_{(i,k_1)}$ and $d_{(j,k_2)}$ be the current assignments to the variables X_i and X_j respectively.
    (b) Let $\{e_1, e_2\}$ be the direction pair associated with the entry $\langle X_i, X_j \rangle = \langle d_{(i,k_1)}, d_{(j,k_2)} \rangle$ in $C(X_i, X_j)$.
    (c) Choose p uniformly at random from $\{e_1, e_2\}$.
    (d) If $p = \mathcal{LF}$: set X_j to $d_{(j,k_2-1)}$.
    (e) If $p = \mathcal{RT}$: set X_j to $d_{(j,k_2+1)}$.
    (f) If $p = \mathcal{DN}$: set X_i to $d_{(i,k_1+1)}$.
    (g) If $p = \mathcal{UP}$: set X_i to $d_{(i,k_1-1)}$.
END ALGORITHM

Fig. 8. A simple randomized algorithm for solving smooth binary CSPs

distance between A_1 and A_2 is defined as $|\mathcal{RN}(d_{(1,i_1)}, D_1) - \mathcal{RN}(d_{(1,j_1)}, D_1)| + |\mathcal{RN}(d_{(2,i_2)}, D_2) - \mathcal{RN}(d_{(2,j_2)}, D_2)| \dots |\mathcal{RN}(d_{(N,i_N)}, D_N) - \mathcal{RN}(d_{(N,j_N)}, D_N)|$. Here, $\mathcal{RN}(d, D)$ denotes the rank of the element d in D (under the chosen ordering for D).

Lemma 3: The ordered distance between two complete assignments A_1 and A_2 is at most $|D_1| + |D_2| \dots |D_N|$, and 0 if and only if $A_1 = A_2$.

Proof: Consider the ordered distance $|\mathcal{RN}(d_{(1,i_1)}, D_1) - \mathcal{RN}(d_{(1,j_1)}, D_1)| + |\mathcal{RN}(d_{(2,i_2)}, D_2) - \mathcal{RN}(d_{(2,j_2)}, D_2)| \dots |\mathcal{RN}(d_{(N,i_N)}, D_N) - \mathcal{RN}(d_{(N,j_N)}, D_N)|$. Inside the ith term, the difference in the ranks of any two elements can be at most $|D_i|$. Hence, the ordered distance is always $\leq |D_1| + |D_2| \dots |D_N|$. Further, since each term can only be ≥ 0, the ordered distance can be 0 only when all the individual terms are 0—which happens only when A_1 and A_2 are identical.

Lemma 4: The expected number of iterations of 'SOLVE-SMTH-BIN-CSP' is only $O((|D_1| + |D_2| \dots |D_N|)^2)$.

Proof: From Lemma 3, the maximum ordered distance between the initial assignment I, and the true assignment A^*, is $|D_1| + |D_2| \dots |D_N|$. Further, in each iteration, we perform a random walk exactly analogous to that in Figure 2(C)— with the left end-point being A^*, I being only as far as the other end-point, and a maximum of $O(|D_1| + |D_2| \dots |D_N|)$ nodes in between. The truth of the Lemma then follows from the properties of random walks on directed graphs.

From Lemma 2 and Lemma 4, the expected running time of 'SOLVE-SMTH-BIN-CSP' is $O(MK^2 + MN^2K^2) = O(MN^2K^2)$. ($N$ is the number of variables, M is the number of constraints, and K is the size of the largest domain.) We note that although Lemma 4 is only an expected case analysis, Markov's inequality yields that the probability that we do not terminate even after k (say a constant 100) times the expected number of iterations is $\leq 1/k$ ($\leq 1/100$). (It is also possible to establish tighter bounds on the tail probabilities when restarts are employed.) We also note that like any Monte-Carlo algorithm, the above algorithm will find A^* (if it exists) with a very high probability, and will not report any solution if in fact none exist.

4.3   Reducing the Running Time Complexity

In the above analysis of the running time, we notice that the factor M arises due to the inner loop of the procedure SOLVE-SMTH-BIN-CSP (see step (3) in Figure 8), where we are required to repeatedly check for the presence of a violated constraint. In this subsection, we will show how we can reduce this factor significantly by employing appropriate data structures. We exploit the fact that it is sufficient for us to consider *any* violated constraint in every iteration. This is because every violated constraint is smooth, and gives us a chance to move closer to the solution with a probability ≥ 0.5.

Figure 9 presents a diagrammatic illustration of the required data structures. A series of doubly linked lists are maintained. The list '*All*' contains all the constraints, and for every variable X_i, a list L_i is maintained. L_i contains exactly

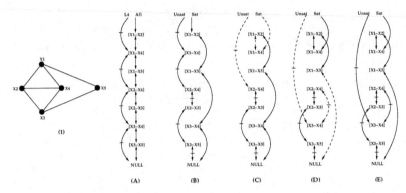

Fig. 9. Illustrates the data structures (and the operations performed on them) to reduce the running time of the randomized algorithm for solving smooth binary CSPs. (1) shows the constraint network of an example binary CSP on 5 variables. (A) shows two doubly linked lists, '*All*' and 'L_4'. The pointers in 'L_4' are distinguished from those of '*All*' by using a small horizontal mark on them. (B) shows the two doubly linked lists, '*Sat*' and '*Unsat*'. The pointers in '*Unsat*' are distinguished from those of '*Sat*' by using a small horizontal mark on them. (C) shows how the lists '*Sat*' and '*Unsat*' are updated when the first unsatisfied constraint in (B) (viz. $C(X_1, X_4)$)) is chosen, and the variable X_4 happens to be reassigned, possibly now satisfying $C(X_1, X_4)$. (D) shows what happens when $C(X_2, X_4)$ remains unsatisfied, but $C(X_3, X_4)$ changes from being satisfied to being unsatisfied. (E) shows the final lists of satisfied and unsatisfied constraints. Note that the ordering of the constraints is inconsequential in all the lists

those constraints that variable X_i participates in. Further, a list of satisfied constraints ('*Sat*'), and a list of unsatisfied constraints ('*Unsat*') are also maintained. These lists are updated incrementally in every iteration, and in the beginning, are built in accordance with the initial assignment I. Additions to '*Sat*' or '*Unsat*' are always made at the beginning of the lists, and therefore take constant time. Similarly, deletion from '*Sat*' or '*Unsat*' (given a pointer to the element to be deleted) takes constant time (because the lists are doubly linked, and deletion can be realized by linking together the neighbors of the element to be deleted).

In every iteration, the first constraint in '*Unsat*' is chosen, and the assignment of one of the two variables participating in it is changed. The only constraints that can be affected by this are the ones in which this variable appears. Walking through the corresponding list of all such constraints, we check each one of them for being satisfied or not. If a constraint under consideration was originally satisfied and is now unsatisfied (or vice-versa), we perform the appropriate addition and deletion operations on the '*Sat*' and '*Unsat*' lists. Both these operations can be done in constant time, and the complexity of the update procedure is therefore equal to the number of elements in the list (that contains all the constraints the chosen variable participates in).

If the maximum number of constraints any variable participates in is d (corresponds to the degree of the constraint network), the running time of the randomized algorithm for solving smooth constraints can now be reduced to $O(N^2 K^2 d)$.

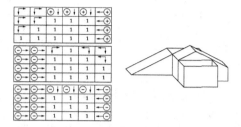

Fig. 10. Shows a 3-ary smooth constraint. The three 2D (0,1)-matrices on the left are respectively the three layers of a 3D (0,1)-matrix, with the top-most matrix corresponding to the bottom-most layer. For clarity, only the '1's are explicitly shown in the matrices. The required pair of directions is indicated against every other entry (which is implicitly a '0'). A circled '+' indicates the outward direction (out of the page), and a circled '-' indicates the inward direction (into the page). The right side of the figure shows the same constraint with all the '1's (its feasible region) appropriately enclosed

This is less than that of the deterministic algorithm for solving smooth binary constraints (see [5]). In the worst case too, d is only as large as N, and the running time is $O(N^3 K^2)$ (equaling that of the deterministic algorithm, but with a much lesser space complexity).

5   Generalization to Non-binary Constraints

Smoothness for general r-ary constraints is defined in a manner very similar to that for binary constraints. In the (0,1)-matrix representation of the constraint, the following property is required to hold: "At every '0', there exist two directions such that with respect to every other '1' in the matrix, moving along at least one of these directions decreases the ordered distance to it".

The only difference is that the number of dimensions in r-ary constraints is r, and checking for smoothness requires us to consider C_2^{2r} possible pairs of directions. Figure 10 shows an example of a 3-ary constraint that is smooth. We also note that convex constraints (like hyperplanes) are not necessarily smooth.

A *smooth CSP* is one in which all the constraints are smooth for chosen orderings on the domains of variables (irrespective of whether or not different constraints are of different arities). The procedure for solving a smooth CSP remains the same as in the binary case, and (as before) is related to random walks on directed graphs.

We note in passing, that the generalization of row-convexity (and its associated implications on tractability) to the non-binary case is rather complicated (see [10]). An r-ary relation R on a set $\mathcal{X}'$ of variables $\{X_1, X_2 \ldots X_r\}$ is *row-convex* if for any subset of $r - 2$ variables $\mathcal{Z} \subseteq \mathcal{X}'$, and for every instantiation z of the variables in $\mathcal{Z}$, the binary relation $\Pi_{(\mathcal{X}'-\mathcal{Z})}(\sigma_z(R))$ is row-convex (σ indicates the *selection* operation—i.e. only those entries consistent with $\mathcal{Z} = z$). If $\mathcal{R}$ is a network of relations whose arity is r or less, and if $\mathcal{R}$ is strongly $2(r - 1) + 1$ consistent, then if there exists an ordering of the domains

$D_1, D_2 \ldots D_N$ such that the relations are row-convex, the network is globally consistent. Enforcing strong $2(r-1)+1$ consistency, however, will not ensure global consistency because we may need to record constraints whose arity is greater than r.

6   Properties and Examples of Smooth CSPs

In this section, we will provide some illustrative examples of smooth CSPs. In particular, we will show that 2-SAT constraints, CRC constraints, implicational constraints and binary integer linear constraints are all smooth.

A 2-SAT instance consists of Boolean variables, and a set of constraints with every constraint relating at most 2 variables. It is easy to see that no matter how the '1's are placed in a 2×2 (0,1)-matrix (and irrespective of how many '1's there are), it will always be smooth—hence proving the tractability of 2-SAT.

In the case of CRC constraints, we argue that they are smooth binary constraints. Clearly, any empty row or column in a CRC constraint can be ignored because the required pair of directions for all '0's in an empty row (column) is just $\{\mathcal{UP}, \mathcal{DN}\}$ ($\{\mathcal{LF}, \mathcal{RT}\}$). Suppose that there were a '0' for which no such pair of directions existed. By the property of connected row-convexity, it must be the case that either the row or the column in which it lies does not have '1's appearing contiguously—hence leading to a contradiction. Therefore, such a pair of directions exists for every '0', and CRC constraints are smooth. Figure 11 shows an example to illustrate the equivalence between CRC constraints and smooth binary constraints. The foregoing arguments suggest an extremely simple randomized algorithm for solving CRC constraints—the time and space complexity of which is much less than that of the corresponding best known deterministic algorithm (see [5]). We also note that the randomized algorithm circumvents the use of complex data structures otherwise required for optimally implementing path-consistency subroutines.

Similarly, a simple way to solve implicational constraints is to first argue that the complete constraints and two-fans are smooth, irrespective of the orderings

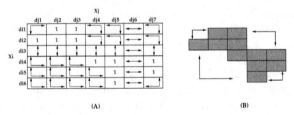

Fig. 11. Shows the equivalence between smooth binary constraints and CRC constraints. The left side of the figure shows an example of a CRC constraint with the required pair of directions marked against each '0' (not shown explicitly for clarity). The right side of the figure illustrates the general pattern of the required pair of directions for '0's in a CRC constraint

chosen for the domains of the variables. Consider complete constraints. By definition, every '0' is either in an empty row or in an empty column—making the required pair of directions to be $\{\mathcal{UP}, \mathcal{DN}\}$ or $\{\mathcal{LF}, \mathcal{RT}\}$ respectively. Now consider two-fans. Ignoring all the empty rows and columns, it is easy to see that the '1's fall along a '+' sign, dividing the '0's into 4 quadrants. The required directions for any '0' in any quadrant are then those that lead towards the "axes" ('1's). Figure 12 presents examples that illustrate the above arguments. Permutation constraints, however, are not necessarily smooth; but can be *dissolved* in a pre-processing step. The idea is to choose orderings on the domain values of all the variables so that all the permutation constraints begin to resemble identity matrices—barring empty rows and columns. Since two-fans and complete constraints remain so for any domain orderings of the participating variables, all the constraints can then be made simultaneously smooth.

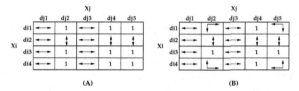

(A)   (B)

Fig. 12. Illustrates why implicational constraints are special cases of smooth constraints. (A) shows the case of a complete constraint, and (B) shows the case of a two-fan constraint. All the permutation constraints are assumed to be *dissolved*

Finally, consider an integer linear program, where the domains of the variables are finite subsets of the integers, and all the linear constraints are binary. That is, the constraints are of the form: $aX_i + bX_j \leq c$ (where a, b and c are integer constants). A network of binary integer linear constraints can be solved in polynomial time (although general integer linear programs are NP-hard to solve). It can be shown that each element in the closure under composition, intersection, and transposition of the resulting set of (0,1)-matrices is row-convex, provided that when an element is removed from a do-

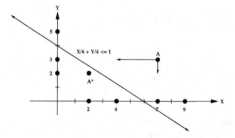

Fig. 13. Illustrates why binary integer linear constraints are smooth. A is the current assignment, and A^* is the true assignment (if it exists). The domains of variables X and Y are indicated using dark circles on the corresponding axes

main by arc-consistency, the associated (0,1)-matrices are "compacted"—i.e. all the empty rows and columns occurring in intermediate matrices (while establishing path-consistency) are removed (see [10]). A much more straightforward approach, however, is to notice that when the domains of the variables are ordered according to their values (i.e. when they are sorted), all the constraints turn out to be convex. Further, since all the constraints are binary, at every infeasible point, there always exist two directions, at least one of which decreases the ordered distance to the true solution (no matter where it is in the feasible region). Figure 13 shows a small example to illustrate this. Note, however, that the ordered distance is still measured using the *ranks* of the domain elements, instead of their absolute values. Similarly, the moves also employ the *ranks* instead of the absolute values, hence explaining why we would randomly choose to decrease X by 3 units, or decrease Y by only 1 unit in Figure 13.

7   Conclusions and Future Work

We identified a property of constraints called smoothness, and presented an extremely simple randomized algorithm for solving smooth constraints. The complexity of the algorithm was shown to be much less than the lower bound for establishing path-consistency (which is employed by most deterministic algorithms that establish the tractability of many different constraint languages). Central to our algorithm was the idea of exploiting the relationship between smooth constraints and random walks on directed graphs. We provided simple deterministic algorithms to test for the smoothness of a given CSP under given domain orderings of the variables. We also showed that some other known tractable constraint languages, like the set of implicational constraints, and the set of binary integer linear constraints, are special cases of smooth constraints, and can therefore be solved much more efficiently than the traditional space-time complexities attached with them. As part of our future work, we are interested in automatically finding domain orderings for the variables in a CSP (if one exists) so as to make it smooth (at least in certain special cases).

References

1. Bulatov A. A., Krokhin A. A. and Jeavons P. G. 2000. Constraint Satisfaction Problems and Finite Algebras. *ICALP'00*.
2. Cooper M. C., Cohen D. A. and Jeavons P. G. 1994. Characterizing Tractable Constraints. *Artificial Intelligence, 65:347-361*.
3. Dechter R. 1992. Constraint Networks. *Encyclopedia of Artificial Intelligence, second edition, Wiley and Sons, pp 276-285, 1992*.
4. Dechter R. and Pearl J. 1991. Directed Constraint Networks: A Relational Framework for Causal Modeling. *Proceedings of the Twelfth International Joint Conference on Artificial Intelligence. IJCAI'91*.

5. Deville Y., Barette O. and Van Hentenryck, P. 1999. Constraint Satisfaction over Connected Row-Convex Constraints. *Artificial Intelligence, 109:243-271.*
6. Doyle P. G. and Snell E. J. 1984. Random walks and Electrical Networks. *Carus Math. Monographs 22, Math. Assoc. Amer., Washington, D. C.*
7. Jeavons P. G., Cohen D. A. and Cooper M. 1998. Constraints, Consistency and Closure. *Artificial Intelligence, 101:251-265.*
8. Jeavons P. G. 1998. On the Algebraic Structure of Combinatorial Problems. *Theoretical Computer Science, 200:185-204.*
9. Mohr R. and Henderson T. C. 1986. Arc and Path Consistency Revisited. *Artificial Intelligence, 28:225-233.*
10. Van Beek P. and Dechter R. 1995. On the Minimality and Global Consistency of Row-Convex Constraint Networks. *Journal of the ACM (JACM) Archive Volume 42, Issue 3 (May 1995) Pages: 543 - 561.*
11. Van Hentenryck P. Deville Y. and Teng C. M. 1992. A Generic Arc-Consistency Algorithm and its Specializations. *Artificial Intelligence, 57:291-321.*

A SAT-Based Decision Procedure for Mixed Logical/Integer Linear Problems

Hossein M. Sheini and Karem A. Sakallah

University of Michigan
Dept. of Electrical Engineering and Computer Science, 1301
Beal Avenue, Ann Arbor, MI 48109-2122, USA
{hsheini, karem}@umich.edu

Abstract. In this paper, we present a method for solving Mixed Logi-cal/Integer Linear Programming (MLILP) problems that integrates a polynomial-time ILP solver for the special class of Unit-Two-Variable-Per-Inequality (unit TVPI or UTVPI) constraints of the form $ax + by \leq d$, where a, $b \in \{-1, 0, 1\}$, into generic Boolean SAT solvers. In our approach the linear constraints are viewed as special literals and replaced by binary "indicator" variables to generate a pure logical problem. The resulting problem is subsequently solved using a SAT search procedure which invokes the linear UTVPI solver to incremen-tally check the consistency of the UTVPI constraints whenever any of the indicator variables are assigned to true. The linear UTVPI solver, on the other hand, can possibly pass *implications* or *no-good con-straints* to the Boolean SAT solver. Checking the consistency of the UTVPI constraints incrementally enables the UTVPI solver to effi-ciently interact with the different components of the SAT solver. Ad-ditionally, several heuristics and encoding methods are proposed to ac-commodate the special circumstances of activating UTVPI constraints by the SAT solver. Empirical evidence is presented that demonstrates the advantages of our combined method for large problems.

1 Introduction

In the past several years, there have been numerous efforts aimed at solving problems of Mixed Logical/Integer Linear Programming (MLILP) or Mixed Integer Nonlinear Programming (MINLP) where continuous and discrete vari-ables and nonlinearities are involved in the objective function and constraints. In practice, these problems are applied in areas as diverse as process design and synthesis, planning and scheduling, process and batch control and, recently, bioinformatics. A complete collection of MINLP applications can be found in [15]. Several different approaches to solve these problems have been widely investigated in operations research and programming, among which are Branch-and-Bound (BB), Outer Approximation (OA), Generalized Benders' Decomposition (GBD), Extended Cutting Plane (ECP) and Branch-and-Cut methods. The reader is referred to [14] for a survey of these various schemes.

R. Barták and M. Milano (Eds.): CPAIOR 2005, LNCS 3524, pp. 320–335, 2005.

On the modeling front, Generalized Disjunctive Programming (GDP) [23] has been recently introduced and studied as an alternative to MINLP. In GDP disjunctions and logic propositions are used to represent discrete decisions in the continuous space and constraints in the discrete space, respectively. This model consists of algebraic constraints, logic disjunctions and logic relations and is generally formulated as follows:

$$\min \quad Z = \sum_k c_k + f(x)$$

$$\text{s.t.} \quad g(x) \leq 0$$

$$\vee_{i \in D_k} \begin{bmatrix} Y_{ik} \\ h_{ik}(x) \leq 0 \\ c_k = \gamma_{ik} \end{bmatrix} ; k \in SD \tag{1}$$

$$\Omega(Y) = True$$

$$x \in \mathbb{R}^n, c \in \mathbb{R}^m, Y \in \{True, False\}^m$$

where Y_{ik} is a Boolean variable that establishes whether constraint $h_{ik}(x) \leq 0$ is true and if so, the variable c_k will be activated to a value γ_{ik}. $\Omega(Y) = True$ corresponds to logic propositions expressed in Conjunctive Normal Form (CNF). Note that this model can always be reformulated back into general Mixed Integer Programming (MIP) by replacing the disjunctions with *Big-M* constraints.

In addition to Big-M conversion followed by standard MINLP methods, other techniques including Disjunctive Branch-and-Bound, logic-based OA and GBD have been proposed for solving GDP problems. An overview of these methods can be found in [16].

On the other hand, many applications in hardware and software verification as well as in artificial intelligence are cast as decision problems which are, basically, special cases of GDP in which both logical and integer/real variables are involved. The ever-increasing strength of SAT solvers in addressing such problems has made SAT-based techniques the method of choice in these applications. These methods are concerned with determining the satisfiability/ unsatisfiability of quantifier-free formula consisting of Boolean variables and Integer Linear Inequality predicates. Recently, there has been a growing interest in applying SAT-based methods to the general class of optimization problems as well. The two common methods for solving such problems are to either integrate the SAT algorithm into a Branch-and-Bound procedure [6] or to encode the objective function into the SAT solver as a constraint with an adjustable right-hand-side and iteratively solve a sequence of increasingly "tighter" SAT problems [1].

In order to be able to utilize SAT-based methods to solve these problems, they are represented in Mixed Logical/Integer Linear Conjunctive Normal Form (MLIL-CNF) where a Boolean variable (or its negation) or an integer inequality predicate constitutes a literal. In particular, Two-Variable-Per-Ine-

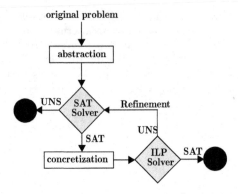

Fig. 1. Solution of MLIL-CNF problems by abstraction/refinement

quality (TVPI) linear constraints have already proven useful in different areas such as program verification [24], buffer over-run vulnerability detection [26], scheduling and several other logic programming applications. In this paper, we address the special class of *unit* TVPI, or UTVPI, constraints of the form $ax + by \leq d$ where $a, b \in \{-1, 0, 1\}$ and propose a hybrid method to solve such problems by integrating Jaffar's polynomial-time algorithm [19] within a modern CNF SAT solver.

Current methods for solving such hybrid problems rely mostly on an *abstraction/refinement* [4] scheme in which each linear constraint is replaced with a Boolean *indicator* variable. As soon as the SAT solver finds a solution to the Boolean problem, an ILP solver is activated to verify the consistency of that solution. This is done by constructing and solving a system of simultaneous linear constraints corresponding to the satisfying assignments found for the indicator variables. The algorithm terminates if the ILP solver proves the consistency of the SAT solution or all the satisfying solutions to the SAT problems are proved to be inconsistent. The overall method is illustrated in Figure 1.

In this paper, we introduce another method where the linear constraints are checked for consistency on demand upon assignment to their corresponding indicator variables. This was made possible by utilizing an incremental algorithm for solving the UTVPI problem. Additionally, our combined method also facilitates an efficient interaction between the two solvers via the indicator variables. In other words, the UTVPI solver can yield implications and generate "no-good" (learned) constraints in the SAT solver as soon as such combinations of UTVPI constraints are detected.

The remainder of this paper is organized as follows. In Section 2 we cover some preliminaries. Section 3 and Section 4 describe our method for encoding and solving the hybrid problem. Experimental results are reported in Section 5 and we conclude in Section 6.

2   Preliminaries

We define an *atom* to be either a Boolean variable or a UTVPI constraint $ax_i + bx_j \leq d$ where $a, b \in \{0, \pm 1\}$ and $x_i, x_j, d \in \mathbb{Z}$[1]. A *literal* is an atom or its negation. A *clause* is a disjunction of literals. Finally, a Mixed Logical/Integer Linear Conjunctive Normal Form (MLIL-CNF) satisfiability instance is a conjunction of clauses. Note that if the set of UTVPI atoms is empty, the MLIL-CNF instance reduces to a pure Boolean CNF instance. On the other hand, if the set of Boolean atoms is empty, it reduces to a UTVPI programming problem with logical combinations.

2.1   Boolean Satisfiability Solver

The Boolean satisfiability problem (SAT for short) is a decision problem that seeks to find a set of truth assignments to the literals in a Boolean CNF formula such that the entire formula evaluates to true, or to prove that no such assignments exist, i.e., the formula is unsatisfiable.

Every Boolean formula can be transformed to CNF in linear time and its satisfiability can be checked using modern backtrack search algorithms based on the DPLL procedure of Davis, Putman, Logemann and Loveland [10,11]. The major computational steps of these algorithms are the following procedures:

Procedure 1: Decision. This procedure is called whenever no assignments (implications) is forced by BCP procedure, explained below. In that case, the next literal to be assigned is decided based on a heuristic. The most common and particularly effective heuristic in most problems is Variable State Independent Decaying Sum (VSIDS) heuristic introduced in [22]. In this method, for each variable, the frequency of occurrence in conflict-induced (learned) clauses or "no-goods" is stored in a counter and the variable with highest counter is selected for next assignment. These counters are periodically scaled down to emphasize recent conflicts.

Procedure 2: Boolean Constraint Propagation (BCP). After each decision, the current assignment is extended to find any clause where at most one of its literals is non-false. If all literals but one in a clause are assigned to false, that is, the condition for that clause to be unit, the remaining literal is implied to true. If all literals are false, the procedure has detected a conflict and should trigger a backtrack to an earlier decision level. Considering that modern conflict-based backtracking SAT solvers spend most of their run-time in this step, a significant optimization in this regard was made by the introduction of watched literals [22]. It was noted that watching just two literals per clause, regardless of clause size, is sufficient to detect when the clause becomes unit. A clause is pro-

---

[1] Note that single-variable constraints are easily accommodated by introducing a dummy variable with a zero coefficient.

```
loop
  propagate()- BCP procedure
  if not conflict then
    if all variables assigned then
      return SAT
    else
      decide() - choose new assignment
  else
    learn() - conflict based learning
    if conflict in root decision level
      return UNSAT
    else
      backtrack() - undo assignments
```

Fig. 2. Modern SAT algorithm

cessed only when either of its two watched literals is set to false; such an assignment triggers the search for another unassigned literal to replace the one that just became false. The clause becomes unit if the only remaining unassigned literal is the other watched literal which must now be implied to true to satisfy the clause.

Procedure 3: Conflict-based Learning. As soon as the BCP procedure detects a conflict in a clause, a so-called conflict-induced clause or a "no-good" is generated and recorded [21]. This enables the solver to prune away a portion of the search space thus avoiding a recurrence of the same conflict. Additionally, conflict-induced clauses enable the solver to backtrack non-chronologically in the search tree without compromising completeness [21]. The conflict-induced clause is obtained by backward traversal of the implication graph starting from the conflict and examining the literals at each cut. The procedure terminates when a unique implication point (UIP) is reached. The UIP is a cut in which only one literal is assigned in the conflict-level and the rest were assigned in previous decision levels. The learned clause will be the disjunction of literals on the UIP cut with opposite polarities.

Procedure 4: Random Restarts and Backtracking. Recent studies [22] show that using random restarts and backtracking can be very effective in helping the SAT solver extricate from hard regions in the search space and exploring a different subtree. This is achieved by either periodically resetting all variable assignments and randomly selecting a new sequence of decisions, or randomly backtracking to a decision level involving any literal in the conflict-induced clause. (The overall SAT algorithm is depicted in Figure 2)

2.2   UTVPI Integer Constraint Solver

Problems consisting of conjunctions of UTVPI constraints can be decided using generic ILP solvers such as CPLEX [18] or XPRESS-MP [9]. However, the full-scale Simplex techniques adopted in these solvers do not take advantage of the simple structure of UTVPI constraints and cannot be efficiently integrated within the backtrack search process of modern SAT solvers.

The solution method we adopt in this paper for deciding systems of UTVPI integer linear constraints is a polynomial-time transitive closure algorithm proposed by Jaffar et al. [19] which, in turn, is an extension of Shostak's method for TVPI *real* constraints [24].

A set of UTVPI constraints is said to be *transitively closed* if for each pair of constraints sharing a variable with opposite signs there exists an inequality constraint between the two remaining variables. For instance, the transitive closure of $\{x - y \leq d , y + z \leq d'\}$ is $\{x - y \leq d , y + z \leq d' , x + z \leq d + d'\}$ and we say that $x + z \leq d + d'$ is *implied by* $x - y \leq d$ and $x , y + z \leq d'$. In the MLIL-CNF context, we will be interested in a dynamically-changing set of UTVPI constraints. To keep such a set transitively closed, whenever a new constraint is added to the set, all its implied constraints must be derived and added. When an implied constraint ends up involving a single variable, it may need to be *tightened* in order to maintain the unit coefficient property. Specifically, the constraint implied by $ax + by \leq d$ and $ax - by \leq d'$ (recall that $a, b \in \{0, \pm 1\}$) is $2ax \leq d + d'$ whose tightening yields $ax \leq \lfloor (d + d')/2 \rfloor$. It is easy to show that the overall complexity of maintaining a tightened and transitively-closed set of UTVPI constraints is cubic in time and quadratic in space.

To illustrate, consider the following set of UTVPI constraints:

$$C = \{y + z \leq 4 , x - y \leq 5 , x + y \leq 2\} \tag{2}$$

The transitively-closed and tightened set of constraints derived from (2) is easily shown to be:

$$\text{Trans}(C) = C \cup \{x + z \leq 9 , 2x \leq 7\}$$
$$\tag{3}$$
$$\text{Tighten}(\text{Trans}(C)) = C \cup \{x + z \leq 9 , x \leq 3\}$$

Jaffar et al. [19] showed that a set of UTVPI integer constraints C is satisfiable iff $\text{Tighten}(\text{Trans}(C))$ does not contain a constraint of the form $0 \leq d$ where $d < 0$. An example of an unsatisfiable constraint set is:

$$C = \{y - z \leq 1 , x - y \leq 1 , z - x \leq -3\}$$
$$\tag{4}$$
$$\text{Tighten}(\text{Trans}(C)) = C \cup \{x - z \leq 2 , 0 \leq -1\}$$

Our MLIL-CNF SAT solver maintains a database of transitively closed and tightened constraint sets. Specifically, suppose that in the course of searching for a solution the sequence of UTVPI constraint sets $C_1 , C_2 , C_3 , \dots$ is generated. As each such set C_i is produced, the corresponding $\text{Trans}(\text{Tighten}(C_i))$ set is computed incrementally by adding/removing any implied constraints when a new constraint is added/removed.

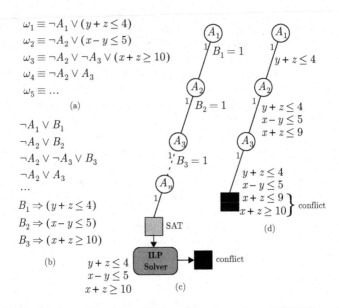

$$\omega_1 \equiv \neg A_1 \vee (y + z \leq 4)$$
$$\omega_2 \equiv \neg A_2 \vee (x - y \leq 5)$$
$$\omega_3 \equiv \neg A_2 \vee \neg A_3 \vee (x + z \geq 10)$$
$$\omega_4 \equiv \neg A_2 \vee A_3$$
$$\omega_5 \equiv \ldots$$

(a)

$$\neg A_1 \vee B_1$$
$$\neg A_2 \vee B_2$$
$$\neg A_2 \neg A_3 \vee B_3$$
$$\neg A_2 \vee A_3$$
$$\ldots$$
$$B_1 \Rightarrow (y + z \leq 4)$$
$$B_2 \Rightarrow (x - y \leq 5)$$
$$B_3 \Rightarrow (x + z \geq 10)$$

(b)

Fig. 3. (a) Simple hybrid problem (b) Lazy representation of UTVPI constraints (c) Assignments without instant UTVPI constraint satisfiability checking (d) Assignments with instant UTVPI constraint satisfiability checking

3   Encoding UTVPI Constraints into the SAT Solver

Our framework for solving the Mixed Logical/Integer Linear Programming (MLILP) problem is similar to the one proposed by Hooker et al. [17] where the continuous and discrete elements of the problem are linked by conditionals in each constraint. Specifically, let

$$\varphi = g(A_1, \ldots, A_n, C_1(X), \ldots, C_k(X)) \qquad (5)$$

denote a Boolean constraint function defined over n Boolean variables $A_1, \ldots, A_n$ and k UTVPI constraints $C_1(X), \ldots, C_k(X)$. Therefore, an MLIL SAT problem would have the following form:

$$\hat{\varphi} = g(A_1, \ldots, A_n, B_1, \ldots, B_k) \wedge \bigwedge_{i=1}^{k} (B_i \Rightarrow C_i) \qquad (6)$$

Boolean *Indicator* variables, B_i's, are used to replace each linear inequality in the Boolean representation of the problem and consequently produce a pure logical problem of the form: $g(A_i, \ldots, A_n, B_1, \ldots, B_k)$. The model of the form (6) is a conservative abstraction of the more straightforward encoding of the linear constraint where they are linked to indicator variables by equality relation and therefore resulting in the following model:

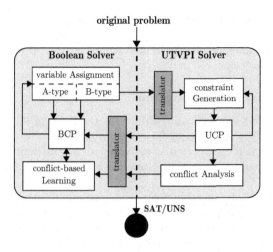

Fig. 4. Hybrid Solver

$$\hat{\hat{\varphi}} = g(A_1,...,A_n,B_1,...,B_k) \wedge \bigwedge_{i=1}^{k}(B_i = C_i) \tag{7}$$

Obviously, forms (6) and (7) are equi-satisfiable due to the fact that g is positive unate in all indicator variables, B_i. In other words, when $\hat{\varphi}$ is unsat, so is $\hat{\hat{\varphi}}$, and when $\hat{\varphi}$ is sat, it is possible that $\hat{\hat{\varphi}}$ is unsat; however, the only situation in which this happens is when at least one B_i is assigned to false and the corresponding linear constraint is forced to be satisfied. Such a solution can be changed so that $B_i = true$ restoring consistency between the linear constraint and its indicator variable without affecting the satisfiability of the original formula. A linear constraint which is enforced unconditionally will be written with a true indicator variable and can be put in the conditional form $true \Rightarrow ax + by \leq d$.

The practical effect of this optimization is that the linear solver need only process those constraints that are "active," i.e. those whose indicator variables are true. We also never have to consider constraints whose indicator is false. Figure 3 demonstrates an example of this process, comparing with abstraction/refinement technique of Figure 1.

4   Solving the Hybrid Satisfiability Problem

The linear constraint solver or in this case the UTVPI solver works in close conjunction with the Boolean SAT solver, as shown in Figure 4. The communication between the two solvers is through the indicator variables which are linked to linear constraints and recognized by both solvers. The Boolean SAT solver performs as the primary solvers and communicates with the linear solver every time an indicator variable is assigned to true. The linear solver, on the

other hand, performs independently from the Boolean SAT solver and has the
capability to produce implications due to current assignments to indicator variables and/or produce no-goods in case of detecting a conflict. Both these procedures are explained below.

4.1   Hybrid Constraint Propagation

Upon true assignment to an indicator variable by the Boolean solver, if no conflict is detected in the Boolean problem, the assignment is passed to the
UTVPI solver where it is interpreted as activation of a constraint which should
be added to the already transitively-closed and tightened set of UTVPI constraints.

The role of UTVPI solver at this point is to generate all implied constraints
to maintain the transitively-closed and tightened property of its set of constraints. An indicator variable is implied in case of generating a constraint
equal to or stronger than its corresponding UTVPI constraint or its complement. For instance, if B is a Boolean indicator variable for $x - y \leq 5$, generating $x - y \leq 3$ will imply B to true and generating $y - x \leq -6$ implies B to false.
In case of detecting infeasibility, the Boolean solver is notified of a conflict.
Figure 3d is an instance of the collaboration between the two solvers. Upon
assigning A_2 to true, the corresponding UTVPI constraint, $x - y \leq 5$, is added
to the set of activated constraints which at that point only contains $y + z \leq 4$.
In order to maintain the transitively closure of the constraint set, the new constraint $x + z \leq 9$ is also generated. As soon as A_3 is implied to true by the
Boolean solver and consequently $x + z \geq 10$ is activated, the UTVPI solver
detects a conflict and forces the Boolean solver to backtrack.

4.2   Hybrid Learning

The learning algorithm is invoked as soon as the hybrid propagation detects a
conflict in a Boolean clause or in the set of activated UTVPI constraints. The
UTVPI solver participates in the learning process in two different cases: first
the case when a conflict is the result of activating a new UTVPI constraint,
and second when an indicator variable needed for conflict-analysis is implied by
the UTVPI solver. In both cases, the reason should be prepared in terms of
indicator variables to be passed to the learning engine of the Boolean solver.

Therefore, the complete learning process in case of a UTVPI inconsistency
detection can be divided into two separate procedures, as follows:.

Finding Infeasible Set of UTVPI Constraints. The incremental property of
the UTVPI satisfiability checking algorithm makes it straight-forward to generate the exact subset of inconsistent constraints causing the conflict. Together
with each constraint, either the reason for its generation being two other constraints each sharing one variable with the current constraint, or an assignment
to its corresponding Boolean indicator variable is stored. Upon detecting an
inconsistent UTVPI constraint, the reason for each of the two conflicting con-

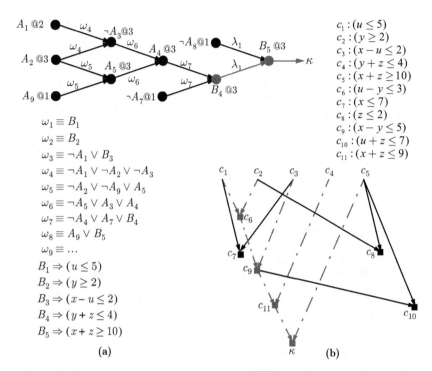

$c_1 : (u \leq 5)$
$c_2 : (y \geq 2)$
$c_3 : (x - u \leq 2)$
$c_4 : (y + z \leq 4)$
$c_5 : (x + z \geq 10)$
$c_6 : (u - y \leq 3)$
$c_7 : (x \leq 7)$
$c_8 : (z \leq 2)$
$c_9 : (x - y \leq 5)$
$c_{10} : (u + z \leq 7)$
$c_{11} : (x + z \leq 9)$

$\omega_1 \equiv B_1$
$\omega_2 \equiv B_2$
$\omega_3 \equiv \neg A_1 \vee B_3$
$\omega_4 \equiv \neg A_1 \vee \neg A_2 \vee \neg A_3$
$\omega_5 \equiv \neg A_2 \vee \neg A_9 \vee A_5$
$\omega_6 \equiv \neg A_5 \vee A_3 \vee A_4$
$\omega_7 \equiv \neg A_4 \vee A_7 \vee B_4$
$\omega_8 \equiv A_9 \vee B_5$
$\omega_9 \equiv \ldots$

$B_1 \Rightarrow (u \leq 5)$
$B_2 \Rightarrow (y \geq 2)$
$B_3 \Rightarrow (x - u \leq 2)$
$B_4 \Rightarrow (y + z \leq 4)$
$B_5 \Rightarrow (x + z \geq 10)$

(a)

(b)

Fig. 5. (a) Implication graph at decision level (@)3 for Boolean problem: $\wedge \, \omega_i$ in which $\lambda_1 \equiv \neg B_4 \vee B_5 \vee A_8$ is already learned. At decision level 3, A_2 is assigned (decided) to false which would ultimately result in conflict detected by UTPVI solver. (b) resolution graph demonstrating UTVPI solver incremental procedure to detect a conflict in set of integer constraints

straints is traced back until a set of Boolean indicator variables is obtained and transferred to the Boolean SAT Solver learning engine as the conflicting assignment. The same procedure will be adopted for finding the implying assignment sequence for an indicator variable, which is implied by the UTVPI solver. The learning engine of the Boolean solver uses this information to construct the implication graph.

Generating Conflict-Induced Learned Clause. Upon receiving a set of conflicting assignments to indicator variables from the UTVPI solver, the Boolean solver performs the exact learning procedure according to [21]. The conflict-induced clause is generated by traversing Boolean implication graph and is recorded to prune current conflict. Backtrack level is computed by analyzing the learned clause. The backtrack level should be also passed to the UTVPI solver to update its data structure and remove deactivated constraints and their children.

An example of the complete procedure is demonstrated in Figure 5. In this example, constraints c_1, c_2 and c_3 are activated upon true assignments to their indicator variables (B_1, B_2 and B_3 at decision levels 0, 0 and 2, respectively). Constraints c_6, c_7 and c_8 are generated and recorded as soon as both their parent constraints are added. At decision level 3, implying indicator variables B_4 and B_5 adds constraints c_4, c_5 to set of UTVPI constraints which ultimately results in a conflict. The reason for this conflict can easily be obtained by tracing the conflict back in the implication graph which is recorded by the UTVPI solver (dotted lines in Figure 5b). The conflicting assignment, being $B_1 \wedge B_2 \wedge B_3 \wedge B_4 \wedge B_5$, is returned to the Boolean solver which by backward traversal of implication graph (Figure 5a) learns the following clause at the UIP: $\neg B_1 \vee \neg B_2 \vee \neg B_3 \vee A_8 \vee A_7 \vee \neg A_4$

5   Experimental Results

We conducted several experiments to evaluate our hybrid method using our SAT solver, Pueblo, augmented with a UTVPI solver engine. Pueblo is built on top of MiniSAT [13] and inherits its strategy for random restarts, VSIDS and clause removal. All experiments were conducted on a Pentium-IV 2800MHz machine with 1 GB of RAM running Linux 2.4.20.

5.1   Timing Analysis of Combinational Circuits

The first set of benchmarks [7] used for this evaluation is the problem of finding the worst-case propagation delay of a circuit, that is, the maximum elapsed time between a change in the inputs and stabilization of the outputs. In this model, for each gate in the combinational model of the circuit, a gate-delay is also taken into account and therefore, the gates, rather than performing as an instantaneous Boolean functions, are viewed as functions whose inputs are gradually propagated to the outputs.

Determining the worst-case delay has several important practical application in industry, for instance in calculating the frequency of the clock with which the circuit can operate. The width of the clock needs to be large enough to cover the maximum stabilization time of the circuit in order to guarantee successful propagation of all admissible inputs to the outputs. Current methods practiced in industry, rely on calculating the path with maximum accumulated delay from the inputs to the outputs without considering the logical relations between circuit components. The result of such methods would be over pessimistic due to the fact that some paths in the circuit might be logically impossible to be exercised.

We consider the class of bi-bounded inertial delay models [8] where each propagation delay associated with a gate is characterized by lower and upper bounds as illustrated in the example of Figure 6. The delay operator D_I where $I = [l, u]$ on each gate is a non-deterministic function with following characteristics:

1. changes in the input should be propagated to the output if they persist for u time
2. changes in the input that persist for less than l time, are not propagated at all to the output

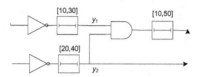

Fig. 6. A timed Boolean circuit

A separate clock variable is associated with each delay element (gate) and will reset by its state transition. The system's behavior is modeled using a Timed Automaton (TA) [2] which is a tuple $\mathcal{A} = (Q, C, I, \Delta)$ where Q is a finite set of states, C is a finite set of clock variables, I is a staying condition, assigning a conjunction of inequalities over C to every state, and Δ is a state-transition relation. The basic verification problem is to determine whether a state with given properties is reachable within exactly k transitions. There are two types of transitions: *Elapse of time* when only the clock variable increments, and *Location switch* when a new state is reached and its clock is reset to zero. The set of states, state-transition relation and the properties in the TA can be modelled by Boolean functions together with *Difference Linear Inequalities* of the form $x - y \leq d$ that represent time pass and therefore, the model checking problem will reduce to checking the satisfiability of the combined formula. For details on model checking of timed automata, the reader is referred to [2].

To compute the maximum stabilization time in the circuit, an auxiliary clock is considered which is never reset to zero and at each reachable configuration, represents the total time elapsed. The circuits modeled in these benchmarks are combinational bit-adder circuits constructed from gates and the satisfiable problem is concerned with the question that "whether there is a run of the automaton with k transitions which remains in an unstable state after d time". The problem would be checking the satisfiability of a logical combination of a set of Boolean variables and difference linear inequalities. Three different methods/solvers are used to solve this problem:

1. Hybrid online method used in Pueblo as described in this paper and adopting encoding methods of (6) and (7) separately,
2. Translation to MIP using *Big-M* method and solving it using XPRESS-MP [9]. In this method, each difference inequality of the form $x - y \leq d$ is replaced by a fresh binary variable B_k in its corresponding Boolean clause and linear constraint $x - y \leq d + M(1 - B_k)$, where M is the Big-M parameter, is added to the set of constraints. Note that each clause can be transformed to a linear constraint using $\sum_{i=1}^{n} \dot{x}_i \geq 1 \Leftrightarrow \bigvee_{i=1}^{n} \dot{x}_i$ where all complemented literals are replaced by $\dot{x}_i' = 1 - \dot{x}_i$,

Table 1. Comparisons of different methods on bit-adder benchmarks

benchmark (bits, k, d)	SAT/ UNS	Variables num/bin	Run-Time (sec.)			
			Pueblo(6)	Pueblo(7)	MIP	SVC
ba(4, 5, 5)	SAT	151/3104	0.3	2.43	6.88	328.18
ba(4, 5, 10)	UNS	151/3104	0.3	1.01	236.66	769.02
ba(4, 6, 5)	SAT	176/3710	0.42	4.41	27.96	>1000
ba(4, 6, 10)	UNS	176/3710	0.4	0.96	361.28	>1000
ba(4, 7, 5)	SAT	201/4306	1.36	14.37	15.17	>1000
ba(4, 7, 10)	UNS	201/4306	0.6	6.11	>1000	>1000
ba(4, 8, 5)	SAT	226/4922	3.78	35.23	199.26	>1000
ba(4, 8, 10)	UNS	226/4922	1.94	5.52	>1000	>1000
ba(4, 9, 5)	SAT	251/5528	3.22	69.19	62.38	>1000
ba(4, 9, 10)	UNS	251/5528	5.03	17.11	>1000	>1000
ba(4, 10, 5)	SAT	276/6134	2.99	51.89	212.4	>1000
ba(4, 10, 10)	UNS	276/6134	9.71	20.72	>1000	>1000

3. Stanford Validity Checker [5] using Shostak-style [24] method in combination with Binary Decision Diagrams (BDD).

Table 1 shows a comparison of these methods conducted on 4-bit adder benchmarks with different numbers of state transitions and duration. As it is obvious, mainly due to dominance of binary variables in these benchmarks, the performance of SAT-based methods are much better than LP relaxation method of XPRESS-MP and BDD-based method of SVC. It is also clear that for UNSAT problems, the advantages of SAT-based methods is more evident over MIP method of XPRESS-MP. The better performance due to encoding based on (6) rather than (7) is also evident for these benchmarks.

5.2   Job-Shop Scheduling

The Job-Shop Scheduling Problem (JSSP) consists of a finite job sets to be processed on a finite number of machines. Each job must be processed on every machine and each operation has to be scheduled according to its precedence and capacity constraints. The decision problem in this case is to find "whether there exists a schedule with makespan of L for the problem that satisfies all constraints". It can also be stated as an optimization problem to find the minimum makespan by iteratively solving the decision problem. This problem has been thoroughly studied for the past several decades in both OR and CLP and several efficient algorithms and heuristics have been developed that solve JSSP with high level of performance. Such methods include Edge-Finding techniques [3], and meta heuristics such as Tabu Search [12].

Our hybrid method presented in this paper does not intend to compete with those specialized JSSP techniques and heuristics and rather should be viewed as a framework to benefit current algorithms and make it possible to couple those with advances made in SAT.

A scheduling problem can be stated with two families of constraints, expressed as follows, where $precede(t_1, t_2)$, represents when task t_2 cannot be performed before t_1, $d(t)$ denotes the duration of task t and $r(t)$, the resource that it uses. The problem then can be expressed as follows:

$$\forall t_1, t_2 \in T, \ precede(t_1, t_2) \Rightarrow time(t_2) \geq time(t_1) + d(t_1)$$

$$\forall t_1, t_2 \in T, \ r(t_1) = r(t_2) \Rightarrow time(t_2) \geq time(t_1) + d(t_1)$$

$$\vee \ time(t_1) \geq time(t_2) + d(t_2) \tag{8}$$

The encoding of the problem into our framework will result in Boolean CNF clauses with one or two literals and their corresponding UTVPI constraints in the form of difference inequalities. Table 2 provides the results obtained with our solver on well-known JSSP benchmarks of [20] when proving the satisfiability of their best known makespan. This table clearly illustrates the advantages of main algorithms of SAT solvers, namely the conflict learning, non-chronological backtracking and VSIDS variable ordering, on these benchmarks. Particularly, these methods help the search algorithm to focus on the bottlenecks that is proven to be the best heuristic in solving JSSP.

6   Conclusions and Future Work

In this paper, we introduced a combined method in solving problems of Boolean propositions and a special class of linear inequalities. The integration of logic-based reasoning and linear programming methods promises to be a vibrant area of research for the next several years in various research communities ranging from OR and CP to Artificial Intelligence and Software and Hardware Verification. Recent advances in SAT-based algorithms such as introduction of efficient techniques to compute Minimally Unsatisfiable Subformula (MUS) or several new decision heuristics can be utilized in LP and ILP problems, given an efficient hybridization methodology. As we learn more about the trade-offs involved, we will be able to develop effective integration strategies that outperform individual techniques.

One promising direction of future research involves extending the techniques described above to include a more general class of constraints, specifically integer or real linear or nonlinear constraints. Since in most hybrid problems, integer solvers act as the bottleneck in the SAT-based algorithms and considering that non-UTVPI constraints comprise a small portion of integer constraints in

Table 2. Applying Pueblo on JSSP benchmarks. No VISIDS refers to adopting fixed variable ordering, set at the beginning of the search

benchmark [size]	minimum makespan	Pueblo Run-Time (sec.)		
		with VSIDS and Learning	no VSIDS, with Learning	no VSIDS, no Learning
la01 [10*5]	666	0.79	16.91	>1000
la03 [10*5]	597	7.32	47.79	208.29
la06 [15*5]	926	6.12	>1000	>1000
la07 [15*5]	890	66.34	>1000	>1000
la11 [20*5]	1222	179.04	>1000	>1000
la14 [20*5]	1292	178.25	195.03	>1000
la17 [10*10]	784	13.93	>1000	>1000
la18 [10*10]	848	25.95	>1000	>1000
la19 [10*10]	842	150.84	660.70	>1000
la20 [10*10]	902	35.77	>1000	>1000

those application [25], methods described in this paper when combined with other ILP methods that does not negatively impact performance, sounds a viable option. Specifically, the independent characteristics of our method together with its ability to collaborate with the SAT solver during the search, makes it more practical for large problems and more friendly to parallel algorithms.

Acknowledgment

This work was funded in part by the National Science Foundation (NSF) under ITR grant No. 0205288.

References

[1] F. Aloul, A. Ramani, I. Markov, and K. Sakallah, "Generic ILP versus Specialized 0-1 ILP: an Update," *ICCAD*, pp. 450-457, 2002.
[2] R. Alur and D. L. Dill, "A Theory of Timed Automata," *Theoretical Computer Science*, vol. 126, pp. 183-235, 1994.
[3] D. Applegate and B. Cook "A Computational Study of the Job Shop Scheduling Problem," *ORSA Journal on Computing*, vol.3(2), pp. 149-156, 1991.
[4] G. Audemard, P. Bertoli, A. Cimatti, A. Kornilowics, and R. Sebastiani, "A SAT-Based Approach for Solving Formulas over Boolean and Linear Mathematical Propositions," *CADE*, pp. 193-208, 2002.
[5] C. Barrett, D. Dill, and J. Levitt , "Validity Checking for Combinations of Theories with Equality," *FMCAD, LNCS 1166*, pp. 187-201, 1996

[6] A. Bemporad and N. Giorgetti, "SAT-based Branch & Bound and Optimal Control of Hybrid Dynamical Systems," *CP-AI-OR*, pp. 96-111, 2004.

[7] R. Ben Salah, M. Bozga, and O. Maler, "On Timing Analysis of Combinational Circuits," *FORMATS*, 2004

[8] J. A. Brzozowski and C.J.H. Seger, "Asynchronous Circuits," *Springer*, 1994.

[9] Dash Inc., XPRESS-MP 15.25.03, http://www.dashoptimization.com.

[10] M. Davis and H. Putnam, "A Computing Procedure for Qualification Theory," *Journal of the ACM*, vol. 7, pp. 102-215, 1960.

[11] M. Davis, G. Logemann, and D. Loveland. "A Machine Program for Theorem Proving," *Communications of the ACM*, vol. 7, pp. 394-397, 1962.

[12] M. Dell'Amico and M. Trubian "Applying Tabu Search to the Job-Shop Scheduling Problem" *Annals of Operations Research*, vol. 41, pp. 231-252, 1993.

[13] N. Eén and N. Sörensson, "An Extensible SAT-solver," *SAT*, pp. 502-508, 2003.

[14] I. E. Grossmann and Z. Kravanja, "Mixed Integer Nonlinear Programming: A Survey of Algorithms and Applications," *Large-Scale Optimization with Applications, Part II: Optimal Design Control*, Springer-Verlag, pp. 73-100, 1997.

[15] I. E. Grossmann and N. V. Sahinidis, (eds) "Special Issue on Mixed Integer Programming and its Application to Engineering," *Part I/II, Optim. and Engin.*, 2002.

[16] J. N. Hooker, "Logic-based Methods for Optimization: Combining Optimization and Constraint Satisfaction," Wiley, 2002.

[17] J. N. Hooker, G. Ottosson, E. S. Thorsteinsson, and H. Kim, "On Integrating Constraint Propagation and Linear Programming for Combinatorial Optimization", *AAAI*, pp. 136-141, 1999.

[18] ILOG CPLEX, http://www.ilog.com/products/cplex.

[19] J. Jaffar, M. Maher, P. Suckey, and R. Yap, "Beyond Finite Domains," *Workshop on Principles and Practice of Constraint Programming*, 1994.

[20] S. Lawrence, "Resource Constrained Project Scheduling: An Experimental Investigation of Heuristic Scheduling Techniques (Supplement)," Graduate School of Industrial Administration, Carnegie-Mellon University, Pittsburgh, PA, 1984.

[21] J. P. Marques-Silva and K. A. Sakallah, "GRASP: A Search Algorithm for Propositional Satisfiability," *IEEE Trans. on Computers*, vol. 48(5), pp. 506-521, 1999.

[22] M. Moskewicz, C. Madigan, Y. Zhao, L. Zhang, and S. Malik, "Chaff: Engineering an Efficient SAT Solver," *DAC*, pp. 530-535, 2001.

[23] R. Raman, I. E. Grossmann, "Modelling and Computational Techniques for Logic Based Integer Programming," *Computers and Chemical Engineering*, vol. 18, 1994.

[24] R. Shostak, "Deciding Linear Inequalities by Computing Loop Residues," Journal of the ACM, vol. 28(4) pp. 769-779, 1981.

[25] S. Seshia and R. Bryant, "Deciding Quantifier-Free Presburger Formulas Using Parameterized Solution Bounds," LICS, pp. 100-109, 2004.

[26] D. Wagner, J. S. Foster, E. A. Brewer, and A. Aiken, "A First Step Towards Detection of Buffer Overrun Vulnerabilities," *Network and Distributed System Security Symposium*, Internet Society, 2000.

Symmetry and Search in a Network Design Problem

Barbara M. Smith

Cork Constraint Computation Centre, University College Cork, Ireland
b.smith@4c.ucc.ie

Abstract. An optimization problem arising in the design of optical fibre networks is discussed. A network contains client nodes, each installed on one or more SONET rings. A constraint programming model of the problem is described and compared with a mixed integer programming formulation. In the CP model the search is decomposed into two stages; first partially solving the problem by deciding how many rings each node should be on, and then making specific assignments of nodes to rings. The model includes implied constraints derived by considering optimal solutions to subproblems. In both the MIP and CP models, it is important to deal with the symmetry of the problem. In the CP model, two sources of symmetry are separated; one is eliminated dynamically during search and the other by assigning ranges rather than explicit values to one set of decision variables. The resulting CP model allows optimal solutions to be found easily for benchmark problems.

1   Introduction

In this paper, the development of constraint programming models for an optimization problem arising in the design of optical fibre networks is discussed. The problem was introduced by Sherali, Smith and Lee [8], who discuss the practical scenario that the problem is abstracted from and described mixed integer programming (MIP) formulations. In both the MIP and CP models, it is important to deal with symmetry; Sherali and Smith [7] discuss different ways of dealing with the symmetry in one of the MIP models.

A CP model of a simplified version of the problem (ignoring the demand capacities of the rings), using similar variables and constraints to the MIP model and passed to a CP solver, could solve only small instances. An earlier paper [9] describes how this model evolved into one that could solve larger instances of the full problem; further improvements are presented here. In this paper, Section 2 describes the problem and the MIP model, with the methods in [7] for dealing with symmetry. The following sections describe a CP model for the simplified problem in which the traffic capacity of the rings is ignored; solving this simplified problem is a precursor to solving the full problem. It is useful to separate two sources of symmetry in the problem: Section 7 describes how the symmetry due to the fact that the rings are interchangeable can be eliminated in the CP model. The full problem, taking the traffic capacity into account, is returned to in section 8; this introduces further symmetry from the fact that demand can often be split between rings in different ways. The CP model of the previous sections is adapted, and equivalent solutions are avoided by assigning ranges rather than specific values to the

R. Barták and M. Milano (Eds.): CPAIOR 2005, LNCS 3524, pp. 336–350, 2005.
ⓒ Springer-Verlag Berlin Heidelberg 2005

decision variables. Results are presented showing that optimal solutions for the largest instances in [8, 7] can be found easily. Section 9 draws conclusions.

2   Problem Description and MIP Model

Sherali, Smith and Lee [8] describe a network design problem arising from the deployment of synchronous optical networks (SONET). The network contains a number of client nodes and there are known demands (in terms of numbers of channels) between pairs of nodes. (It is explained in [8] that when the model is used for planning, the value assigned to each demand takes account of the uncertainty in predicting the actual demand over the planning horizon.) A SONET ring joins a number of nodes; a node is installed on a ring using an 'add-drop multiplexer' (ADM) that is capable of adding and dropping the traffic. Each node can be installed on more than one ring, and traffic can be routed between a pair of client nodes only if they are both installed on the same ring. In this scenario, there is no traffic allowed between rings, but the demand between a pair of nodes can be split between two or more rings. There are capacity limits on the rings (in terms of both nodes and channels). The objective is to minimise the total number of ADMs required, while satisfying all the demands.

The largest test instances used in [8] have 13 nodes, with 24 demand pairs. The rings can accommodate 5 nodes and 40 traffic channels, and there are 7 rings available. (It appears that the cost of SONET rings is negligible, so that there is no practical limit on the number of rings used. The limit is specified in order to model the problem.) 80% of the demand pairs were uniformly generated between 1 and 5, and 20% between 1 and 25. There are also two smaller sets of 15 problems each, one set having 7 nodes and 8 demand pairs and the other 10 nodes and 15 demand pairs.

The data for one of the large instances is given below. The first line gives the origin nodes, the second the destination nodes, and the third the demands in terms of number of channels.

```
1  1  2 2 2  2  2  3 4 4  4  4 5 5 7  7  7  8  8  8  9 10 11 12
9 11  3 5 9 10 13 10 5 8 11 12 6 7 9 10 12 10 12 13 12 13 13 13
8  2 25 5 2  3  4  2 4 1  5  2 5 4 5  2  6  1  4  1  5  9  3  2
```

For this instance, an optimal solution ignoring the traffic levels uses 22 ADMs on 5 rings. The sets of nodes installed on the rings are: {4, 8, 10, 12, 13}, {4, 5, 6, 11}, {1, 2, 9, 11, 13}, {2, 3, 5, 7, 10}, {7, 9, 12}. However, this is not a feasible solution if the level of demand is taken into account, because the fourth ring requires 41 channels. The demand pairs on this ring are not on any other ring, so that the demand cannot be split between two rings. An optimal solution respecting the traffic limit uses 23 ADMs.

The Mixed Integer Programming (MIP) model in [8, 7] is as follows. Let $N = \{1, ..., n\}$ be the set of nodes, and d_{ij} be the traffic demand between nodes i and j. This defines a set of edges $E = \{(i, j), i < j, d_{ij} > 0\}$ and a demand graph $G(N, E)$. Let $M = \{1, .., m\}$ be the set of rings, r be the maximum number of nodes that can be installed on any ring, and b be the capacity of a ring in terms of the number of channels. Let S_i be the set of neighbours of i in G, i.e. the set of nodes j such that there is a demand between nodes i and j, so that $S_i = \{\rho \in E : \rho = (i,j) \text{ or } \rho = (j,i) \text{ for some } j\}$.

The model has two sets of decision variables: $x_{ik}, \forall i \in N, k \in M$, where $x_{ik} = 1$ if node i is assigned to ring k, 0 otherwise; $f_{\rho k} \forall \rho = (i,j) \in E, k \in M$, representing the fraction of the demand between the nodes i and j that is assigned to ring k. The model can be stated as:

$$\text{minimize} \quad \sum_{i \in N} \sum_{k \in M} x_{ik}$$

$$\text{subject to:} \quad \sum_{k \in M} f_{\rho k} = 1 \qquad\qquad \forall \rho \in E \qquad (1)$$

$$\sum_{\rho \in E} d_\rho f_{\rho k} \leq b \qquad\qquad \forall k \in M \qquad (2)$$

$$\sum_{i \in N} x_{ik} \leq r \qquad\qquad \forall k \in M \qquad (3)$$

$$0 \leq f_{\rho k} \leq x_{ik} \qquad\qquad \forall i \in N, k \in M, \rho \in S_i \qquad (4)$$

$$x_{ik} \in \{0,1\} \qquad\qquad \forall i \in N, k \in M \qquad (5)$$

This basic model, RD1, has many alternative optimal solutions, because given any feasible network design, equivalent designs can be obtained by "simply reshuffling the demand allocations made to the various individual rings" [7]. Sherali and Smith developed a number of alternative models in which the problem symmetry was reduced by introducing hierarchy into the model. The most successful model, RD3, replaced constraints (3) by:

$$r \geq \sum_{i \in N} x_{i1} \geq \sum_{i \in N} x_{i2} \geq \dots \geq \sum_{i \in N} x_{im}$$

i.e. ring 1 has the largest number of ADMs, followed by ring 2 and so on. In their experiments, this model gave the shortest average run-time, even though it does not eliminate all the symmetry: each ring can accommodate only 5 nodes, and most rings that are used at all will have 4 or 5 nodes allocated to them. The average run-time for the 15 largest instances was just over 21,200 sec., compared with nearly 196,000 sec. for RD1 (on a Sun Ultra 10 Workstation, running CPLEX).

Another model, RD2, which gave an average run-time of over 40,000 sec., replaced constraints (2) by:

$$b \geq \sum_{\rho \in E} d_\rho f_{\rho 1} \geq \sum_{\rho \in E} d_\rho f_{\rho 2} \dots \geq \sum_{\rho \in E} d_\rho f_{\rho m}$$

i.e. ring 1 has the largest demand, followed by ring 2, and so on. Although this model was not as good as the previous model, RD3, by itself, Sherali & Smith found that using the demand allocated to successive rings to break ties in the case that two successive rings have the same number of nodes allocated was promising, and in experiments gave the second smallest average run-time (just under 23,000 sec. on average).

Although even the worst model with symmetry breaking constraints (RD2) did much better than the original model, it is surprising that RD3, using a simple set of constraints which leave much of the symmetry intact, did so well. Sherali and Smith concluded that this was because the constraints added tended to encourage integral solutions, since the nonzero coefficients are all 1s. In CP, on the other hand, experience is that as much as possible of the symmetry in the problem should be eliminated. In CP models of the SONET problem, as shown later, all the symmetry can be eliminated quite easily and this contributes to its successful solution.

3   The Unlimited Traffic Capacity Problem

This section describes a CP model for a simplified version of the problem in which the demand capacity of the rings is ignored. The problem initially proved difficult to solve using CP, so that it was helpful to start with a simplified problem. Moreover, solving the simpler problem gives a good lower bound on the solution to the full problem, in fact often a feasible solution, so that it is useful to solve the simpler problem even now that better models have been developed. Finally, considering the simpler problem and the full problem separately makes explicit the two sources of symmetry in the problem, since one arises only in the full problem. These sources of symmetry can be eliminated in different ways. Solution of the full problem will be described in section 8.

For the simplified problem, the relevant data can be represented by the demand graph, $G(N, E)$. Figure 1 gives an example, representing one of the large instances.

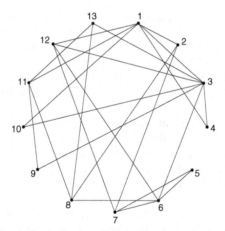

Fig. 1. Demand pairs in a sample instance of the SONET problem

It is possible to build a CP model the unlimited traffic capacity problem using only the x_{ik} variables of the MIP model. However, it was found that the resulting model could only be solved in a reasonable time (using ILOG Solver) for the 15 smallest instances. In any case, the expression of the constraint that if there is a demand between a pair of nodes they must both appear on the same ring is somewhat awkward for CP.

Partly to make it easier to express this constraint, two dual sets of set variables were introduced: variable R_k represents the set of nodes assigned to ring k, $\forall k \in M$, and N_i is the set of rings that node i is assigned to, $\forall i \in N$.

The model can then be written as:

$$\text{minimize} \quad t = \sum_{i \in N} \sum_{k \in M} x_{ik} = \sum_k |R_k| = \sum_i |N_i|$$

$$
\begin{array}{llr}
\text{subject to:} & |N_i \cap N_j| \geq 1 & \forall (i,j) \in E \quad (6) \\
& |R_k| \leq r & \forall k \in M \quad (7) \\
& (x_{ik} = 1) = (i \in R_k) = (k \in N_i) & \forall i \in N, k \in M \quad (8) \\
& x_{ik} \in \{0, 1\} & \forall i \in N, k \in M \quad (9)
\end{array}
$$

The first constraint expresses that if there is a demand between nodes i and j, there must be at least one ring that they are both assigned to. The second could alternatively be expressed as: $\sum_i x_{ik} \leq r$, and so is equivalent to (3) in the MIP model. The third constraint is a set of channelling constraints linking the old and new variables.

The x_{ik} variables are used as the search variables and an optimal solution is found using the branch-and-bound optimization provided in ILOG Solver: whenever a solution is found, Solver adds a constraint that in future solutions, the value of t (the objective variable) must be smaller. Hence, when there is no solution satisfying the current constraint on t, the last solution found is known to be optimal.

4   How Many ADMs for Each Node?

The approach described in the last section can solve the small and medium SONET instances to a limited extent. With suitable variable ordering heuristics, good, and sometimes optimal, solutions for these instances can be found very quickly, but proving optimality is slow; as before, of course, we are still dealing only with the unlimited traffic capacity problem.

Part of the difficulty is that very little can be deduced from the assignment of a value to one of the x_{ik} variables, or even from several such assignments. In particular, when it comes to proving optimality, it is hard to tell from such assignments whether the current bound on the objective can be met. Thus, the search explores many unprofitable branches, with no means of pruning them.

As described in [9], the proof of optimality can be speeded up by focusing on the number of ADMs each node needs. Hence, for each node, an integer variable n_i is introduced, where $n_i = |N_i|$, i.e. n_i is the number of rings that node i is installed on.

Deciding how many rings each node should appear on is not sufficient to solve the problem, so the n_i variables cannot be used alone as search variables. After the n_i variables have been assigned, the search continues by assigning values to the x_{ik} variables. Once it has been decided how many rings each node should appear on, it is easy to find a consistent assignment of nodes to rings, or prove that there is no such assignment. This is still a complete search, since if no assignment to the x_{ik} variables can be found, the search backtracks to the n_i variables.

Hence, the problem is decomposed into first deciding how many rings each node should be on, and then assigning each node to a specific set of rings. A similar approach is common in scheduling and has proved useful for other problems, for instance in [4].

A further improvement comes from assigning increasing values of the objective variable until a solution is found, rather than using the branch-and-bound procedure described earlier. This is easily implemented by making t a search variable and assigning it first. One benefit is that, since the objective t is equal to the sum of the n_i variables, fixing t reduces their domains. Once again, if no feasible assignment to the n_i variables can be found, the search backtracks to try a larger value of t; hence, the first solution found must be optimal.

This would not be a useful solution strategy if finding the optimal solution were very difficult and we were prepared to settle for a good solution and forgo the proof of optimality. Alternatively, if there were a large gap between the smallest value of t found by propagating the constraints and the optimal value, proving that every intervening value is infeasible might take too long. For these instances of the SONET problem, the minimum value of t which need be considered is between 18 and 20 (given the implied constraints on the n_i variables described in the next section) and the optimal value is between 20 and 24, so that there are only a few infeasible values of t to consider before the optimal value is reached.

5   Implied Constraints

It has proved much easier to reason about the number of rings that each node must be installed on than whether or not node i should be installed on ring k. Consequently, the n_i variables have been a fruitful source of implied constraints, allowing infeasible assignments to be detected early.

The simplest constraint is that if the degree δ_i of node $i \geq r$, it must be placed on more than one ring: for instance, in Figure 1, nodes 1, 3 and 6 must each appear on at least 2 rings. In general, $n_i \geq \lceil \frac{\delta_i}{r-1} \rceil$.

A similar constraint was used in [8], but only for the node with largest degree in the demand graph: this node is assigned to the first n_i rings in order to reduce the symmetry in the problem. Here, we do not need such a constraint to deal with symmetry (as discussed below in section 7), but the set of constraints is useful in detecting when the current value of the objective cannot be attained.

The remaining implied constraints have been derived by considering subproblems consisting of pairs of nodes with a demand between them, and their neighbours in the demand graph.

- if two nodes i and j are connected, and each of them is connected to fewer than r other nodes (so that, considered individually, it appears that each could be placed on just one ring), but together they are connected to at least $r - 1$ other nodes, then at least one of them must be on at least two rings i.e. $n_i + n_j \geq 3$.
  In Figure 1, nodes 9 and 11 are connected to nodes 1, 3, 8 and 13, and to each other. There must be a ring that both nodes are installed on, since they are connected, and both sets of neighbours cannot fit onto this ring as well. Hence, one of nodes 9 and 11 must be installed on another ring.

- if two nodes i and j are connected, and one is connected to at least r nodes (so that the last constraint does not apply), but the other is not, and they are connected to a total of more than $2r-3$ other nodes between them, then nodes i and j require at least four ADMs, i.e. $n_i + n_j \geq 4$.

  For instance, for the example of Figure 1, node 3 must be on at least 2 rings, whereas node 13 could be installed on a single ring, with its neighbours. One more of node 3's neighbours could be on the same ring, but this would leave 5 of the demand pairs involving node 3 not yet accommodated, and this would require at least 2 more rings. Hence $n_3 + n_{13} \geq 4$.

- Suppose that two nodes i, j are not connected; that each could be installed on just one ring but they could not both be installed on the same ring (i.e. $\delta_i \leq r - 1$, $\delta_j \leq r - 1$, $\delta_i + \delta_j \geq r - 1$); and they have a mutual neighbour k which has more than $r - 1$ neighbours. (In Figure 1, nodes 2, 11 and 1 meet these conditions.) If nodes i and j appear on just one ring each, then node k must also be installed on these rings; if the three nodes have more than $2r-4$ other distinct neighbours in total, then node k must also appear on a third ring. We can add a constraint to reflect this: if $n_i = 1$ and $n_j = 1$ then $n_k \geq 3$.

  For instance, nodes 2, 11 and 1 have more than $2r-4$ neighbours (nodes 3, 4, 7, 8, 9, 10, 13). The ring with nodes 2 and 1 could also accommodate two more of node 1's neighbours; the ring with nodes 11 and 1 could accommodate another; but this still leaves one of the demand pairs involving node 1 unplaced, requiring another ring. Hence, if $n_2 = 1$ and $n_{11} = 1$ then $n_1 \geq 3$.

In [8], valid inequalities are derived which similarly express a lower bound on the number of ADMs required by a subset of the nodes. The inequalities are generated by selecting subsets of at least r nodes (where r is the maximum number of nodes on any ring), or a set of nodes with total demand between them greater than b (where b is the maximum demand capacity of a ring). For instance, any connected set of $r+1$ nodes requires at least $r+2$ ADMs. In [8], these inequalities are reported to reduce runtime by about one-third, with some care in selecting appropriate sets of nodes. However, in the CP model, they would be weaker than the implied constraints listed earlier.

The foregoing implied constraints would still be valid in the full problem, i.e. taking into account the traffic capacity of the rings. The following constraint would not then be valid, but is useful in the simplified problem:

- the number of rings that a node is installed on should be no more than its degree, i.e. $n_i \leq \delta_i$.

The following dominance rules can also be added. They are not true of every consistent solution, and so are not logical consequences of the existing constraints, as implied constraints are, but they will be satisfied by at least one optimal solution:

- a ring cannot have just one node on it, i.e. $|R_k| \neq 1$, for $1 \leq k \leq m$. In fact, any ring must have two *connected* nodes on it.
- the total number of nodes allocated to two non-empty rings must be more than the number that can be accommodated on one ring, i.e. if $|R_k| > 0$ and $|R_l| > 0$ then $|R_k| + |R_l| > r$, for all k, l with $1 \leq k < l \leq m$. Otherwise there is an equally good solution in which the two rings are combined into one.

All these additional implied constraints and dominance rules are useful: they reduce both search and runtime.

Once a complete assignment to the n_i variables has been found, consistent with all these constraints, the following constraints are useful while the x_{ik} variables are being assigned. They help to complete the solution quickly or decide that it cannot be completed.

- if $n_i = 1$, and node i is installed on ring k, then all the neighbours of node i must also be on ring k, i.e. if S_i is the set of neighbours of node i:

$$\text{if } n_i = 1 \text{ and } x_{ik} = 1 \text{ then } \{i\} \cup S_i \subseteq R_k \quad \text{for } 1 \leq i \leq n, 1 \leq k \leq m$$

- similarly, if $n_i = 2$, once node i has been allocated to two rings, all its neighbours must be on these two rings as well, i.e.

$$\text{if } n_i = 2 \text{ and } x_{ik} = 1 \text{ and } x_{il} = 1 \text{ then } \{i\} \cup S_i \subseteq R_k \cup R_l$$
$$\text{for } 1 \leq i \leq n, 1 \leq k < l \leq m$$

6   Variable Ordering Heuristics

The search strategy described earlier requires two variable ordering heuristics, one for the n_i variables and one for the x_{ik} variables. The heuristics described in this section are used both for the simplified problem and the full problem that includes the traffic capacities of the rings.

Devising variable ordering heuristics for the first stage of search i.e. for assigning the n_i variables, is more complicated than when there is only one set of search variables. The aim is not just to find an assignment to satisfy the direct constraints on the first-stage search variables (which is easily done), but to find one that can be extended to the second-stage variables. This makes it difficult to predict the behaviour of a proposed heuristic.

Three variable ordering heuristics have been compared: smallest domain, minimum degree and maximum degree. (The last two are static orders, using the original degrees in the demand graph.) When the objective variable is assigned first, any variable ordering will generate the same set of sub-optimal complete assignments to the n_i variables (i.e. assignments for which the value of t is less than its optimal value). This is because all sub-optimal assignments satisfying the implied constraints on the n_i variables must be found, whatever heuristic is used, and will fail only when the second stage tries to assign the x_{ik} variables. The different heuristics tested sometimes require slightly different numbers of backtracks to explore the suboptimal assignments; however, they principally differ in the number of complete assignments to the n_i variables that they consider at the optimal value of t, before finding one that leads to a solution. On average, minimum degree is the best of the three heuristics, with smallest domain slightly worse, and maximum degree significantly worse. Clearly, the total number of complete assignments to the n_i variables at the optimal value of t is the same for all three heuristics, but they each consider them in a different order, and hence find a solution earlier or later. Poor performance of a heuristic indicates that the assignments it considers first are those that are less likely to be successfully extended to the x_{ik} variables, but it is difficult to identify the reason.

A number of variable ordering heuristics have been investigated for the x_{ik} variables, which are assigned once a complete assignment to the n_i variables has been found. In all cases, the smallest numbered ring not yet fully occupied is chosen, and then a node chosen to place on this ring. The value ordering is to choose 1 before 0, i.e. when considering variable x_{ik}, choose to place node i on ring k before choosing not to place it. The best heuristic found chooses the node i assigned to fewest rings (i.e. for which n_i is smallest), breaking ties by choosing the node with largest degree. This means that any node that is only on one ring (i.e. $n_i = 1$) is placed first. The constraint in section 5, that if $n_i = 1$ and $x_{ik} = 1$, then every neighbour of node i must be on ring k, will then be triggered. By choosing the node with maximum degree, the largest number of neighbouring nodes will be placed as a result of this constraint. Since all these nodes have to be on the same ring as node i, given the first-stage assignment, postponing placing them is likely to lead to future failure and wasted search.

The heuristics investigated were selected largely by trial and error; although those chosen have been found to be superior to the others considered, it is unlikely that they are the best possible heuristics for these models. Nevertheless, the selected heuristics are giving good results.

7   Symmetry Breaking

Sherali & Smith's aim in [7] was to investigate ways of dealing with symmetry in MIP models in order to speed up solution time: the SONET problem was one of several considered. Symmetry also causes difficulties for search in constraint programming. The symmetry in the SONET problem partly arises because the available rings are indistinguishable: a solution is unchanged by permuting the rings, with their associated nodes. This symmetry could be eliminated in a similar way to [7] by adding constraints to the model to distinguish them; for instance, constraints that the set of nodes installed on ring k is lexicographically greater than the set of nodes on ring $k+1$ (so that any empty rings are the largest numbered). However, it is important to ensure that symmetry-breaking constraints do not conflict with the variable ordering. If they do, it can happen that the solutions in a symmetry equivalence class which the constraints eliminate are those that would be found earlier, given the variable and value ordering, than the solutions allowed by the constraints. Thus, finding a solution can be hindered rather than helped by the symmetry breaking. As discussed earlier, a dynamic variable ordering heuristic is used when assigning nodes to rings, and this makes it difficult to ensure that symmetry-breaking constraints are consistent with the variable ordering.

An alternative approach is to eliminate the symmetry using SBDS (Symmetry Breaking During Search) [5]. This adds constraints during search, on backtracking to a choice point, to ensure that no partial assignments symmetrically equivalent to those already considered will be considered in future. Since the constraints are added dynamically, depending on the choices already made, SBDS is compatible with any variable ordering, including dynamic orderings. SBDS requires the user to supply a function describing the effect of each symmetry on the assignment of a value to a variable. It is particularly simple to use when the symmetry is equivalent to permuting the rows or columns of a matrix of variables, since in that case the symmetry functions need only describe the transpositions of pairs of rows/columns, not all their permutations. Here, if the variables x_{ik} are thought of as corresponding to the elements of a 2-dimensional matrix, the columns of the matrix (corresponding to the 2nd subscript) can be permuted and hence, SBDS only requires functions describing the effects of transposing pairs of rings.

The symmetry breaking functions required by SBDS were first written for the original CP model, using only the x_{ik} variables. With the two-stage search described in section 4, the symmetry between the rings can be broken in the same way, using the same SBDS functions. Symmetry breaking is only needed when assigning values to the x_{ik} variables; the symmetry does not affect the n_i variables. This demonstrates once again the usefulness of having a symmetry breaking method that does not depend on the search order: with the two-stage search, it would be even more difficult than before to use symmetry-breaking constraints and ensure compatibility with the variable ordering, since the order in which the x_{ik} variables are assigned depends on the previous assignments to the n_i variables, as described in section 6.

The effect of using SBDS on the effort required to solve the SONET problem (ignoring the traffic capacity) is shown in Table 1. The model used includes the search strategy and variable ordering heuristics described earlier. The 15 medium-sized instances (10 nodes, 15 demand pairs, 6 available rings) are trivial if the symmetry is eliminated; if not, some of them take much longer to solve. For the larger instances, with 13 nodes,

Table 1. Solving medium-sized SONET instances, with and without symmetry-breaking, using ILOG Solver 6.0. 'Value' is the minimum number of ADMs required. 'Backtracks' is the number of times the search backtracks following failure, reported by Solver. 'Time' is the cpu time in seconds on a 1.7GHz Pentium M PC

Instance	Value	With SBDS		No symmetry breaking	
		Backtracks	Time	Backtracks	Time
1	14	14	0.03	596	0.21
2	14	11	0.02	27	0.03
3	14	25	0.03	292	0.12
4	13	43	0.04	917	0.31
5	15	49	0.04	7,423	2.35
6	14	19	0.02	20	0.03
7	13	9	0.03	17	0.02
8	14	12	0.02	13	0.02
9	15	33	0.03	499	0.18
10	14	50	0.05	2,270	0.70
11	12	1	0.02	9	0.03
12	15	50	0.03	776	0.27
13	15	104	0.05	2,361	0.74
14	15	270	0.18	28,960	9.14
15	15	28	0.02	667	0.21
Average		47.86	0.04	2,989.13	0.96

24 demand pairs and 7 available rings, it is essential to break the symmetry in order to solve them in a reasonable time: being larger, these problems are more difficult to solve even when the symmetry is eliminated, and the larger number of available rings means that the number of symmetrically equivalent solutions is up to 7!, rather than 6!. (Given m available rings, there are fewer than $m!$ symmetrically equivalent solutions, since swapping a pair of empty rings does not change the solution.)

8   Modelling Traffic Capacity and Demand Splitting

As stated earlier, in solving the problem ignoring the traffic capacity we can at least get a lower bound on the number of ADMs required in the full problem. In fact, several of the solutions found for the large instances are still feasible, and therefore optimal, when the demand levels between pairs of nodes and the traffic capacities of the rings are added. Hence, developing a model to solve the simpler problem is not simply a stage on the way to solving the full problem: the solutions found can be useful.

As in the MIP model, the demand capacity of the rings can be modelled by introducing variables $f_{\rho k}$ for $\rho = (i, j) \in E$ and $k \in M$. $f_{\rho k}$ is the fraction of the demand between nodes i and j that is assigned to ring k, and $0 \leq f_{\rho k} \leq 1$.

The constraints of these variables from the MIP model can be used without change. In the CP model, most $f_{\rho k}$ variables are set to 0 or 1 by these constraints as the x_{ik} variables are assigned. If neither node in a demand pair, corresponding to $\rho \in E$, appears on ring k, the corresponding $f_{\rho k}$ variable will be set to 0; if a demand pair appears on only one ring, the fraction of the demand on that ring is 1. Relatively few demand pairs appear on more than one ring, and so require reasoning about the fraction of the demand allocated to each.

In the MIP models, the variables $f_{\rho k}$ are decision variables. However, this creates another source of alternative equivalent solutions, since there are usually several ways of splitting the demand between rings, given a feasible allocation of nodes to rings. For instance, the optimal solution for the example in section 2, taking the demand levels into account, uses 23 ADMs. The sets of nodes on each ring are: {4,8,10,12,13}, {1,4,9,11,13},{4,5,6}, {2,3,7,10,13}, {2,5,7,9,12}. The total traffic demand between the nodes on the fourth ring is 45 channels, whereas the capacity of a ring is 40. However, this solution is feasible, because nodes 10 and 13, with a demand of 9 units between them, are also both installed on the first ring, which has enough spare capacity to take all the demand between these two nodes. Hence, a feasible solution can allocate 0, 1, 2, 3 or 4 channels on the first ring to the demand between nodes 10 and 13 and the remaining demand from this pair to the fourth ring.

Some of the formulations in [7] attempt to reduce the effects of this form of symmetry. For instance, RD2, discussed in section 2, has constraints to ensure that the total demand on successive rings is non-increasing, and these constraints reduce both forms of symmetry in the problem, to some extent. However, this model does not give as good performance as RD3, which only reduces the symmetry between the rings.

Constraint programming offers a different way to avoid considering equivalent alternative solutions in this case; it is not necessary to decide how the demand between a pair of nodes is to be split between rings, but simply to ensure that it can be done feasibly. This can be done in ILOG Solver: the bounds on the $f_{\rho k}$ variables can be reduced until they are consistent with the constraints acting on them. For instance, in the example given earlier where the nine channels required between nodes 10 and 13 can be split between two rings, with up to four channels on one ring and the remainder on the other, the domain of the relevant variables would be reduced to [0, .., 0.4444] and [0.5555, .., 1]. To take another simple case, if two nodes forming a demand pair are installed on two rings, both of which can accommodate all the demand between the pair, the ranges of the demand fraction variables will both be left as [0, .., 1], indicating that any fraction of the demand can be accommodated on one ring and the rest on the other. In this way, the alternative solutions from splitting the demand in different ways are not explicitly considered. When the bounds of the variables have been reduced, any value remaining (that gives an integral number of channels as a fraction of the total demand between a pair of nodes) can form part of a feasible solution consistent with the traffic capacity of the rings. It is straightforward then to construct a solution by choosing say the smallest value in some variable's domain (which must be one of the allowed fractions), re-establishing consistency and continuing in this fashion until every variable has been assigned.

A few other minor changes to the previous model are required to deal with the traffic capacity. The constraint that any two rings must have a total of more than r nodes installed on them is no longer correct, but can be modified so that two rings *either* have more than r nodes *or* more than b channels between them. Further, it is no longer necessarily true that a node with degree δ_i should appear on at most δ_i rings, and this constraint is dropped.

Table 2. Solving the SONET problem for the large instances, either ignoring the traffic capacity of the rings, or taking it into account. 'Optimum' is the minimum number of ADMs in each case

	Without traffic capacity			With traffic capacity		
Instance	Backtracks	Time	Optimum	Backtracks	Time	Optimum
1	543	0.41	22	990	0.95	22
2	445	0.42	20	451	0.65	20
3	401	0.42	22	417	0.62	22
4	680	0.55	23	1,419	1.52	23
5	150	0.07	20	922	0.70	22
6	948	0.85	22	306	0.29	22
7	86	0.11	20	982	1.15	22
8	40	0.07	20	34	0.09	20
9	2,280	2.11	22	35,359	45.13	23
10	2,280	1.77	23	4,620	6.75	24
11	349	0.35	22	352	0.54	22
12	48	0.05	20	1,038	1.09	22
13	88	0.09	21	105	0.14	21
14	558	0.45	23	1,487	1.66	23
15	1,038	0.97	22	13,662	19.59	23
Average	662.27	0.58		4142.93	5.39	

Table 2 shows the results with this strategy. When taking the traffic capacity into account, the revised model was used to find a new optimal solution from scratch in all cases, even when the solution already found was still feasible and therefore optimal. The average solution time for the instances used in the Sherali & Smith paper is now only a few seconds. The average is dominated by just two of the problems (9 and 15) for which satisfying the traffic capacity requires a larger number of ADMs than the simpler problem; further improvements to the CP model will focus on problems of this kind. In [7], the average solution time (with the level of symmetry breaking giving best performance) was 21,220 sec., although on Sun Ultra 10 (a much slower machine), and the corresponding average number of nodes in the branch and bound tree using CPLEX was about 600. It is hard to compare the performance of the CP model and the MIP model, but it is clear that these instances are now easy to solve using the CP model and search strategy described.

9   Conclusions

Sherali and Smith [7] investigated different possible ways of reducing the symmetry in the MIP model of the SONET problem. The most successful reformulation of those that they compared (RD3) leaves much of the symmetry intact: it imposes constraints that the number of nodes installed on successive rings is non-increasing. Optimal solutions for their sample instances have several rings with the same number of nodes, which will not be distinguished by these constraints. Furthermore, the RD3 mode does not address the symmetry due to the fact that when a demand is split between two or more rings it can often be done in different ways. In the CP model, the symmetry due to the rings being interchangeable can be eliminated using SBDS: this can reduce the run-time to solve the simplified problem for the medium-sized instances by more than an order of magnitude, as shown in Table 1. In view of the importance of eliminating the symmetry in the CP model, it is surprising that in the MIP approach reducing the symmetry, rather than eliminating it completely, is sufficient to achieve the best results. It is not clear what the implications are for hybrid MIP/CP algorithms for problems with symmetry.

The remaining symmetry in the problem, from splitting the demand in different ways, can also easily be eliminated in the CP model by finding feasible ranges of values for the fraction of demand on each ring, rather than choosing specific values. For a demand fraction variable, $f_{\rho k}$, whose domain is still an interval after the domains are made consistent with the constraints, the values in the interval that correspond to integral numbers of channels can be seen as *partially interchangeable*, as defined by Choueiry and Noubir [2]. In their definition, two values for a (discrete-valued) variable in a constraint satisfaction problem are partially interchangeable with respect to a subset A of variables if and only if any solution involving one value implies a solution involving the other, with possibly different values for variables in A. The set A for a variable $f_{\rho k}$ consists of the variables $f_{\rho k'}$ for $k' \in M$, $k' \neq k$, whose domain is an interval rather than a fixed value after its bounds have been reduced, i.e. A consists of the fraction variables for the same demand pair on other rings whose values are neither 0 nor 1. Choueiry and Noubir give an algorithm to identify and exploit partially interchangeable values, but in this case, it is not required: the method described in section 8 is sufficient. This demonstrates that partially interchangeable values can in some problems be easily exploited to allow the possible alternative solutions to be implicitly found, without enumerating them or choosing between them.

The CP model explicitly represents more features of the problem than the MIP models (the set of nodes on each ring, the set of rings each node is on, the number of rings each node is on), using multiple sets of variables, linked by channelling constraints. This richer model allows the decomposition of the search into two stages: first deciding how many rings each node should be on, and then assigning the nodes to specific rings, backtracking to the first stage if the assignment in the second stage fails. Although a similar decomposition has been used in other problems, it is worth noting that much previous work in CP using redundant models linked by channelling constraints [1] has used one model as a primal model, providing the search variables. The CP model of the SONET problem is an example of the more complex possibilities in using redundant models.

Modelling the problem in this way also allows implied constraints to be derived on the variables representing the number of rings each node is on. The implied constraints have been derived by hand from considering subgraphs of the demand graph, consisting of pairs of connected nodes and their neighbours, and finding lower bounds on the number of rings the pair of nodes must be installed on. It would be possible to extend these constraints to take account of the level of demand between the nodes, and not just the existence of a demand. This was done, for instance, in deriving valid inequalities for the MIP model [8]. It would also be possible to derive further implied constraints systematically by solving subproblems consisting of two or three connected nodes and their neighbours in the demand graph. The implied constraints already used have proved invaluable, and such a systematic approach, taking the demand levels into account, could lead to further improvements. Generating implied constraints from solutions to subproblems could also be a useful approach in CP models for other problems.

Considerable effort has been put into developing the CP model from the initial simple model based on the MIP model. Remodelling currently requires expertise in constraint programming, and it would be preferable if the process could be automated or at least supported. A proposal for a system to do this is presented by Frisch et al. [3], using models of the SONET problem as an illustration.

The SONET problem initially appeared intractable for constraint programming; remodelling has eventually resulted in a CP model that appears competitive with Sherali & Smith's MIP model. The largest instances that they solved can be solved with the CP model in an average of 30 sec. on a 600 MHz Celeron PC. Optimization is generally considered to be difficult for CP, but this experience suggests that successful CP models could be developed for other initially unpromising optimization problems too. Régin [6], for instance, has developed a CP approach to solving the maximum clique problem and has shown that it is competitive with existing methods, achieving new solutions to benchmark problems. Improved CP models might improve overall performance even for problems that are eventually solved using a CP/IP hybrid.

Acknowledgments

I should like to thank other members of the APES and CPPod groups (http://www. dcs.st-and.ac.uk/~apes, http://www. dcs.st-and.ac.uk/~cppod), especially Ian Miguel, for their encouragement and interest through several iterations of work on this problem. I am also grateful to Hanif Sherali and Cole Smith for sending me their sample instances, and to Sarah Fores and Les Proll for bringing the problem to my attention. This material is based in part on work done while the author was employed at the University of Huddersfield, U.K., and in part on works supported by the Science Foundation Ireland under Grant No. 00/PI.1/C075.

References

1. B. M. W. Cheng, K. M. F. Choi, J. H. M. Lee, and J. C. K. Wu. Increasing constraint propagation by redundant modeling: an experience report. *Constraints*, 4:167–192, 1999.
2. B. Y. Choueiry and G. Noubir. On the Computation of Local Interchangeability in Discrete Constraint Satisfaction Problems. In *AAAI-98*, pages 326–333, 1998.
3. A. M. Frisch, B. Hnich, I. Miguel, B. M. Smith, and T. Walsh. Towards Model Reformulation at Multiple Levels of Abstraction. In A. M. Frisch, editor, *Proceedings of the International Workshop on Reformulating Constraint Satisfaction Problems: Towards Systematisation and Automation*, 2002.
4. A. M. Frisch, I. Miguel, and T. Walsh. Modelling a Steel Mill Slab Design Problem. In *Proceedings of the IJCAI'01 Workshop on Modelling and Solving Problems with Constraints*, pages 39–45, 2001.
5. I. P. Gent and B. M. Smith. Symmetry Breaking During Search in Constraint Programming. In W. Horn, editor, *Proceedings ECAI'2000*, pages 599–603, 2000.
6. J.-C. Régin. Using Constraint Programming to Solve the Maximum Clique Problem. In F. Rossi, editor, *Principles and Practice of Constraint Programming - CP 2003*, LNCS 2833, pages 634–648. Springer, 2003.
7. H. D. Sherali and J. C. Smith. Improving Discrete Model Representations Via Symmetry Considerations. *Management Science*, 47:1396–1407, 2001.
8. H. D. Sherali, J. C. Smith, and Y. Lee. A branch-and-cut algorithm for solving an intra-ring synchronous optical network design problem allowing demand splitting. *INFORMS J. on Computing*, 12:284–298, 2000.
9. B. M. Smith. Search Strategies for Optimization: Modelling the SONET Problem. Report APES-70-2003, Aug. 2003. Presented at the CP'03 Workshop on Modelling and Reformulating Constraint Satisfaction Problems.

Integrating CSP Decomposition Techniques and BDDs for Compiling Configuration Problems*

Sathiamoorthy Subbarayan

Department of Innovation,
The IT University of Copenhagen,
Denmark
sathi@itu.dk

Abstract. We present the tree-of-BDDs approach, a decomposition scheme for compiling configuration problems. Methods for minimum explanations and full interchangeable value sets detection are also given. Experiments show that the techniques presented here can drastically reduce the time and space requirements for interactive configurators.

1 Introduction

A *configuration problem* (CP) will list the number of parameters (variables) defining a product, their possible values and the rules by which those values can be chosen. It can be viewed as a *Constraint Satisfaction Problem* (CSP), where the solutions to the CSP are equivalent to valid configurations. An *interactive configurator* is a tool, that takes a CP as input and interactively helps the user to choose his preferred *valid* configuration. The Binary Decision Diagram (BDD) [1] based symbolic CSP compilation technique [2] can be used to compile all solutions of a CP into a single (monolithic) BDD. Once a BDD is obtained, the functions required for interactive configuration can be efficiently implemented. Such approaches do not exploit the fact that CPs are specified in hierarchies. Due to this the BDD could be unnecessarily large. Such hierarchies are close to trees in shape. Hence, the tree decomposition techniques for CSPs could be used to enhance compilation. The contributions of this work are

1. A compilation scheme using the hinge decomposition technique [3] and BDDs. In the experiments the scheme results in upto 96% reduction in space.
2. Efficient minimum explanation (see Section 2) algorithms.
3. Methods to detect full interchangeable values for problem reformulations.
4. Efficient implementation and evaluation of the above techniques.

The space reduction is of great importance in online configuration applications, and also in embedded configuration, where one needs to embed the configuration details on a product itself, so that it could be reconfigured as and when required. Reductions in space and time enable the tree-of-BDDs scheme to scale high.

---

* This work was supported by the FIRST Graduate School (http://first.dk/).

R. Barták and M. Milano (Eds.): CPAIOR 2005, LNCS 3524, pp. 351–365, 2005.

The necessary background is given in Section 2. The tree-of-BDDs compilation scheme, and the algorithms for explanations and detection of full interchangeable values are presented in the subsequent sections. Discussion on experiments and related work, followed by concluding remarks, finish this paper.

2   Background

Some of the definitions in this section are based on the definitions in [2, 4, 5]. Let X be a set of variables $\{x_1, x_2, \ldots, x_n\}$ and D be the set $\{D_1, D_2, \ldots, D_n\}$, where D_i is the domain of values for variable x_i. A *relation* R over the variables in M, $M \subseteq X$, is a set of allowed combinations of values for the variables in M. Let $M = \{x_{m1}, x_{m2}, \ldots, x_{mk}\}$, then $R \subseteq (D_{m1} \times D_{m2} \times \ldots D_{mk})$. R restricts the ways in which the variables in M could be assigned values. A *constraint satisfaction problem instance* CSP is a triplet (X, D, C), where $C = \{c_1, c_2, \ldots, c_m\}$ is a set of constraints. Each constraint, c_i, is a pair (S_i, R_i), where $S_i \subseteq X$ is the scope of the constraint and R_i is a relation over the variables in S_i. Without loss of generality, we assume that the variables whenever grouped in a set are ordered in a fixed sequence. The same ordering is assumed on any set of the values of variables and the pairs with variables in them.

An *assignment* is a pair (x_i, v), where $x_i \in X$ and $v \in D_i$. The assignment (x_i, v) binds the value of x_i to v. A *partial assignment*, PA, is a set of assignments for all the variables in Y, where $Y \subseteq X$. A partial assignment is *complete* when $Y = X$. The notation $PA_{|xs}$, where xs is a set of variables, means the restriction of the elements in PA to the variables in xs. Similarly, $R_{|xs}$, where R is a relation, means the restriction of the tuples in R to the variables in xs. Let $var(PA) = \{x | (x, v) \in PA\}$, the set of variables assigned values by a PA. Let $val(PA) = \{v | (x, v) \in PA\}$, the set of values assigned for $var(PA)$. A partial assignment PA *satisfies* a constraint c_i, when $val(PA_{|var(PA) \cap S_i}) \in R_{i|var(PA) \cap S_i}$. A complete assignment CA is a *solution* S for the CSP when CA satisfies all the constraints in C. Let SOL denote the set of all solutions of the CSP.

A *configuration problem instance* CP is given by the available options in a product and the rules in which the choices for those options can be selected. In this paper, we only consider the static CP. In case of dynamic CP [6], some of the choices for the available options might add new options and rules to the CP. Methods for static CP can be easily extended to dynamic CP [7]. A CP can be modelled as a CSP instance, in which options, choices, and rules, in the CP, will correspond to variables, domains, and constraints, in the CSP. The SOL of the CSP will then denote the all valid-configurations of the CP. Hereafter, the terms CSP, and SOL may be used directly in place of the corresponding configuration terms. Fig. 1 shows a T-shirt configuration problem [7]. The first constraint in it implies that a T-Shirt with *MIB* (Men-in-Black) print is *black* in color. The second constraint implies that a T-shirt with *STW* (Save-the-Whales) print is not *small* in size.

A *configurator* is a tool that helps an user in selecting his preferred *valid* product. An *interative configurator* IC is a configurator which interacts with the

$X = \{color, size, print\}$

$D = \{\{black, white, red, blue\}, \{small, medium, large\}, \{MIB, STW\}\}$

$C = \{(\{color, print\}, \{(black, MIB), (black, STW), (white, STW), (red, STW),$
$\quad (blue, STW)\}) + (\{size, print\}, \{(small, MIB), (medium, MIB), (medium, STW),$
$\quad (large, MIB), (large, STW)\})\}$

Fig. 1. A T-shirt configuration problem as a CSP instance

user as and when a choice is made by the user. After each and every choice selection by the user the IC shows a list of unselected options and the valid choices for each of them. The IC only shows the list of valid choices to the user. This prevents the user from selecting a choice, which along with the previous choices made by the user, if any, will result in no valid product according to the specification. The configurator, hence, automatically hides the invalid choices from the user. The *hidden choices* will still be visible to the user, but with a tag that they are inconsistent with the current PA. When the current PA is extended by the user, some of the choices might be implied for consistency. Such choices are automatically selected by the configurator and they are called *implied choices*. Let us assume that the SOL of a CP can be obtained. Given SOL, the three functionalities required for interactive configuration are: *Display, Propagate,* and *Explain.*

  Display is the function, which given a CSP and a corresponding SOL', SOL' $\subseteq$ SOL, lists X, the options in CSP and CD_i, the available valid choices, for each option $x_i \in X$, where $CD_i = \{v | (x_i, v) \in S, S \in SOL'\}$. CD_i is the current valid domain for the variable x_i. *Propagate* is the function, which given a CSP, a corresponding SOL', and (x_i, v), where $v \in CD_i$, restricts SOL' to $\{S | (x_i, v) \in S, S \in SOL'\}$. Propagate could also be written as restricting SOL' to SOL'$|_{(x_i,v)}$. Sometimes the restriction might involve a set of assignments, which is equivalent to making each assignment in the set one by one. Let (x_{ih}, v_{ih}) be an implied or hidden choice. *Explain(x_{ih}, v_{ih})* is the process of generating, E, a set of one or more selections made by the user, which implies or hides (x_{ih}, v_{ih}). The E is called an *explanation* for (x_{ih}, v_{ih}). An explanation facility is required when the user wants to know why a choice is implied or hidden. Let PA be the current partial assignment that has been made by the user. By the definition of explanation, $Display(CSP, SOL_{|PA \setminus E})$ will list v_{ih} as a choice for the unselected option x_{ih}. Each selection, (x_i, v), made by the user could be attached a non-negative value as its priority, $P(x_i, v)$, and the explain function can be required to find a minimum explanation. The cost of an explanation is $Cost(E) = \sum_{(x_i,v) \in E} P(x_i, v)$. An explanation E is *minimum*, if there does not exist an explanation E' for (x_{ih}, v_{ih}), such that $Cost(E') < Cost(E)$. Minimum explanations are useful when different options in a product model have different priorities. For example, in a car configuration problem the main options like engine could be given high priority. Once the user decides on an option of high priority, minimum explanations will try to protect the high priority decision as much as possible.

INTERACTIVECONFIGURATOR(CP)
1   SOL:=COMPILE(CP)
2   SOL':=SOL, PA:={ }
3   **while** |SOL'| > 1
4   Display(CP, SOL')
5   $(x_i,v) :=$ 'User input choice'
6   **if** $(x_i,v) \in CD_i$
7   SOL':=Propagate(CP,SOL',(x_i,v))
8   PA:=PA$\cup\{(x_i,v)\}$
9   **else**
10   E:=Explain(CP, SOL', (x_i,v))
11   **if** 'User prefers (x_i,v)'
12   PA:=(PA\ E)$\cup\{(x_i,v)\}$
13   SOL':=SOL$_{|PA}$
14 **return** PA

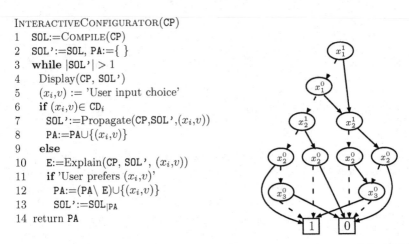

Fig. 2. The Interactive Configurator   **Fig. 3.** BDD for SOL of the T-shirt

The algorithm for interactive configuration is given in Fig. 2. The COMPILE function will return a BDD encoding all the valid solutions for the input CP. Since a CP will not change quite often, the COMPILE function need not have to do the entire compilation everytime the interactive configurator is used.

A BDD [1] is a rooted directed acyclic graph with two terminal nodes marked 1 and 0, respectively. All the non-terminal nodes will be associated with a Boolean variable. Each non-terminal node will have two outgoing edges: *low* and *high*. The nodes in a BDD will be ordered based on a linear variable order. BDDs can be used to represent Boolean functions. A BDD for a given function can be obtained by standard composition functions on BDDs representing atomic elements of the function. Given an assignment to the variables in the function, there exists a unique path from the root node to one of the terminal nodes, defined by recursively following the high edge, when the associated variable is assigned true, and the low edge, when the associated variable is assigned false. If the path leads to the 1-terminal, then the assignment is a solution, otherwise not. Although the size of a BDD can be exponential in the worst case, the BDDs are small for many practical functions. Due to this BDDs have been successfully used in several research areas, including: verification, CSP, and planning. The size of the BDDs are very sensitive to the used variable ordering. Further details on BDDs can be obtained from [1]. When solution space of a non-Boolean function needs to be represented by a BDD, each variable x_i with domain D_i will be represented by l_i Boolean variables, where $l_i = \lceil \lg |D_i| \rceil$. Each value of x_i will be represented by an unique combination of Boolean values for the corresponding Boolean variables. A BDD corresponding to any constraint can be obtained by the composition function on BDDs representing the atomic elements of the constraint. In case of CSP, the conjunction of the constraints represent SOL, and hence the conjunction function when applied to the BDDs obtained for all the constraints in the CSP will give a monolithic-BDD, representing the SOL. As l_i Boolean variables could represent 2^{l_i} values, in cases where $|D_i| < 2^{l_i}$, additional rules need to be added to maintain domain integrity. The BDD representing the solution space of the T-shirt example is shown in Fig. 3. In the example there are three variables: color (x_1), size (x_2), and print (x_3). As explained, each variable is represented by a list of Boolean variables. For example, the variable x_2 is represented by the Boolean variables (x_2^1, x_2^0). In the BDD any path from the root node to the terminal node 1, corresponds to one or more valid configurations. For example, the path from the root node to the terminal node 1, with all the variables taking low values represents the valid configuration $(black, small, MIB)$. Another path with x_1^1, x_1^0, and x_2^1 taking low values, and x_2^0 taking high value represents two valid configurations: $(black, medium, MIB)$ and $(black, medium, STW)$, respectively. In this path the variable x_3^0 is a don't care variable, and hence leads to two valid configurations. Any path from the root node to the terminal node 0 corresponds to one or more invalid configurations.

Let b be a BDD. $Var(b)$ denotes the set of variables in b. The notation b might also denote the function represented by b. The three operations required for interactive configuration – display, propagate, and explain – are implemented using the following BDD-operations:

1. Restrict($b,x_i=v$): Restrict will return a function obtained by substituting each occurence of x_i in b by v. The complexity of restrict is linear in the size of the given BDD. A non-Boolean variable restriction is done by a sequence of restrictions on the Boolean variables that encode the non-Boolean variable.
2. Exist(b,x_i): Exist(b,x_i)=Restrict($b,x_i=0$)$\lor$ Restrict($b,x_i=1$). The exist function will existentially quantify x_i from b. The complexity of exist is quadratic. Usually existential quantifications result in a smaller BDD. When a non-Boolean variable or a set of them is given as second argument, all the corresponding Boolean variables will be existentially quantified out.
3. Conjoin(b_1,b_2): Conjoin operation will give the result of AND-operation between b_1 and b_2.
4. Proj(b_1,b_2): Proj(b_1,b_2)=Exist(Conjoin(b_1,b_2),Var(b_2)\Var(b_1)).
5. ($b_1=b_2$): Equality testing of BDDs could be done in constant time.

Display is implemented by a polynomial algorithm [2] that traverses the paths from the root to the 1-terminal. Propagate is implemented by restrictions. In Section 5, algorithms for minimum explanations will be presented.

3   Tree Decomposition for CSP Compilation

The *constraint graph* (V,E) of a given CSP will contain a node for each constraint in the CSP and an edge between two nodes if their corresponding constraints share at least one variable in their scopes. Each edge will be labelled by the variables that are shared by the scope of the corresponding constraints. A subset of the edges, $(E' \subseteq E)$, in a constraint graph is said to satisfy the *connectedness* property, if for any variable v shared by two constraints, there exists a path between the corresponding nodes in the constraint graph, using only the edges in E' with v in their labels. A *join graph* of a constraint graph contains all the nodes in the constraint graph, but the set of edges in the join graph is a subset of the edges in the constraint graph such that the connectedness property is satisfied. If the join graph does not have any cycles, then it is called a *join tree*. Given a constraint graph of a CSP, it has a join tree if its maximum spanning tree, when the edges are weighted by the number of shared variables, satisfies the connectedness property [8]. A CSP with a join tree is called an *acyclic CSP*. Several tree decomposition techniques [8, 3] convert any CSP into an acyclic one: CSP'. All the tree decomposition techniques create clusters of constraints in a CSP, such that all the constraints in the CSP will be in at least one of the clusters. In the CSP', the conjunction of the original constraints in each of the clusters will be a constraint. A solution for CSP or CSP' will also be a solution for the other. A constraint of a CSP is said to be *minimal* when all the solutions of the constraint can be extended to a solution for the CSP. Acyclic CSPs can be efficiently solved. The nice property of an acyclic CSP is that, mutual projections (*arc-consistency*) between constraints adjacent in its join tree, makes all the constraints minimal. Minimality of constraints is enough to obtain the efficient functions required for interactive configuration. Since each constraint is minimal, a valid choice for a variable in a constraint will also be a valid choice for the variable in the entire CSP, and the display function will use this. Given an assignment, the propagate function just needs to restrict all the constraints in which the variable is present, and propagate the effect to other constraints through projections. Hence, instead of compiling the conjunction of all the constraints in a CSP into a monolithic-BDD, the tree decomposition based compilation scheme just requires a BDD for each constraint in the corresponding CSP'. We call the BDDs representing the constraints of CSP' as the *tree-of-BDDs*.

Since the configuration specifications have a hierarchical description, they might be amenable for CSP decomposition techniques. As the sum of sizes of the BDDs in a tree-of-BDDs is potentially smaller than the size of the corresponding monolithic BDD, decomposition schemes might lead to significant space savings in CSP compilations.

4   Configuration Functions on Tree-of-BDDs

Given a CP, we obtain an equivalent acyclic CP' using the hinge decomposition algorithm [3]. Then, we compile each constraint in CP' into a BDD. The resulting set of BDDs form the tree-of-BDDs. The tree-of-BDDs is made minimal by imposing arc-consistency. Due to acyclicity of the CP', this could be done efficiently. The data-structures used for implementing the display and propagate functions, *DisplayDe* and *PropagateDe*, on a tree-of-BDDs are:

1. TreeNodes(TN): A list of nodes (BDDs) in the join tree.
2. TreeNodeVars(TV): A list of sets of variables. One set for the variables that occur in each node of TN.

PROPAGATEDE$(AL, VNL, (x_i,v))$
  MQ.Clear()
  $\forall N \in VNL[x_i]$
    N:=Propagate$(N,(x_i,v))$
    MQ.Push(N)
  while MQ.NotEmpty()
    $CURR$:=MQ.Pop()
    $\forall N \in AL[CURR]$
      N':=Proj$(N,CURR)$
      if $(N' \neq N)$
        N:=N'
        MQ.Push(N)
  return

DISPLAYDE(TN, TV, R)
  VD:={}
  $\forall N \in TN$
    if $(R[N]$=false$)$
      VD:=$VD \cup Display(N)_{|TV[N]}$
  print VD
  return

Fig. 4. The PropagateDe   **Fig. 5.** The DisplayDe

3. AdjList (AL): A list of adjacency list for each node in TN. Adjacency list of a node n is the list of nodes that are adjacent to n in the join tree.
4. VarNodeList(VNL): A list of list of nodes, where each list of nodes corresponds to a variable, and stores the nodes in which the variable occurs.
5. ModifiedQueue(MQ): A queue to store the nodes whose BDDs were changed due to a propagation. The queue is maintained, so that the changes are propagated to the neighbours of the corresponding node.
6. ValidChoices(VC): A set of pairs. Each pair is made up of a variable x_i and its CD_i consistent with a **PA**.
7. Redundant(R): A list of Boolean values, one for each node in TN. Since a variable, x_i, could be present in more than one node of the tree, the *display* procedure on one among the nodes in which x_i is present is enough to obtain CD_i. The values in R are such that calling the display function on the nodes with R value set false is enough to obtain VC. The R is heuristically populated to maximize the number of nodes with R value set true. Experiments showed that several nodes could be redundant.

The *PropagateDe* function, listed in Fig. 4, makes a tree-of-BDDs consistent with an assignment (x_i,v). Due to acyclicity of CP', once a node is changed during an assignment, it will not change again due to projections on it by its neighbours. Hence, the propagation is one way, and the complexity of the function is polynomial. The *DisplayDe* function, listed in Fig. 5, calls the Display function on all non-redundant nodes in the tree-of-BDDs, and prints the valid choices. The complexity of DisplayDe is also polynomial.

5   Algorithms for Minimum Explanations

Explanations have been of interest to both AI and CSP fields [9, 10, 11, 4]. In this section we present two algorithms for minimum explanations: one each for the monolithic-BDD and the tree-of-BDDs schemes.

5.1   Minimum Explanations in the Monolithic-BDD Scheme

Let the *Explain* function be called with (x_{ih}, v_{ih}) as argument. Let b be the BDD representing SOL. An explanation would correspond to a path from the root node of the BDD b to the 1-terminal node. Such a path must contain the assignment (x_{ih}, v_{ih}), and the path is said to violate an assignment $(x_i, v) \in$ PA, if $\psi \not\Rightarrow (x_i, v)$, where ψ is the conjunction of the literals designating the path. A minimum explanation would then correspond to such a path in which the sum of the costs of the violated assignments in PA is minimum. A minimum cost path could be obtained by using a variant of a shortest path algorithm for directed acyclic graphs(DAG) [12].

For simplicity the cost for violating each user assignments is assumed to be one. The general case is discussed later. We also assume that all the Boolean variables representing a variable in a CSP are placed consequently in the variable order of the BDDs. This is a meaningful assumption, as such Boolean variables are highly related, and it is advisable to keep them together in the variable order for minimizing the size of the BDDs. Let n be a node in the BDD b, Var(n) denote the Boolean variable x_i^k associated with the node, and CSPVar(n) denote the CSP variable x_i corresponding to Var(n). The CSPVar of terminal nodes will be a null value. If the variable x_i^k is assigned true (false) in PA, then the high edge of node n will have a cost one (zero), and the low edge cost zero (one). If x_i^k is unassigned in PA, then both of its outgoing edges will have cost zero. Given a path from the root node to the 1-terminal node, the cost for violating a variable x_i needs to be counted only once. But, when two Boolean variables corresponding to x_i differs from the corresponding values assigned in PA, then the cost will be counted twice in a path. To prevent this, we use a flag, *is_added*, associated with each node in the BDD. Once a cost is added for violating a variable x_i, then *is_added* flag of the subsequent nodes will be set to *true*, until the path moves into a node belonging to a variable other than x_i. As in the classical shortest path algorithms, we also have two more values associated with each node in the BDD: *cost* and *parent*.

Now, the problem resembles closely to the shortest path problem in a DAG. Such an algorithm is presented in Fig. 6. The algorithm uses *Relax*, an algorithm listed in Fig. 8. Given a node n, if either of the child nodes could be reached in lesser cost through n, the values of the corresponding child node will be changed by the relax procedure. The conjunction in line 9 of the *Relax* procedure is necessary, because if the $left$.is_added is set to $false$ and both the costs are the same, the current path to $left$ could potentially be costlier than the path through the corresponding n. AssignedVal (AV) is an array containing a value for all the Boolean variables in the BDD b. The values in AV will be true (false) if the corresponding variable is assigned true (false) in PA. Otherwise d, a don't care value. The *TopologicalSort* function returns a topologically sorted ordering of the nodes in an input BDD. In a topologically sorted order, if a node a appears before node b, then there will not be any path from b to a in the input BDD. The *Initialize* function will assign an infinite cost to all the nodes in the BDD. For the root node, it will assign a zero cost, a null parent, and set is_added flag to false.

EXPLAIN(b,PA,(x_{ih},v_{ih}))
  $b := b_{|(x_{ih},v_{ih})}$
  $NQ :=$ TopologicalSort(b)
  Initialize(b)
  while NQ.NotEmpty()
    $n := NQ$.Pop()
    Relax(n)
  $E :=$ 'Explanation corresponding
    to the path to 1-terminal'
  return E

Fig. 6. The Explain

EXPLAINDE(PA, (x_{ih},v_{ih}))
  $TN' :=$ SOL$_{|(x_{ih},v_{ih})}$
  while (TN'.size()$\neq1$)
    $b_1 := TN'$.Pop()
    $b_2 := TN'$.Pop()
    $b_{12} :=$ Conjoin(b_1,b_2)
    QuantifyUnnecessaryVars(b_{12})
    TN'.Push(b_{12})
  $E :=$ Explain(TN'.Pop(),PA,(x_{ih},v_{ih}))
  return E

Fig. 7. Sketch of the ExplainDe

RELAX(n)
```
1     if (n.is_added=true ∨ AV[Var(n)]=d)
2        leftcost := 0, rightcost := 0
3     else if  (AV[Var(n)]≠ d ∧ AV[Var(n)]=true)
4        leftcost := 1, rightcost := 0
5     else
6        leftcost := 0, rightcost:=1
7     left:=LOWCHILD(n), right:=HIGHCHILD(n)
8     if ( (left.cost > n.cost+leftcost) ∨
9          (left.cost=n.cost+leftcost ∧ left.is_added=false) )
10       left.cost := n.cost+leftcost
11       left.parent := n
12       if (leftcost=1 ∧ CSPVar(left)=CSPVar(n))
13          left.is_added := true
14       else if (leftcost=0 ∧ CSPVar(left)=CSPVar(n))
15          left.is_added := n.is_added
16       else
17          left.is_added := false
18    if ( (right.cost > n.cost+rightcost) ∨
19         (right.cost=n.cost+rightcost ∧ right.is_added=false) )
20       right.cost:=n.cost+rightcost
21       right.parent := n
22       if (rightcost=1 ∧ CSPVar(right)=CSPVar(n))
23          right.is_added := true
24       else if (rightcost=0 ∧ CSPVar(right)=CSPVar(n))
25          right.is_added := n.is_added
26       else
27          right.is_added := false
28    return
```

Fig. 8. The Relax procedure used by the Explain algorithm

Since each node in a BDD has exactly two outgoing edges, the complexity of topological sorting is linear in the BDD size. The relax procedure is also called linear number of times. Hence, the complexity of the explanation algorithm is linear. In the general case, when the cost need not be one and may be any positive value, then two paths might have to be remembered by each node. When the relax procedue is called for a node, a path to one of its child nodes with *is_added* flag set to *true* may be costlier than another path with the flag set *false*. But still, the cheaper path with flag set *false* could potentially violate the PA in the consecutive nodes belonging to the same CSPVar. Until a CSPVar boundary is crossed, or the difference between the cost of two such paths is larger than the cost of the corresponding assignment, both the paths have to be remembered. To handle this problem two paths, one with *is_added* flag set *false*, and another with *is_added* set *true*, needs to be remembered. When the paths cross a CSPVar boundary, only the best one among them will be selected. The algorithm would still remain linear.

This algorithm was inspired by a similar algorithm discussed in [13]. However, the problem in [13] was made easier by adding an additional Boolean variable for each dynamic constraint added to a BDD. The additional variables will be used as switches to turn on or off the corresponding dynamic constraints. The additional variables will significantly increase the size of the BDD.

5.2   Minimum Explanations in the Tree-of-BDDs Scheme

Given a tree-of-BDDs representing $SOL_{|(x_{ih}, v_{ih})}$, adjacent BDDs in the tree-of-BDDs can be conjoined together until a single BDD is obtained, which can be given as input to the Explain algorithm to obtain a minimum explanation. But the problem with this approach is that the BDD obtained at the end will be as large as the corresponding BDD in the monolithic case, and hence, we lose the benefit of using decompositions for reducing space usage. But we know that:

1. In CPs, just assigning values for less than half of the variables, will imply a value for the rest, and result in a CA. We have observed this during our previous experiments [2].
2. When all the variables other than those in the $var(PA)$ are existentially quantified from the BDD representing $SOL_{|(x_{ih}, v_{ih})}$, the Explain algorithm will still be able to produce a minimum explanation. Hence, the unnecessary variables $(X \backslash var(PA))$ can be abstracted away.

Using the above listed facts, unnecessary variables could be abstracted away by existential quantifications, as and when they appear in only one BDD of the tree-of-BDDs. The sketch of the explanation procedure, *ExplainDe*, based on this technique is listed in Fig. 7. For simplicity, the changes made to the other datastructures, like TV and VNL, are not shown in the algorithm.

6   Methods for Full Interchangeability Detection

In this section, we present methods to detect full interchangeable (FI) values of a CSP in both the monolithic-BDD and the tree-of-BDDs scheme. A value a for a variable x_i is *fully interchangeable* [14] with a value b if and only if every solution in which x_i=a remains a solution when b is substituted for a and vice-versa. In a monolithic-BDD representing SOL, two values of a variable x_i are fully inter-changeable, when all the paths from the root node of SOL to its 1-terminal node, containing either one of the values, has a *companion* path containing the other value. The later path shares the first path in nodes corresponding to all vari-ables other than x_i. Based on this observation, a polynomial procedure similar to the *Display* function, which traverses all solution paths in a BDD can be used to find FI values. But, we describe here a simpler algorithm using existential quantifications.

Theorem 1. *A value a for a variable x_i is FI with a value b if and only if $Exist(SOL_{|(x_i,a)},x_i)=Exist(SOL_{|(x_i,b)},x_i)$ and vice-versa.*

The proof of the above theorem is based on the definition of restrict function and existential quantification. Given a variable x_i with domain $\{a_1,\ldots,a_n\}$, the BDDs corresponding to $\forall_{k=1}^{n} Exist(SOL_{|(x_i,a_k)},x_i)$ can be used to find FI value sets of the variable x_i. After that, the corresponding CSP could be reformulated by choosing a representative value for each FI value sets. Such reformulations could simplify the problem. Reformulations could be made transparent to the user by appropriately changing the *Display* function. This method could be extended to the tree-of-BDDs scheme as follows. A value a for a variable x_i is *neighbourhood interchangeable* (NI) [14] with a value b if and only if for every constraint on x_i, the values compatible with x_i=a are exactly those compatible with x_i=b. By the definitions of FI and NI, NI⇒FI, while FI⇏NI.

Theorem 2. *When all the constraints in a CSP are minimal, FI⇔NI*

The proof of the above theorem is simple as any solution of a minimal constraint can be extended to a solution for the entire CSP. In the tree-of-BDDs scheme the minimality of a join tree is maintained. Hence, for a given variable x_i, the NI value sets can be deduced by using the algorithm, explained for finding FI value sets in the monolithic-BDD scheme, on all the nodes in which x_i is present. By the above theorem, the obtained NI value sets are equivalent to the FI value sets. An example illustrating the benefit of FI value sets detection and reformulation is shown in Fig. 9. This example is taken from a *PC* configuration problem [15]. In this example, the variable *GraphicsCardId* has six values in its domain. But after FI value sets detection and reformulation, the domain size reduces to two, and hence just a single Boolean variable is enough to encode the variable. This will reduce the size of the BDD. The extent of the space reduction depends on the position of the variable in the BDD variable order and the structure of the PC instance.

7   Experimental Results

Experiments are done by implementing the techniques presented so far, on top of CLab [16], an open source interactive configurator based on the monolithic-BDD

$GraphicsCardId = \{Asus, Diamond, Creative, ATIRage, Matrox, ATIBulk\}$
FI value sets for $GraphicsCardId$ are $\{Asus, Diamond, Creative, Matrox\}$
and $\{ATIRage, ATIBulk\}$

Domain of $GraphicsCardId$ in the reformulated model is $\{Asus, ATIRage\}$

Fig. 9. An example for FI value sets detection and reformulation

scheme. CLab does not have an explanation facility, and the linear minimum explanation algorithm was implemented in it. The tree-of-BDDs scheme is implemented as a tool called iCoDE (interactive Configurator with Decompositions and Explanations). The iCoDE source code, which includes CLab with explanations, is available at [17]. A Pentium Xeon machine with 4GB RAM and 1MB L2 Cache is used in the experiments. The default variable ordering, the order in which variables appear in the input file, is used. Four configuration instances are used in the experiments: PC, Renault [4], psr-1-32-1 and psr-1-32-2. First two of them are publicly available on the web [15]. PC instance is a personal computer configuration problem. Renault instance is a car configuration problem, and it was the only instance used in the experiments of [4]. Renault instance is quite large, the input file size is around 23 Megabytes, and it has totally around 200,000 tuples in 113 constraints of it.The other two instances are power supply restoration problems modelled as configuration problems. The characteristics of the benchmarks are listed in Table 1. $\sum |D_i|$ refers to the sum of the domain size of the variables in the instance. Arity of a constraint is the size of its scope. Max a refers to maximum arity of the constraints. $\sum a_i$ refers to the sum of the constraint arities.

The compilation details are listed in Table 2. CLab and iCoDE refer to the monolithic-BDD and the tree-of-BDDs scheme, respectively. The column "Hinge(sec)" refers to the time taken by the Hinge CSP Decomposition algorithm. Cyclicity is the size of the maximum-sized cluster in the output of the hinge decomposition. Lower cyclicity is preferable and higher cyclicity means a large cluster size and it will require more space. As specified in [3], cyclicity is an invariant for a given instance. Shuffling the constraints sequence in an input instance will not change the cyclicity of the instance. The following columns list the time taken for compilation in both the schemes. In the larger instance, Renault, iCoDE takes only around 50% of the time taken by CLab. The iCoDE timings could be considerably improved by making the BDDs corresponding to the original constraints arc-consistent, before building the BDDs for each cluster in the join tree. In the present implementation, arc-consistency is done only after the BDDs for each cluster is obtained. The compilation timings include the time taken for finding FI values. Reformulating the instances, by removing FI values, resulted in decreasing the space requirement by around 10% in the instances. In case of iCoDE, the instance will be reformulated by exploiting the fully interchangeable values. BDD-nodes for CLab, refers to the number of nodes in the monolithic-BDD obtained after compilation. In case of

Table 1. Configuration Benchmarks

Benchmark	Variables	$\sum |D_i|$	Constraints	Max a	$\sum a_i$	$|SOL|$
PC	45	383	36	16	107	1.19×10^6
Renault	99	402	113	10	588	2.84×10^{12}
psr-1-32-1	110	258	222	9	913	1.72×10^{11}
psr-1-32-2	110	296	222	9	913	1.76×10^{13}

Table 2. Compilation Details

Benchmark	Hinge (sec)	Cyclicity	$|TN|$	Compile (sec)		BDD-nodes		Peak BDD-nodes	
				CLab	iCoDE	CLab	iCoDE	CLab	iCoDE
PC	0.04	21	16	0.11	0.19	16494	4458	0.08M	0.04M
Renault	0.25	25	73	119	77	455796	17602	2.5M	0.08M
psr-1-32-1	2	74	114	0.46	4	56923	8917	0.6M	0.6M
psr-1-32-2	2	74	114	2	9	246775	22101	1.2M	1.2M

Table 3. Response and explanation time comparison

Benchmark	ART(sec)		WRT(sec)		AET (sec)		WET (sec)	
	CLab	iCoDE	CLab	iCoDE	CLab	iCoDE	CLab	iCoDE
PC	0.004	0.0001	0.050	0.006	0.004	0.010	0.010	0.030
Renault	0.070	0.0003	0.452	0.020	0.160	0.088	0.440	0.921
psr-1-32-1	0.016	0.0010	0.057	0.038	0.031	1.208	0.067	2.155
psr-1-32-2	0.037	0.0002	0.618	0.107	0.122	5.38	0.329	10.78

iCoDE, BDD-nodes refers to the total number of nodes in the tree-of-BDDs. As in any standard BDD package, there is a chance for two BDDs in the hinge tree to share some BDD-nodes, and such shared nodes will be counted only once. Hinge decomposition(iCoDE) for Renault instance results in 96% decrease in the number of BDD-nodes required. Peak BDD-nodes refers to the number of nodes used by CLab (iCoDE), to finish the compilation process. Even if the final BDD is small, the intermediate BDDs required during the compilation process may be very large, and hence the interest in Peak BDD-nodes. For the Renault instance, iCoDE uses only 80,000 BDD-nodes to finish the compilation process, which is 97% less than that used by CLab. Response and explanation times of both the schemes are compared in Table 3. The values listed are obtained from 10,000 random interactions. Each response is the sum of the time taken for a call to the *Propagate* function and a subsequent call to the *Display* function. During the random interactions, the time taken for calls to *Explain* function was also measured. ART and WRT refers to the average and worst response time, respectively. AET and WET refers to the average and worst time taken for minimum explanations, respectively. In case of Renault instance, iCoDE results in 200X speedup for ART and 20X speedup for WRT. In explanation results for Renault instance, even though the *ExplainDe* function is not linear like *Explain*, in average case it takes less time to generate an explanation. This is due to 96% decrease in the BDD-nodes value. The WET by iCoDE is around twice the time taken by CLab. We believe that WET of iCoDE could be improved well, as the quantification scheduling of *ExplainDe* is naive. Several non-naive approaches like [18], could be used to improve the WET for iCoDE. Efficient techniques for building BDDs [19] could also be used to reduce the WET of iCoDE. Also, the explanation function in CLab is a contribution of this work. Results on the last two instances show the same phenomenon as the Renault.

8   Related Work

We are not aware of any work detecting FI values for problem reformulation. In [20], a scheme without decomposition, and with NI detection, was presented. An automaton-based scheme, similar to the monolithic-BDD scheme, without decomposition was presented in [4]. In [5], a scheme combining tree cluster-ing [8] and cartesian product representation was presented. That work did not have minimum explanations. In [21], Tree-Driven Automata, a scheme combining automata and tree decomposition, was presented without experimental results. It did not focus on the required functionalities, like explanations.

9   Conclusion

A decomposition scheme, tree-of-BDDs, for compiling models for interactive con-figurators was presented. The decomposition scheme results in a drastic reduc-tion in space required for storing the compiled solutions. A linear algorithm for minimum explanations in the monolithic-BDD scheme was given. Using ab-stractions, the explanation algorithm was extended to work in the tree-of-BDDs scheme. Procedures for exploiting full interchangeable values was given. All the techniques presented here were experimentally evaluated as useful. Altogether, we believe that this work improves the state-of-the-art in configurators and CSP compilation techniques.

Future work include: precise complexity analysis of the techniques presented here, experiments on BDD variable orderings, hybrid representations instead of just BDDs, using multi-valued decision diagrams instead of BDDs [22], efficient quantification scheduling for explanations in the tree-of-BDDs scheme, and com-paring other tree decomposition techniques with hinge decomposition technique.

Acknowledgements

Special thanks to Henrik Reif Andersen, Rune M. Jensen, Erik R. van der Meer, Tarik Hadzic, and the anonymous reviewers for commenting on this work.

References

1. Bryant, R.E.: Graph-based algorithms for boolean function manipulation. IEEE Transactions on Computers **8** (1986) 677–691
2. Subbarayan, S., Jensen, R.M., Hadzic, T., Andersen, H.R., Hulgaard, H., Møller, J.: Comparing two implementations of a complete and backtrack-free interactive configurator. In: CP'04 CSPIA Workshop. (2004) 97–111
3. Gyssens, M., Jeavons, P.G., Cohen, D.A.: Decomposing constraint satisfaction problems using database techniques. Artificial Intelligence **66** (1994) 57–89
4. Amilhastre, J., Fargier, H., Marquis, P.: Consistency restoration and explanations in dynamic CSPs-application to configuration. Artificial Intelligence **1-2** (2002) 199–234
5. Madsen, J.N.: Methods for interactive constraint satisfaction. Master's thesis, Department of Computer Science, University of Copenhagen (2003)
6. Mittal, S., Falkenhainer, B.: Dynamic constraint satisfaction problems. In: AAAI. (1989) 25–32
7. van der Meer, E.R., Andersen, H.R.: BDD-based recursive and conditional modular interactive product configuration. In: CP'04 CSPIA Workshop. (2004) 112–126
8. Dechter, R.: Constraint Processing. Morgan Kaufmann (2003)
9. Junker, U.: Quickxplain: Conflict detection for arbitrary constraint propagation algorithms. In: IJCAI Workshop on Modelling and Solving with constraints. (2001)
10. Jussien, N.: e-constraints: explanation-based constraint programming. In: CP01 Workshop on User-Interaction in Constraint Satisfaction. (2001)
11. Freuder, E.C., Likitvivatanavong, C., Moretti, M., Rossi, F., Wallace, R.J.: Explanations and optimization in preference-based configurators, LNCS 2627, Recent Advances in Constraints, Springer (2002) 58–71
12. Cormen, T.H., Leiserson, C.E., Rivest, R.L., Stein, C.: Introduction to Algorithms (Second Edition). MIT Press/McGraw-Hill (2001)
13. Bouquet, F., Jégou, P.: Using OBDDs to handle dynamic constraints. Information Processing Letters **62** (1997) 111–120
14. Freuder, E.C.: Eliminating interchangeable values in constraint satisfaction problems. In: AAAI. (1991) 227–233
15. CLib: Configuration library. http://www.itu.dk/doi/VeCoS/clib/ (online)
16. Jensen, R.M.: CLab: A C++ library for fast backtrack-free interactive product configuration. In: Proceedings of CP, Springer LNCS (2004) 816 http://www.itu.dk/people/rmj/clab/.
17. iCoDE. http://www.itu.dk/people/sathi/icode/ (online)
18. Chauhan, P., Clarke, E.M., Jha, S., Kukula, J.H., Shiple, T.R., Veith, H., Wang, D.: Non-linear quantification scheduling in image computation. In: Proceedings of ICCAD. (2001) 293–299
19. Cabodi, G., Camurati, P., Quer, S.: Dynamic scheduling and clustering in symbolic image computation. In: Proceedings of DATE. (2002) 150–157
20. Weigel, R., Faltings, B.: Compiling constraint satisfaction problems. Artificial Intelligence **115** (1999) 257–287
21. Fargier, H., Vilarem, M.C.: Compiling CSPs into tree-driven automata for interactive solving. Constraints **9** (2004) 263–287
22. Kam, T., Villa, T., Brayton, R.K., Sangiovanni-Vincentelli, A.L.: Multi-valued decision diagrams: Theory and applications. Multiple-Valued Logic: An International Journal **4** (1998) 9–62

Formulations and Reformulations in Integer Programming

Michael Trick

Tepper School of Business, Carnegie Mellon,
Pittsburgh, PA USA 15213
trick@cmu.edu*

Abstract. Creating good integer programming formulations had, as a basic axiom, the rule "Find formulations with tighter linear relaxations". This rule, while useful when using unsophisticated branch-and-bound codes,is insufficient when using state-of-the-art codes that understand and embed many of the obvious formulation improvements. As these optimization codes become more sophisticated it is important to have finer control over their operation. Modelers need to be even more creative in reformulating their integer programs in order to improve on the automatic reformulations of the optimization codes.

1 Introduction

Integer programming has shown itself to be an effective mechanism for solving a wide variety of difficult combinatorial optimization problems of practical interest. While no technique can solve every instance of such problems quickly, integer programming has been robust and effective enough to play a key role in solving problems in applications such as airline crew scheduling, combinatorial auction winner determination, telecommunications network design, sports scheduling and many other applications.

Despite the practical success of integer programming, initial forays into this area are often full of frustration: seemingly obvious formulations "don't work", leading to excessive computation time for even small instances. Success with integer programmming seems to be a hit-or-miss proposition, with more misses than hits.

In this note, I examine two problems of practical interest: a transportation design problem and a sports scheduling problem. We will show that key to the successful application of integer programming to these problems is the choice of formulation. In both cases, initial formulations lead to intractible instances, while "good" formulations can be solved very quickly with modern software. However,

---

* Thanks to DASH Optimization who provided the XPRESS-MP software under their Academic Partner Program. A previous version of this paper was presented in the FORMUL'04 workshop held in conjunction with Constraint Programming 2004, Toronto. The author thanks the participants of that workshop for their comments.

R. Barták and M. Milano (Eds.): CPAIOR 2005, LNCS 3524, pp. 366–379, 2005.

the "good" formulations have to be very creative, since modern software embeds most of the obvious formulation improvements.

The general issue of formulations in integer programming has been little studied. Textbooks generally provide lots of examples in the hope that readers will be able to find generalizations. One exception is the book by Williams [6], which does concentrate on formulations and provides some broad perspective on integer programming formulations. Otherwise the integer programming literature contains a vast number of formulations, many with computational experience, with few generalizations on what leads to a successful formulation. The few generalizations we have are so well understood that they are included in modern software (as we will see) to the extent that model formulations do not need to include the "improvements": the software will generate them itself.

So, integer programming formulations often "don't work", taking excessive time to find and prove optimal solutions, but modern software already includes some of the obvious improvements. What is a modeler to do?

By closely examining these two cases, I believe that there are general things to be learned. First, I think the integer programming paradigm where models are given by the variables, objective, and linear constraints can be greatly enhanced by learning from the constraint programming field whereby models are often given by higher-level constructs. As we will see, within integer programming, there is a huge difference between the linear constraint

$$4x_1 + 10x_2 + 7x_3 + 5x_4 + 8x_5 \leq 17$$

and what might be denoted the "knapsack" constraint

$$\texttt{knapsack}([4, 10, 7, 5, 8], x, 17)$$

with all the implications that come from our understanding of knapsack constraints. Constraint programmers understand this difference, while integer programmers tend to muddy up the distinction (or leave it to software to handle).

Second, there is still room to provide better formulations to software in the standard integer programming sense: formulations with better relaxations. It must be understood, however, that software is already pretty good at "tightening" formulations, so the modeler has to be quite creative to get beyond what the software can do. This leads to the interesting question of what can be embedded in software: in the race between modelers and software, will there always be a role for modelers or will software be able to include everything a modeler can think of?

Third, one area where modelers have a advantage is in the creation of problems with a huge number of constraints or variables. Such formulations can be very powerful, but are difficult for integer programming codes to generate since they involve the concept of a cut (or variable) generation algorithm, rather than generating the cuts or variables themselves. Given the power of such formulations, is there any mechanism for automatically generating these models, or will human modelers always be required to provide guidance here?

2   Example 1: Transportation Design

My first example is a transportation design problem that came from a consulting project I did a year or two ago. The company sends packages between pairs of cities. The amount it sends is high volume (multiple trucks per day), so its trucks go simply between pairs (there are no complicated routing issues) in one-way trips. For a pair of cities (A,B), the company has a set of packages to be sent from A to B. Each package has a size, a time for which is available to load at A, and a time for which is needed at B. Trucking firms have provided the company with a set of truck choices. Each "choice" consists of a truck of a particular size, leaving A at a particular time, and arriving at B at a particular time. Each choice has a cost. The goal of the company is to choose a set of trucks that can hold all of the packages and gets them to B on time. A package cannot be split among multiple trucks.

Naturally the real problem is more complex, with more cities, complicated routing, multiple capacity constraints, splittable packages, and other aspects, but this simplified model has most of the critical features.

The natural integer programming formulation for this has a set of binary (0-1) variables for the decision on whether to use a particular truck (indexed by i) and a binary variable for whether package j goes onto truck i. We handle the timing issues by an array can_use(i,j) which is 1 if truck i can handle package j (that is, j is available at A before i departs, and i arrives at B before j is required there). This results in the formulation in Figure 1 (written in the language Mosel [7]).

Don't worry if you are not familiar with Mosel: this is a straightforward integer programming formulation. Constraints (1) ensure that the total size of the packages assigned to a truck is no more than the capacity of the truck. Constraints (2) ensure that x(j,i) is 0 whenever y(i) is (NUM_PACKAGE is the number of packages in the instance). While (2) might look to be a strange formulation of the constraint, this is a "standard" integer programming approach to handling this requirement. Constraints (3) for every package to go on some truck. (4) and (5) enforce the integrality restrictions.

I will illustrate the effect of various formulations with a single 10 truck, 20 package instance (the real examples are at least an order of magnitude larger). The formulation above with this instance solved with XPRESS-MP [7] (Optimizer version 15.20.05) results in 11.2 seconds of computation time (3Gz, Intel/Windows machine, 2Gb memory, default settings), with 31,825 nodes in the branch-and-bound tree. This time is not extreme, but it is much larger than we would want with such a small instance.

Now, it is a fundamental tenet of integer programming that the key to a successful formulation is a "tight" linear relaxation. The linear relaxation of the above model replaces (4) and (5) with

```
forall (i in TRUCKS)
        y(i) <= 1                          ! (4')
    forall (i in TRUCKS, j in PACKAGES)
```

```
model "Transportation Planning"
    uses "mmxprs"

    declarations
        TRUCKS = 1..10
        PACKAGES = 1..20
        capacity: array(TRUCKS) of real
        size: array(PACKAGES) of real
        cost: array(TRUCKS) of real
        can_use: array(PACKAGES,TRUCKS) of real
        x: array(PACKAGES,TRUCKS) of mpvar
        y: array(TRUCKS) of mpvar
    end-declarations

    capacity:= [100,200,100,200,100,200,100,200,100,200]
    size := [17,21,54,45,87,34,23,45,12,43,
             54,39,31,26,75,48,16,32,45,55]
    cost := [1,1.8,1,1.8,1,1.8,1,1.8,1,1.8]
    can_use:=[1,1,1,1,1,1,0,0,0,0,   1,1,1,1,0,0,0,0,0,0,
        1,1,1,1,1,1,1,1,0,0,   1,1,1,1,1,1,1,0,0,0,
        0,1,1,1,1,0,0,0,0,0,   0,1,1,1,1,1,1,1,0,0,0,
        0,0,1,1,1,1,1,1,1,1,1,   0,0,1,1,1,1,1,1,1,0,0,
        0,0,1,1,1,1,0,0,0,0,   0,0,0,1,1,1,1,1,1,1,0,
        0,0,0,1,1,1,1,0,0,0,   0,0,0,1,1,1,0,0,0,0,
        0,0,0,0,1,1,1,1,1,1,0,   0,0,0,0,1,1,1,1,0,0,
        0,0,0,0,1,1,1,1,1,1,1,   0,0,0,0,0,1,1,1,1,1,
        0,0,0,0,0,1,1,1,1,0,   0,0,0,0,0,0,1,1,1,1,
        0,0,0,0,0,0,0,1,1,1,   0,0,0,0,0,0,0,0,0,1,1]


    Total := sum(i in TRUCKS) cost(i)*y(i)
    forall(i in TRUCKS)
            sum(j in PACKAGES) size(j)*x(j,i) <= capacity(i)
                                        ! (1) Packages fit
    forall (i in TRUCKS)
            sum (j in PACKAGES) x(j,i) <= NUM_PACKAGE*y(i)
                                        ! (2) use only
                                        ! paid for trucks
    forall (j in PACKAGES)
            sum(i in TRUCKS) can_use(j,i)*x(j,i) = 1
                                        ! (3) every
                                        ! package on truck
    forall (i in TRUCKS)
            y(i) is_binary              ! (4) no partial trucks
    forall (i in TRUCKS, j in PACKAGES)
            x(j,i) is_binary            ! (5) no package splitting

    minimize(Total)
end-model
```

Fig. 1. Transportation formulation

```
x(j,i) <= 1                              ! (5')
```

(Note that nonnegativity of the variables is assumed). This results in a linear program (the x and y variables can take on fractional values). A formulation with linear relaxation F_1 is tighter than another with relaxation F_2 if every fractional feasible solution to F_1 is also a fractional feasible solution to F_2 and the reverse is not true. Note that tightness is a property of the linear relaxation of a formulation.

Every integer programmer will look at the formulation given and immediately identify improvements. The main issue is in the constraints (2). These are well-known "weak" constraints. For example, it is straightforward to see that a package can be assigned to a truck whose corresponding y value is as small as $\frac{1}{\text{NUM\_PACKAGE}}$. We can "cut off" this sort of solution by replacing (2) with the constraints

```
forall (i in TRUCKS, j in PACKAGES) x(j,i) <= y(i)
                              !(2') tighter formulation
```

Now, if x(j,i) = 1 for a particular i,j then the corresponding y(i) must also be 1. The addition of these constraints leads to a tighter formulation. Note that the new formulation is quite a bit larger: instead of one constraint for every truck, we have a constraint for every (truck,package) pair. While this makes it slower to solve the linear program at each node of the branch-and-bound tree, the resulting decrease in size of the tree far outways this.

Further improvements can be had by replacing

```
forall(i in TRUCKS)
        sum(j in PACKAGES) size(j)*x(j,i) <= capacity(i)
                              ! (1) Packages fit
```

with

```
forall(i in TRUCKS)
        sum(j in PACKAGES) size(j)*x(j,i) <= capacity(i)*y(i)
                              ! (1') Packages fit
```

Again, depending on the exact coefficients, this can lead to a tighter formulation.

At this point, integer programmers step back, look self-satisfied, and move on to other problems.

Unfortunately, when put into XPRESS-MP (other sophisticated codes will work similarly), the results are not what was expected. The time for our instance goes *up*, doubling to 22.1 seconds with 50,631 nodes in the branch-and-bound tree.

What has happened? The primary point is that the relaxation solved by XPRESS-MP or any other top-quality code is not the same as the naive relaxation. The code already has the ability to identify "obvious" tightenings. In the case of constraints (2), there is a technique during preprocessing that sets each binary variable to 1 and determines any variable fixing that might occur (exactly as would happen with constraint propagation in constraint programming). If variable y must be 1 once x is 1, then the constraint $y \geq x$ can be added. This will generate constraints (2') from (2) automatically. From a modeler's point of view, there is no need to add (2'): that "trick" is already known to the software.

Now, if the code is unsophisticated, it is important to add tightenings such as 2'. Turning off preprocessing and cut generation from XPRESS-MP leads to a formulation and solution code that takes 1851 seconds and 2.4 million nodes with 2'. Without 2', after the same 1851 seconds, branch-and-bound has taken 5 million nodes (since the linear program is smaller) but still has a duality gap with a lower bound of 1.22 and an upper bound (feasible solution) of 8.4 (8.2 is optimal). Time to optimality is measured in days.

Why does adding 2' to the sophisticated code actually slow things down for this instance? Only some of the 2' constraints are relevant, while the others simply make the instance larger. XPRESS-MP is able to generate only the relevant constraints by including only those that are violated by the linear relaxation. This leads to smaller, equally tight, formulations. Such formulations solve quicker than the larger formulation.

This interaction of formulation with solution code is shown even stronger with the addition of the constraint:

```
sum(i in TRUCKS)
        capacity(i)*y(i) >= sum (j in PACKAGES)size(j)
                         ! (6) Have sufficient capacity
```

This constraint says simply: the total capacity of the trucks chosen must be sufficient to handle the total size of the packages to be transported.

This constraint does not tighten the linear relaxation: it is a linear combination of previous constraints, so it cannot improve the relaxation. Standard IP formulation approaches would therefore not include the constraint.

Aardal [1] noted the surprising result that if you include this redundant constraint into the formulation, sophisticated codes solve instances much faster (they worked on a closely related location problem, where the timing aspects of the packages do not come into play).

For our instance, solution is instantaneous, and no branching is done: the problem is solved at the initial relaxation. How can adding a constraint that does not improve the relaxation affect the solution process to such an extreme extent?

Again, the key is that XPRESS-MP (or any other sophisticated code) does not solve the naive relaxation. In this case, the constraint (6) is recognized as a specially structured constraint, called a knapsack constraint. A tremendous amount is known about knapsack constraints (they form the basis for the ground-breaking work of Crowder, Johnson and Padberg [3] who used an understanding of knapsack constraints to solve general integer programs, a fundamental breakthrough in computational integer programming), and that knowledge is embedded in current codes. In particular, a set of constraints called *cover inequalities* are known to provide a much tighter formulation than just the linear knapsack constraint alone. The constraints are added "automatically" by XPRESS-MP, resulting in an extremely tight formulation that is solved without branching.

This is an example of an interaction between the human modeler and the software. The modeler is needed to identify the knapsack inequality (at least current software does not automatically identify the redundant knapsack) but then the software is able to bring in all of its knowledge about knapsack constraints.

This development is not surprising (I believe) for constraint programmers. In constraint programming, it is common to add redundant constraints in order to improve propagation. In integer programming, however, it is unusual to add a constraint that doesn't improve the linear relaxation in order to take advantage of the automatic cut generation available in the software.

This would be more obvious to integer programmers if the solution codes offered more flexibility in the handling of cover and other inequalities. Currently, for every code I am aware of, you can do no more than set a level of aggressiveness in searching for cover inequalities (0=no inequalities, 1=1 round of search, etc.). It is not possible to identify some constraints as good prospects for finding cover constraints and others as poor areas. If integer programming codes were like constraint programming codes, then it would be possible to write (6) as something like

```
knapsack(capacity,y,'>',sum(j in PACKAGES) size(j))
            with STRONGCUTS
```

or

```
knapsack(capacity,y,'>',sum(j in PACKAGES) size(j))
            with FASTCUTS
```

or

```
knapsack(capacity,y,'>',sum(j in PACKAGES) size(j))
            with NOCUTS
```

where the '>' denotes a "$\geq$" knapsack and STRONGCUTS, FASTCUTS, and NOCUTS tell the optimizer which cut generation routine to us. This will guide software in the amount of cuts to generate for this particular constraint, rather than the current approach of setting the generation for all constraints at once.

To review for this problem, a naive software implementation of branch-and-bound for a simple formulation doesn't work: solutions take hours or days. Tightening the formulation in the traditional sense only works for simple codes: sophisticated codes already include standard tightening. The best formulation requires understanding the capabilities of the software and adding a seemingly irrelevant constraint.

3   Example 2: Sports Scheduling

For our second problem, I will discuss some experiments on the Traveling Tournament Problem. The Traveling Tournament Problem (TTP), introduced by Easton, Nemhauser, and Trick [4], is a simplification of a sports scheduling problem that arose in the scheduling of Major League Baseball (MLB). The requirements for MLB take many pages to describe, but the key aspects of a "good" MLB schedule is flow (the number of consecutive home or away series a team plays) and distance traveled (how far teams must fly in the schedule). Additional "real world" constraints include stadium availability, the scheduling of key rivals, holiday requirements and much more. The TTP ignores most of these requirements and concentrates on flow and distance.

Given n teams with n even, a double round robin tournament is a set of games in which every team plays every other team exactly once at home and once away. A game is specified by an ordered pair of opponents. Exactly $2(n-1)$ slots or time periods are required to play a double round robin tournament. Distances between team sites are given by an n by n distance matrix D. Each team begins at its home site and travels to play its games at the chosen venues. Each team then returns (if necessary) to its home base at the end of the schedule.

Consecutive away games for a team constitute a road trip; consecutive home games are a home stand. The length of a road trip or home stand is the number of opponents played (not the travel distance).

The Traveling Tournament Problem is defined as:

Input: n, the number of teams; D an n by n integer distance matrix; L, U integer parameters.

Output: A double round robin tournament on the n teams such that

- The length of every home stand and road trip is between L and U inclusive, and
- The total distance traveled by the teams is minimized.

The parameters L and U define the trade off between distance and pattern considerations. For $L = 1$ and $U = n - 1$, a team may take a trip equivalent to a traveling salesman tour. For small U, teams must return home often, so the distance traveled will increase. In this paper, we will concentrate on $L = 1$ and $U = 3$, which corresponds with the MLB ideal.

In addition, a "no-repeaters" constraint can be added: if team A plays at team B in slot t, then B does not play at A in slot $t + 1$.

Instances of the TTP seem very difficult, even for relatively small n. For $n = 4$, optimal solutions are relatively easy to find, but even $n = 6$ is nontrivial. The largest instances solved to optimality are at $n = 8$ (in contrast, MLB has two leagues: one with $n = 14$ and one with $n = 16$).

Many researchers have worked on heuristics for this problem, but there has been relatively little work on complete (or provably optimal) approaches.

The most direct formulation for this problem as an integer program defines a variable plays(i,j,t) which equals 1 if team i plays at team j in slot t. In this

way, we can ensure a double-round robin structure with constraints like (where
TEAMS is defined to be the range $1\dots n$ and SLOTS is $1\dots 2n-2$:

```
forall (i in TEAMS, t in SLOTS)
 plays(i,i,t) = 0 ! (1) no team plays itself

forall (i in TEAMS, t in SLOTS)
 sum(j in TEAMS) (plays(i,j,t)+plays(j,i,t)) = 1
                            ! (2) team i plays one team in each slot

forall (i,j in TEAMS | i <> j)
 sum (t in SLOTS) plays(i,j,t) = 1
                            ! (3) team i plays at team j exactly once
```

Handling the "no more than 3 home or away in a row" can be handled with
constraints like

```
forall (i in TEAMS, t in 1..2*n-5)
  1 <= sum(j in TEAMS) (plays(i,j,t)+plays(i,j,t+1)+
                    plays(i,j,t+2)+plays(i,j,t+3)) <= 3
                        ! (4) no more than 3 away in a row
```

These variables are not sufficient for the objective function, however. To get
the distance traveled, additional variables are needed. Define `location(i,j,t)`
to be 1 if team i is in location j in slot t (so `location(i,i,t)=1` implies i is home
in slot t. Define `follows(i,i1,i2,t)` to be 1 if team i travels from location $i1$
to location $i2$ between slots t and $t+1$. Then the following constraints links all
the variables together:

```
forall (i,j in TEAMS, t in SLOTS)
 if (i=j) then
   location(i,i,t) = sum(k in TEAMS) plays(k,i,t)
 else
   location(i,j,t) = plays(i,j,t)
 end-if
 ! (5) define location in terms of plays

forall (i in TEAMS)
  forall (j1,j2 in TEAMS, t in 1..2*n-3)
    follows(i,j1,j2,t) >=
        location(i,j1,t)+location(i,j2,t+1) - 1
 ! (6) define follows in terms of location
```

Now the total distance traveled is

```
Total := sum(i,j,k in TEAMS, t in 1..2*n-3) DIST(j,k)*follows
(i,j,k,t)+
          sum(i,j in TEAMS) DIST(i,j)*location(i,j,1)+
          sum(i,j in TEAMS) DIST(j,i)*location(i,j,9)
  ! (7) Distance traveled
```

The "no-repeaters" requirement is

```
forall (i,j in TEAMS, t in 1..2*n-3)
  plays(i,j,t)+plays(j,i,t)+
        plays(i,j,t+1)+plays(j,i,t+1) <= 1
  ! (8) no repeaters
```

This gives a complete formulation for the TTP. Unfortunately, putting this formulation into XPRESS-MP gives very poor result, even for $n = 6$. The initial relaxation value for that instance (letting XPRESS-MP be aggressive in adding initial cuts) is only 2186, while the optimal value is 23,916. After 1800 seconds, the lower bound has improved to only 5434 , while the best feasible solution found is 25650 . Again, running to optimal takes days.

To improve this formulation, it might be possible to add constraints to give a better linear relaxation. For instance, since the assignment for every week corresponds to a matching problem, Trick [5] suggests adding the "odd-set" constraints for each week.

An alternative (and better) approach is to reformulate by redefining the variables. The formulation given seems quite complicated because multiple types of variables are needed to correctly model the "distance traveled" aspects. Instead of using plays(i,j,t) as a fundamental variable, we can formulate this problem using variables corresponding to each road trip and home stand. Define trips1(i,i1,t) to be 1 if team i makes a trip to team $i1$ in slot t, and then returns home. Let trips2(i,i1,i2,t) be 1 if team i makes a trip to $i1$ in slot t then on to team $i2$ in slot $t+1$ and then returns home. trips3(i,i1,i2,i3,t) is the corresponding variable for length-3 trips: first to $i1$, then $i2$, then $i3$ before returning home. Similarly, home1(i,t) corresponds to a length-1 homestand in slot t; home2(i,t) a length-2 homestand in t and $t + 1$; and home3(i,t) a length-3 homestand beginning at t.

Each road-trip variable has a cost, corresponding to the distance traveled. This gives an objective function of

```
Total := sum(i,i1 in TEAMS,t in SLOTS) cost1(i,i1,t)*trips1(i,i1,t)+
    sum(i,i1,i2 in TEAMS, t in SLOTS) cost2(i,i1,i2,t)*trips2(i,i1,i2,t)+
    sum(i,i1,i2,i3 in TEAMS, t in SLOTS) cost3(i,i1,i2,i3,t)*trips3
    (i,i1,i2,i3,t)
```

For constraints, we still have constraints that require each team to play at most one game in each slot. This looks like the following (the constraint for slots 1 and 2 is slightly different, based on which trips are feasible for the slot):

```
forall (i in TEAMS, t in 3..10)
  sum(i1 in TEAMS) trips1(i,i1,t) +
  sum (i1,i2 in TEAMS) (trips2(i,i1,i2,t)+trips2(i,i1,i2,t-1)) +
  sum (i1,i2,i3 in TEAMS) (trips3(i,i1,i2,i3,t)+trips3(i,i1,i2,i3,t-1)+
                        trips3(i,i1,i2,i3,t-2)) +
  home1(i,t) +
  home2(i,t)+home2(i,t-1) +
  home3(i,t)+home3(i,t-1)+home3(i,t-2) = 1
```

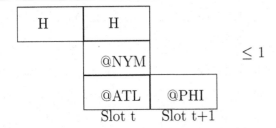

Fig. 2. Constraint: One Game per Slot

This is illustrated in Figure 2.

There are also constraints that either i is away in slot t or some team is away and playing i in that slot (the subscripts get a little messy):

```
forall (i in TEAMS, t in 3..10)
  sum(i1 in TEAMS) trips1(i,i1,t) +
  sum (i1,i2 in TEAMS) (trips2(i,i1,i2,t)+trips2(i,i1,i2,t-1)) +
  sum (i1,i2,i3 in TEAMS) (trips3(i,i1,i2,i3,t)+trips3(i,i1,i2,i3,t-1)+
                          trips3(i,i1,i2,i3,t-2)) +
  sum(i1 in TEAMS) trips1(i1,i,t)+
  sum(i1,i2 in TEAMS) trips2(i1,i2,i,t-1)+
  sum(i1,i2 in TEAMS) trips2(i1,i,i2,t)+
  sum(i1,i2,i3 in TEAMS) trips3(i1,i2,i3,i,t-2) +
  sum(i1,i2,i3 in TEAMS) trips3(i1,i2,i,i3,t-1) +
  sum(i1,i2,i3 in TEAMS) trips3 (i1,i,i2,i3,t)= 1
```

It is also necessary to ensure that no away trip for team i is followed immediately by another away trip:

```
forall (i,i1 in TEAMS, t in 3..2*n-3)
  trips1(i,i1,t)+trips1(i1,i,t)+
  trips1(i,i1,t+1)+trips1(i1,i,t+1)+
  sum(i2 in TEAMS) (trips2(i,i2,i1,t-1)+trips2(i1,i2,i,t-1)) +
  sum(i2 in TEAMS) (trips2(i,i1,i2,t+1)+trips2(i1,i,i2,t+1)) +
  sum(i2,i3 in TEAMS) (trips3(i,i2,i3,i1,t-2)+trips3(i1,i2,i3,i,t-2)) +
  sum(i2,i3 in TEAMS)(trips3(i,i1,i2,i3,t+1)+trips3(i1,i,i2,i3,t+1))<= 1
```

Figure 3 illustrates this constraint.

Additional constraints preclude a home-stand after a home-stand and repeaters.

This formulation is inspired by the "variable generation" formulations useful in airline crew scheduling and many other applications (see Barhart et al. [2] for a fine survey). By encapsulated complicated structure (in this case, the distance traveled) in an expanded variable definition, we can create formulations with tight relaxations. In this case, we do not have to resort to branch-and-price since the number of variables is still relatively small (4400 for the $n = 6$ case).

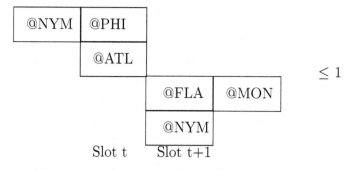

Fig. 3. Constraint: No away after away

The strength of this formulation is shown immediately by XPRESS-MP. The initial relaxation value for $n = 6$ is 21624.7, an order of magnitude larger than that of our initial formulation. In fact, we obtain the optimal solution for this instance after "merely" 4136 seconds and 66,000 nodes in the tree.

Despite the improvement, the time required is still quite long, and does not bode well for solving larger instances. There are some "obvious" strengthenings available. For instance, for the "no away trip after away trip" constraint, it is possible to add more variables to the constraint. This is illustrated in Figure 4.

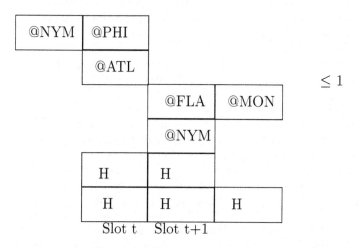

Fig. 4. Constraint: No away after away (strengthened)

This clearly is a strengthening, and resulted in significant improvement in previous versions of XPRESS-MP. Putting this constraint in the current version of XPRESS-MP, however, leads to another nasty surprise: the initial relaxation value is identical, and the overall solution trajectory is a little worse (taking 15 seconds longer to find and prove an optimal solution).

What has happened? Again, I have added a strengthening that the system already knows about: the "strengthened" constraint is known as a "clique inequality" and is part of the XPRESS-MP repertoire. XPRESS-MP can generate that strengthening on its own: my strengthening of the constraint did not help. In fact, it slightly slowed the solution, for reasons that are unclear. If I want to improve my formulation, I need to find constraints or other reformulations that the sophisticated software package does not know about.

To summarize this example, again we have an initial formulation that is hopeless, and XPRESS-MP (other any other package, I believe) cannot improve on it. By reformulating the model using different variables, we ended up with a much better formulation. Trying to improve that model, however, let to overlap with the optimization package's knowledge, and led to no improvement.

4   Conclusions

Through two examples, I have argued that traditional approaches to "reformulation" in integer programming are not practical, since modern, sophisticated software already understands and implements obvious modeling "tricks". In order to improve on a formulation it is necessary to understand what the software knows and to provide insight beyond that knowledge base. For the transportation problem, this insight was in the form of a "redundant" but very important knapsack constraint that was "hidden" in the formulation. Adding this constraint allowed the software to add additional constraints, greatly improving the formulation. For the sports scheduling problem, the added knowledge was in the form of reformulating the variables of the problem to better encapsulate complicated structure. This reformulation was much better than the initial approach, though still not sufficient to solve even small instances (like the $n = 8$ instance).

These experiences suggest that, at least for integer programs, the art of improving formulations is getting more complicated: the simple rules of the past ("find formulations with better relaxations") are becoming less relevant as the relaxation used by the software is often not the relaxation given by the model. Understanding the software sufficiently to provide improved relaxation relative to the solved-relaxation requires highly sophisticated knowledge, and knowledge that can go out of date with every version released of the software.

But there still is room for the modeler to improve the formulations. Is it possible that the software packages will eventually "find" the knapsack constraint needed for the transportation problem? Probably. Can the software do the variable reformulation needed for the sports scheduling problem? Probably not, and almost certainly not if the integer program is only given the formulation in terms of variables, linear constraints, and linear objective. This sort of reformulation requires a deeper understanding of the problem structure.

In order to further develop "reformulations" as a research area and an area of practical interest, it would be useful to have more control over the solving of models. It is in that spirit that I proposed the concept of defining some linear constraints as knapsack constraints while others are just linear constraints: this would define to the solver where the modeler thinks it is likely there are useful strengthenings (such as cover constraints).

Further, while most work in integer programming formulations has tried to find *one* integer programming formulation based on a higher level description of a problem, perhaps it is useful to come up with approaches that can generate multiple formulations for experimentation. Can we create a system that begins with a high level description of a problem and generates a series (or continuum) of formulations, perhaps based on the number of variables or constraints?

At this point, sophisticated software has embedded a lot of the simple reformulation rules integer programmers have developed. We now are challenged to find more sophisticated approaches to spur on the software.

References

1. Aardal, K. (1998). Reformulation of capacitated facility location problems: how redundant information can help, *Annals of Operations Research* **82** 289-308.
2. Barnhart C., Johnson E.L., Nemhauser G.L., Savelsbergh M.W.P. and Vance P.H. (1998), Branch-and-Price: Column Generation for Huge Integer Programs, Operations Research 46, 316.
3. Crowder H., Johnson E.L. and Padberg M.W. (1983), Solving Large Scale Zero-One Linear Programming Problems, *Operations Research* **31**: 803-834.
4. Easton, K., G.L. Nemhauser, and M.A. Trick (2003), Solving the Traveling Tournament Problem: A Combined Integer and Constraint Programming Approach, in *PATAT'2002*, E. Burke and P. Causmaecker (eds), Springer Lecture Notes in Computer Science **2740**, 63–77.
5. Trick, M.A. (2003). Integer and Constraint Programming Approaches for Round Robin Tournament Scheduling, in *PATAT'2002*, E. Burke and P. Causmaecker (eds), Springer Lecture Notes in Computer Science **2740**, 63–77 (2003).
6. Williams H.P. (1999), *Model Building in Mathematical Programming*, Wiley, New York.
7. XPRESS-MP Extended Modeling and Optimisation Subroutine Library, Reference Manual (2004), Dash Associates, Blisworth House, Blisworth, Northants.

Nondeterministic Control for Hybrid Search

Pascal Van Hentenryck[1] and Laurent Michel[2]

[1] Brown University, Box 1910, Providence, RI 02912
[2] University of Connecticut, Storrs, CT 06269-3155

Abstract. Hybrid algorithms combining local and systematic search often use nondeterminism in fundamentally different ways. They may differ in the strategy to explore the search tree and/or in how computation states are represented. This paper presents nondeterministic control structures to express a variety of hybrid search algorithms concisely and elegantly. These nondeterministic abstractions describe the search tree and are compiled in terms of first-class continuations. They are also parameterized by search controllers that are under user control and specify the state representation and the exploration strategy. The resulting search language is thus high-level, flexible, and directly extensible. The abstractions are illustrated on a jobshop scheduling algorithm that combines tabu search and a limited form of backtracking. Preliminary experimental results indicate that the control structures induce small, often negligible, overheads.

1   Introduction

In the last decade, hybridizations between local and systematic search have received increased attention and contributed many interesting results. Such hybridizations include the use of limited backtracking to intensify local search algorithms around elite solutions [7], variable-depth search procedures that explore trees of moves to select neighbors [1], as well as large neighborhood search where systematic search performs the neighborhood exploration (e.g., [2, 9, 12]).

These hybridizations often lead to fundamentally different search algorithms which may use trailing, copying, incremental checkpointing, or a combination of them in order to restore computation states appropriately. It is therefore a challenge to design flexible, efficient, and elegant search languages to support local and systematic search, as well as their hybridizations.

This paper originated as an attempt to address this challenge in COMET, an object-oriented programming language supporting a constraint-based architecture for local search [5, 14, 15, 16]. It led to the design and implementation of novel nondeterministic abstractions addressing the specificities of hybridizations between local and systematic search, while encompassing the wealth of results in search languages (e.g., [4, 8, 10, 13, 17]).

From a programming standpoint, the nondeterministic control structures of COMET specify the search tree to explore and closely resemble those of OPL. However, these control structures are also parameterized by a search controller

R. Barták and M. Milano (Eds.): CPAIOR 2005, LNCS 3524, pp. 380–395, 2005.

that specifies both the search strategy, i.e., how the search tree should be explored, and how search nodes must be stored/restored. Hence, together with other control abstractions of COMET, they provide a rich and flexible search language with several desirable properties.

Perhaps the most significant property is the novel separation between control and state. Indeed, *the control structures only specify nondeterminism. How to explore the resulting search tree and how to store/restore the search nodes are decisions left to the search controller and thus under programmers' control.* This functionality is critical for hybrid search where the state restoration is not necessarily performed using trailing. Instead, state restoration may be based on concepts such as solutions and checkpoints that restore previously saved states with various degrees of incrementality. Note that the separation between control and state also enables different implementation technologies for systematic search (e.g., [11]) to coexist in the same system.

Equally important is the implementation of the nondeterministic control structures which are compiled into continuations in COMET. First-class continuations provide an elegant and efficient abstraction to specify the control flow of nondeterministic abstractions. Moreover, continuations may be implemented to induce no overhead when nondeterminism is not used, which is important for pure local search applications.

Finally, the search language is open and extensible, thanks to continuations and the separation between control and state. Search controllers elegantly implement a variety of systematic search procedures, as well as incomplete search algorithms typically found in hybridizations. Moreover, different state representations, such as trailing, copying, and checkpointing, can be encapsulated inside search controllers, allowing the same nondeterministic abstractions to be used for fundamentally different search procedures.

The rest of this paper introduces the nondeterministic abstractions of COMET. The goal is to convey the rationale underlying their design and implementation, and to illustrate them on a complex application. Section 2 recalls the concept of continuations and illustrates their use in COMET. The nondeterministic abstractions are described in Section 3 and various search controllers are presented in Section 4. The abstractions are illustrated on a hybrid algorithm for jobshop scheduling in Section 5. The last two sections present the experimental results and conclude the paper.

2   Continuations

Continuations provide a flexible control structure to implement several higher-level abstractions such as exceptions, coroutines, and nondeterminism. Informally speaking, a continuation is a snapshot of the runtime data structures that allows the execution to restart from this point at a later stage of the computation. More precisely, a continuation is a pair $\langle I, S \rangle$, where I is an instruction pointer and S is a stack to execute the code starting at I. In COMET, continuations are obtained through instructions of the form

```
continuation c ⟨body⟩
```

```
0.   function int fact(int n) {if (n==0) return 1;else return n*fact(n-1);}
1.   int i = 4;
2.   continuation c { i = 5; }
3.   int r = fact(i);
4.   cout << "fact(" << i << ") = " << r << endl;
5.   if (i == 5) call(c);
```

Fig. 1. Continuations in COMET

that binds c to a continuation $\langle I, S \rangle$, where I is the next instruction in the code and S is the stack when the continuation is captured. It then executes its body and continues in sequence. The resulting continuation can be invoked with the call `call(c)` that restores the stack S and restarts execution from I. Consider the code displayed in Figure 1. The code outputs

```
fact(5) = 120
fact(4) = 24
```

Indeed, the continuation c in line 2 consists of an instruction pointer to line 3 and a stack whose entry for i stores the value 4. The COMET implementation first calls the factorial function with argument 5 (since i = 5 is executed when the continuation is taken). Since i has value 5, the implementation calls the continuation (line 5), which restarts execution in line 3 with a stack whose entry for i has value 4. The COMET implementation thus calls `fact(4)`, displays its result, and terminates (since i is 4).

Consider now the code displayed in Figure 2. The code has the same effect but it clearly illustrates the complex control/stack patterns that may be induced by continuations. Indeed, the continuation is taken in line 3, i.e., inside the function `getContinuation` that returns the continuation. The instruction pointer is on line 4 (the `return` instruction) and the stack contains two frames for the global and the function scopes. When the continuation is called on line 9, the stack is restored, the execution restarts in line 4, returns the correct continuation c, and proceeds to compute `fact(4)`, displays its results, and terminates. Note that continuations, like closures, are first-class objects that can be stored in data structures, used as arguments, and returned as values.

```
0.   function int fact(int n) {if (n==0) return 1;else return n*fact(n-1);}
1.   int i = 4;
2.   function Continuation getContinuation() {
3.      continuation c { i = 5; }
4.      return c;
5.   }
6.   Continuation c = getContinuation();
7.   int r = fact(i);
8.   cout << "fact(" << i << ") = " << r << endl;
9.   if (i == 5) call(c);
```

Fig. 2. Continuation in COMET Again

3   Nondeterminism

This section describes some of the nondeterministic abstractions of COMET. It assumes initially that nondeterminism is implemented through depth-first search before relaxing this assumption.

The Try Instruction. Figure 3 depicts a nondeterministic program that generates all binary arrays of size 4 and displays their decimal values, i.e., 0 1 2 3 4 5 6 7 8 9 10 11 12 13 14 15. Line 1 simply declares the array and lines 2-7 specify the core of the nondeterministic search. A depth-first search controller is created in line 2 and used in all subsequent nondeterministic control structures. Lines 4-5 specify the nondeterministic choices: They iterate over all variables and assign them nondeterministically either to 0 or 1. These lines, as well as the output instruction, are encapsulated into an `exploreall` instruction (line 3) in order to produce all solutions. These solutions are obtained by depth-first search, since this is the search controller used in the instruction.

```
0.    include "SearchController";
1.    int x[1..4] = 0;
2.    DFS sc();
3.    exploreall<sc> {
4.       forall(i in 1..4)
5.          try<sc> x[i] = 0; | x[i] = 1;
6.       cout << 8 * x[1] + 4 * x[2] + 2 * x[3] + x[4] << " ";
7.    }
```

Fig. 3. A Simple Nondeterministic Program in Comet

The Exploreall Instruction. The `exploreall` instruction is used to find all solutions to a nondeterministic program. Its implementation simply fails each time it finds a new solution. It is possible to exit an `exploreall` instruction early by using the `exit` method on the search controller.

The Tryall Instruction. Consider now Figure 4 that features a simple backtracking program for the 8-queens problem. The COMET program declares the queens array and the depth-first search controller (lines 1-3) and specifies the search using a `tryall` instruction (lines 4-6). The instructions

```
forall(q in 1..8)
   tryall<dfs>(v in R: !attack(queen,q,v))
      queen[q] = v;
```

iterate over all variables and nondeterministically assign them a value so that the queen in column q does not attack the queens in columns 1..q-1. Observe the iterative style for nondeterminism that is traditionally appreciated by programmers [3]. Observe that the `tryall` instruction performs a nondeterministic choice and then continues the execution normally. Hence it is not only the body of the `tryall` but also the "continuation" of the execution that is nondeterministic.

```
0.    include "SearchController";
1.    int queen[1..8];
2.    DFS dfs();
3.    forall(q in 1..8)
4.      tryall<dfs>(v in 1..8: !attack(queen,q,v))
5.        queen[q] = v;
6.    function bool attack(int[] queen,int i, int v) {
7.      forall(k in 1..i-1)
8.        if (queen[k]==v || queen[k]+k==v+i || queen[k]-k==v-i)
9.          return true;
10.     return false;
11.   }
```

Fig. 4. A Comet Program for the Queens Problem

4   Search Controllers

The nondeterministic instructions only define the search tree to explore. It is the role of the search controller to specify how to explore it, including how to store/restore computation states. This section reviews the interface of search controllers, shows how to compile nondeterminism in terms of the interface and continuations, and reviews a variety of controllers.

The Interface of Search Controllers. Search controllers in COMET are subclasses of SearchController which is (partially) described in Figure 5. Several of the methods were informally described earlier. Method start is called by exploreall to specify what to do when no choice points are left to explore. It receives a continuation that is generally executed in method exit that terminates the search. Methods addChoice and fail constitute the core of the interface and are typically overridden in specific controllers. Method addChoice adds a new choice point, while method fail restarts execution from an earlier choice point. Observe that choice points are continuations: They primarily specify the control flow, not the computation states.

```
class SearchController {
  Continuation _exit;
  Event closeChoice;
  SearchController() { _exit = null; }
  void start(Continuation e) { _exit = e; }
  void exit() { call(_exit); }
  void addChoice(Continuation c) {}
  void fail() { exit(); }
  ...
}
```

Fig. 5. The Search Controller Interface in COMET (Partial Description)

Compiling Nondeterminism. It is interesting to sketch how nondeterminism is implemented in terms of search controllers and continuations. The try instructions are compiled in terms of continuations. A try instruction

```
try<sc> ⟨left⟩ | ⟨right⟩
```

is compiled into

```
bool rightBranch = true;
continuation c { sc.addChoice(c); rightBranch = false; ⟨left⟩ }
if (rightBranch) ⟨right⟩
```

The implementation creates a continuation c, transfers it to the search controller as a new choice and then executes the left branch of the choice point. On backtracking, i.e., when the continuation c is invoked, the right branch is executed since rightBranch is true in that context. The compilation of a tryall instruction is essentially similar but uses iterators to assign its parameter that is represented as a local variable. An exploreall instruction

```
exploreall<sc> ⟨body⟩
```

is compiled into

```
continuation c { sc.start(c); ⟨body⟩; sc.fail(); }
```

The implementation takes a continuation c representing what must be executed when no more choice points are left unexplored, i.e., when all the solutions of its body have been explored. It stores the continuation in the search controller, executes the body of the instruction, and fails, which induces the search controller to consider unexplored choices. Note also that method exit on the controller invokes the continuation c by default.

A Simple Search Controller. Figure 6 depicts the depth-first search controller used so far. The controller maintains a stack of continuations. Method addChoice pushes the continuation on the stack, while method fail pops and invokes the top continuation. Nothing else is necessary for solving the queens problem. Indeed, the continuation automatically saves the parameters of the forall and tryall instructions as they are stored on the stack. Moreover, the values of the queens do not need to be restored because of the depth-first strategy.

```
0.    class DFS extends SearchController {
1.        Stack{Continuation} stack;
2.        DFS(): SearchController() { stack = new Stack{Continuation};}
3.        void addChoice(Continuation f) { stack.push(f); }
4.        void fail() { if (stack.empty()) exit() else call(stack.pop()); }
5.    }
```

Fig. 6. The Depth-First Search Controller

(Re)storing Search Nodes. In more complex applications or with other strategies, the search controllers must save and restore additional data structures. Figure 7 revisits the simple nondeterministic program presented earlier. It now declares a local solver (line 1) and incremental variables (line 2), since this is the underlying technology for the jobshop scheduling application described subsequently. Note the try instruction that assigns incremental variables using := instead of =.

```
0.    include "LocalSolver";
1.    LocalSolver mgr();
2.    var{int} x[1..3](mgr) := 0;
3.    SDFS sc(mgr);
4.    exploreall<sc> {
5.      forall(i in 1..3)
6.        try<sc> x[i] := 0; | x[i] := 1;
7.      cout << 8 * x[1] + 4 * x[2] + 2 * x[3] + x[4] << endl;
8.    }
```

Fig. 7. The Simple Nondeterministic Program Revisited

The program also declares a depth-first search controller whose implementation is depicted in Figure 8. The controller stores, not only continuations, but also the states of the incremental variables. Line 2 in Figure 8 declares a stack of continuations and a stack of solutions. Solutions in COMET capture a snapshot of the incremental variables, which can then be restored at a later computation stage [5]. When a choice point is created (method addChoice), the controller captures a solution (new Solution(m)) and pushes it onto the solution stack. On backtracking, the instruction sol.pop().restore() restores the solution.

Observe the decoupling between the control and data aspects of choice points. The control flow is abstracted by continuations that are used by the controller to implement the search strategy. The representation of search nodes is under user control and may use abstractions such as solutions, checkpoints, and computation spaces [10]. In other words, the controller describes how to save and restore the nodes independently of the specifications of search tree and the search strategy. As a consequence, the node representation can be changed by replacing or modifying the controller without affecting the rest of the program. For instance, to use checkpoints [14] instead of solutions, it suffices to store checkpoints on the stack using instructions such as new Checkpoint(m). Checkpoint restorations undo and, possibly reexecute, operations on incremental variables as described in [14]. Similar issues arise in CP systems. Trailing-based systems may

```
0.    class SDFS extends SearchController {
1.      LocalSolver m; Stack{Continuation} cont; Stack{Solution} sol;
2.      SDFS(LocalSolver mgr): SearchController() {
3.        m = mgr;
4.        cont = new Stack{Continuation}; sol = new Stack{Solution};
5.      }
6.      void addChoice(Continuation f) {
7.        cont.push(f); sol.push(new Solution(m)); }
8.      void fail() {
9.        if (cont.empty()) exit();
10.       else { sol.pop().restore(); call(cont.pop()); }
11.     }
12.   }
```

Fig. 8. The Depth-First Search Controller with Solutions

```
0.    class SDS extends SearchController {
1.       LocalSolver m; Queue{Continuation} cont; Queue{Solution} sol;
2.       SDS(LocalSolver mgr): SearchController() {
3.          m = mgr;
4.          cont = new Queue{Continuation}; sol = new Queue{Solution};
5.       }
6.       void addChoice(Continuation f) {
7.          cont.push(f); sol.push(new Solution(m)); }
8.       void fail() {
9.          if (cont.empty()) exit();
10.         else { sol.pop().restore(); call(cont.pop()); }
11.      }
12.   }
```

Fig. 9. The Discrepancy Search Controller

use CP checkpoints that capture trail pointers and use semantic decomposition for strategies [6], while copy-based systems only save the state of the solver.

A Simple Discrepancy Controller. Figure 9 describes a controller where the stack was replaced by queue, implementing a search strategy where the choices are explored by increasing number of discrepancies. By replacing SDFS by SDS in line 3 of Figure 7, the program displays 0 8 4 2 1 12 10 9 6 5 3 14 13 11 7 15. *Observe the simplicity of moving from depth-first search to this new search strategy thanks to the high-level control and state abstractions of* COMET. The control and the states are fully abstracted by continuations and solutions, providing a concise and elegant specification of the search strategy.

An Iterative Discrepancy Controller. Figure 10 depicts a search controller for an iterative implementation of limited discrepancy search. The exploration strategy consists of a sequence of searches, where the i-th search allows at most i discrepancies. The controller maintains a variable _discr to count the number of discrepancies that is incremented in method fail each time a new choice is explored. Method fail only calls a continuation whenever the maximum number of discrepancies is not exceeded (line 22). Otherwise, the controller recursively fails to explore another choice with fewer discrepancies. Note that line 20 pops the continuation and restores the values of the incremental variables, including the number of discrepancies.

The discrepancy phases are initiated in the overridden start method. Interestingly, it also uses a tryall instruction to explore all the discrepancies up to the maximum depth (line 14). Observe that, like in the queens problem, the nondeterminism operates not only on the body of the tryall but also on whatever follows the start method (e.g., the body of an exploreall instruction). The first two phases of the resulting program display 0 1 2 4 8 0 1 2 3 4 5 6 8 9 10 12.

A Memento Controller. Figure 11 depicts the search controller to be used in jobshop scheduling. The key idea underlying the controller is to only store the

```
0.   class IDS extends SearchController {
1.     LocalSolver m;
2.     Stack{Continuation} cont;
3.     Stack{Solution} sol;
4.     var{int} _discr;
5.     int _maxDiscr;
6.     int _maxDepth;
7.     SIDS(LocalSolver mgr,int d) : SearchController() {
8.       m = mgr; _maxDepth = d;
9.       cont = new Stack{Continuation}(); sol = new Stack{Solution}();
10.      _discr = new var{int}(mgr) := 0;
11.    }
12.    void start(Continuation e) {
13.      super.start(e);
14.      tryall<this>(d in 1.._maxDepth) { _discr := 0; _maxDiscr = d; }
15.    }
16.    void addChoice(Continuation f) { cont.push(f); sol.push(f); }
17.    void fail() {
18.      if (cont.empty()) exit();
19.      else {
20.        Continuation c = cont.pop(); sol.pop().restore();
21.        _discr := _discr + 1;
22.        if (_discr > _maxDiscr) fail(); else call(c);
23.      }
24.    }
25. }
```

Fig. 10. The Iterative Discrepancy Search Controller

```
0.   class Memento extends SearchController {
1.     LocalSolver m;
2.     Stack{Continuation} cont;
3.     Stack{Solution} sol;
4.     int _maxSize;
5.     Memento(LocalSolver mgr,int maxSize) : SearchController() {
6.       m = mgr; _maxSize = maxSize;
7.       cont = new Stack{Continuation}; sol = new Stack{Solution};
8.     }
9.     void addChoice(Continuation f) {
10.      if (cont.getSize() == _maxSize) { cont.drop(); sol.drop(); }
11.      cont.push(f); sol.push(new Solution(m));
12.    }
13.    void fail() {
14.      if (cont.empty()) exit();
15.      else { sol.pop().restore(); call(cont.pop()); }
16.    }
17. }
```

Fig. 11. The Memento Search Controller

last k choice points. If a new choice point f is available, and there are already k choice points, the controller first drops the earliest stored choice point before pushing f onto the stack. This controller, called Memento in the following, is a simple modification to the depth-first controller. Indeed, method addChoice now tests whether the maximum number of choices is reached, in which case the earliest choice is dropped from the stack before the push. Once again, observe the simplicity of the controller and the flexibility of the abstractions to implement incomplete search procedures in a natural fashion.

5   An Hybrid Search for Jobshop Scheduling

This section presents an implementation in COMET of the hybrid algorithm of Nowicki and Smutnicki for jobshop scheduling [7]. The algorithm is a tabu-search procedure with a very interesting intensification component based on a limited form of backtracking. Informally speaking, the algorithm maintains the k best solutions found so far. Whenever the tabu search completes, it backtracks to one of these k solutions and explores all its neighbors by restarting a tabu search from each of them. Of course, these new tabu searches may introduce new choice points that will be explored subsequently on backtracking. The rest of this section presents the core of this hybrid search, using the control and scheduling abstractions presented in [14, 16].

Figure 12 presents the core of the algorithm. The search procedure is organized as a series of phases (lines 6-16). Each phase terminates after curIter iterations (line 16) or whenever a new best solution is found (line 11). A phase (lines 7-15) consists in exploring the neighborhood and selecting the best move (lines 3, 8-9). Observe the declaration of a MinNeighborSelector object (line

```
1.    void JobshopAlgorithm::search() {
2.       memento = new Memento(mgr,mementoSize);
3.       MinNeighborSelector N();
4.       int li = 0;
5.       exploreall<memento> {
6.          do {
7.             int oldBest = bestSoFar;
8.             if (exploreNeighborhood(N)) {
9.                call(N.getMove());
10.               if (oldBest > makespan.value()) {
11.                  li = 0; bestSoFar = makespan.value(); curIter = maxIter;
12.                  visitAllNeighbors();
13.               }
14.            } else memento.exit();
16.         } while (li++ < curIter);
17.      }
18.   }
```

Fig. 12. Hybrid Search for Jobshop Scheduling

```
1.    void JobshopAlgorithm::visitAllNeighbors() {
2.      AllNeighborsSelector neighborhood();
3.      exploreNeighborhood(neighborhood);
4.      tryall<memento>(i in 1..neighborhood.getSize()-1)
5.        call(neighborhood.getMove(i));
6.    }
```

Fig. 13. Jobshop Scheduling: The Intensification

3) that is passed to the neighborhood exploration (line 8) to return the best move. The selected move, which is a closure, is executed in line 9: It performs the moves and updates the tabu list. If it improves the best solution, the phase is terminated in line 11 (ignore line 12 for the time being). The neighborhood exploration may return false, meaning that the current solution is optimal (i.e., it satisfies a necessary condition for optimality). When this is the case, the search terminates by calling method exit on the memento declared in line 2.

As mentioned earlier, one of the most innovative aspects of this algorithm is its intensification: Its goal is to explore the search space around elite solutions more extensively. The algorithm maintains a stack of the best k solutions found during the search. When the tabu search completes, the algorithm pops the best solution from the stack and restarts a phase of the tabu search from each of its neighbors. The algorithm terminates when the stack is empty. Observe however that each additional phase may find new best solutions that are themselves pushed onto the stack. The intensification is featured in line 12 of Figure 12 and the implementation is depicted in Figure 13.

The implementation declares another neighbor selector to retrieve all neighbors (line 2). Method exploreNeighborhood, called with this selector, collects all neighbors sorted by decreasing quality (line 3). Once the neighbors are available, the nondeterministic instruction tryall explores all of them nondeterministically, except the best one that has been explored by the tabu search previously.

This implementation is particularly elegant for several reasons. First, it captures the essence of the intensification concisely and naturally. It only needs four lines of code to express a sophisticated intensification component. Second, the search strategy is completely disconnected from the intensification. If a different search strategy (e.g., depth-first search) is desired, it suffices to replace the memento by a depth-first search controller. Third, the implementation is completely generic: It does not explicitly refer to the neighborhood and changes to the neighborhood (in method exploreNeighborhood) do not affect the intensification. Finally, observe that method exploreNeighborhood is used both to find the best neighbor (line 8 in Figure 12) and all the neighbors (line 3 in Figure 13) by passing different neighbor selectors.

The original algorithm in [7] also includes another interesting feature: It reduces the length of the phases each time all neighbors of a choice point have been explored. This functionality can be elegantly accommodated in the search procedure by adding the instruction

```
whenever memento@closeChoice() curIter = curIter - 400;
```

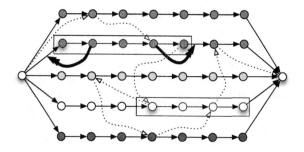

Fig. 14. The Neighborhood of the Jobshop Algorithm

between lines 4 and 5 in Figure 12. This instruction features an event [14] that specifies that, whenever all alternatives of a `tryall` instruction are exhausted (event `closeChoice` of the search controller class), `curIter` must be reduced by 400. Observe the compositionality of the search language and the synergy between the existing and novel control abstractions of COMET.

For completeness, it is useful to discuss the neighborhood briefly. The neighborhood focuses on one critical path from the source to the sink only. Moreover, only the critical arcs at the start or at the end of a critical block are considered for swapping. In other words, the neighborhood identifies the activities at the start (resp. at the end) of a critical block on the selected path and considers the moves that swap such activities with their successors (resp. predecessors). The

```
1.    bool JobshopAlgorithm::exploreNeighborhood(NeighborSelector N) {
2.      set{Activity} SB();
3.      set{Activity} EB();
4.      bool optimal = collectBlockEdges(SB,EB);
5.      if (!optimal)
8.        forall(v in SB) {
9.          Activity s = v.getSucc();
10.         int eval = makespan.estimateMoveForward(v);
11.         if (acceptMove(v,s,eval)) {
12.           found = true;
13.           neighbor(eval,N) { v.moveForward(); tl.makeTabu(s,v); }
17.         }
18.       }
19.       forall(v in EB) {
20.         Activity p = v.getPred();
21.         int eval = makespan.estimateMoveBackward(v);
22.         if (acceptMove(p,v,eval)) {
23.           found = true;
24.           neighbor(eval,N) { v.moveBackward(); tl.makeTabu(v,p); }
28.         }
29.       }
32.     return !optimal;
33.   }
```

Fig. 15. Jobshop Scheduling: The Neighborhood Exploration

resulting neighborhood is not connected, although it is very effective in practice. Moreover, if there is only one critical block on the selected path, then the moves cannot decrease the length of the makespan and the solution can be shown to be optimal. The neighborhood is described visually in Figure 14. In the figure, the machines are depicted horizontally and the job precedences are shown by dashed arcs. The large bold arrows show the two moves associated with the first block. Figure 15 describes parts of the neighborhood exploration which specifies what the neighborhood is, not how to use it. The key for this separation of concerns is the **neighbor** construct [14] which uses closures to represent moves (lines 13-16).

6   Experimental Results

This section presents some preliminary results on the performance of nondeterministic control structures. The first test compares the nondeterministic program in Figure 4 (N in the following) with the "traditional" recursive algorithm R depicted in Figure 16 (the function **noattack** is similar and not shown here). Since no state information is saved in the traditional implementation, this experiment captures the cost of the control structures. It is a worst-case scenario since, in sophisticated applications, the control cost is typically amortized by the state saving and restoration, as well as by propagation and/or the maintenance of incremental data structures. Table 1 depicts the CPU times of the recursive and nondeterministic algorithms for various n. The results indicate that the overhead of the nondeterministic implementation is small and ranges from 7 to 25% which is very reasonable for such a worst case scenario. Observe also the contrast between the iterative style of the nondeterministic program and the recursive style of the traditional program.

```
0.    int n = 8;
1.    range R = 1..n;
2.    int queen[R];
3.    function bool search(int[] queen,int i) {
4.      if (i > n) return true;
5.      forall(v in R: !attack(queen,i,v)) {
6.        queen[i] = v;
7.        if (search(queen,i+1)) return true;
8.      }
9.      return false;
10.   }
11.   search(queen,1);
```

Fig. 16. A Recursive Backtracking Algorithm for the Queens Problem

Table 2 presents some experimental results on jobshop scheduling. It reports the quality and performance of the algorithm on the LA instances. Each line correspond to 50 runs of the algorithm and report statistics on the best, worst,

Table 1. Performance Evaluation of the Nondeterministic Control Instructions

n	16	18	20	22	24	26
R		0.18	0.79	4.78	51.71	14.11	15.52
N		0.20	0.98	5.57	56.80	15.20	16.76
$(N-R)/R$		11.11	24.05	16.53	9.84	7.73	7.99

Table 2. Performance Results on Jobshop Scheduling

Bench	$B(V)$	$m(V)$	$M(V)$	$\mu(V)$	$\sigma(V)$	$m(B)$	$M(B)$	$\mu(B)$	$\sigma(B)$
LA19	842	842 (40)	846	842.4	1.0	0.31	13.45	4.20	3.19
LA20	902	902 (50)	902	902.0	0.0	0.17	2.97	0.95	0.62
LA21	1047	1047 (4)	1061	1052.8	2.7	1.05	19.29	6.56	4.39
LA22	927	930 (6)	939	934.4	2.2	1.53	11.26	5.45	2.72
LA23	1032	1032 (50)	1032	1032.0	0.0	0.34	1.01	0.60	0.14
LA24	935	938 (1)	944	943.3	1.5	0.93	12.39	4.54	2.96
LA25	977	977 (8)	986	980.0	2.6	1.42	17.59	6.14	4.01
LA26	1218	1218 (50)	1218	1218.0	0.0	1.61	6.55	3.24	1.21
LA27	1235	1236 (1)	1269	1251.2	6.9	3.30	23.37	9.66	4.64
LA28	1216	1216 (46)	1225	1216.5	1.9	1.95	23.07	7.46	4.07
LA29	1157	1163 (3)	1190	1174.8	7.1	3.12	33.78	11.77	5.68
LA30	1355	1355 (50)	1355	1355.0	0.0	0.68	2.01	1.40	0.36
LA31	1784	1784 (50)	1784	1784.0	0.0	1.59	5.80	3.26	0.78
LA32	1850	1850 (50)	1850	1850.0	0.0	1.41	6.23	3.66	1.00
LA33	1719	1719 (50)	1719	1719.0	0.0	0.27	3.58	1.19	0.69
LA34	1721	1721 (50)	1721	1721.0	0.0	2.73	7.29	4.91	1.32
LA35	1888	1888 (50)	1888	1888.0	0.0	0.27	4.24	2.03	0.73
LA36	1268	1268 (24)	1291	1272.4	5.7	2.25	20.51	9.53	4.55
LA37	1397	1402 (3)	1428	1412.8	5.6	3.28	32.31	13.35	6.55
LA38	1196	1196 (13)	1208	1200.7	3.1	3.14	29.83	9.39	6.12
LA39	1233	1233 (23)	1251	1237.0	5.2	3.66	28.50	13.49	6.74
LA40	1222	1226 (5)	1234	1231.2	2.8	2.93	24.24	10.36	5.89

average, and standard deviation of the solution quality and CPU times. The
number of times the best solution was found is also reported. On this algorithm,
the overhead of nondeterminism is not noticeable, since restoring a solution
amounts to recomputing the makespan and critical arcs, which is much more
costly than creating and restoring continuations.

7   Conclusion

This paper presented nondeterministic control structures for hybrid search procedures which often differ in their underlying node selection strategies and their
implementation of search nodes. From a modeling standpoint, the main contribution of the abstraction is to decouple the specification of the search tree, the
node selection, and the node representation. In particular, the nondeterministic

abstractions separate the specification of the search tree (i.e., the computations to be explored), the control flow (i.e., how the computations are actually explored), and the node representation (i.e., how the search nodes are stored and restored). All these aspects of search procedures remain under programmers' control, combining a high-level iterative style with the flexibility and extensibility necessary to implement a variety of search procedures. From an implementation standpoint, the nondeterministic control structures are compiled into first-order continuations, inducing no overhead when nondeterminism is not used. The expressiveness and practicability of the abstractions was demonstrated by presenting several search controllers, a tabu procedure for job-shop scheduling featuring an intensification based on backtracking search, and unit performance tests to estimate the cost of continuations.

Acknowledgments. This work was partially supported by NSF ITR Awards ACI-0121497.

References

1. A. Balas, E.; Vazacopoulos. Guided Local Search with Shifting Bottleneck for Job Shop Scheduling. *Management Science*, 44(2):262–275, 1998.
2. R. Bent and P. Van Hentenryck. A Two-Stage Hybrid Local Search for the Vehicle Routing Problem with Time Windows. *Transportation Science*, 8(4):515–530, 2004.
3. de Givry, S. and Jeannin, L. Tools: A library for partial and hybrid search methods. In *CP-AI-OR'03*, Montreal, Canada, 2003.
4. F. Laburthe and Y. Caseau. SALSA: A Language for Search Algorithms. In *CP'98*, Pisa, Italy, October 1998.
5. L. Michel and P. Van Hentenryck. A Constraint-Based Architecture for Local Search. In *OOPSLA'02.*, pages 101–110, Seattle, WA, USA, November 4-8 2002.
6. L. Michel and P. Van Hentenryck. A Decomposition-Based Implementation of Search Strategies. *ACM Transactions on Computational Logic*, 5(2):351-383, 2004.
7. E. Nowicki and C. Smutnicki. A Fast Taboo Search Algorithm for the Job Shop Problem. *Management Science*, 42(6):797–813, 1996.
8. L. Perron. Search Procedures and Parallelism in Constraint Programming. In *CP'99*, pages 346–360, Alexandra, VA, October 1999.
9. L.M. Rousseau, M. Gendreau, and G. Pesant. Using Constraint-Based Operators to Solve the Vehicle Routing Problem with Time Windows. *Journal of Heuristics*, 8:43–58, 2002.
10. C. Schulte. Programming Constraint Inference Engines. In *CP'97*, 519–533, Linz, Austria, October 1997.
11. Christian Schulte. Comparing Trailing and Copying for Constraint Programming. In *ICLP-99*, pages 275–289, Las Cruces, NM, USA, November 1999.
12. P. Shaw. Using Constraint Programming and Local Search Methods to Solve Vehicle Routing Problems. In *CP'98*, pages 417–431, Pisa, Italy, October 1998.
13. P. Van Hentenryck. *The OPL Optimization Programming Language*. The MIT Press, Cambridge, Mass., 1999.
14. P. Van Hentenryck and L. Michel. Control Abstractions for Local Search. In *CP'03*, pages 65–80, Cork, Ireland, 2003.

15. P. Van Hentenryck and L. Michel. Scheduling Abstractions for Local Search. In *CP-AI-OR'04*, pages 319–334, Nice, France, 2004.
16. P. Van Hentenryck, L. Michel, and L. Liu. Constraint-based Combinators for Local Search. In *CP'04*, Toronto, Canada, 2004.
17. P. Van Hentenryck, L. Perron, and J-F. Puget. Search and Strategies in OPL. *ACM Transactions on Computational Logic*, 1(2):1–36, October 2000.

Computing Explanations for the Unary Resource Constraint

Petr Vilím

Charles University, Faculty of Mathematics and Physics,
Malostranské náměstí 2/25, Praha 1, Czech Republic
vilim@kti.mff.cuni.cz

Abstract. Integration of explanations into a CSP solver is a technique addressing difficult question *"why my problem has no solution"*. Moreover, explanations together with advanced search methods like directed backjumping can effectively cut off parts of the search tree and thus speed up the search.

In order to use explanations, propagation algorithms must provide some sort of reasons *(justifications)* for their actions. For binary constraints it is mostly easy. In the case of global constraints computation of factual justifications can be tricky and/or computationally expensive.

This paper shows how to effectively compute explanations for the unary resource constraint. The explanations are computed in a lazy way. The technique is experimentally demonstrated on job-shop benchmark problems. The following propagation algorithms are considered: edge-finding, not-first/not-last and detectable precedences. Speed of these filtering algorithms and speed of the explanation computation is the main interest.

1   Introduction

To show a typical usage of the unary resource constraint, let us consider the following *shop-scheduling* problem. We are given a set of machines and a set of *jobs* which must be processed. A job consists of a set of *operations*, each operation requires exclusive usage exactly one machine. Processing of an operation cannot be interrupted by any other operation. Exact processing time of each operation is known in advance. In the case of *jobshop* problem, operations in a job must be processed in a certain order. In *openshop* an order of operations in a job is arbitrary. The problem is to find a schedule with minimal completion time of all jobs, i.e. a schedule with minimal *makespan*.

Shop-scheduling problems can be modeled as a constraint satisfaction problem (CSP). In this case unary resource[1] constraints are typically used as abstractions of machines. In case of openshop, unary resource constraints are also used to model jobs.

---

[1] In this paper, a resource always denotes a unary resource.

R. Barták and M. Milano (Eds.): CPAIOR 2005, LNCS 3524, pp. 396–409, 2005.

In relation to a resource operations are called *activities*. Each activity i has following requirements:

- earliest possible starting time est_i
- latest possible completion time lct_i
- processing time p_i

A purpose of the resource constraint is to reduce a search space by tightening the time bounds est_i and lct_i. This process of elimination of unfeasible values is called *propagation*, an actual propagation algorithm is often called *filtering* algorithm. Interval $\langle est_i, lct_i \rangle$ is called a *time window* of the activity i. Thus the role of the resource constraint can be seen as a process of tightening of these time windows.

There are several filtering algorithms for unary resources, in this paper we focus on the edge-finding [7, 8], not-first/not-last [10, 9, 3] and detectable precedences [10]. Each of these algorithms filters out different inconsistent values, therefore these algorithms can be used together to achieve better pruning.

There are only a few attempts to combine explanations with unary resource constraint. The author is aware of the paper [6] where Guéret et. al. solved several open openshop problems using explanations. This result was achieved by very simple explanations for unary resource constraints. This paper focuses on computing more accurate explanations. In paper [2] explanations are used for solving dynamic schedule problems.

This paper differs from the previous work in two main aspects: explanations are computed in a lazy way and justifications are very tightly connected with filtering algorithms. This way the computation is very fast and resulting explanations are more accurate.

2   Explanations

The purpose of the explanation is to capture a *reason* why a search (sub)tree failed. Advanced search methods (directed backjumping, dynamic backtracking [5]) can exploit such information and speed up the search. The idea is to identify a reason of fail and cut off other branches of the search tree which are known to fail for the same reason.

An explanation has to describe all properties of the subproblem which leads to the infeasibility. This way the explanation can be seen as a *relaxation* of the original unfeasible subproblem. The important point is that this relaxation remains unfeasible. Our intention is to find as general relaxation (explanation) as possible. More general explanation can cover more subproblems and dismiss them as unsolvable.

Let us precisely define a specific type of explanations which is used in this paper:

Definition 1. *An fail explanation is an unfeasible CSP[2] which is a relaxation of the current search node. The explanation consists of:*

1. *A subset Υ of initial constraints and search decision constraints valid in the current search node.*
2. *Conflict windows $\langle \underline{est}_i, \overline{lct}_i \rangle$ for activities. A conflict window for an activity i is a superset of the current time window: $\langle est_i, lct_i \rangle \subseteq \langle \underline{est}_i, \overline{lct}_i \rangle$. I.e. the conflict window is a relaxation of the current time window.*
   If no conflict window is given for an activity i then we consider the time window to be $\langle -\infty, \infty \rangle$.

The idea of this definition follows. In explanation we relax constraints which are not in the set Υ. We also relax domains by replacing time windows by conflict windows. And still the problem remains unfeasible. Conflict window $\langle -\infty, \infty \rangle$ is special, it says that the activity is irrelevant: can be processed at any time and yet the problem has no solution.

The explanation can be compared with state in any other search node. A problem has no solution as long as all constraints from the set Υ remain in the system and all time windows are covered by associated conflict windows.

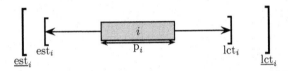

Fig. 1. Activity i, its time window and conflict window

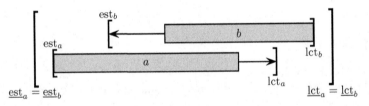

Fig. 2. Two activities a and b in conflict, their time windows and conflict windows

2.1   Initial Explanation

When propagation comes to a dead end, an initial explanation must be computed. This initial explanation simply describe the reason of the fail which was found.

For shop-scheduling problems, the usual reason why propagation generates fail is *overloading*. Let us consider a subset $\Omega \subseteq T$ of activities on one resource.

---

[2] Constraint Satisfaction Problem.

We can define processing time, earliest starting time and latest completion time of the set Ω as:

$$\mathrm{est}_\Omega = \min\{\mathrm{est}_i, \ i \in \Omega\}$$
$$\mathrm{lct}_\Omega = \max\{\mathrm{lct}_i, \ i \in \Omega\}$$
$$\mathrm{p}_\Omega = \sum_{i \in \Omega} \mathrm{p}_i$$

All activities from the set Ω must be processed during the interval $\langle \mathrm{est}_\Omega, \mathrm{lct}_\Omega \rangle$. However, if $\mathrm{p}_\Omega > \mathrm{lct}_\Omega - \mathrm{est}_\Omega$ then no solution exists. Empty domain for an activity i is a special case of overloading for $\Omega = \{i\}$.

The explanation for overloading consists of the unary resource constraint and conflict windows for activities $i \in \Omega$. These conflict windows can be $\langle \mathrm{est}_\Omega, \mathrm{lct}_\Omega \rangle$. However conflict windows can be little bit wider, as long as $\underline{\mathrm{lct}}_i - \underline{\mathrm{est}}_i \leq \mathrm{p}_\Omega - 1$. Let Δ is defined as:

$$\Delta = \mathrm{p}_\Omega - (\mathrm{lct}_\Omega - \mathrm{est}_\Omega) - 1$$

Conflict windows for $i \in \Omega$ can be set the following way:

$$\left\langle \mathrm{est}_\Omega - \left\lfloor \frac{\Delta}{2} \right\rfloor, \ \mathrm{lct}_\Omega + \left\lceil \frac{\Delta}{2} \right\rceil \right\rangle$$

2.2   Justifications

It is likely that the infeasibility of the problem cannot be simply detected by the overloading. Some propagation or even search must be done first. Initial explanation provided by overloading is just a beginning. The explanation must be refined during the way back in the search tree.

For this purpose, a *justification* must be remembered for each domain reduction. The justification captures the reason which justifies the realized reduction:

Definition 2. *Justification is a CSP which is a relaxation of the state just before the reduction. Filtering algorithm would generate exactly the same reduction for the relaxed CSP as for the original one.*

Justification consists of the propagated constraint and a set of conflict windows.

Justifications are written on the stack during constraint propagations and used for explanation (re)computation during way back in the search tree. Naturally, we are looking for as general justification as possible – more general justifications result in the more general explanation.

Let us describe more formally how justifications are used to refine explanations during way back in the backtrack:

1. Once a fail is found, an initial explanation is created.
2. One by one the reductions made by the constraint propagation are undone in the reverse order than they were originally made. For that, all realized reductions and their justifications are stored in a stack.

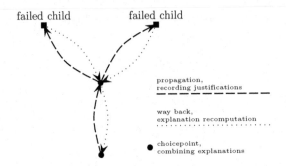

Fig. 3. Operations on the search tree

3. After undoing a particular reduction, it can happen that the explanation is not the relaxation of the current problem any more. For example est_i may become greater than est_i and thus the conflict window does not cover the time window any more. In that case, the explanation must be repaired using the justification associated with the undone reduction. It is done in the following way:

    i. est_i is set to $-\infty$.

    ii. Constraint which generated the reduction is added into the explanation: $\Upsilon := \Upsilon \cup \{c\}$.

    iii. Conflict windows from the justification are "merged" into the conflict windows of the explanation. Let $\langle est'_k, lct'_k \rangle$ be the conflict window for the activity k in the justification. Then the resulting conflict window for the activity k in the explanation is:

$$\langle est_k, lct_k \rangle := \langle \max\{est_k, est'_k\}, \min\{lct_k, lct'_k\} \rangle$$

4. In a choicepoint, the explanation is a combination of explanations from all child nodes. For example, let us suppose that the branching was done by addition of a constraint **a** in the first child node and the negation of this constraint ¬**a** in the second child node. The resulting explanation consists of:

    i. Subset of current constraints Υ:

$$\Upsilon = (\Upsilon^{\mathbf{a}} \setminus \{\mathbf{a}\}) \cup (\Upsilon^{\neg \mathbf{a}} \setminus \{\neg \mathbf{a}\})$$

    ii. Conflict windows for activities $\langle est_k, lct_k \rangle$:

$$\langle est_k, lct_k \rangle := \langle \max\{est_k^{\mathbf{a}}, est_k^{\neg \mathbf{a}}\}, \min\{lct_k^{\mathbf{a}}, lct_k^{\neg \mathbf{a}}\} \rangle$$

Justifications are similar to explanations, however justifications are much more simple. For each particular filtering algorithm, justification can be held in a specialized data structure which exploit a particular method of filtering. Detailed descriptions of justifications for different algorithms are provided in following sections 4–7.

3   Directed Backjumping

Before going into details about justifications, let us show how to implement directed backjumping using explanations.

Consider a the choicepoint from the item 4 above. If $\mathbf{a} \notin \varUpsilon^{\mathbf{a}}$ then the explanation for the first branch $\mathbf{a}$ is valid also for the second branch $\neg\mathbf{a}$. And so the second branch can be skipped because it would fail anyway. This way, explanations can speed up the search.

4   Precedence Justification

Together with resource constraint, binary precedence constraints are used to model shop-scheduling problems. A precedence constraint $i \ll j$ assures that the activity i finish before the activity j starts. Precedence constraints can be used to model ordering of operations within a job in jobshop. They are also often used as search decisions.

Let us introduce a notation convention. Whenever a reduction of a domain is made (i.e. est_i is increased or lct_i is decreased), then est_i and lct_i denote values before the reduction, est_i' and lct_i' denote values after the reduction.

Propagation of the precedence constraint $i \ll j$ is quite simple: whenever est_i is increased, the constraint propagates this change into the value est_j:

$$est_j' := \max\{est_j,\ est_i + p_i\} \tag{1}$$

Similarly, when lct_j is decreased, lct_i can be adjusted:

$$lct_i' := \min\{lct_i,\ lct_j - p_j\} \tag{2}$$

All propagation algorithms considered in this paper have two symmetric versions. One of them increase values est_i (in this case the rule (1)), the second one decrease values lct_i (the rule (2)). Since propagation algorithms and their justifications are symmetrical, we will always consider only one of these symmetric versions – the one which changes est_i.

Justification for a reduction made by the rule (1) is quite simple: precedence constraint itself and the conflict window $\langle est_i, \infty \rangle$ for the activity i.

Now let us now focus on the usage of such justification. Explanation must be recomputed only if[3] $est_j < \underline{est}_j$. In that case, explanation is recomputed the following way:

i. Precedence constraint $i \ll j$ is added into the explanation.
ii. $\underline{est}_i$ must be changed. According to justification it is enough to set $\underline{est}_i := est_i$. However, it is possible that $\underline{est}_j < est_j'$, i.e. to put the activity j into the conflict, it is enough to increase est_j to $\underline{est}_j$. Thus it is enough to set $\underline{est}_i := \underline{est}_j - p_i$.
iii. Conflict window of the activity j is enlarged: $\underline{est}_j := -\infty$.

---

[3] Note that est_j is value before the reduction, i.e. the value after undo of this reduction.

Recording and using one justification for a precedence constraint has both time complexity and space complexity $\mathcal{O}(1)$.

5   Not-First/Not-Last Justifications

Filtering algorithm not-first/not-last [10, 3, 9] is based on the following rule *not-first* and its symmetric variant *not-last*. Let us consider an activity i and a set $\Omega \subset T$ such that the activity i cannot start before the set Ω. We denote such property $i \not\ll \Omega$:

$$\forall \Omega \subset T, \forall i \in (T \setminus \Omega): \quad \text{lct}_\Omega - \text{est}_i < p_\Omega + p_i \quad \Rightarrow \quad i \not\ll \Omega \qquad (3)$$

If $i \not\ll \Omega$, some activity j from the set Ω must finish before the activity i can start. This allows to increase est_i:

$$i \not\ll \Omega \quad \Rightarrow \quad \text{est}'_i := \max \left\{ \text{est}_i, \ \min \left\{ \text{est}_j + p_j, \ j \in \Omega \right\} \right\} \qquad (4)$$

A justification for such change of est_i must guarantee that if all activities remain inside conflict windows, inequality (3) remains valid and the value est'_i in the rule (4) remains the same or it is even greater.

The inequality (3) remains valid as long as lct_Ω does not increase too much. Hence for each activity j in the set Ω, the bound of the conflict window $\underline{\text{lct}_j}$ must fulfill the following inequality:

$$\forall j \in \Omega: \quad \underline{\text{lct}_j} < \text{est}_i + p_\Omega + p_i$$

Similarly, as long as $\min\{\text{est}_j + p_j, \ j \in \Omega\}$ does not decrease, the rule (4) still justifies the reduction. Therefore:

$$\forall j \in \Omega: \quad \underline{\text{est}_j} \geq \text{est}'_i - p_j$$

To fulfill both last inequalities, the conflict windows for activities from the set Ω are assigned in the following way:

$$\forall j \in \Omega: \quad \langle \text{est}'_i - p_j, \ \text{est}_i + p_\Omega + p_i - 1 \rangle$$

Also, the conflict window $\langle \text{est}_i, \infty \rangle$ must be assigned to the activity i.

Just constructed justification has time and space complexity $\mathcal{O}(n)$ because all activities from the set Ω must be enumerated into the justification. However, only some special types of sets Ω can be considered in order to find all reductions resulting from the rule not-first.

Let us consider one particular reduction according to the rule not-first (3), (4). Let Ψ be the set constructed the following way:

$$\Psi = \{j, \ j \in T \ \& \ \text{est}_j + p_j \geq \text{est}'_i \ \& \ \text{lct}_j \leq \text{lct}_\Omega \ \& \ j \neq i\} \qquad (5)$$

If we exchange the set Ω by the set Ψ in the rules (3) and (4), these rules would raise exactly the same change of the est_i. In fact, all not-first algorithms [10, 3, 9] consider only sets in this form. I.e. whenever a change of a est_i is made, $\Omega = \Psi$.

Thanks to this special form of the set Ω, the set Ω can be characterized only by the values est_i' and lct_Ω. Using the rule (5), the set Ω can be reconstructed in time $\mathcal{O}(n)$. Thus the justification has size only $\mathcal{O}(1)$ and can be recorded in the time $\mathcal{O}(1)$. All three not-first/not-last algorithms [10, 3, 9] can be easily modified to record such justifications without changing their time complexities, i.e. $\mathcal{O}(n \log n)$ for [10] and $\mathcal{O}(n^2)$ for [9, 3].

Usage of each one not-first justification takes time $\mathcal{O}(n)$. One run of the not-first algorithm can generate only n changes maximum. Thus the way back "through" the not-first/not-last propagation takes $\mathcal{O}(n^2)$ maximum. In addition, a lot of justifications can be skipped because they do not interfere with the current explanation (i.e. $est_i \geq \underline{est_i}$).

Finally, let us consider usage of a not-first justification. Let us suppose that $est_i < \underline{est_i}$. I.e. we are in a situation when the current explanation is valid after the change (i.e. $\underline{est_i} \leq est_i'$), but not before it. Thus some repair of the current explanation is necessary. The justification captures the reason why est_i was increased to est_i'. However in order to make the current explanation valid, it may not be necessary to increase the est_i so much. It is enough to increase est_i to $\underline{est_i}$. I.e. the justification can be weakened before it is merged into the current explanation. This weakening can be achieved by using $\underline{est_i}$ instead of est_i' in all conflict windows.

6   Edge-Finding Justifications

Edge-finding is well known filtering algorithm for unary resource constraint. The algorithm is based on the following rules (6), (7) and their symmetric versions [3, 7]. Consider a set $\Omega \subseteq T$ and an activity $i \notin \Omega$. The activity i has to be scheduled after all activities from the set Ω if:

$$\forall \Omega \subset T, \ \forall i \in (T \setminus \Omega) : \ \min\{est_\Omega, \ est_i\} + p_\Omega + p_i > lct_\Omega \ \Rightarrow \ \Omega \ll i \quad (6)$$

The reason follows: if the activity i is not scheduled after the set Ω then the last activity from the set Ω cannot finish before $\min\{est_\Omega, \ est_i\} + p_\Omega + p_i$, what is more than the allowed maximum lct_Ω.

Once it is known that the activity i must be scheduled after the set Ω, est_i can be adjusted:

$$\Omega \ll i \ \Rightarrow \ est_i' := \max\{est_i, \ ECT_\Omega\} \quad (7)$$

Where ECT_Ω denotes a lower bound of the earliest completion time of a set Ω. ECT_Ω is defined by the following formula:

$$ECT_\Omega = \max\{est_{\Omega'} + p_{\Omega'}, \ \Omega' \subseteq \Omega\} \quad (8)$$

There are several implementations of edge-finding algorithm, [4] presents a $\mathcal{O}(n \log n)$ algorithm, another two $\mathcal{O}(n^2)$ algorithms can be found in [7,8].

We are interested in providing justifications for reductions made by edge-finding. Naturally, a justification consists of the unary resource constraint itself and some set of conflict windows. These conflict windows have to assure, that while the time windows of activities remain inside the conflict windows, the reduction made according to the rules (6) and (7) would be at least the same.

Lets start with the inequality (6). There are several ways how to extent time windows to conflict windows. Lets look at one of them: we allow an extension only on the right side of the time windows. Thus to satisfy the inequality (6), the conflict windows can be:

$$\forall j \in \Omega : \quad \langle r,\, r + p_\Omega + p_i - 1 \rangle \qquad (9)$$
$$i : \quad \langle r,\, \infty \rangle$$

where $r = \min\{\text{est}_\Omega,\, \text{est}_i\}$.

Sure, such conflict windows are not sufficient for the rule (7). This rule demands that for one particular set $\Omega' \subseteq \Omega$, the value $\text{est}_{\Omega'}$ remains the same: $\text{est}_{\Omega'} = \text{est}'_i - p_{\Omega'}$. Putting that together with previous conflict windows (9), the final conflict windows are:

$$\forall j \in \Omega' : \quad \langle \text{est}'_i - p_{\Omega'},\, r + p_\Omega + p_i - 1 \rangle$$
$$\forall j \in (\Omega \setminus \Omega') : \quad \langle r,\, r + p_\Omega + p_i - 1 \rangle$$
$$i : \quad \langle r,\, \infty \rangle$$

These conflict windows are sufficient for both rules (6) and (7).

Enumeration all activities from the set Ω in the explanation would again slow down the justification generation to $\mathcal{O}(n)$. Fortunately, a trick similar to not-first justification can be used here. Let us consider one particular reduction of est_i. Let the set Ω' be such a subset of the set Ω that $\text{ECT}_\Omega = \text{est}_{\Omega'} + p_{\Omega'}$. Note that the set Ω' must exists thanks to the definition (8) of the ECT_Ω. Further, let the sets Φ and Θ are defined the following way:

$$\Phi = \{j,\ j \in T\ \&\ \text{est}_\Omega \le \text{est}_j\ \&\ \text{lct}_j \le \text{lct}_\Omega\}$$
$$\Theta = \{j,\ j \in T\ \&\ \text{est}_{\Omega'} \le \text{est}_j\ \&\ \text{lct}_j \le \text{lct}_\Omega\}$$

These sets are in the form of so called *task intervals*. We can use the set Φ in the rule (6) instead of the set Ω and the inequality stays holding. The set Θ can be used to estimate a lower bound of ECT_Φ:

$$\Theta \subseteq \Phi \quad \Rightarrow \quad \text{ECT}_\Phi \ge \text{est}_\Theta + p_\Theta \quad \Rightarrow \quad \text{est}'_i \ge \max\{\text{est}_i,\, \text{est}_\Theta + p_\Theta\}$$

Because $\text{est}_\Theta = \text{est}_{\Omega'}$ and $p_\Theta \ge p_{\Omega'}$:

$$\text{ECT}_\Omega = \text{est}_{\Omega'} + p_{\Omega'} \le \text{est}_\Theta + p_\Theta$$

Therefore the set Ω can be replaced by the set Φ in the justification and the set Ω' can be replaced by the set Θ. In fact, edge-finding algorithms consider only sets Ω and Ω' in a form of task intervals, i.e. $\Omega = \Phi$ and $\Omega' = \Theta$.

Thus the result is very similar to the not-first justification. Instead of the enumeration of the sets Ω and Ω' in the justification, it is sufficient to record only the values est_Ω, lct_Ω and $\text{est}_{\Omega'}$. Both the sets can be reconstructed from these values within the time complexity $\mathcal{O}(n)$. The justification has size only $\mathcal{O}(1)$ and can be recorded in time $\mathcal{O}(1)$. Both edge-finding algorithm [7, 8] can be easily modified to generate justifications without changing their time complexity $\mathcal{O}(n^2)$.

Before using the justification, we can weaken it the same way as not-first justification: it is not necessary to increase est_i to est'_i to reach the infeasibility. Sufficient value of est'_i is $\underline{\text{est}}_i$. However this time we must be more careful. Simple replacement of est'_i by $\underline{\text{est}}_i$ in the definition of the conflict windows leads to invalid justifications. Conflict windows for the activities $j \in \Omega'$ should be:

$$\forall j \in \Omega' : \quad \langle \max\{\underline{\text{est}}_i - p_{\Omega'}, r\}, \; r + p_\Omega + p_i - 1 \rangle$$

This way the conflict window cannot run out from the conflict interval (9).

7   Justifications for Detectable Precedences

Detectable precedences is another propagation algorithm which can be used together with the edge-finding and not-first/not-last [10]. Let i and j be two different activities on the same resource. The precedence $j \ll i$ is said to be *detectable*, if the following inequality holds:

$$\text{est}_i + p_i > \text{lct}_j - p_j \tag{10}$$

Simply when the previous inequality holds then it is not possible to schedule the activity i before the activity j. The filtering algorithm builds a set Θ of all activities j, which precede the activity i according to detectable precedences:

$$\Theta = \{j, \; j \in T \; \& \; j \ll i \text{ is detectable}\}$$

The activity i cannot start until all of the activities from the set Θ finish, thus est_i can be adjusted:

$$\text{est}'_i := \max\{\text{est}_i, \; \text{ECT}_\Theta\}$$

Let Ω' be a subset of the set Ω such that $\text{ECT}_\Omega = \text{est}_{\Omega'} + p_{\Omega'}$. The justification has to assure two things: that $\Omega' \ll i$ and that the value $\text{est}_{\Omega'} + p_{\Omega'}$ does not decrease. Note that activities from the set $\Omega \setminus \Omega'$ are not included in the justification at all.

Lets start with $\text{est}_{\Omega'} + p_{\Omega'}$. Because this value cannot decrease, $\underline{\text{est}}_j$ must fulfill the following inequality:

$$\forall j \in \Omega' : \quad \underline{\text{est}}_j \geq \text{est}_{\Omega'}$$

Also it has to be assured that $\Omega' \ll i$. For each $j \in \Omega'$ the precedence $j \ll i$ is detectable. And it remains detectable as long as the inequality (10) remains valid:

$$\text{est}_i + p_i > \text{lct}_j - p_j$$

To assure that, conflict windows can be set the following way:

$$i: \quad \left\langle \mathrm{est}_i - \left\lceil \frac{\Delta}{2} \right\rceil, \ \infty \right\rangle$$

$$\forall j \in \Omega': \quad \left\langle \mathrm{est}_{\Omega'}, \ \mathrm{est}_i + \mathrm{p}_i + \mathrm{p}_j - \left\lceil \frac{\Delta}{2} \right\rceil - 1 \right\rangle$$

$$\text{where } \Delta = \mathrm{est}_i + \mathrm{p}_i - \max\left\{ \mathrm{lct}_j - \mathrm{p}_j, \ j \in \Omega' \right\} - 1$$

Again, we do not have to enumerate all activities from the set Ω' in the explanation. The set Ω' can be easily reconstructed using value $\mathrm{est}_{\Omega'}$. The justification has space complexity $\mathcal{O}(1)$ and it can be recorded within time $\mathcal{O}(1)$. Therefore recording of justifications do not change time complexity of the filtering algorithm. Processing of each relevant justification during way back takes time $\mathcal{O}(n)$.

Like other justifications, a justification for the detectable precedences can be weakened before it is used. The idea is still the same: to reach the conflict, it is not necessary to increase est_i to est'_i, value $\underline{\mathrm{est}}_i$ is enough. However this time est'_i does not occur in the conflict window definition directly. But $\mathrm{est}_{\Omega'} = \mathrm{est}'_i - \mathrm{p}'_\Omega$. Hence conflict windows for the activities j from the set Ω' can be enlarged the following way:

$$\forall j \in \Omega': \quad \left\langle \underline{\mathrm{est}}_i - \mathrm{p}_{\Omega'}, \ \mathrm{est}_i + \mathrm{p}_i + \mathrm{p}_j - \left\lceil \frac{\Delta}{2} \right\rceil - 1 \right\rangle$$

8   Experimental Results

The ideas presented in this paper were implemented in a C++ jobshop solver. Several jobshop problems of sizes 10x10 to 15x15 from the OR library [1] were used as a benchmark problems. The task is to find and prove the minimal makespan. Problems were solved using backtracking with directed backjumping based on explanations.

In order to make number of backtracks small, initial upper bound was set to the known optimal makespan. The solver has to find a solution first and then prove that there is no better solution.

The experiments shows that computation of explanation is very fast, it takes only 3–5% of the CPU time. The reason follows: propagation algorithms find a reduction only in 4–31% of runs (exact ratio depends on the filtering algorithm and an order in which the algorithms are called). From the recorded justifications, only 25–80% are really used (again, the ratio depends on the filtering algorithm).

The problems were solved twice. First time using explanations, second time without it. Tables 2 and 1 show the results. Columns CH1 and CH2 are number of choicepoints (i.e. nodes of a search tree without leaves), columns T1 and T2 shows the computation time.

Note that not all explanation computation was excluded in the second run. However, as said before, no more than 3–5% time could be saved by that.

Table 1. Jobshop instances: first branching strategy

Problem	Size	Makespan	CH1	CH2	CH Saving	T1	T2	T Saving
ft10	10x10	930	11478	14192	19.12%	13.112s	16.068s	18.40%
abz5	10x10	1234	5294	6445	17.86%	5.358s	6.538s	18.05%
abz6	10x10	943	3033	3807	20.33%	3.025s	4.077s	25.80%
la16	10x10	945	119	151	21.19%	0.147s	0.189s	22.22%
la17	10x10	784	26	27	3.70%	0.045s	0.045s	0.00%
la18	10x10	848	2429	2641	8.00%	2.422s	2.697s	10.20%
la19	10x10	842	14107	14989	5.89%	14.910s	16.058s	7.15%
la20	10x10	902	2623	2912	9.92%	2.812s	3.220s	12.67%
la36	15x15	1268	962	33749	97.15%	2.556s	40.214s	93.64%
la37	15x15	1397	61	61	0.00%	0.158s	0.157s	-0.64%
orb01	10x10	1059	16268	17137	5.07%	19.808s	20.927s	5.35%
orb02	10x10	888	10937	13818	20.85%	11.987s	15.348s	21.90%
orb03	10x10	1005	44152	50820	13.12%	48.051s	55.384s	13.24%
orb04	10x10	1005	1220	1319	7.50%	1.525s	1.618s	5.75%
orb05	10x10	887	2587	3312	21.89%	2.717s	3.587s	24.26%
orb06	10x10	1010	9838	10397	5.38%	11.709s	12.337s	5.09%
orb07	10x10	397	16476	20745	20.58%	16.251s	21.231s	23.46%
orb08	10x10	899	15	15	0.00%	0.039s	0.037s	-5.41%
orb09	10x10	934	491	515	4.66%	0.615s	0.641s	4.06%

Table 2. Jobshop instances: second branching strategy

Problem	Size	Makespan	CH1	CH2	CH Saving	T1	T2	T Saving
ft10	10 x 10	930	5931	6246	5.05%	4.928s	5.255s	6.23%
abz5	10 x 10	1234	2188	2963	26.16%	1.500s	2.093s	28.34%
abz6	10 x 10	943	840	863	2.67%	0.618s	0.659s	6.23%
la16	10 x 10	945	1025	1231	16.74%	0.604s	0.786s	23.16%
la17	10 x 10	784	45	45	0%	0.037s	0.037s	0%
la18	10 x 10	848	828	838	1.20%	0.606s	0.626s	3.20%
la19	10 x 10	842	5088	5447	6.60%	3.783s	4.085s	7.40%
la20	10 x 10	902	1353	1369	1.17%	1.076s	1.096s	1.83%
la36	15 x 15	1268	2636	2890	8.79%	5.477s	6.123s	10.56%
la37	15 x 15	1397	2554	6398	60.09%	2.869s	6.763s	57.58%
la39	15 x 15	1233	251	276	9.06%	0.554s	0.597s	7.21%
la40	15 x 15	1222	23606	26408	10.62%	49.816s	56.019s	11.08%
orb01	10 x 10	1059	5214	5220	0.12%	4.903s	4.931s	0.57%
orb02	10 x 10	888	3200	3448	7.20%	2.339s	2.620s	10.73%
orb03	10 x 10	1005	12603	12699	0.76%	10.214s	10.422s	2.00%
orb04	10 x 10	1005	1938	1969	1.58%	1.584s	1.658s	4.47%
orb05	10 x 10	887	1625	1771	8.25%	1.134s	1.248s	9.14%
orb06	10 x 10	1010	6145	6618	7.15%	4.493s	4.867s	7.69%
orb07	10 x 10	397	2066	2190	5.67%	1.552s	1.673s	7.24%
orb08	10 x 10	899	45	45	0%	0.040s	0.041s	2.44%
orb09	10 x 10	934	532	535	0.57%	0.443s	0.451s	1.78%
orb10	10 x 10	944	146	146	0%	0.157s	0.160s	1.88%
ta04	15 x 15	1175	115525	185278	37.65%	2m 33s	4m 12s	39.42%
ta07	15 x 15	1227	763719	1290715	40.83%	20m 10s	35m 47s	43.67%
la38	15 x 15	1196	1381989	1596715	13.45%	37m 57s	45m 34s	16.71%

Two different branching schemes were used to show the influence of the branching strategy to the directed backjumping:

1. The first branching strategy finds a resource with a smallest slack time. Then all yet unscheduled activities on the resource are taken and branching is done on a decision which of them will be the first. For results, see table 1.

2. The second branching strategy is taken from [11]. The resource with the relatively smallest slack is taken and branching is done by ordering two longest unordered activities on this resource. The results are in the table 2.

As can be seen, in case of quickly solvable instances ($\sim$ 1000 choicepoints) backjumping does not significantly improve the performance. However for harder instances, the savings of time and choicepoints reach 20% for the first branching strategy and 40% for the second branching strategy. The problem la36 in table 1 is quite exceptional: 97.15% of choicepoints are eliminated.

9   Conclusions and Further Work

Experimental results shows that explanations can significantly speed up the search, especially for hard problems. Also lazy computation of explanations seams to be quite effective.

In the future work we would like to explore more advanced search methods: dynamic backtracking, MAC-DBT or decision-repair.

Another technique often used to prune a search space is *shaving* [7]. For scheduling problems, shaving turned out to be quite effective. It could be interesting to combine explanations with shaving.

References

[1] OR library. URL http://mscmga.ms.ic.ac.uk/info.html.
[2] Narendra Jussien Abdallah Elkhyari, Chrstelle Guéret. Conflict-based repair techniques for solving dynamic scheduling problems. In *Principles and Prictice of Constraint Programming (CP 2002)*, pages 702–707, Ithaca, USA, 2002. Springer-Verlag.
[3] Philippe Baptiste and Claude Le Pape. Edge-finding constraint propagation algorithms for disjunctive and cumulative scheduling. In *Proceedings of the Fifteenth Workshop of the U.K. Planning Special Interest Group*, 1996.
[4] Jacques Carlier and Eric Pinson. Adjustments of head and tails for the job-shop problem. *European Journal of Operational Research*, 78:146–161, 1994.
[5] Matthew L. Ginsberg, James M. Crawford, and David W. Etherington. Dynamic backtracking, 1996. URL http://citeseer.ist.psu.edu/ginsberg96dynamic.html.
[6] Christelle Guéret, Narendra Jussien, and Christian Prins. Using intelligent backtracking to improve branch and bound methods: an application to openshop problems. *European Journal of Operational Research*, 127(2):344–354, 2000. ISSN 0377-2217. URL http://www.emn.fr/jussien/publications/gueret-EJOR00.pdf.
[7] Paul Martin and David B. Shmoys. A new approach to computing optimal schedules for the job-shop scheduling problem. In W. H. Cunningham, S. T. McCormick, and M. Queyranne, editors, *Proceedings of the 5th International Conference on Integer Programming and Combinatorial Optimization, IPCO'96*, pages 389–403, Vancouver, British Columbia, Canada, 1996.

[8] Claude Le Pape Philippe Baptiste and Wim Nuijten. *Constraint-Based Scheduling: Applying Constraint Programming to Scheduling Problems.* Kluwer Academic Publishers, 2001.

[9] Philippe Torres and Pierre Lopez. On not-first/not-last conditions in disjunctive scheduling. *European Journal of Operational Research*, 1999.

[10] Petr Vilím. $O(n \log n)$ filtering algorithms for unary resource constraint. In *Proceedings of CP-AI-OR 2004*. Springer-Verlag, 2004.

[11] Armin Wolf. Better propagation for non-preemptive single-resource constraint problems. In *Proceedings of the ERCIM/CoLogNet workshop 2004*, 2004.

Author Index